Handbook of Polymer Reaction Engineering
Volume 2

Edited by
Th. Meyer, J. Keurentjes

Further Titles of Interest

E. S. Wilks (Ed.)

Industrial Polymers Handbook

Products, Processes, Applications

2000

ISBN 3-527-30260-3

H.-G. Elias (Ed.)

Macromolecules

Vols. 1–4

2005

ISBN 3-527-31172-6, 3-527-31173-4, 3-527-31174-2,
3-527-31175-0

M. F. Kemmere, Th. Meyer (Eds.)

Supercritical Carbon Dioxide

in Polymer Reaction Engineering

2005

ISBN 3-527-31092-4

M. Xanthos (Ed.)

Functional Fillers for Plastics

2005

ISBN 3-527-31054-1

R. C. Advincula, W. J. Brittain, K. C. Caster, J. Rühe (Eds.)

Polymer Brushes

2004

ISBN 3-527-31033-9

H.-G. Elias (Ed.)

An Introduction to Plastics

Second, Completely Revised Edition

2003

ISBN 3-527-29602-6

Handbook of Polymer Reaction Engineering

Edited by
Thierry Meyer, Jos Keurentjes

Volume 2

WILEY-VCH

WILEY-VCH Verlag GmbH & Co. KGaA

Editors

Dr. Thierry Meyer
Swiss Federal Institute of Technology
Institute of Process Science
EPFL, ISP-GPM
1015 Lausanne
Switzerland

Prof. Jos T. F. Keurentjes
Process Development Group
Eindhoven University of Technology
P.O. Box 513
5600 MB Eindhoven
The Netherlands

Library of Congress Card No.: Applied for

British Library Cataloging-in-Publication Data: A catalogue record for this book is available from the British Library

Bibliographic information published by Die Deutsche Bibliothek
Die Deutsche Bibliothek lists this publication in the Deutsche Nationalbibliografie; detailed bibliographic data is available in the Internet at ⟨http://dnb.ddb.de⟩.

Printed in the Federal Republic of Germany
Printed on acid-free paper

Composition Asco Typesetters, Hong Kong
Printing betz-druck gmbh, Darmstadt
Bookbinding Litges & Dopf Buchbinderei GmbH, Heppenheim

ISBN-13 978-3-527-31014-2
ISBN-10 3-527-31014-2

Foreword

A principal difference between science and engineering is intent. Science is used to bring understanding and order to a specific object of study – to build a body of knowledge with truth and observable laws. Engineering is more applied and practical, focused on using and exploiting scientific understanding and scientific principles to make products to benefit mankind. A polymer reaction engineer seeks the applied and practical as the title implies, but the path to success is most often through polymer science. This truth is steeped in history – there are many examples of polymeric products commercialized without adequate understanding of the chemistry and physics of the underlying polymerization. Polymer reaction engineers, faced with detriments in process safety, product quality or product cost, become the driving force behind many polymer science developments. As such, polymer reaction engineering is more a collaboration of polymer science and reaction engineering. A collaboration where polymer reaction engineers develop a firm understanding of the many aspects of polymer chemistry and physics to successfully apply chemical engineering principles to new product developments. Only through the integration of science and engineering are such products realized.

This handbook is a testimony to this melding of polymer science and chemical engineering that defines polymer reaction engineering. Thierry Meyer and Jos Keurentjes have compiled a strong list of contributors with an even balance from academia and industry. The text offers a comprehensive view of polymer reaction engineering. The text starts with an overview describing the important integration of science and engineering in polymer reaction engineering and ends with recent and potential breakthrough developments in polymer processing. The middle chapters are divided into three sections. The first section is devoted to the science and chemistry of the major types of polymerization. Included are step and chain growth polymerizations with chapters devoted specifically to several different chain growth methods. The central section of the middle chapters is dedicated to polymer reaction engineering tools and methods. The very important topics of safety and process control are detailed and help frame the conditions through which successful scale-ups are achieved. The last section of the middle chapters is focused on the physics and physical nature of formed polymers including their physical and mechanical structure. In these chapters, many of the processes that modify polymers through man-made and natural change are discussed. The details of polymer end use are also presented.

This tome represents the first published handbook on polymer reaction engineering and should be well received in academia and industry. Polymer reaction engineering became recognized as a separate engineering discipline in the 1970's. It is well past due that such a handbook be published. The broad scope and depth of coverage should make it an important reference for years to come.

Michael C. Grady, ScD
Senior Engineering Associate
DuPont
Philadelphia, Pennsylvania

Contents

Volume 1

Preface

Freshly started as chairman and secretary of the Working Party on Polymer Reaction Engineering it never crossed our mind to edit a book on this subject. This changed when Wiley-VCH asked if the working party would be able to provide a translation of the Handbuch der Technischen Polymerchemie, written in 1993 by Adolf Echte. We decided to do so, but not exactly. Very rapidly we were convinced that we needed a completely new book, covering the field of polymer reaction engineering in a modern, broad and multidisciplinary approach. Many of the working party members directly agreed to participate, others needed somewhat stronger persuasion techniques, and for some chapters we hired authors from other institutions. In June 2003 we had completed the list of contributors, coming from Europe, Canada and the USA. Now, roughly one year later, the new handbook is there.

The quality an edited book like this very much depends on the quality of the individual contributions. It has been a great pleasure for us to see that all authors have taken their writing jobs very seriously. With these contributions, we are sure that this book represents the state of the art in polymer reaction engineering. It is intended to attract equally readers that are new in the field as well as readers that may be considered expert in some of the topics but want to broaden their knowledge. We are convinced that the multidisciplinary and synergetic approach presented in this book may act as an eye-opener for research and development activities going on in strongly related areas. We hope the reader will take advantage of this approach, where the references given in the various chapters may be a starting point for further reading.

Reading books, you often read the preface as well. We have seen numerous examples from which the frustration is quite obvious. Of course things may not always work out the way you plan, that has also been the case for this book. Maybe we were just lucky, but we have greatly enjoyed doing this. Editing this book has also been a starting point for the editors to become friends, including Swiss cheese fondue and Dutch "Friese nagelkaastaart" in a friendly home setting. From that perspective also Francine and Maartje have had their part both of the workload but also of the fun of all this.

Finally, we would like to thank Karin Sora and Renate Doetzer from Wiley-VCH for their help with the editing process. They really know to find the balance between waiting and pushing in order not to diverge too far from the schedule.

Lausanne & Eindhoven, fall 2004
Thierry Meyer & Jos Keurentjes

List of Contributors

P. Aarnoutse
Polymer-Analysis Group
Department of Chemical Engineering (ITS)
Faculty of Science, University of Amsterdam
Nieuwe Achtergracht 166
1018 WV Amsterdam
The Netherlands

Dr. U. S. Agarwal
Polymer Technology Group
Department of Chemical Engineering and Chemistry
Eindhoven University of Technology
P.O. Box 513
5600 MB Eindhoven
The Netherlands

Prof. J. M. Asua
The University of the Basque Country
Institute for Polymer Materials (POLYMAT)
Paseo Manuel Lardizabal 3
20018 Donostia-San Sebastián
Spain

Dr. R. Bachmann
Bayer AG
ZT-TE-SVT
51368 Leverkusen
Germany

Prof. M. J. Barandiaran
The University of the Basque Country
Institute for Polymer Materials (POLYMAT)
Manuel Lardizabal, 3
20018 Donostia-San Sebastián
Spain

Dr. C. W. M. Bastiaansen
Eindhoven University of Technology
Den Dolech 2
5600 MB Eindhoven
The Netherlands

Prof. D. J. Broer
Philips Research Laboratories
Prof. Holstlaan 4
5656 AA Eindhoven
The Netherlands

and

Eindhoven University of Technology
Den Dolech 2
5600 MB Eindhoven
The Netherlands

and

Dutch Polymer Institute (DPI)
P.O. Box 902
5600 AX Eindhoven
The Netherlands

Polymer Technology
Department of Chemical Engineering and Chemistry
Eindhoven University of Technology
P.O. Box 513
5600 MB Eindhoven
The Netherlands

Prof. B. W. Brooks
Loughborough University
Department of Chemical Engineering
Loughborough
Leicestershire, LE11 3TU
United Kingdom

Dr. A. Buttè
Swiss Federal Institute of Technology
Zurich, ETHZ
Institut für Chemie- und Bioingenieurwissenschafften
Gruppe Morbidelli
ETH Hoenggerberg/HCI F135
8093 Zurich
Switzerland

Dr. J. P. Congalidis
E.I. du Pont de Nemours and Company
DuPont Central Research and Development
Experimental Station
Wilmington, DE 19880
USA

Dr. M. R. P. F. N. Costa
Faculty of Engineering
University of Porto
Rua Roberto Frias, s/n
4200-465 Porto
Portugal

Prof. J. C. de la Cal
The University of the Basque Country
Institute for Polymer Materials (POLYMAT)
Paseo Manuel Lardizabal, 3
20018 Donostia-San Sebastián
Spain

Dr. A. J. de Koning
DSM Research
Oude Postbaan 1
6167 RG Geleen
The Netherlands

Dr. T. W. de Loos
Delft University of Technology
Faculty of Applied Sciences
Department Chemical Technology
Julianalaan 136
2628 BL Delft
The Netherlands

C. de Witz
Philips Research Laboratories
Prof. Holstlaan 4
5656 AA Eindhoven
The Netherlands

Dr. L. J. Evers
DSM Melamine
Oude Postbaan 1
6167 RG Geleen
The Netherlands

Dr. A. Hofland
DSM Coating Resins
Ceintuurbaan 5
8022 AW Zwolle
The Netherlands

Dr. K.-D. Hungenberg
BASF AG
Polymer Technology, B1
67056 Ludwigshafen
Germany

Prof. R. A. Hutchinson
Department of Chemical Engineering
Queen's University
Dupuis Hall, 19 Division St.
Kingston, Ontario K7M 2G9
Canada

Dr. P. D. Iedema
Department of Chemical Engineering
University of Amsterdam
Nieuwe Achtergracht 166
1018 WV Amsterdam
The Netherlands

Dr. J. F. G. A. Jansen
DSM Research
Oude Postbaan 1
6167 RG Geleen
The Netherlands

Dr. J. A. Juijn
Research Institute
Department QRI
P.O. Box 9600
6800 TC Arnheim
The Netherlands

Dr. M. F. Kemmere
Process Development Group
Department of Chemical Engineering and Chemistry
Eindhoven University of Technology
P.O. Box 513
5600 MB Eindhoven
The Netherlands

Prof. J. T. F. Keurentjes
Process Development Group
Eindhoven University of Technology
P.O. Box 513
5600 MB Eindhoven
The Netherlands

N. H. Kolhapure
DuPont Engineering Research and Technology
1007 N. Market St.
Wilmington, DE 19898-0001
USA

Prof. J. R. Leiza
The University of the Basque Country
Institute for Polymer Materials (POLYMAT)
Paseo Manuel Lardizabal 3
20018 Donostia-San Sebastián
Spain

J. Mattheij
DSM Melamine
Oude Postbaan 1
6167 RG Geleen
The Netherlands

Dr. T. Meyer
Swiss Federal Institute of Technology
Institute of Process Science
EPFL, ISP-GPM
1015 Lausanne
Switzerland

L. J. Molhoek
DSM Coating Resins
Ceintuurbaan 5
8022 AW Zwolle
The Netherlands

Prof. M. Morbidelli
Swiss Federal Institute of Technology
Zurich, ETHZ
Institut für Chemie- und Bioingenieurwissenschaften
Gruppe Morbidelli
ETH Hoenggerberg/HCI F135
8093 Zurich
Switzerland

Prof. E. B. Nauman
The Isermann Department of Chemical and Biological Engineering
Rensselaer Polytechnic Institute
Troy, NY 12180
USA

Dr. Q. T. Nguyen
Laboratory of Polymers (LP)
Ecole Polytechnique Fédérale de Lausanne
1015 Lausanne
Switzerland

J. C. Plummer
Laboratory of Composite and Polymer Technology (LTC)
Ecole Polytechnique Fédérale de Lausanne
1015 Lausanne
Switzerland

Dr. J. R. Richards
E. I. du pont de Nemours and Company
DuPont Engineering and Research Technology
Experimental Station
Wilmington, DE 19880
USA

Dr. C. Sánchez
Eindhoven University of Technology
Den Dolech 2
5600 MB Eindhoven
The Netherlands

and

Dutch Polymer Institute (DPI)
P.O. Box 902
5600 AX Eindhoven
The Netherlands

Prof. P. J. Schoenmakers
Polymer-Analysis Group
Department of Chemical Engineering (ITS)
Faculty of Science, University of Amsterdam
Nieuwe Achtergracht 166
1018 WV Amsterdam
The Netherlands

Prof. L. C. Simon
Department of Chemical Engineering
University of Waterloo
200 University Avenue West
Waterloo, Ontario N2L 3G1
Canada

Prof. J. B. P. Soares
Department of Chemical Engineering
University of Waterloo
200 University Avenue West
Waterloo, Ontario N2L 3G1
Canada

Prof. F. Stoessel
Swiss Institute for the Promotion of Safety and Security
Chemical Process Safety Consulting
Klybeckstrasse 141
WKL-32.322
4002 Basel
Switzerland

Prof. G. Storti
Swiss Federal Institute of Technology Zurich, ETHZ
Institut für Chemie- und Bioingenieurwissenschaften
Gruppe Morbidelli
ETH Hoenggerberg/HCI F125
8093 Zurich
Switzerland

Prof. R. A. T .M. van Benthem
Coating Technology
Department of Chemical Engineering
and Chemistry
Eindhoven University of Technology
P.O. Box 513
5600 MB Eindhoven
The Netherlands

Dr. M. van Duin
DSM Research
Oude Postbaan 1
6167 RG Geleen
The Netherlands

11
Safety of Polymerization Processes[1]

Francis Stoessel

11.1
Introduction

Polymerization – that is, polycondensation and polyaddition – performed on an industrial scale presents a number of specific reaction and process engineering aspects which differentiates these processes from reactions of low molecular weight molecules. This is also true for aspects of the dynamic control of polymerization reactors. Therefore the concepts developed for low-molecular-weight chemistry must be adapted to the specific problems of polymerization reactions. The scope of the present chapter is the assessment of risks linked with the performance of polymerization reactions on an industrial scale. Moreover, the focus is on the control of the course of reaction by means of chemical reaction engineering [1]. Risks linked with handling of raw material or products are not treated in this chapter. Part of the chapter is based on a comprehensive text by Moritz [2], who is acknowledged for his authorization to use it here.

Barton and Nolan [3] and later Maddison and Rogers [4] analyzed the causes of chemical runaway incidents based on statistics from the British Health and Safety Executive. A classification by reaction type (Figure 11.1) reveals how prominent polymerizations are. From 134 cases where the causes could be identified, 64 were due to polymerization reactions, including polycondensations and polyadditions. From these 64 polymerization incidents, 13 (20%) involved phenol–formaldehyde resin production. This led the British Plastics Federation (BPF) to publish a booklet, *Guidance for the Safe Production of Phenolic Resins* [5]. A further investigation among more recent incidents involving polymer reactions showed that most of the incidents occur in vinyl chloride, vinyl acetate, and polyester resins. Since polymerization is the type of reaction where most runaway situations occur, it is worth devoting special attention to process safety. Understanding how a reaction may leave its "normal" course and enter a runaway course is essential for the design of safe and economic processes. This behavior has its roots in the thermodynamics

1) The symbols used in this chapter are listed at the end of the text, under "Notation".

Handbook of Polymer Reaction Engineering. Edited by T. Meyer, J. Keurentjes

ISBN: 3-527-31014-2

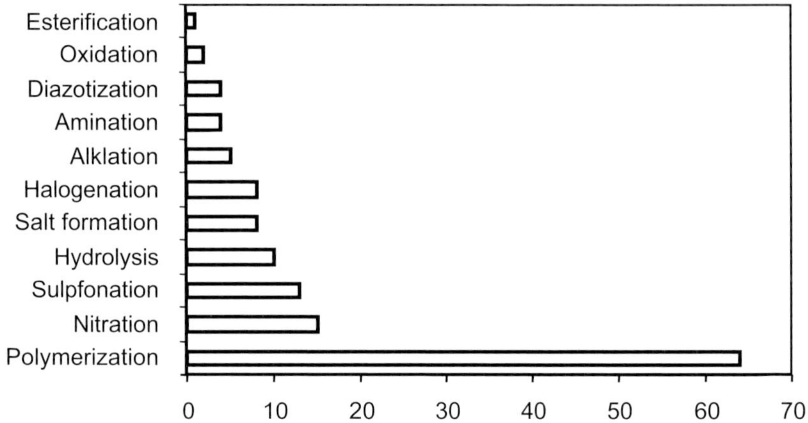

Fig. 11.1. Number of runaway incidents, classified by reaction type.

and chemical engineering aspects of the reactions. This will be explained in the present chapter.

In Section 11.2, general principles of reactor safety and heat balance of reactors are presented, with an emphasis on specific aspects of polymerizations. Section 11.3 is devoted to safety-related thermodynamics and reaction engineering aspects of polymerization reactions. In Section 11.4, cooling of polymerization reactors is reviewed. The chapter is concluded by a section describing safety aspects of industrial processes, together with technical risk-reducing solutions.

11.2
Principles of Chemical Reactor Safety Applied to Polymerization

Accidents happening in polymerization reactors are practically always due to a lack of control of the course of reaction caused by a disturbance of the heat balance, which results in a temperature increase leading to loss of control of the reactor and a runaway reaction. In this section a systematic procedure based on a failure scenario with six key questions, allowing assessment of the criticality of a process, is presented. Since the heat balance is at the center of our concerns in matters of thermal control of reactors, the different terms of the heat balance will be examined. Finally, aspects of the dynamic stability of reactors and of the thermal stability of reaction masses are analyzed.

11.2.1
Cooling Failure Scenario

A common practice in the assessment of risks due to runaway reactions is to use the cooling failure scenario as developed by Gygax [1, 6]. This is a worst-case as-

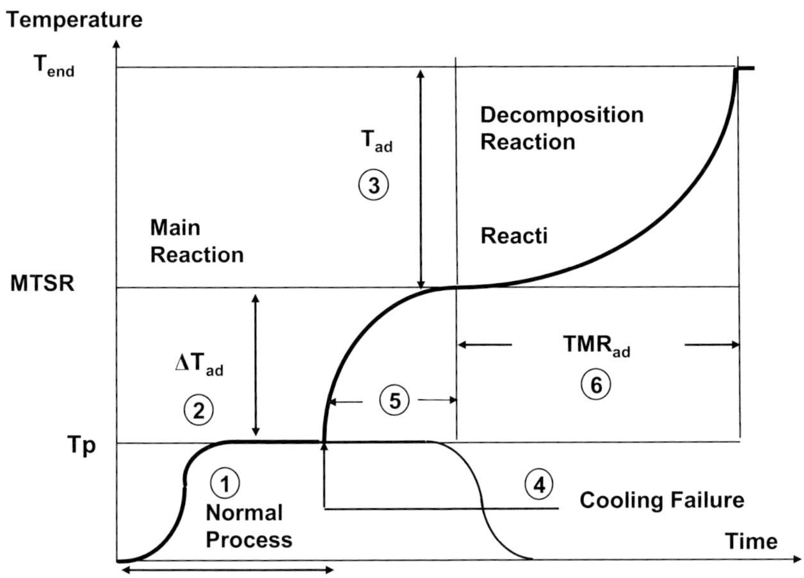

Fig. 11.2. Cooling failure scenario, presenting the consequences to the desired reaction of loss of cooling and triggering of a secondary decomposition reaction. The numbers represent the key questions used in the assessment of thermal risks (see text).

sumption that is useful for the risk assessment. It is assumed that, while the reactor is at reaction temperature, a cooling failure occurs. If at this instant unconverted monomer is still present in the reactor, the temperature will increase due to the completion of the reaction. This temperature increase is proportional to the amount of nonreacted material. At the temperature reached at the end of this period, a secondary decomposition reaction may be triggered and the heat produced by this reaction may lead to a further increase in temperature. The following questions help to develop the runaway scenario and to determine the data required for the risk assessment. This scenario (Figure 11.2) was developed for chemical reactions in general and is well suited to polymerization reactions.

1. Can the process temperature be controlled by the cooling system? This is a typical question which should be answered during process development. To ensure thermal control of the reaction, the power of the cooling system must be sufficient to dissipate the heat produced in the reactor at any time. Special attention must be devoted to the strong change in the viscosity of the reaction mass during polymerization and to possible fouling at the reactor wall (see Section 11.4). An additional condition, which must be fulfilled, is that the reactor is operated in the dynamic stability region as described in Section 11.2.4.
2. What temperature can be attained after runaway of the desired reaction? The

answer to this question necessitates a study of the kinetics of the reaction, in order to determine the degree of accumulation of monomer in the reactor as a function of time. The concept of MTSR (maximum temperature of the synthesis reaction) was developed for this purpose. In the case of polymerization reactions, the accumulation of monomer is not the only factor which must be known; the number of living chains must also be considered.

3. What temperature can be attained after runaway by decomposition? The thermal data of the secondary decomposition reactions allow calculation of the adiabatic temperature rise and determination of the final temperature, starting from the level of the MTSR. This temperature gives a direct indication of the possible consequences of a runaway. In polymerization reactions, the temperature reached after loss of control of the reaction itself will often determine the consequences. Polymerization reaction masses are often thermally stable, so this question is often not relevant. Exceptions are polycondensation reactions, especially hot melts or reactive resins.
4. At which moment does the cooling failure have the worst consequences? Since the amount of unconverted reactants and the thermal stability of the reaction mass may vary with time, it is important to know at which instant the accumulation, and therefore the thermal potential, is highest. This will be the worst case, and obviously the safety measures have to account for it.
5. How fast is the runaway of the desired reaction? Generally, industrial reactors are operated at temperatures where the desired reaction is fast. Hence, a temperature increase above the normal process temperature will cause a significant acceleration of the reaction: therefore, in most cases, this period of time is short. For polymerization reactions, where decomposition of the reaction mass is not critical, this time will determine the choice of technical risk reduction measures. The concept of time to maximum rate under adiabatic conditions (TMR_{ad}) as used for decomposition reactions can be applied to the polymerization itself, starting from the process temperature. It allows estimation of the probability of entering a runaway situation, as explained below for decomposition reactions.
6. How fast is the runaway of the decomposition starting at the MTSR? The dynamics of the decomposition reaction play an important role in the determination of the probability of an incident. The TMR_{ad} concept was developed for that purpose [7, 8]. The principles of using the time to explosion (TMR_{ad}) is that the longer the time to explosion, the greater is the chance of recovering the situation. This means that for a longer time to explosion, the probability of triggering the runaway is lower and, conversely, for a shorter TMR_{ad} this probability is higher.

The questions mentioned in this scenario can be answered using the results of calorimetric experiments which can be directly used in the determination of the characteristic temperature levels: after a cooling failure they will give us first the temperature due to the runaway of the desired reaction, and then the temperature reached after the runaway of the decomposition reaction (T_{end}).

11.2.2 Criticality Classes Applied to Polymerization Reactors

For reactions presenting a thermal potential we can consider the relative position of four temperature levels [9]:

- The process temperature ($T_{Process}$) is the initial temperature in the cooling failure scenario. In the case of a non-isothermal process, the initial temperature will be taken at the instant when the cooling failure has the heaviest consequences (worst case) according to Question 4 in the scenario, above.
- The maximum temperature of synthesis reaction (MTSR) depends essentially on the degree of accumulation of unconverted monomer and catalytic activity. Therefore it is strongly dependent on the process design. In polymerization processes, this temperature is often high enough to become critical.
- The maximum temperature for technical reasons (MTT) represents the technical limit of the equipment. In an open system, as for solution and emulsion polymerizations, it can be the boiling point. For a closed system, it is the temperature at which the pressure reaches the maximum permissible, that is, the set pressure of a safety valve or bursting disk. This could be the case for mass polymerization. This temperature (MTT) is often the critical factor for polymerization reactions.
- The temperature at which the TMR_{ad} is 24 h (T_{D24}): this temperature is defined by the thermal stability of the reaction mixture. It is the highest temperature at which the thermal stability of the reaction mass is unproblematic. This is often a secondary factor since polymerization masses are usually thermally stable.

These four temperature levels allow division of the scenarios into five different classes, going from the least critical (1) to the most critical (Figure 11.3).

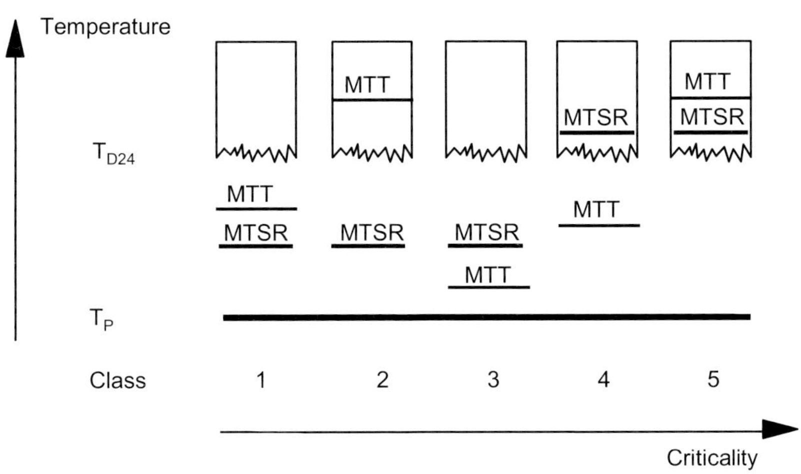

Fig. 11.3. Criticality classes of failure scenarios.

11.2.2.1 Description of the Criticality Classes

- Class 1: After loss of control of the synthesis reaction, the *MTT* cannot be reached and the decomposition reaction cannot be triggered. Only if the reaction mass was maintained for a long time under heat accumulation conditions, could the *MTT* be reached. Then the evaporative cooling may serve as an additional safety barrier. The process is thermally safe.
- Class 2: After loss of control of the synthesis reaction, the MTT cannot be reached and the decomposition reaction cannot be triggered. The situation is very similar to Class 1, but if the reaction mass is maintained for a longer time under heat accumulation conditions, the decomposition reaction could be triggered and reach the MTT. In this case, reaching this temperature could be a hazard if the heat release rate at the MTT is too high with respect to vaporization or pressure rise. If the reaction mass is not kept for a longer time under heat accumulation conditions, the process is thermally safe.
- Class 3: After loss of control of the synthesis reaction, the MTT will be reached, but the decomposition reaction cannot be triggered. In this situation, the safety of the process depends on the heat release rate of the synthesis reaction and on the pressure, respectively, at the MTT.
- Class 4: After loss of control of the synthesis reaction, the MTT will be reached and the decomposition reaction could theoretically be triggered. In this situation, the safety of the process depends on the heat release rate of *both* the synthesis reaction and the decomposition reaction at the MTT. Evaporative cooling or the emergency pressure relief may serve as a safety barrier.
- Class 5: After loss of control of the synthesis reaction, the decomposition reaction will be triggered and the MTT will be reached during the runaway of the decomposition reaction. It is very unlikely that the evaporative cooling or the emergency pressure relief can serve as a safety barrier in this case. The heat release rate of the decomposition at the MTT determines the thermal safety of the process. This is the most critical of all the scenarios.

For polymerization reactions the relative positions of the MTSR and the MTT are dominant: decomposition reactions play a role only in specific cases, such as resins. Thus the most important classes are 1 to 3, whereas the consequences of exceeding the MTT level may be dramatic.

These criticality classes are very useful in the decision making process of choosing the protection strategy for a reactor. Depending on the criticality class of the scenario, different measures can be applied to avoid, to control, or to stop the runaway:

- Class 1: No special measure is required for this class of scenario. The evaporative cooling or the emergency pressure relief could serve as a barrier.
- Class 2: No special measure is required. The reaction mass should not be held for longer time under heat accumulation conditions. The evaporative cooling or the emergency pressure relief could eventually serve as a barrier.

- Class 3: The reaction medium temperature exceeds the technical possibilities of the reactor. Thus technical measures are required. In a first approach one should try to avoid conditions in which a runaway situation develops. If this cannot be ensured, then emergency measures are required.
- Class 4: Similar to Class 3. The same measures apply here but the additional heat release rate due to the secondary reaction has also to be taken into account.
- Class 5: In this class the MTT is very unlikely to serve as a safety barrier. Therefore only emergency measures can be used. Since in most cases the decomposition reactions release very high energies, particular attention has to be paid to the design of safety measures. It is worthwhile to consider an alternative design of the process in order to reduce the severity, or at least the probability, of triggering a runaway reaction.

11.2.3
Heat Balance of Reactors

The heat balance is important as well for the design of reactors, for their scaleup, for the risk assessment, and especially for the assessment of the reactor stability. The heat balance is also at the center of the evaluation of calorimetric experiments as used for safety studies. Thus understanding the heat balance of a reactor is essential for the design of safe processes. Hereafter the different contributions to the heat balance, such as the heat release rate of the reaction, the heat exchange at the wall of the reactor, the heat dissipated by the stirrer, the heat accumulation in the reactor, the effects of the sensible heat of the feed, and the heat losses, will be discussed in detail. The different terms of the heat balance are expressed as heat release rates or thermal power.

11.2.3.1 Heat Production

The heat production corresponds to the rate of heat release by the reaction. Therefore it is proportional to the reaction enthalpy and to the reaction rate [Eq. (1)].

$$q_{RX} = -r_A \cdot (-\Delta H_R) \cdot V \tag{1}$$

This term will be of primary importance with respect to reactor safety: mastering the heat release by the reaction is the key to reactor safety. For a single reaction of order n, the reaction rate can be expressed as Equation (2).

$$-r_A = k_0 \cdot \exp\left[\frac{-E_A}{RT}\right] \cdot C_A^n \tag{2}$$

Two features of these expressions are important for safety purposes: the heat release rate of a reaction is an exponential function of temperature. Secondly the heat release rate is proportional to the volume of the reaction mass. Therefore it will vary with the cube of the linear dimension of the vessel (L^3) containing the reacting mass. It can also be expressed in a form allowing enhancement of the conversion rate [Eq. (3)].

$$q_{RX} = N_{A0} \cdot (-\Delta H_R) \frac{dX}{dt} \tag{3}$$

Consequently, the heat of reaction is obtained by integration, which allows one to express the conversion as thermal conversion in Equation (4).

$$X_{th} = \frac{\int_0^t q_{RX} \cdot d\tau}{\int_0^\infty q_{RX} \cdot d\tau} = \frac{\int_0^t q_{RX} \cdot d\tau}{Q_{RX}} \tag{4}$$

The thermal conversion is often used in the kinetic evaluation of calorimetric experiments. It is also very useful for safety assessments, since it gives direct information on the accumulation of reactants that could react even after a cooling failure. The thermal conversion is used for the calculation of the MTSR.

11.2.3.2 **Heat Exchange**

There are several mechanisms for heat exchange between a reacting medium and a heat carrier: radiation, conduction, and forced or natural convection. Here we shall consider convection only.

The heat exchanged with a heat carrier across the reactor wall by forced convection is proportional to the heat exchange area and to the driving force, that is, the temperature difference between the reaction medium and the heat carrier. The proportionality coefficient is the overall heat transfer coefficient [Eq. (5)]

$$q_{EX} = U \cdot A \cdot (T - T_C) \tag{5}$$

In the case of significant change in the physical chemical properties of the reaction mixture, which is quite common with polymerization reactions, the overall heat exchange coefficient U will also become a function of time. This is essentially due to the viscosity change during polymerization and to possible fouling at the reactor wall, which may sometimes become important. With respect to safety, two important features must be considered here: the heat removal is a linear function of temperature and since it is proportional to the heat exchange area, it varies as the square of the linear dimension of the equipment (L^2). This means that when the dimensions of a reactor have to be changed, as for scaleup, the heat removal capacity increases more slowly than the heat production rate. Therefore the heat balance becomes more critical for larger reactors. The contribution of heat exchange to the heat balance is examined in more detail in Section 11.4.1.

11.2.3.3 Heat Accumulation

Heat accumulation represents the variation of the energy content of a system with temperature [Eq. (6)].

$$q_{AC} = M_R \cdot c_p \cdot \frac{dT}{dt} \tag{6}$$

This includes all system compounds, the reaction mass as well as the reactor itself, or at least the parts directly in contact with the reacting system. However, in safety assessments, and as a worst case assumption, the heat capacity of the reactor itself is often neglected. Since heat accumulation is the consequence of a difference between heat production rate and cooling rate, it results in a variation of the temperature of the reactor contents.

11.2.3.4 Convective Heat Transport due to Feed

If a feed stream to a reactor is at a different temperature from the contents of the reactor, there is convective heat transport to the reactor. The thermal effect of the feed stream must be accounted for in the heat balance [Eq. (7)].

$$q_{Feed} = \dot{m}_{Feed} \cdot c_{p,\,Feed} \cdot (T_{Feed} - T) \tag{7}$$

This effect is also called "sensible heat". It can also be used as an additional means of cooling in semi-batch and in continuous reactors. When the temperature difference between reactor and feed is important and/or the feed rate is high, this term may play an important role. In such cases, when the feed is stopped an abrupt increase in the reactor temperature may result. This term is also important in calorimetric measurements, where the appropriate correction must be performed.

11.2.3.5 Stirrer

The mechanical energy dissipated by the agitator is converted into viscous friction energy and finally altered into thermal energy. In most cases this term may be neglected when compared to the heat released by a chemical reaction. But with viscous reaction masses, as for example with polymerization reactions, this term must be integrated in the heat balance. It can be estimated from Equation (8).

$$q_S = Ne \cdot \rho \cdot n^3 \cdot d_S^5 \tag{8}$$

Computation of the thermal energy dissipated by a stirrer requires knowledge of the Newton number, also called the power number, which depends on the stirrer type and the flow regime characterized by the Reynolds number. The contribution of the stirrer power to the heat balance is examined in more detail in Section 4.3.

11.2.3.6 Heat Losses

Industrial reactors are thermally insulated for safety reasons (hot surfaces) and for economic reasons (heat losses). Nevertheless, at higher temperatures heat losses may become important. Their calculation may become tedious, since heat losses are often due to a combination of losses by radiation and by natural convection. If estimation is required, a simplified expression using a global overall heat transfer coefficient may be useful [Eq. (9)].

$$q_{Loss} = h \cdot (T - T_{Amb}) \tag{9}$$

The simplest way of estimating the overall heat transfer coefficient (h) is by direct measurement at plant scale.

11.2.3.7 Simplified Expression of the Heat Balance

An overall heat balance taking all the terms explained above into account can be established [Eq. (10)].

$$q_{AC} = q_{RX} + q_{Feed} + q_S - q_{EX} - q_{Loss} \tag{10}$$

But in most cases a simplified heat balance, which comprises only the most important terms on the right-hand side of Equation (10), is sufficient for safety purposes. Such an expression will be used in Section 11.2.4. The problems linked with the cooling of polymerization reactors are presented in detail in Section 11.4.

11.2.4 Dynamic Control of Reactors

In fact a chemical reactor is governed simultaneously by its heat balance and its mass balance. If we consider a single reaction of order n, we obtain a set of equations [Eqs. (11)].

$$\begin{cases} \dfrac{dT}{dt} = \Delta T_{ad} \cdot \dfrac{dX_A}{dt} - \dfrac{U \cdot A \cdot (T - T_C)}{M_R \cdot c_p} \\ \dfrac{dX_A}{dt} = k_0 \cdot \exp\left[\dfrac{-E_A}{RT}\right] \cdot C_{A0}^n \cdot (1 - X_A)^n \end{cases} \tag{11}$$

In this system of coupled differential equations, the mass balance corresponds to the reaction rate and the heat balance is a simplified version showing only the heat production by the reaction and the heat removal by the cooling system, both terms resulting in heat accumulation. This system presents the property of parametric sensitivity, meaning that a small change in one of the parameters may lead to dramatic changes in the solution of the system of equations, that is, in the behavior of the reactor. This is an old [10–12], but always real, problem [13, 14]. This behavior may be observed for batch reactors and for tubular reactors (plug flow reactors) and also for bed reactors [15, 16]. Calorimetric methods make it possible to

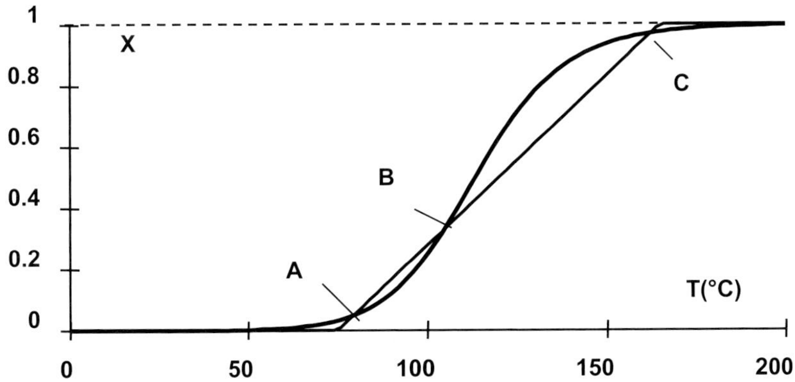

Fig. 11.4. Multiplicity of solutions for an exothermal reaction performed in an adiabatic CSTR. The straight line represents the heat balance and the S-shaped line the mass balance. A: cold branch operating point. B: instable operating point. C: hot branch operating point.

determine the data used to assess the dynamic stability of a reactor as a function of the production scale [17–19]. When performing this type of assessment, it is important to consider also variations in the overall heat transfer coefficient and in the temperature of the cooling system, which may vary with time or even with season.

A further specific aspect of the reactor heat balance is the multiplicity of solutions to the system of equations. This situation may arise with a CSTR in which an exothermal reaction is performed. The mass balance [Eq. (12)] is coupled with the heat balance [Eq. (13)], which gives a system of equations [Eq. (14)] that is represented graphically in Figure 11.4.

$$F_{A0} \cdot X_A = -r_A \cdot V \tag{12}$$

$$q_{RX} = -r_A \cdot V \cdot (-\Delta H_R) = F_{A0} \cdot c_p \cdot (T_R - T_0) \tag{13}$$

$$X_A = \frac{c_p}{-\Delta H_R} \cdot (T_R - T_0) = \frac{1}{\Delta T_{ad}} \cdot (T_R - T_0) \tag{14}$$

In such cases, the temperature of the reactor may jump from the low-temperature solution to the high-temperature solution, which can be catastrophic if the reactor is not designed to be operated under these conditions. This problem was extensively studied by Ray and co-workers [20].

11.2.5
Thermal Stability of Polymerization Reaction Masses

Since polymerizations are very exothermal, the potential temperature increase in the case of loss of control of the reaction itself is often great enough to cause evaporation of some components of the reaction mixture or even a partial cracking of the polymer chains. Hence the pressure of the system may increase, and cause the rupture of the vessel and consequently heavy damage to the plant and/or its environment. In most cases the thermal stability of polymerization reaction masses is not critical by itself: the main risk stems from the intended polymerization reaction. In polyaddition reactions as used for the production of resins, and especially of hot melts, the reaction mass remains reactive even after the end of the polymerization itself. Often curing is followed by a decomposition of the products that is accompanied by large gas releases, which means that decomposition may easily result in the explosion of the vessel containing the resin. This kind of product may present critical situations in the case of storage or transportation in large amounts. This problem goes beyond the scope of this handbook, however.

11.3 Specific Safety Aspects of Polymerization Reactions

Polymerization reactions form a well defined and specific class of reactions. They present some special features linked to the reaction kinetics and also to the thermodynamic aspects. These topics are presented in this section together with the factors that may affect the reaction rate.

11.3.1 Kinetic Aspects

The reactions taking place during the synthesis of a polymer are rather complex in nature. The description of the chemistry of a polymerization reaction often involves over 20 different elementary reactions. This means that control of the overall reaction rate that governs the process safety may be rather complicated. Nevertheless the kinetically determining step in polymerization reactions is the chain growth reaction.

With respect to chain growth, two main types of reaction may be distinguished (Table 11.1). One type is the addition of one monomer molecule at a time at the

Tab. 11.1. Polymerization mechanisms.

Type of chain growth[a]	*Reaction steps*	*Terminology for reaction*
Monomer addition $P_n^* + M \rightarrow P_{n+1}^*$	initiation growth termination transfer	polymerization (radical, ionic, coordination)
Polymer addition $P_n + P_m \rightleftarrows P_{m+n} + X$	growth	polycondensation polyaddition

[a] P_n^*: active monomer molecule with chain length n; M: monomer; P_n: bifunctional polymer molecule with length n; X: low molecular condensation product.

end of a growing chain until the chain is terminated. This is called a chain growth polymerization. The active center may be a free radical, an ion, or an available coordination bond on a transition metal complex of the catalyst. Once the active center is created, the growth of the chain is very fast, which results in a very short living time for the active chain – in the order of seconds – until it is deactivated by termination. Thus the concentration of active chains is extremely low, at approximately 10^{-8} mol L^{-1}. This renders these polymerizations very sensitive to impurities. After termination a dead polymer chain remains, but in the case of free-radical polymerization the dead chain may react again with further radicals.

The other possible growth mechanism involves two molecules carrying reactive groups, which react together leading to the addition of monomers, oligomers, or polymers. In this case the growth is achieved by the addition of longer elements than in chain growth. This type of reaction is called step growth polymerization, polyaddition, or polycondensation. With this mechanism, the growth of the chain proceeds more slowly, so the lifetime of the active chain and therefore the polymerization time are longer. These are often equilibrated reactions, so that the low molecular by-products must be removed from the reaction mass in order to shift the equilibrium toward the products.

These different chain growth mechanisms result also in different activation energies. These are in the order of magnitude of only 20 kJ mol^{-1} for chain growth polymerizations. But polycondensations and polyaddition present high activation energies in the order of 100 kJ mol^{-1}. Consequently the reaction temperatures are very different: whereas polymerizations are often performed at modest temperature levels from 50 to 100°C, polycondensations require higher temperatures between 150 and 250°C. Since the activation energy dictates also the variation of the reaction rate with temperature, chain growth polymerizations are less sensitive to temperature excursions than step growth polymerizations are.

11.3.2 Thermochemical Aspects

Polymerizations are generally exothermal reactions with specific energies up to 3600 kJ kg^{-1}, corresponding to an adiabatic temperature rise of up to 1800 K. Some typical reaction enthalpies are presented in Table 11.2, together with the specific heat of reaction and adiabatic temperature rise obtained for mass polymerization. Most free-radical and ionic polymerizations have negative standard enthalpies and standard entropies; thus at higher temperatures these reactions must be considered reversible [Eq. (15)].

$$P_n^* + M \rightleftharpoons P_{n+1}^* \tag{15}$$

As for every reversible exothermal reaction, there is an equilibrium temperature at which the chain growth and depolymerization rates are equal. This temperature is called the ceiling temperature and is related to the equilibrium concentration of the monomer according to Eq. (16).

Tab. 11.2. Typical reaction enthalpies and corresponding adiabatic temperature rises.

Monomer	Physical state[a]	Temperature [°C]	$-\Delta H_R$ [kJ mol^{-1}]	$-\Delta H_R$ [kJ kg^{-1}]	$-\Delta T_{ad}$ [°C]
Ethylene	g c	25	101.5	3620	1810
Propene	l c	25	84	2000	1000
Isobutene	l c	25	48	855	428
Butadiene	l c	25	73	1350	676
Isoprene	l c	25	75	1100	559
Chloroprene	l c	61.3	68	768	384
Acrylamide[b]	s	74.5	81.5	1147	574
Acrylonitrile	l c′	74.5	76.5	1423	721
Acrylic acid	l c	74.5	67	930	465
Methyl acrylate	l c	76.8	78	1100	435
Methyl methacrylate	l c	74.5	55.5	550	277
Vinyl acetate	l c	74.5	88	1022	511
Vinyl propionate	l c	74.5	86	860	430
Vinyl chloride	l c	25	71	1135	542
Vinylidene chloride	l c′	25	75.5	780	390
Styrene	l c	25	70	672	336
α-Methylstyrene	l c	25	35	296	148
Ethylene oxide	l c′	25	94.5	2145	1073
Propylene oxide	g	25	75.5	1300	650
Trioxane[c]	s c′	30	19.5	216	108
ε-Caprolactam	l c	200	15.5	137	68
2-Pyrrolidone	l c	75	4.5	53	26

[a] Physical state: l: liquid; g: gas; s: in solution, c: condensed amorphous, c′: crystalline or part crystalline.
[b] Solvent is water.
[c] Solvent is methylene chloride.

$$T_C = \frac{\Delta H_R}{\Delta S^0 + R \ln C_{M,C}} \tag{16}$$

It is important to realize that the ceiling temperature is not a constant, but it is a function of the monomer concentration. For most monomers, this equilibrium concentration is lower than the detection limit of common analytical methods. An exception is α-methylstyrene, with a ceiling temperature of 61°C for 100% monomer and a concentration of 2.2 mol L^{-1} at 25°C.

In the runaway of a polymerization, if a temperature in the range of the ceiling temperature may be reached, or in other words the MTSR may be close to the ceiling temperature ($MTSR \approx T_C$), then the safety analysis must account for the contribution of the depolymerization reaction, which produces low molecular species and may result in a pressure increase. This can easily be realized with the criticality classes presented above, by choosing the temperature T_{D24} on the basis of the depolymerization reaction.

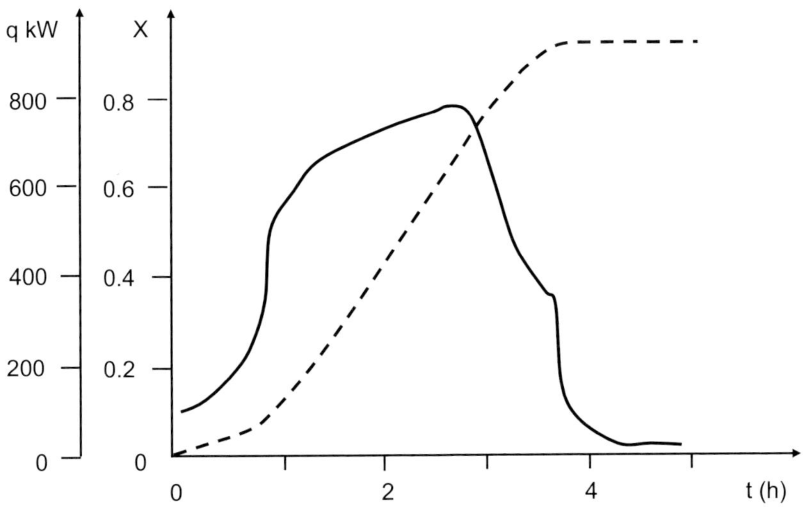

Fig. 11.5. Heat release rate (solid line) and monomer conversion (broken line) during the polymerization of 5000 kg vinyl chloride and 5000 kg water as a suspension. (after Hamielec [21]).

The high enthalpy of reaction and the high concentrations often used in industrial polymerization processes lead to very high heat release rates. As an example [21], the polymerization of vinyl chloride as a 50% (by weight) suspension in water with a charge of 5000 kg vinyl chloride and 5000 kg water releases an energy of 5.7 GJ and the heat release rate is about 780 kW (Figure 11.5). This high heat flow presents an engineering challenge in order to ensure a sufficient heat transfer coefficient.

The high heat release rates observed in polymerization reactions do not represent by themselves the only safety engineering problem. A further issue is the fact that the reaction dynamics leads to fast changes of the heat release rate during the process. This in turn requires a fast-acting temperature control system, which may be difficult to achieve for large reactor volumes. In the example given in Figure 11.5, after about 40 to 50 minutes of reaction time, corresponding to 5 to 10% conversion, the heat release rate increases from 100 to over 700 kW. This is even more difficult, in the sense that the reaction temperature should be maintained constant in order to obtain the required properties of the final product. The accumulated heat at this stage represents approximately 90%, or 5.4 GJ. This energy could lead to an adiabatic temperature increase of over 1900 K if not controlled by the temperature control system. The thermodynamic characteristics of the reaction and the heat transfer capability of the reactor may lead to a parameter combination giving rise to the parametric sensitivity as explained in Section 11.2.4 and Eq. (11).

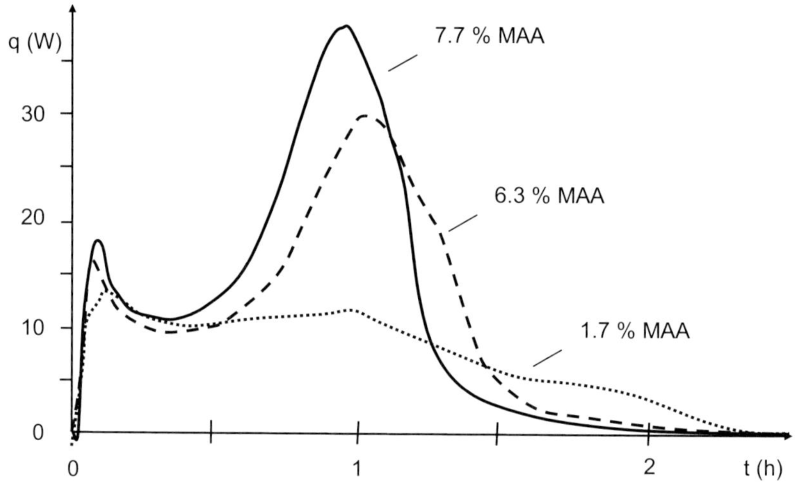

Fig. 11.6. Heat release rate as a function of time for the copolymerization of methacrylic acid (MAA) with styrene in emulsion with different mass fractions of methacrylic acid. The self-accelerating behavior corresponding to the gel effect is evident after 20 min. Composition of the reaction mass: 418 g water, 120 g styrene, 10/8/2 g methacrylic acid, 0.505 g $K_2S_2O_8$, 7 g Na dodecylbenzenesulfonate.

11.3.3
Factors Leading to Changing Heat Release Rates

Some polymerization processes present a particular kinetic behavior: after a certain time the reaction accelerates itself, leading to an increase in the heat release rate. This can be exemplified with the copolymerization of styrene and methacrylic acid in emulsion [22]. At the beginning of the polymerization, a fast increase of the heat release rate is observed (Figure 11.6). This is due to the formation of latex particles that are essentially nucleated during the initial reaction phase: the reaction rate is often proportional to the number of latex particles. For higher concentrations of methacrylic acid, a second acceleration of the reaction can be observed: this self-acceleration is also called gel effect or the Tromsdorff effect. It is due to the fact that the viscosity of the reaction mixture increases and the termination reactions of radical chains become diffusion controlled. Thus the concentration of living chains increases and so does the heat release of the reaction. This phenomenon appears as a self-acceleration, and like other types of self-accelerated reactions, this is also sensitive to small changes in the process conditions. This is shown in Figure 11.6 by the changes in the initial concentration of methacrylic acid, which cause amplified changes in the maximum heat release rate and also in the time point where this maximum occurs. The maximum heat release rate increases from 14 W kg^{-1} with 1.7% methacrylic acid to 30 or 40 W kg^{-1} when the concentration is increased to 6.3 or 7.7%.

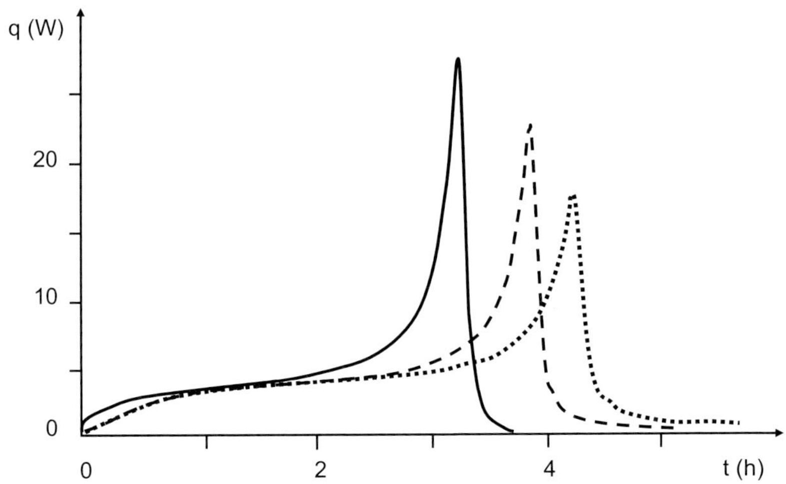

Fig. 11.7. Heat release rate as a function of time for the copolymerization of methacrylic acid (MAA) with styrene in emulsion following a seeding process. The mass fraction of methacrylic acid was varied: solid line, 10%; broken line, 4.2%; dotted line, 1.7%. Composition of the reaction mass: 451 g water, 9 g styrene as seed ($d_P = 112$ nm), 108/115/118 g styrene, 12/5/2 g methacrylic acid, 0.505 g $K_2S_2O_8$, 0.26 g Na dodecylbenzenesulfonate.

The same copolymerization performed as a seeding process shows a totally different behavior that implies dramatic changes in the heat release rate as a function of time. In a seeding polymerization, the time required for the formation of the latex particles disappears because a given amount of latex particles are added at the beginning of the process [22]. Therefore the particles swell in the presence of the monomer and only the phases of particle growth and monomer depletion can be observed (Figure 11.7). This leads to an initial heat release rate, which remains practically constant during particle growth. The reaction seems to follow a zero-order rate law. Then after a period of time depending on the initial concentration of methacrylic acid, the gel effect leads to an increasing reaction rate. Consequently the heat release rate increases practically exponentially, even under isothermal conditions. The time interval after which this increase occurs and the maximum heat release rate depend on the concentration of methacrylic acid. Such a behavior may easily lead to a runaway situation if not anticipated. Additionally the particle size of the seed affects the dynamic behavior of the reaction.

Therefore it is essential to study the thermal behavior of a polymerization reaction during process development. A safe process may only be designed if these phenomena are thoroughly understood and if the corresponding engineering means are used for scaleup. A well suited tool for this kind of study is reaction calorimetry [23–26].

11.4 Cooling of Polymerization Reactors

Different technical means may be used for cooling industrial reactors. Among them we consider three methods: direct cooling using the sensible heat of a feed stream, which was treated in Section 11.2.3; indirect cooling using heat exchange across the reactor wall or with internal cooling systems; and hot cooling using the latent heat of evaporation. These latter two cooling techniques are discussed in Sections 11.4.1 and 11.4.2. In Section 11.4.3, the problem of viscosity change is examined.

11.4.1 Indirect Cooling: Heat Exchange Across the Reactor Wall

The heat transfer across a wall can be expressed by the general equation, Eq. (5). In practice, the temperature difference between the reaction medium and the cooling system must remain within reasonable limits. One limitation is given by the dynamic stability in the case of a batch reaction. But another limitation is due to the fact that too low a wall temperature may cause the buildup of a polymer film, or fouling at the inner side of the wall. In a homogeneous reaction mixture, even without fouling, the resistance of the laminar heat transfer film increases as the wall temperature decreases. The result is that too low a cooling medium temperature may have adverse results, since the reduction of the heat transfer coefficient may overcome the benefits of the higher temperature gradient. Increasing the heat transfer area by using inserts such as cooling coils or other shapes of built-in heat exchangers is often problematic. These inserts may enhance the formation of a polymer film at the wall and they also may have a negative impact on the agitation of the reaction mass. Further, they render the cleaning operations of the reactor more difficult. Another method, which is often used to increase the heat exchange area in low molecular reactions, is loop circulation through an external heat exchanger. This technique may be inadequate if the polymers are sensitive to the shear forces that will arise in a circulation pump. In certain cases reactors with an increased height to diameter ratio are used in order to increase the specific cooling area.

The problem of the determination of the overall heat transfer coefficient can be simplified if one considers that the overall resistance to heat transfer consists of several resistances in series [Eq. (17)].

$$\frac{1}{U} = \underbrace{\frac{1}{h_R} + \frac{d_p}{\lambda_p}}_{\substack{\textit{depends on}\\ \textit{reaction mass}}} + \underbrace{\frac{d_w}{\lambda_w} + \frac{d_f}{\lambda_f} + \frac{1}{h_C}}_{\substack{\textit{depends}\\ \textit{on reactor}}} = \frac{1}{h_R} + \frac{d_p}{\lambda_p} + \frac{1}{\varphi} \tag{17}$$

The first two terms, representing the liquid film at the inner wall of the reactor and any polymer deposit, only depend on the reactor contents, that is, on the agitation and on the physical properties of the reaction mass. Here the resistance of the internal laminar film and of the polymer deposit at the wall play a key role. Therefore the reactor should be cleaned regularly with high-pressure cleaner. The last three terms depend on the reactor itself, especially on its heat exchange system, that is, on the reactor wall, on fouling in the jacket and on external liquid film. They are often grouped into one term: the equipment heat transfer coefficient (φ). For the description of the heat transfer coefficient of the internal film, there are several correlations available, the most popular of which is presented in Eq. (18) [27].

$$Nu = C^{te} \cdot \mathrm{Re}^{2/3} \cdot \mathrm{Pr}^{1/3} \cdot \left(\frac{\mu}{\mu_W}\right)^{0.14} \qquad (18)$$

This expression is valid for Newtonian fluids; therefore with polymers its validity must be verified. In this correlation, the dimensionless numbers are defined in Eq. (19). Here the Reynolds number (Re) is expressed for a stirred tank where the flow rate corresponds to the tip speed of the agitator.

$$Nu = \frac{h_R \cdot d_R}{\lambda} \quad \mathrm{Re} = \frac{n \cdot d_S^2 \cdot \rho}{\mu} \quad \mathrm{Pr} = \frac{\mu \cdot c_p}{\lambda} \qquad (19)$$

The last term in Eq. (18) represents the ratio of the viscosity of the reaction mass at reaction (bulk) temperature to its viscosity at wall temperature. It accounts for the changes of the heat transfer coefficient, when switching from heating to cooling. This produces an inversion of the temperature gradient and therefore affects the viscosity of the product close to the reactor wall. With reactions performed in solvents it can generally be neglected, but may become important in the case of polymers: the viscosity of the reaction mass is often important and its temperature dependence may give this term a value that cannot be neglected. Equation (18) can be solved for h_R, the internal film heat transfer coefficient. By grouping the terms in an appropriate way, the heat transfer coefficient of the reaction mass can be split into two parts, one (z) representing the technical data of the reactor and the other (γ) grouping the physical properties of the reaction mass [Eq. (20)]. The gravity constant g is introduced to rend the first term dimensionless and to give the second the dimensions of a heat transfer coefficient.

$$h_R = \underbrace{C^{te} \frac{n^{2/3} d_S^{4/3}}{d_R g^{1/3}}}_{\substack{\text{technical data}\\ \text{of the reactor}}} \underbrace{\sqrt[3]{\frac{\rho^2 \lambda^2 c_p g}{\mu}}}_{\substack{\text{physical-chemical data}\\ \text{of the reaction mass}}} = z.\gamma \qquad (20)$$

Thus, for a given reaction mass, the heat transfer coefficient of the internal film can be influenced by the revolution speed of the agitator and its diameter. The value of z, characterizing the internal part of the equipment factor, can be calculated using the geometric characteristics of the reactor. Some typical values of the agitator constant are given in Table 11.3. The value of γ can either be calculated

Tab. 11.3. Typical agitator constants.

Agitator	*Constant*
Plate stirrer	0.36
Rushton turbine	0.54
Rushton turbine with pitched blades	0.53
Propeller	0.54
Anchor	0.36
Impeller	0.33
Intermig (Ekato)	0.54

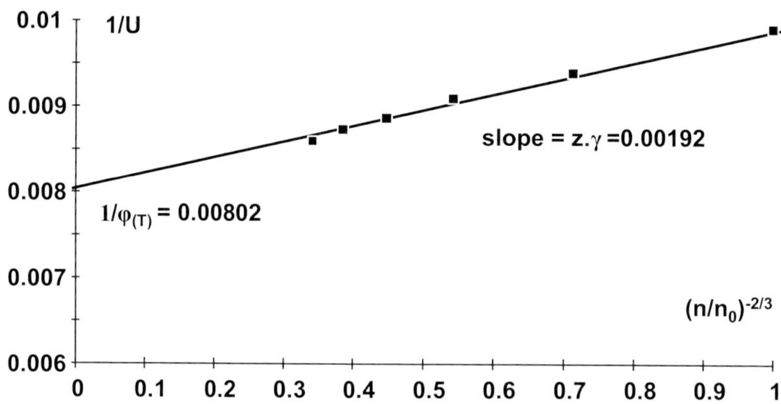

Fig. 11.8. Wilson plot measured with toluene in a reaction calorimeter. The reciprocal heat transfer coefficient is plotted against the agitator revolution speed to the power −2/3. The intercept with the ordinate represents the reciprocal heat transfer coefficient of the equipment and either z or γ can be determined from the slope of the straight line.

from the physical properties of the reactor contents – as far as they are known – or measured in a reaction calorimeter by the Wilson plot method [28, 29]. This parameter is independent of the geometry or size of the reactor. Thus it can be determined at laboratory scale and used at industrial scale. The Wilson plot consists of the determination of the overall heat transfer coefficient as a function of the agitator revolution speed in a reaction calorimeter. The Wilson plot (Figure 11.8) makes it possible to verify that the correlation in Eq. (18) is valid: if the measures fit on a straight line, a validation is built into the method. The intercept with the ordinate represents the reciprocal heat transfer coefficient of the equipment, that is, the wall and the external cooling system of the calorimeter. The slope is the product of z and γ, which allows the determination of either one of these parameters. In a first stage, z is determined by a calibration performed using a solvent with known physical properties. In a second stage γ is determined during the actual measurement with the reaction mixture.

Moreover, the contribution of polymer deposits to the overall resistance to heat transfer may be important. Since the thermal conductivity of polymers is low, they

Tab. 11.4. Factors influencing the heat transfer with some typical values of heat transfer coefficients in an agitated reactor.

Type	*Influencing factors*	*Typical values* $[W\,m^{-2}\,K^{-1}]$	
Internal film	stirrer: speed and type	water	1000
h_R forced convection	reaction mass c_p, λ, ρ, η	toluene	300
	physical data, especially $\rho = f(T)$	glycerol	50
h_R natural convection		water	100
(stirrer failure)		gases	10
Polymer deposit	thermal conductivity (λ)	With $d = 1$ mm,	
	thickness of deposit	PE	300
		PVC, PS	170
Reactor wall λ/d	construction	With $d = 10$ mm,	
	wall thickness (d)	iron	4800
	construction material	stainless steel	1600
	coating	glass	100
		glass lined	800
Fouling at external wall	thermal conductivity (λ)	With $d = 0.1$ mm,	
	thickness of deposit	Gallert	3000
		Kesselstein	5000
External film h_c	jacket	water	
	construction, flow rate	with flow	1000
	heat carrier, physical properties, phase change	no flow	100
		evaporation	3000
	welded half coil	water	
	construction, flow rate	with flow	2000
	physical properties	no flow	200

act as insulators, and thus even thin deposits may affect strongly the heat transfer coefficient, as shown in Table 11.4.

The resistance of the reactor wall and external film (h_c) can be determined in a cooling experiment realized directly with the production reactor filled with a known amount of a substance (M) with known physical chemical properties. The temperature T_R of the contents of the reactor and the average temperature T_C of the cooling system are recorded during this experiment. A heat balance can be calculated between two instants t_1 and t_2: the heat removed from the contents of the reactor is given by Eq. (21), and this is realized with an average cooling power given by Eq. (22) with the average temperature difference as expressed in Eq. (23).

$$Q = M \cdot c_p \cdot (T_{R1} - T_{R2}) \tag{21}$$

$$q_{EX} = U \cdot A \cdot \overline{\Delta T} \tag{22}$$

$$\overline{\Delta T} = \frac{(T_{R1} - T_{C2}) - (T_{R2} - T_{C2})}{\ln(T_{R1} - T_{C1}) - \ln(T_{R2} - T_{C2})} \tag{23}$$

The overall heat transfer coefficient can then be obtained from the heat balance equation [Eq. (24)].

$$U = \frac{M \cdot c_P \cdot (T_{R1} - T_{R2})}{A \cdot \overline{\Delta T} \cdot (t_2 - t_1)} \tag{24}$$

This is a simplified method using only two points. A more accurate method is to use more points by applying the differential equation of Newtonian cooling, which expresses the variation with time of the temperature difference between reactor contents and cooling system [Eq. (25)].

$$-M \cdot c_P \cdot \frac{d(\Delta T)}{\Delta T} = U \cdot A \cdot dt \tag{25}$$

By integration, with the initial condition: $t = 0$, $\Delta T = \Delta T_0 = T_0 - T_C$, one obtains Eq. (26) with the thermal time constant τ_c of the reactor given by Eq. (27).

$$-\ln\left(\frac{\Delta T}{\Delta T_0}\right) = \frac{t}{\tau_C} \quad \text{or} \quad \left(\frac{\Delta T}{\Delta T_0}\right) = \exp\left(-\frac{t}{\tau_C}\right) \tag{26}$$

$$\tau_C = \frac{M \cdot c_P}{U \cdot A} \tag{27}$$

Thus a plot of the logarithm of the ratio of the instantaneous temperature difference to the initial temperature difference, as a function of time, should give a straight line, as far as cooling is Newtonian. The slope of this line is the thermal time constant of the reactor, from which the overall heat transfer coefficient U can be calculated.

Then the equipment heat transfer coefficient φ can be calculated from Eq. (28) written for a clean reactor. If there was some polymer deposit at the reactor wall during the cooling experiment, its resistance must be taken into account.

$$\frac{1}{\varphi} = \frac{1}{U} - \frac{1}{z\gamma} \tag{28}$$

For the calculation of the heat transfer coefficient of the external film some models are also available. These models describe the hydraulics of the flow in the jacket or in the half-welded coils. The results depend strongly on the technical design of these parts of the equipment. Direct measurement is mostly preferred.

11.4.2
Hot Cooling: Cooling by Evaporation

Hot cooling, which uses the latent heat of evaporation of a solvent, is a very efficient technique: on one hand it is independent of the heat transfer at the reactor wall, and on the other hand the condenser can be designed independently of the reactor's geometry. This allows relatively high specific cooling powers to be reached. In case a reaction cannot be performed at boiling temperature, it is possible to work under partial vacuum in order to decrease the boiling point and to work at reflux. Hot cooling can be used as the main cooling system for a reactor working under normal operating conditions; it can also be used as an emergency cooling system, in cases when the boiling point is reached during the temperature increase following a failure of the main cooling system. Obviously, this is only possible provided the condenser is equipped with an independent cooling system and the equipment has been designed for this purpose.

The boiling rate of the solvent, depending on the instantaneous heat release rate of the reaction, governs the whole design of the reflux system. Some technical aspects and limitations must be considered in the design of such cooling systems. Too high a boiling rate could lead to flooding of the vapor tube, when the condensate flows down in countercurrent to the rising vapor. Further, the presence of vapor bubbles in the reaction mass increases its apparent volume. The reaction mass swells, and if its level is high enough to enter the vapor tube, again flooding will occur. A method was developed for predicting these limitations [30, 31]. If the boiling point is reached during runaway, like in scenario of criticality classes 3 or 4, a possible secondary effect of the evaporation is the formation of an explosive vapor cloud, which in turn can lead to a severe room explosion if ignited. In some cases, there is enough solvent present in the reaction mixture to compensate for the energy release, allowing the temperature to be stabilized at the boiling point. This is only possible if the solvent can be distilled off in a safe way, to a catch pot or a scrubber. The thermal stability of the concentrated reaction mixture must also be ensured. In most cases, however, the condensed solvent is refluxed to the reactor. If the capacity of the reflux system is sufficient, all of the vapor produced by an exothermal reaction can be conducted from the reactor to the condenser, where it is entirely condensed. In such a case, the boiling point may act as a safety barrier. In the opposite situation, if the flow rate is too high with respect to the capacity of the system, a pressure increase will result. This may be due to flooding of the vapor tube or to swelling of the reaction mass. This pressure increase may in turn lead to the rupture of the weakest element in the distillation system, followed by the release of a vapor cloud, which could be ignited and cause a vapor cloud explosion. Further, if the condenser capacity is too short, uncondensed vapor will pass to the ventilation system, where it could cause a secondary incident, if it is not compatible with the design. For these reasons, it is essential to know the vapor flow rate due to an exothermic reaction, which depends on the latent heat of evaporation of the solvent. The capacity of the reflux system will be assessed using the vapor velocity in the different parts of the equipment [Eq. (29)].

$$u = \frac{q_{RX}}{\Delta H_V \rho_G S} \tag{29}$$

If the diameter of the vapor tube is insufficient for a given vapor release rate, the

Tab. 11.5. Limiting vapour velocity for different solvents.

	Solvent							
	water	***methanol***	***ethanol***	***acetone***	***dichloro-methane***	***chlorobenzene***	***toluene***	***m-xylene***
ΔH_v [kJ kg^{-1}]	2260	1100	846	523	329	325	356	343
T_b [°C]	100	65	78	56	40	132	111	139
M_w [g mol^{-1}]	18	32	46	58	85	112	92	106
P [mbar]	1013	1013	1013	1013	1013	1013	1013	1013
ρ_g [kg m^{-3}]	0.59	1.15	1.60	2.15	3.31	3.37	2.92	3.13
u_{max} [m s^{-1}]	10.2	6.6	5.3	5.1	4.5	4.4	4.8	4.6

high vapor velocity results in a pressure increase in the reactor leading to a temperature increase and a further acceleration of the reaction. The consequence will be a thermal explosion until the rupture of weak parts of the equipment allows pressure relief. In order to avoid this type of reaction course, it is important to know the maximum vapor velocity admissible in a given tube and consequently the maximum admissible heat release rate for the reaction. To predict whether flooding will occur in existing equipment, an empirical correlation was established experimentally [31]. The experimental study was performed in the laboratory, in the pilot plant and on an industrial scale with various organic solvents and water for tubes with an inside diameter between 6 and 141 mm. The maximum allowable heat release rate is obviously a function of the latent heat of evaporation and of the tube cross-section. It can be calculated by Eq. (30).

$$q_{\max} = (4.52\Delta H_V + 3.37 \cdot 10^6)S \tag{30}$$

Calculations performed for different common solvents show that the limiting velocity remains relatively constant for different classes of solvents. Some values are shown in Table 11.5. This allows the required vapor tube diameter d for a given heat release rate for different solvents to be calculated (Figure 11.9).

The relative volume increase due to swelling of the reaction mass can be estimated using the Wilson correlation (Eqs. (31), (32) with the conditions: if $\tilde{u} < 2$ then $K = 0.68$, $a = 0.62$; or if $\tilde{u} \geq 2$ then $K = 0.88$, $a = 0.40$) [32, 33]. This correlation was first established for air in water; it is easy to use and was found to describe with enough accuracy the swelling of a liquid by bubbles of its vapor.

$$\alpha = K\left(\frac{\rho_V}{\rho_L - \rho_V}\right)^{0.17} \tilde{D}_H^{-0.1}\tilde{u}^a \tag{31}$$

$$\alpha = \frac{H_B - H_0}{H_B} \qquad \tilde{D}_H = \frac{D_H}{\sqrt{\frac{\sigma}{g \cdot (\rho_L - \rho_V)}}} \qquad \tilde{u} = \frac{u}{\sqrt{g\sqrt{\frac{\sigma}{g \cdot (\rho_L - \rho_V)}}}} \tag{32}$$

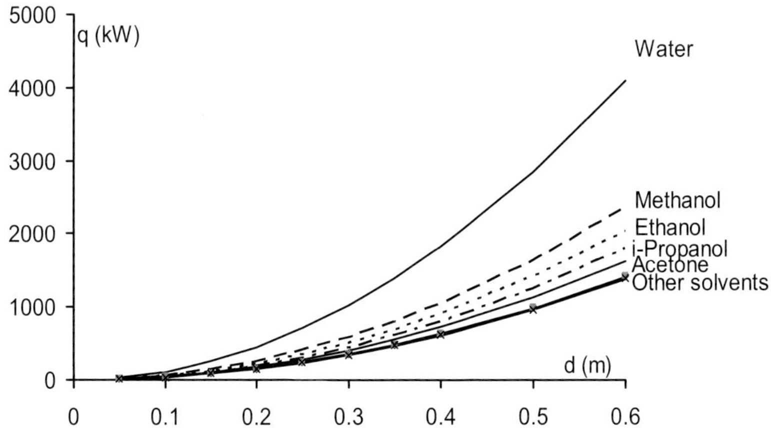

Fig. 11.9. Heat release rate at flooding as a function of the vapor tube (riser) diameter for different solvents.

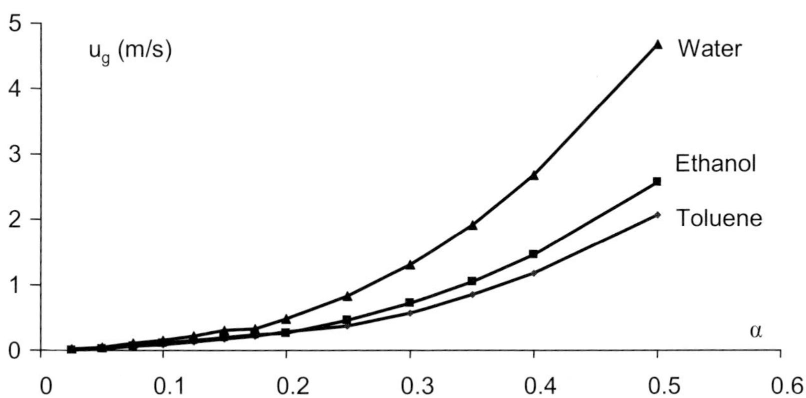

Fig. 11.10. Maximum allowed vapor velocity u_g across the surface of the reaction mass as a function of its degree of swelling α.

The correlation allows the maximum admissible heat release rate for given plant equipment to be calculated, or equipment for the requirements of a given process to be designed (Figure 11.10). They are both based on easily accessible physical-chemical properties of the boiling solvent and on geometric data of the reactor. Such considerations make it possible to adapt the equipment or the process, that is, the degree of filling of the reactor, to the safety requirements. This kind of measure often allows processes to be run under safe conditions, whereas a classical assessment would consider them to be critical.

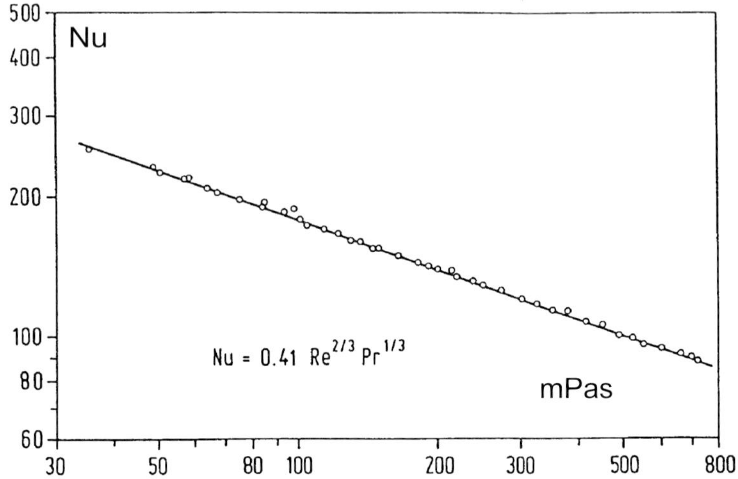

Fig. 11.11. Nusselt number as a function of the apparent viscosity during reticulation of poly(vinyl alcohol) in aqueous solution in a laboratory reactor with an anchor stirrer. Geometry $V_R = 1.3$ L, $D = 0.1$ m, $d_R/D = 0.95$, $H/D = 1.9$, $\rho = 1000$ kg m^{-3}, $\lambda = 0.65$ W m^{-1} K^{-1}. After Schmidt and Reichert [38].

11.4.3
Importance of the Viscosity

In polymerizations the fluid often does not behave in a Newtonian manner. In such cases Eqs. (18)–(20) do not apply as such, and modified equations must be used. It could be shown experimentally [34–36] that the shear forces are proportional to the agitator revolution speed, meaning that an average shear force in the reactor can be used and the dynamic viscosity can be replaced by an apparent viscosity [37]. Such an example is illustrated in Figure 11.11, where the variation of the Nusselt number *Nu* during a polymerization shows the decrease in the overall heat transfer by a factor of three when the viscosity increases by over one order of magnitude [38]. In mass polymerization an increase of the viscosity by six or seven orders of magnitude is common. This means that the viscosity is a key factor for the control of a polymerization reactor, that is, for its safety.

In fact the viscosity influences both the heat balance and the mass balance. It has been shown how the heat transfer coefficient is affected by the viscosity. But the energy dissipation by the stirrer is also strongly dependent on viscosity (see Section 11.4.4). Furthermore, viscosity affects the molecular diffusion, the mass transport, the mixing time, or the residence time distribution, and therefore the reaction rate. Since the reaction rates influence the chain length and particle sizes, they have a direct effect on the polymer properties. In turn they affect the viscosity and the shear forces – there is a feedback effect. Such complex interactions cannot be described by analytical equations, so empirical models must be used. Often these models are very specific for a given reaction system and even for a reactor geometry. Their use affords deep chemical and engineering knowledge and experience. To give an example, increasing the stirrer speed in order to improve the heat transfer may result in an adverse situation: one reason is that the heat dissipation by the stirrer is increased. But a further effect may arise at the same time: the mixing time is reduced, enhancing the mass transfer and resulting in an increase in the reaction rate that leads to an effect opposite to the initial intention.

11.5 Chemical Engineering for the Safety of Polymerization Processes

Chemical reaction engineering traditionally addresses the design of processes operated under nominal operating conditions. But if the same principles are used to design the process to remain safe even under conditions deviating from the intended conditions, it becomes a strong tool for improving process safety [1]. For this reason, in this section the different reactor types are analyzed with regard to their safe operation and especially to the design in order to prevent runaway. Some further risk-reducing techniques for stopping a commencing runaway or mitigating its consequences are also presented.

11.5.1 Batch Processes

Batch processes are characterized by a closed mass balance. This means that during the reaction nothing is added to or withdrawn from the reactor. In fact we consider as batch processes those where the monomer and the solvent, if there is one, are added at the beginning of the reaction. The initiator or catalyst is also added at the beginning of an operation. We extend this definition to processes where a volatile compound is evaporated in order to shift the equilibrium to the products, as is often the case with polyadditions or polycondensations. In these processes the control of the temperature is the unique way of controlling the reaction course. Different strategies of temperature control may be used in this situation.

An isothermal operation requires the cooling capacity to be sufficient at any time to compensate for the heat release rate of the reaction. Moreover, the dynamics of the cooling system must allow sharp changes in the reaction rate, which are often observed in polymerization reactions (see Section 3.3), to be followed. Due to the high energy of reaction, this kind of process is only possible in diluted systems as emulsion polymerization or polymerization in solution.

In most cases non-isothermal processes are used. A part of the energy of the reaction is used to heat the reactor to the desired process temperature in the so-called polytropic reaction mode. A typical temperature control strategy is to use an adiabatic phase, during which there is no cooling, so that the correct initiation of the reaction can be checked. Then, in a second stage, after reaching a specified temperature, the cooling system is switched. The correct choice of the initial temperature and of the cooling system switching temperature is critical for the process safety. This type of strategy is often used in polyaddition or polycondensation reactions, where high reaction temperatures are desired. Advanced temperature control strategies allowing the optimization of productivity, but taking technical limitations of the plant equipment into account, are proposed [39]. Great care must be used to avoid working in the parametrically sensitive field. Thus a thorough study of the process by reaction calorimetry is essential for the safe design of such processes. The aim of process design should be to develop processes that are tolerant to a failure of the cooling system. This is often a utopian dream, due to the high energy of polymerization reactions.

In batch operation, the correct charge of the reactor is essential, and requires great care by the operators, because this is often a manual operation. In every case a batch process requires emergency measures to be taken in order to recover control of the reactor when the cooling system fails, or at least to mitigate the consequences of a runaway. The scenario presented in Section 11.2 is a great help for this purpose.

11.5.2 Semi-batch Processes

In semi-batch operation only a part of the monomer, or no monomer at all, is initially charged. The main part of the monomer is added to the reactor during the process, allowing additional control of the reaction course and of the polymer properties by adjusting the feed rate. In such processes the accumulation of monomer and therefore of heat is limited to a fraction of the overall heat. Such processes are often used for emulsion polymerizations. The continuous feed also reduces the instantaneous heat release by the reaction and requires less cooling capacity. A correct choice of the concentrations and of the temperature even makes it possible to achieve a so-called feed-controlled process. This means that the monomer entering the reactor is immediately converted; that is, there is no monomer accumulation. In such a situation the reaction rate is equal to the feed rate, which can easily be adapted to the cooling capacity of the reactor (Figure 11.12). In the case of any failure, stopping the feed immediately stops the reaction and the heat production. Thus the temperature can be stabilized, making it possible to design fail safe processes.

The design of a safe semi-batch process requires a strict control of the accumulation of unconverted monomer. In fact the accumulation results from a balance between monomer addition by the feed, and monomer conversion by the reaction. Thus it can be influenced by the feed rate, but also by the reaction rate, which in turn can be governed by the temperature and the initiator or catalyst concentration. These parameters must be optimized during process development. Here again, reaction calorimetry is of great help for the determination of the accumulation [40]. It can be calculated from thermal conversion by means of Eq. (4) obtained directly from the calorimetric measurement and the actual amount fed. Just by varying the

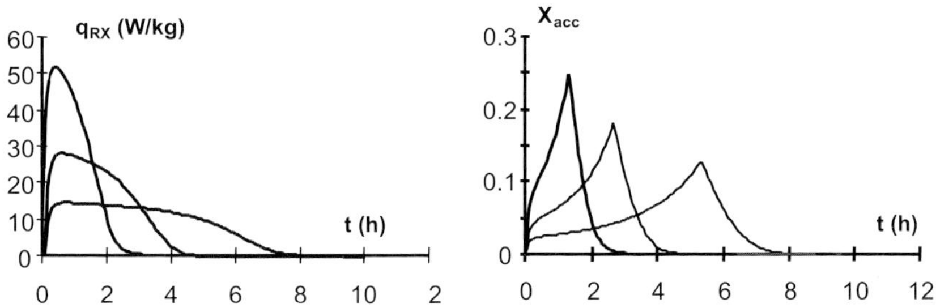

Fig. 11.12. Heat release rate and accumulation obtained in a semi-batch reaction for different feed rates: 2, 4 and 6 h.

feed rate and the process temperature, one can verify that the heat release rate of the reaction never exceeds the cooling capacity of the reactor. A second constraint is the maximum allowed temperature in the case of a cooling failure, that is, the accumulation, and therefore the MTSR should remain below a given level (Figure 11.12) [41, 42].

In practice, emulsion polymerizations are often performed according to the monomer feed process with an initial charge of the monomer, the remaining being fed over time. Figure 11.13 represents a schematic example of an emulsion polymerization according to the monomer feed process [2]. In such processes three phases may be distinguished: the initiation, the feed and the end of reaction. In this example, the delay between initiator addition and continuous feed was varied.

11.5.2.1 **Initiation**

In a first stage, an initial amount of monomer is added to the reactor. This addition is immediately followed by the addition of the initiator. During the delay between initiator addition and monomer feed, inhibition of the polymerization may occur. This depends on the purity of the reactants and on the oxygen concentration remaining in the reactor. Then a fast increase in the thermal power follows: it is due to formation of the particles. On an industrial scale, such a fast increase of heat release rates may generally not be compensated by the comparatively slow cooling system, and results in an accumulation of heat, that is, an increase in the reactor temperature. But at this stage, the monomer concentration is still low. Often the total amount of water was charged initially, conferring a high heat capacity on the system. Thus the resulting temperature increase during this stage remains low. In fact the initial charge must be optimized in such a way that the initiation remains safe. In the case of emulsion feed polymerizations, the initial contents of the reactor represent an aliquot of the total batch. The situation is similar to the monomer feed process, since the initial charge may be designed in such a way that the temperature increase remains within critical limits. The cooling capacity, and sometimes also the heat capacity, of the reactor are high enough to limit the initial temperature excursion.

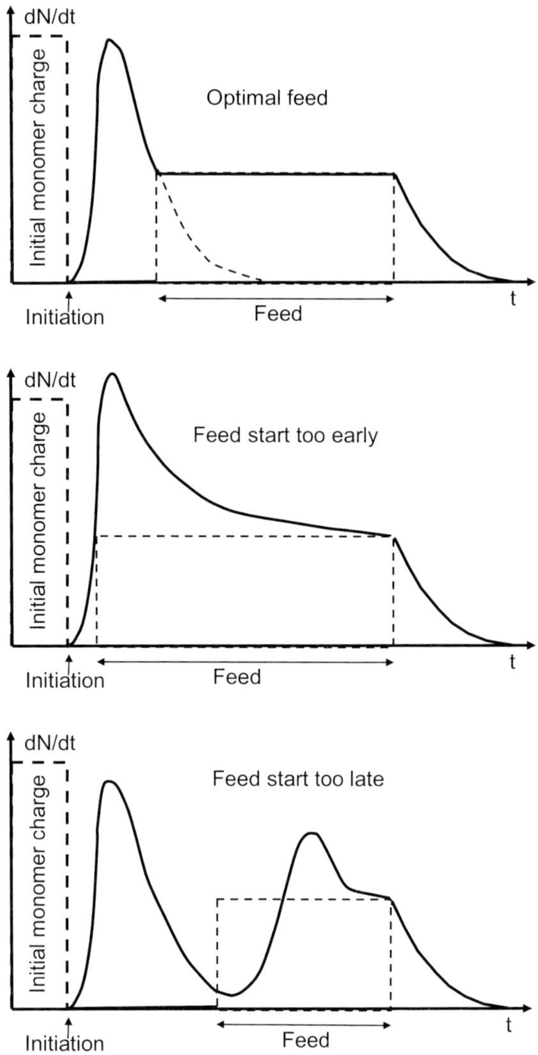

Fig. 11.13. Schematic representations of the molar flow rates of monomer in a semi-batch emulsion polymerization. Top graph: the feed is optimal; it is immediately converted. Middle graph: the feed was started too early; the reaction course is no longer controlled by the feed. Bottom graph: the feed was started too late; the accumulation results in a quasi runaway situation.

11.5.2.2 **Feed**

After the particle growth stage of the emulsion polymerization is terminated, the monomer concentration within the latex particles decreases and so does the heat production rate. In the meantime, the temperature control system is able to reduce the temperature of the jacket. When the molar flow rate (mol s^{-1}) of the monomer depletion reaches a value corresponding approximately to the molar flow rate of the intended feed, the monomer feed should be started (Figure 11.13, top graph). This allows compensation in order to maintain the concentration constant; that is, a quasi steady state can be achieved. The reactor is cooled by two methods: heat exchange across the wall, and via the sensible heat of the cold feed. Here, the emulsion feed processes obviously show an advantage due to the higher heat capacity of the emulsion compared to the monomer.

The right time point for starting the feed is an important safety-relevant parameter. If the feed is started too early during the particle formation period (Figure 11.13, central graph), an accumulation of monomer may result. This will cause an increase in the reaction rate and result in an uncontrollable temperature increase – even a runaway reaction. Matching of feed rate and reaction rate, allowing control of the course of the polymerization, is only achieved in the very late stages of the feed time, if ever. If the feed is started too late, the situation may become even worse (Figure 11.13, bottom graph): the monomer initially present is already converted and the heat release rate is low. But the jacket may still be at a low temperature due to the previous exotherm, and undercool the reactor. Since the feed is increasing the monomer concentration at too low a reaction rate, an accumulation will occur and may lead to a sudden "wake up" of the reaction, which in turn could result in a runaway situation.

11.5.2.3 Final Stage

After the end of the feed period, the heat release rate decreases as a consequence of the monomer depletion. This third stage is very similar to a batch reaction, because the reaction rate can no longer be influenced by the feed.

11.5.2.4 Practical Aspects

The difficulty concerning the initiation on an industrial scale is that the actual initiation of the polymerization must be recognized before a great amount of monomer is added to the reactor. This can be realized either by chemical in-line analysis, or by a heat balance on the industrial reactor [43]. If the initiation fails and the feed is continued, it may lead to an accumulation of monomer that could react spontaneously in an uncontrolled way. Thus the process becomes a batch reaction, for which the reactor is not designed.

In a semi-batch reaction, overcooling, or too low a temperature, may be as critical as too high a temperature, because the reaction rate is lower and the resulting accumulation may be important. If the temperature is adjusted to its nominal value, the accumulated monomer may also react in an uncontrolled way, leading to a runaway reaction. This may be due to degradation of the initiator which slows down the reaction. Such effects can be detected by a heat balance on industrial reactors.

For polymerization or copolymerization processes performed in semi-batch reactors, the feed rate can also be adjusted to meet quality criteria [44–48]. Advanced techniques are proposed to control the feed rate by taking into account simultaneously productivity and safety criteria [49, 50].

Thus not only does a safe process result from sound development, working out the right process parameters, but on an industrial scale it must be checked that the polymerization remains on the "right path".

11.5.3
Continuous Processes

In continuous polymerizations, the main problem is the dynamic stability of the reactor. The stability problems have various different aspects: the thermal stability as introduced above (see Section 11.2.4), the concentration stability, the particle number stability, and the viscosity stability. Even under isothermal conditions these problems may lead to multiplicity or oscillatory behavior. It is worth emphasizing the fact that stability and safety are in no case synonymous: a reactor may be unsafe even if working at a stable working point, or conversely it may be run safely at an instable working point. But knowledge of stability limits of the reactor is essential for the design of a safe process.

11.5.3.1 Concentration Stability

In a continuous stirred tank reactor (CSTR) where a homogeneous polymerization is being performed, the mass balance or the performance equation can be written in a very simple way [Eq. (33)].

$$\frac{dX}{dt} = \frac{X}{\tau} \tag{33}$$

If the reaction follows an autocatalytic kinetic equation, like polymerizations exhibiting the Tromsdorff effect, then the conversion rate as a function of conversion, $dX/dt = f(X)$, shows a typical maximum curve [51]. The operating point is found at the intercept of the kinetics line and a straight line with the space time reciprocal as the slope. This configuration leads to a multiplicity of solutions as represented in Figure 11.14 for several values of the space time. In a range of space time between τ_1 and τ_3, there are three solutions. The upper and lower solutions are stable, whereas the intermediate one is instable. In such a situation, the CSTR should be designed to be operated at the high conversion point (hot branch, or ignited reactor). This allows for a higher productivity and a safer behavior if the high cooling capacity required can be achieved. If the reactor is operated at the lower conversion point (cold branch), a small perturbation, for example of a feed pump, may "ignite" the reactor, which suddenly jumps to the hot branch, with high conversion. This could lead to a runaway.

11.5.3.2 Particle Number Stability

The particle size stability is a special form of the dynamic concentration stability [52]. If the emulsifier is present at a concentration above the critical concentration, new particles will be created. Thus the area growth rate, and therefore the emulsifier consumption, are increased. Then the emulsifier supply by the feed may be in-

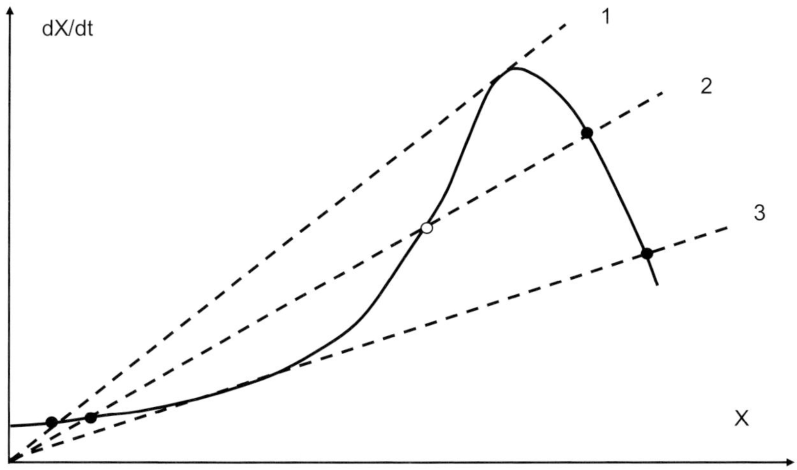

Fig. 11.14. Schematic representation of the conversion rate as a function of conversion (solid line) in an isothermal CSTR with a polymerization presenting a gel effect. Three different space times are represented by the three operating lines (broken lines). Filled circles represent stable operating points and the white circle an unstable operating point.

sufficient to ensure coverage of the particle's surface. This stops the nucleation process of new particles. Since the particles are washed out by the reactor outlet, only fewer, but larger, particles remain in the reactor. The specific interfacial area decreases until the emulsifier consumption becomes lower than the supply. Then the concentration increases, and a new cycle starts. These oscillations may occur especially during start-up of the reactor. They result in oscillations of the conversion that impinge on the quality of the product. They can be avoided by using a adequate seeding policy for the reactor.

11.5.4
Design Measures for Safety

Safety measures or risk-reducing measures must be considered under the aspect of inherent safety. Kletz [53], who was the promoter of these ideas for a long time, formulated some principles for reduction of risks:

- intensification, using so little hazardous material that it will not matter if it all leaks out;
- substitution, using a safer material instead;
- attenuation, using a hazardous material in a safer form;
- limitation of the effects of failures, not by adding on protective equipment but by equipment design or changing the conditions of use;
- simplification, as complex plants provide more opportunities for human error and contain more equipment that can go wrong;

- avoiding knock-on or domino effects;
- making incorrect assembly difficult or impossible;
- making the state of equipment, such as open or shut, clear;
- designing equipment that is able to withstand incorrect installation or operation;
- making equipment easy to control.

Following these principles in a more specific way for polymerization reactions, three levels of priority can be defined: in decreasing order, the first priority is the reduction of severity by design. As a second priority, technical measures for control of the reaction to avoid runaway should be considered. The aim is to obtain a fail safe process by reduction of the probability of occurrence of an incident. As last resort only, emergency measures should be taken in order to mitigate the consequences of runaway. In any case, the basic principle remains: "Avoid runaway rather than mitigate its consequences."

With reference to this principle, different types of risk-reducing measures can be considered for polymerization reactions (Sections 11.5.4.1–11.5.4.10).

11.5.4.1 Process Design

Semi-batch processes are preferred to batch processes where possible. This point was made sbove (see Section 11.4). Continuous processes allow the inventory of reaction mass to be reduced, and represent an elegant way of reducing the severity of a process.

11.5.4.2 Reactor Design

Many design elements are in fact preventive measures. Some of them will be mentioned here. One may consider building a reactor resistant to the maximum pressure even in the case of loss of control of the polymerization. Often the final pressure obtained with polymerization reactions is not so high. Another important element is the agitation. The agitator ensures uniform mixing over a large range of viscosities, avoids local concentration of reactants that could lead to hot spots, and ensures heat transfer. The agitation is also required for temperature measurement, that in turn ensures control of the reactor. In this context it is important to keep heat exchange surfaces (and also the thermometer) clean.

Temperature is often the key parameter for process safety; thus redundant systems should be used, since failure rates in the order of magnitude of one per year are reported for them. The temperature probe must be correctly located and the temperature recorded so that trends can be followed. This is essential for safety as well as for product quality. The rate of temperature change can also represent very important information for process control. Utility-independent thermometers are also very useful. The temperature of the cooling system (if possible at the inlet and outlet of the jacket) should be monitored.

The process should be designed in order to provide adequate heating profiles, especially to avoid inadvertent fast heating. An interlock may stop heating in the case of malfunction.

When reflux cooling is used, one must avoid deposits inside the condenser tubes. To do so, the condenser can be pitched to assist drainage, and must be sized to prevent flooding. The process must be designed to avoid foam-up of the reaction mixture (swelling; see Section 11.4.2).

Services are very important for the control of the reaction. Redundant systems should be provided for electrical power, as well as compressed air or inert gases. The electricity supply must be examined in detail during risk analysis: motors – that is, feeding and cooling pumps, agitators, vacuum, and feed – will be stopped if the electrical power fails. Moreover, instruments dependent on compressed air become inoperative. For this reason, the fail safe position of valves (open for cooling, closed for heating) should be considered.

Failure of vacuum at reflux may lead to the development of a hazardous situation: the boiling point, and thus the reaction temperature, increase and may end up in a runaway situation.

11.5.4.3 Control of Feed

In semi-batch or continuous operation, the feed rate allows control of the reaction course. Hence it plays an important role concerning the safety of the process. With an exothermal reaction, it is important to be able to limit the feed rate by technical means. One possibility is feed by portions, a method that is only applicable for semi-batch reactors. This mode of addition is the traditional way of limiting accumulation. In this case, the addition must be controlled by the conversion: that is, the next portion is added only if the previous portion has been consumed by the reaction. Different criteria can be used to follow the reaction: the temperature, the appearance of the reaction mass, chemical analysis, and so on. For a well designed process, the additions can also be performed on a time basis.

Another possibility is to use a feed tank with a control valve and gravimetric flow. The valve can be controlled by the weight of the reactor, by the weight of the tank, by the level in the tank, or by a flow meter. The maximal feed rate can be limited by the clearance of the valve or by a calibrated orifice. If a centrifugal pump is used, it is necessary to provide also a control valve to limit the flow rate. The flow control strategies are the same as those described above. With a metering pump, the through flow rate can be controlled by a stroke adjusting mechanism or a variable speed drive acting on the stoke frequency. The control can be realized by a fixed adjustment or by through a flow meter.

11.5.4.4 Emergency Cooling

If the cooling system fails, water can be supplied from a hydrant to the cooling jacket or coil, or an independent emergency cooling system can be provided. With such a system, it is critical that the temperature does not fall below the solidification point of the reaction mass, or otherwise a crust would form, resulting in reduced heat transfer, which again may favor the runaway situation. The agitation of the reaction mass is also critical in such a situation: large reaction masses behave adiabatically in practice, even if cooled on the outside. Here the injection of nitrogen into the bottom of the reaction mass has proven to be helpful for an emergency mixing. Such equipment must be tested under practical conditions.

11.5.4.5 **Inhibition**

This technique may be applied to catalytic reactions where an inhibitor can be added in small amounts. Mixing is an especially critical factor here, since a small amount of inhibitor has to be added and mixed into a large volume of reaction mixture [54, 55].

11.5.4.6 **Quenching**

Some reactions can be stopped by the addition of a suitable component. Dilution by an inert and cold material may lower the temperature to slow down the reaction. For this type of measure, the critical factors are the amount and rate of addition and the temperature of the quenching material. The required empty volume must be also be available in the reactor. Calorimetric methods are of great help in the design of such measures, because they allow measurement of the heat of mixing, which is often important, and the thermal stability of the resulting mixture.

11.5.4.7 **Dumping**

This measure is similar to quenching, with the difference that the reaction mass is not kept within the reactor, but transferred into a vessel containing the inhibitor or the diluting compound. This vessel must be prepared to receive the reaction mass at any instant during the process. The transfer line is critical for the success of this measure. It must be designed to allow an emergency transfer even in the case of breakdown of the utilities. This measure is particularly suitable in cases where the reaction mass must be transferred for workup after a normal operation.

11.5.4.8 **Controlled Depressurization**

If a runaway is detected in an early stage, where the temperature and the pressure increase is slow, a controlled depressurization of the reactor can be considered. This is only suitable for cases where a volatile solvent is present; the reactor is slowly depressurized until evaporative cooling occurs. Obviously, a scrubber and/or a reflux condenser must be installed and designed to work with independent utilities.

11.5.4.9 **Pressure Relief**

This measure will not be described in detail in this chapter. The design of venting lines for reactions with thermal potential is a complex matter. There are examples where pressure release was able to protect reactors from an explosion, but also cases where a reactor exploded, even with an open manhole. This measure only applies with reaction systems where the pressure still increases significantly for small temperature increases above the normal operating level. In addition, the discharge line must end in a catch tank or in a scrubber to avoid spillage with possibly toxic or flammable material. The behavior of the reaction mass – that is, the temperature and pressure increase – under runaway conditions must be known. Critical factors are foaming and two-phase flow, which require specific design methods [56–59]. Special care must be taken to define the runaway scenario against which the reactor has to be protected. Emergency vent systems may also be designed for liquid discharge (dumping).

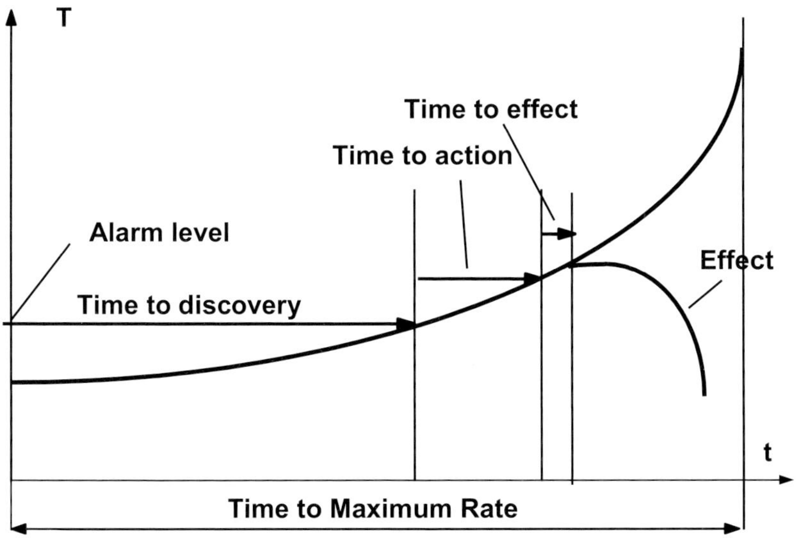

Fig. 11.15. Chronological development of contingency measures.

11.5.4.10 Time Factor

Time plays a primary role in the efficiency of a measure. The following steps must be taken from the instant a failure occurs up to recovery of the control of the process (see Figure 11.15): First, when a failure or a malfunction occurs, it must be detected. The detection time can be influenced by the choice of appropriate alarm settings and instrumentation. But the most important is the choice of an adequate parameter which must be monitored to detect a malfunction. The design of alarms, interlocks, and control strategies is an important part of process design. Once the alarm is switched on, some time is required for the measure to be applied. Quenching or dumping requires some time for the transfer of the mass, an emergency cooling system must be switched on, and the cooling medium must flow at the required temperature with the required flow rate. Finally, the measure must become effective: Some time also elapses from the instant when the measure was applied until its effects become sensitive to the process. This is especially true with large reactors such as those used in polymerization processes. This time factor must be estimated for an effective design of safety measures, and compared with the dynamics of the runaway of the desired reaction and of decomposition reactions. Here the TMR_{ad} concept (Section 11.2.1) is of great help.

11.6 Conclusion

Even if polymerization reactions are critical from the point of view of thermal process safety, there are means to systematically identify the risks and to design the reactors in order to minimize the risks. The systematic evaluation scheme presented in this text has proven to be well suited for risk identification. A method based on the criticality classes and on calorimetric tools for process development has been presented. Despite the fact that more advanced methods are available, these find only few applications in the industrial practice. This is probably not due to their inherent quality, but may be an effect of long-term traditions in the polymer industry, which render innovation more difficult.

References

1 R. Gygax, *Chemical Engineering Science*, **1988**, *43*(8), 1759.
2 H. U. Moritz, *Sichere Handhabung chemischer Reaktionen*, Dechema, Frankfurt a.M., **1995**, Vol. 3, p. 115.
3 J. A. Barton, P. F. Nolan, *Safety of chemical batch reactors and storage tanks*, ECSC, EEC, EAEC, Brussels, **1991**.
4 N. Maddison, R. L. Rogers, *Chemical Technology Europe*, **1994**, *11–12*, 28.
5 P. A. Dunn, *Guidance for the safe production of phenolic resins*, British Plastics Federation, London, **1979**.
6 R. Gygax, *ESCIS (Expert Commission for Safety in the Swiss Chemical Industry), Lucerne*, **1993**, Vol. 8.
7 R. Gygax, *Chemical Engineering Progress*, **1990**, 53.
8 D. I. Townsend, J. C. Tou, *Thermochimica Acta*, **1980**, *37*, 1.
9 F. Stoessel, *Chemical Engineering Progress*, **1993**, *10*, 68.
10 C. Van-Heerden, *Industrial Engineering Chemistry*, **1953**, *45*(6), 1242.
11 N. R. Amundson, O. Bilous, *AIChE Journal*, **1955**, *1*(4), 513.
12 R. Aris, *Introduction to the analysis of chemical reactors*, Prentice-Hall, Englewood Cliffs, **1965**.
13 A. Varma, M. Morbidelli, H. Wu, Cambridge University Press, Cambridge, UK, **1999**.
14 P. Hugo, *Chemie Ingenieur Technik*, **1980**, *52*(9), 712.
15 K. Y. Choi, W. H. Ray, *Chemical Engineering Science*, **1988**, *43*(10), 2587.
16 K. Y. Choi, W. H. Ray, *Chemical Engineering Science*, **1985**, *40*(12), 2261.
17 B. Hentschel, *Chemie Ingenieur Technik*, **1979**, *51*(8), 823.
18 H. U. Moritz, *Third Berlin International Workshop on Polymer Reaction Engineering*, VCH, Weinheim, **1989**.
19 H. G. Poersch-Parke, A. Avela, K. H. Reichert, *Chemie Ingenieur Technik*, **1989**, *61*(10), 808.
20 A. Uppal, W. H. Ray, A. B. Poore, *Chemical Engineering Science*, **1974**, *29*(4), 967.
21 A. E. Hamielec, *Course Notes*, Course on "Polymer Reaction Engineering", McMaster University, Hamilton, Ontario, **1980**.
22 M. Klocke, *Thesis*, Universität-GH, Paderborn, **1995**.
23 F. Stoessel, *Journal of Thermal Analysis*, **1997**, *49*, 1677.
24 R. N. Landau, L. R. Williams, *Chemical Engineering Progress*, **1991**, *12*, 65.
25 W. Regenass, *8th RC1 User Forum Hilton Head SC, USA*, Mettler-Toledo, **1996**.
26 W. Regenass, *Chimia*, **1997**, *51*, 189.
27 *VDI-Wärmeatlas*, VDI-Verlag, Düsseldorf, **1974**.
28 J. R. Bourne, M. Buerli, W. Regenass, *Chemical Engineering Science*, **1981**, *36*, 347.
29 S. Choudhury, L. Utiger, *Chemie Ingenieur Technik*, **1990**, *62*(2), 154.
30 J. Wiss, F. Stoessel, G. Killé, *Chimia*, **1990**, *44*(12), 401.
31 J. Wiss, F. Stoessel, G. Killé, *Chimia*, **1993**, *47*(11), 417.
32 J. F. Wilson, R. J. Grenda, J. F. Patterson, *Trans. Am. Nucl. Soc.*, **1961**, *4*(37), 356.
33 J. F. Wilson, R. J. Grenda, J. F. Patterson, *Trans. Am. Nucl. Soc.*, **1962**, *5*(25), 151.

34 M. F. Edwards, W. L. Wilkinson, *Chemical Engineering (London)*, **1972**, *265*, 328.
35 A. B. Metzner, R. E. Otto, *AIChE Journal*, **1957**, *3*, 3.
36 H. U. Moritz, *Chemical Engineering Technology*, **1989**, 71.
37 A. Steiff, R. Poggemann, P. M. Weinspach, *Chemie Ingenieur Technik*, **1980**, *52*(6), 492.
38 C. U. Schmidt, K. H. Reichert, *Chemie Ingenieur Technik*, **1987**, *59*, 739.
39 C. Gentric et al., *Chemical Engineering Journal*, **1999**, *75*, 31.
40 F. Stoessel, *Chemical Engineering Progress*, **1995**, *9*, 46.
41 P. Hugo, J. Steinbach, F. Stoessel, *Chemical Engineering Science*, **1988**, *43*(8), 2147.
42 F. Stoessel, O. Ubrich, *Journal of Thermal Analysis and Calorimetry*, **2001**, *64*, 61.
43 J. Schmidt, H. Giesbrecht, *Chemie Ingenieur Technik*, **1996**, *68*(9), 1144.
44 I. S. d. Buruaga, M. Arotçarena, P. D. Armitage, L. M. Gugliotta, *Chemical Engineering Science*, **1996**, *51*(11), 2781.
45 G. Arzamendi, J. Asua, *Ind. Eng. Chem. Res.*, **1991**, *30*, 1342.
46 A. Urretabizkaia, J. R. Leiza, J. M. Asua, *AIChE Journal*, **1994**, *40*(11), 1850.
47 I. S. d. Buruaga, A. Echevarria, P. D. Armitage, J. C. D. L. Cal, R. Leiza, J. M. Asua, *AIChE Journal*, **1997**, *43*(4), 1069.
48 L. M. Gugliotta, J. R. Leiza, M. Arotçarena, P. D. Armitage, J. M. Asua, *Ind. Eng. Chem. Res.*, **1995**, *34*, 3899.
49 T. J. Crowley, K. Y. Choi, *Journal of Process Control*, **1996**, *6*(2–3), 119.
50 O. Abel et al., *Journal of Process Control*, **2000**, *10*(4), 351.
51 R. S. Knorr, K. F. O'Driscoll, *Journal of Applied Polymer Science*, **1970**, *14*, 2683.
52 G. Ley, H. Gerrens, *Makromolekulare Chemie*, **1974**, *75*, 563.
53 T. A. Kletz, *Process Safety Progress*, **1996**, *15*(1), 5.
54 S. Kammel, U. Schlüter, A. Steiff, P. M. Weinspach, *Chemical Engineering Science*, **1996**, *51*(10), 2253.
55 H. Niemann, S. Lighart, S. Duffield, A. Schipper, *PPSS96*, Houston, TX, **1996**.
56 J. Etchells, J. Wilday, *Workbook for chemical reactor relief system sizing*, HSE, Norwich, UK, **1998**.
57 *Guidelines for pressure relief and effluent handling systems*, CCPS-AICHE, New York, **1998**.
58 J. Schmidt, F. Westphal, *Chemie Ingenieur Technik*, **1997**, *69*(6), 776.
59 J. Schmidt, F. Westphal, *Chemie Ingenieur Technik*, **1997**, *69*(8), 1074.

Notation

Symbol	*Name*	*Practical unit*	*SI unit*
A	heat exchange area	m^2	m^2
A, P	chemical compounds		
C	concentration	$mol\ L^{-1}$	$mol\ m^{-3}$
c_P	specific heat capacity	$J\ kg^{-1}\ K^{-1}$	$J\ g^{-1}\ K^{-1}$
d	diameter or thickness	m	m
D	diameter	m	m
E_a	activation energy	$J\ mol^{-1}$	$J\ mol^{-1}$

F	molar flow rate	$mol\ h^{-1}$	$mol\ s^{-1}$
g	acceleration due to gravity	$m\ s^{-2}$	$m\ s^{-2}$
h	heat transfer coefficient	$W\ m^{-2}\ K^{-1}$	$W\ m^{-2}\ K^{-1}$
H	height	m	m
ΔH_R	molar enthalpy of reaction	$kJ\ mol^{-1}$	$J\ mol^{-1}$
ΔH_V	specific enthalpy of vaporization	$kJ\ kg^{-1}$	$J\ g^{-1}$
k	rate constant	function of rate law	function of rate law
k_0	frequency factor	function of rate law	function of rate law
$\dot{m}$	mass flow rate	$kg\ h^{-1}$	$g\ s^{-1}$
M	monomer		
M_R	mass of reaction mixture	kg	g
n	order of reaction	–	–
n	revolution frequency	rpm	s^{-1}
N	number of moles	–	–
q	heat release rate	W	W
Q	thermal energy (heat)	kJ	J
r	rate of reaction	$mol\ m^{-3}\ h^{-1}$	$mol\ m^{-3}\ s^{-1}$
R	universal gas constant	$J\ mol^{-1}\ K^{-1}$ $L\,mbar\,mol^{-1}\ K^{-1}$	$J\ mol^{-1}\ K^{-1}$
S	cross-section	m^2	m^2
S	entropy	$J\ K^{-1}$	
t	time	h	s
T	temperature	°C	K
u	linear velocity	$m\ s^{-1}$	$m\ s^{-1}$
U	overall heat transfer coefficient	$W\ m^{-2}\ K^{-1}$	$W\ m^{-2}\ K^{-1}$
V	volume	m^3	m^3
X	conversion	–	–
z	equipment constant (stirred tank)	–	–

Subscripts

Subscript	***Meaning***	***Example***	
0	initial value	T_0	initial temperature
A, B, P, R, S	chemical compounds	C_A	concentration of A

AC	accumulation	X_{ac}	degree of accumulation
ad	adiabatic	ΔT_{ad}	adiabatic temperature rise
C	coolant	T_{Amb}	ambient temperature
ex	exchange (heat exchange)	T_C	temperature of coolant
f	fouling	q_{ex}	heat dissipation rate
Feed	feed	d_f	thickness of fouling film
G	gas	T_{Feed}	
Loss	loss	ρ_G	specific weight of gas
M	monomer	q_{Loss}	
P	process	C_M	Monomer concentration
P	polymer	T_P	process temperature
R	reactor, reaction mass	λ_P	thermal conductivity of polymer
RX	reaction	d_R	diameter of reactor
S	stirrer	q_{RX}	heat release rate of reaction
TAmb	ambient	d_S	stirrer diameter
th	thermal	X_{th}	thermal conversion
V	vapor	m^*_V	mass flow rate of vapor
W	wall	d_W	wall thickness

Greek

Symbol	*Name*	*Practical unit*	*SI unit*
α	relative volume increase	–	–
γ	material constant for heat transfer	$W\ m^{-2}\ K^{-1}$	$W\ m^{-2}\ K^{-1}$
Δ	difference (used as prefix)	–	–
φ	heat transfer coefficient of equipment	$W\ m^{-2}\ K^{-1}$	$W\ m^{-2}\ K^{-1}$
λ	thermal conductivity	$W\ m^{-1}\ K^{-1}$	$W\ m^{-1}\ K^{-1}$
μ	dynamic viscosity	cP = mPa s	Pa s
ρ	specific weight	$kg\ m^{-3}$	$g\ m^{-3}$
σ	surface tension	$N\ m^{-1} = 10^3\ dyn\ cm^{-1}$	$N\ m^{-1}\ (= kg\ s^{-2})$
τ	time constant	h	s
τ	space time in continuous reactor	h	s

Acronyms

MTSR maximum temperature of synthesis reaction
MTT maximum temperature for technical reasons
TMR_{ad} time to maximum rate under adiabatic conditions (time to explosion)

Dimensionless groups

Symbol	*Name*	*Expression*	*Parameters*
Ne	Newton (power) number	$Ne = \frac{P}{\rho \cdot n_S^3 \cdot d_S^5}$	P = power of stirrer ρ = specific weight of fluid n_S = revolution speed d_S = stirrer diameter
Nu	Nusselt number	$Nu = \frac{h \cdot d}{\lambda}$	h = film heat transfer coefficient d = characteristic length λ = thermal conductivity
Pr	Prandtl number	$\mathrm{Pr} = \frac{\mu \cdot C_p}{\lambda}$	μ = dynamic viscosity C_p = specific heat capacity λ = thermal conductivity
Re	Reynolds number (stirred tank)	$\mathrm{Re} = \frac{n \cdot d^2 \cdot \rho}{\mu}$	n = stirrer frequency d = diameter of agitator ρ = specific weight μ = dynamic viscosity

12 Measurement and Control of Polymerization Reactors[1]

John R. R. Richards and John P. C. Congalidis

12.1 Introduction

Consistent polymer properties are of paramount importance to end-user manufacturers who must produce the polymer in its final form and shape for the intended application. These properties are the result of complex polymer architecture and composition formed in reaction and perhaps further influenced in isolation and extrusion processes. Producing consistent, uniform, and in-specification polymer for the end-user are the tasks of the polymer process measurement and control systems.

Polymer processes, whether batch or continuous, rarely run under stable equilibrium conditions. However, in order to operate such processes safely and in order to set the characteristics of the products optimally, a set of process manipulated variables must be kept constant or systematically modified over the duration of the reaction or in the course of the various reaction steps. The main process variables of this type are temperature, pressure, concentration, amount, flow, and level. Speeds (agitator, gear pumps, extruders), power input and viscosity can be of substantial importance also.

This chapter will discuss various measurement techniques of importance to engineers and scientists designing and operating polymer reactors and associated equipment. We will then discuss basic control concepts and conclude with more advanced control strategies. We have attempted to expand and update the fine chapter from the previous edition of this book [1].

12.1.1 Definitions

The first step in the objective of obtaining constant or systematically adjusted operating conditions is the proper measurement of process variables. There are vari-

1) The symbols used in this chapter are listed at the end of the text, under "Notation".

Handbook of Polymer Reaction Engineering. Edited by T. Meyer, J. Keurentjes

ISBN: 3-527-31014-2

ous organizations developing standards pertaining to measurement and control, including the US National Institute of Standards and Technology (NIST), the Deutsches Institut für Normung (DIN), the International Organization for Standardization (ISO), and the International Electrotechnical Commission (IEC).

In measurement technology there are a number of standard measurement concepts [2–8]:

- A measurement installation translates the measurement quantity to a measured value.
- The measured value consists of a numerical value and a unit of measurement.
- The result of measurement consists of several measured values, which can be reproduced as a value or a curve of values.
- The measurement follows certain measurement principles whose application results in measurement methods.
- The measuring method is executed with a measuring installation, which consists of different measuring instruments and auxiliary devices.
- Measuring instruments are situated in the signal flow and are essential to the measuring installation.
- Auxiliary devices usually serve to provide the auxiliary energy which is needed to provide measuring signals.
- If the measuring signals pass through several measuring devices connected in series, then a measurement chain is present.

According to Figure 12.1 a measurement installation translates the measurement quantity to a measured value. Here we follow the connections of the sensor, transmitter, and transducer as described in Ref. 7:

- The chain consists of a sensing element (sensor), a measuring instrument whose input consists of the measured quantity and its output consists of the corresponding measurement signal.
- At the end of the chain lies the output unit, which provides the desired measurement value.

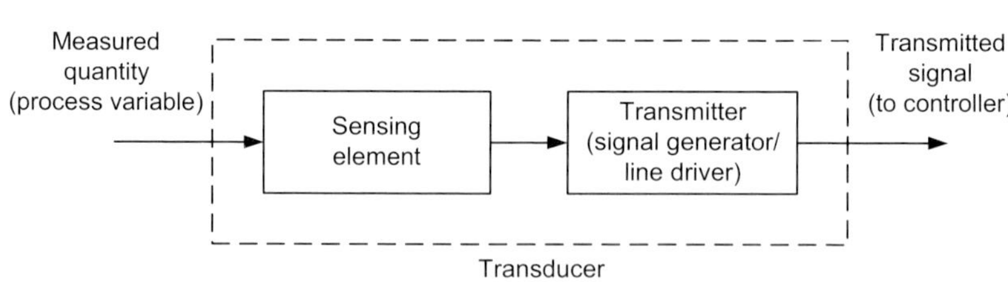

Fig. 12.1. A typical process transducer [7]. D. E. Seborg, T. F. Edgar, D. A. Mellichamp, *Process Dynamics and Control*, Copyright © 2003 John Wiley & Sons, Inc. This material is used by permission of John Wiley & Sons, Inc.

- Between the measured quantity and the measured variable to the output unit lie transducers (see Figure 12.1), which process (transform and amplify) the signal from the measurement instruments ahead of them and feed their signal to the output measurement device.
- Measurement transmitters are those devices which convert an analog input signal to an unambiguous standardized analog output signal [4]. The transmitter is usually required to convert the sensor element output to the standardized form compatible with, for example, a controller input.
- Measurement transducers are devices which have input and output signals in the same or different structure (analog/digital or digital/analog). Sometimes the terms "transmitter" and "transducer" are used interchangeably.
- One calls the measuring device, with its case and accessories, a measuring instrument.
- The instrument gauge or indicator is the part of the measurement device whose movement allows the reading of the measurement value.
- To the measurement instrument also belongs a scale. Between the end points of the scale lies the range, or span, of the measurement device within which individual values of the measured quantity are obtained.
- In order to calibrate and adjust the scale standards are used, which associate a precisely defined output measurement value to a measured quantity.

12.1.2 Measurement Error

Accuracy of a measurement is the degree of conformity of an indicated value to an accepted standard or ideal value [4]. Repeatability is the closeness of agreement among a number of consecutive measurements of the output for the same value of the input. Each individual measurement x_i deviates from the correct or ideal measured value μ [2, 8]. This difference is called the measuring error or deviation d_i [Eq. (1)].

$$d_i = x_i - \mu \tag{1}$$

Measurement errors can be systematic errors, which are found in the measuring method or in the equipment; with knowledge of the correct measured value these errors can be compensated by a correction term or a correction factor. Random errors, which oscillate in sign and amount, cannot be corrected. Averaging eliminates these errors. The simplest mean is the arithmetic average value $\bar{x}$, which is an estimate of the true mean value μ [Eq. (2)].

$$\bar{x} = \frac{1}{n}\sum_{i=1}^{n} x_i \tag{2}$$

The smaller the random errors of the individual measurements and the larger the number of measurements, n, the smaller the deviation of the arithmetic average from the true mean. The empirical sample standard deviation, s, is a measure of the measurement error [Eq. (3)].

$$s = \sqrt{\frac{\sum_{i=1}^{n}(x_i - \bar{x})^2}{n-1}} \tag{3}$$

With very large n, the discrete measurements x_i can be thought of as continuous measurements x and the sample standard deviation s becomes the true standard deviation σ of a Gauss normal differential distribution function [Eq. (4)].

$$f(x) = \frac{1}{\sqrt{2\pi}\sigma} \exp\left[-\frac{1}{2}\left(\frac{x-\mu}{\sigma}\right)^2\right] \tag{4}$$

A tabulation or histogram of how often the various values of x_i occur in replication is approximated by this distribution. From the Gaussian distribution the statistical confidence of the individual values is obtained. From the cumulative values of this distribution, 68.3% of all measurement values are within the interval $\bar{x} \pm \sigma$, 95.4% are within $\bar{x} \pm 2\sigma$, and 99.7% are within $\bar{x} \pm 3\sigma$. Further statistical concepts can be found in Refs. 5 and 8.

12.2 Measurement Techniques

The measurement technique to be chosen is principally determined by the measured quantity and by the accuracy with which the variable must be measured. The measuring instrument produces a signal, which must be transformed in such a way that it can be registered by an indicator or recorder and further processed. This requirement is fulfilled directly by some measuring methods; however, in most cases a measurement transmitter is operated between the sensor and the measurement device. Pneumatic pressures between 3 and 15 psig, or between 0.2 and 1.0 bar, or electrical currents between 4 and 20 mA are used [7] as standard signals. Pneumatic signals are transmitted in plastic tubing and a distance of approximately 300 m between the sensor and the transducer can be accommodated. Electrical direct current signals can be transferred over a distance of a few kilometers. Electrical signals are much more commonly used today than pneumatic signals. To accommodate larger distances the analog signals must be converted to digital signals; in that case the distance between measurement transducer and receiver can be expanded practically at will. In the Sections 12.2.1–12.2.11 some particularly common measurement methods from operational practice will be discussed.

12.2.1
Temperature

In polymer technology temperature is one of the very important variables of state, probably the most important. Depending on the measurement range, it is measured with electrical or mechanical contact thermometers or with optical methods.

With contact thermometers one attempts to keep the thermal capacity of the measuring device and heat transfer as small as possible so that actual temperature changes can be obtained as soon as possible. Often the sensitive measurement sensors are protectively shielded. Here compromises must be made which are appropriate for individual cases.

The most widely used thermometer in polymer technology is the resistance temperature detector (RTD). For higher temperatures or with point measurements, the thermocouple is often preferred. Mechanical contact thermometers (expansion thermometers, rod thermometers, or bimetallic thermometers) are nearly always used only for local measurements. For optical thermometers, radiation pyrometers are used.

12.2.1.1 **Resistance Thermometers**

In resistance thermometers (or RTDs) one uses the dependency of electrical resistance on the temperature [2]. The resistance of metals almost always rises with temperature. In practice one uses platinum thermometers. A light platinum wire on a ceramic support is covered with a protective layer of glass, enamel, or magnesium oxide. This "measurement resistance" is the end of one so-called measurement element, which ends in connecting terminals. The whole assembly is in a sleeve, which is solidly connected to the measuring point in the apparatus with flanges or screws (Figure 12.2).

For pure metals, resistance thermometry is governed by Equation (5), where R_t is the resistance at temperature t, R_0 is the resistance at the reference temperature, usually 0°C, and a, b, c are coefficients based on calibration points [3].

$$R_t = R_0(1 + at + bt^2 + ct^3 + \cdots) \tag{5}$$

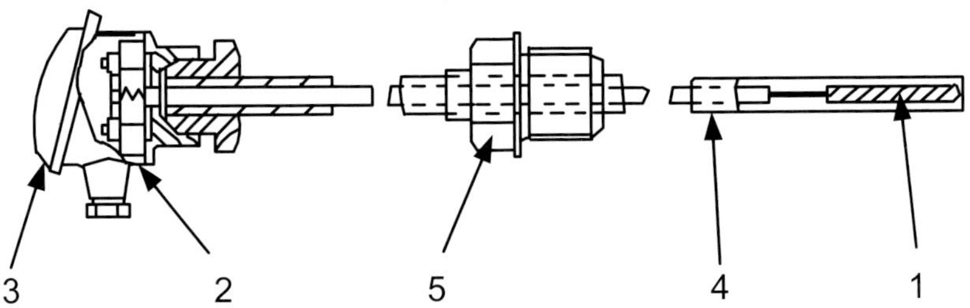

Fig. 12.2. Resistance thermometer thermowell assembly.
1: Measuring resistor; 2: terminal block; 3: connection head; 4: protective pipe; 5: screw tap [1].

For example, platinum has a temperature coefficient of resistance $a = 0.00385\ \Omega/(\Omega°C)$ in the range 0 to 100°C. The measuring range of commercial-purity platinum thermometers is situated between about 200 and 850°C and the accuracy of these thermometers is ±0.15% of span over the full range [4]. Resistance versus temperature can usually be described by just a quadratic function, which can be linearized in measurement amplifiers. If large temperature ranges are not present, the linearization can often be omitted. The resistance is usually measured in a bridge circuit [2].

12.2.1.2 Thermocouples

A thermocouple consists in principle of two wires made of dissimilar metals soldered together in two junctions, which form a closed circuit (Figure 12.3). At the contact junctions a voltage gradient is produced and a current, which depends on the temperature, flows around the circuit. So long as both junctions have the same temperature the voltages compensate and no current flows; if the junctions have different temperatures, then a voltage gradient is produced, which can be measured in an open circuit and which is approximately proportional to the temperature (although the relationship is nonlinear over large temperature scales). If one of the two soldered connections is held at a constant reference temperature, then the other connection can serve as a temperature probe.

Suitable metal pairs, which supply sufficiently large thermal voltages to deliver

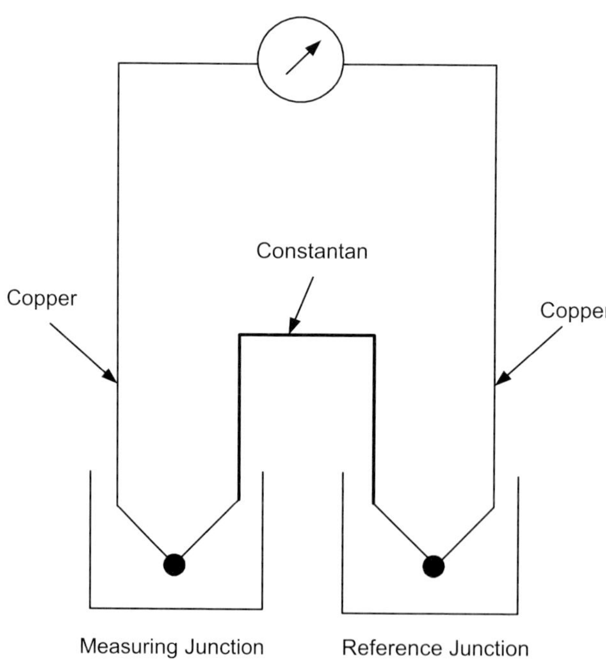

Fig. 12.3. Thermocouple wiring diagram [1].

reliable measured values, are copper/constantan (type T), iron/constantan (type J), chromel/alumel (type K) and platinum (70%)–rhodium/platinum (94%)–rhodium (type B). Their thermal voltages permit measurements up to 371°C, 760°C, 1149°C, and 1700°C respectively [3]. The standard-grade error limits for the first three types are ±0.8°C, ±2.2°C, and ±2.2°C respectively.

Thermocouples offer the benefit that one can obtain point measurements and they are also suitable for measurements of high temperature. The outside shape resembles a resistance thermometer. The usually inevitable shield tube makes the benefit of the point measurement often invalid. A possibility to prevent this is to solder together the head of the shield tube with a soldered joint; at the very least, it is then possible that the components are less sluggish (with time constants of a few seconds). If that is not possible, one must accept time constants of some minutes.

12.2.1.3 **Expansion Thermometers**

The thermal expansion of gases, liquids, and solids is still used to build simple thermometers, which indicate the temperature in the direct proximity of the measurement location. The most usual measuring media are alcohol and mercury. However, mercury poses a health hazard and should only be used when necessary. The measuring range with liquid-filled thermometers (mercury with a glass stem) is situated between −200 and +370°C, with an accuracy of ±1% of full scale, usually less than ±0.5°C [4].

The principle of the thermal expansion of solids is used in bimetallic thermometers, which consist of two metal strips of different materials with different coefficients of expansion firmly connected together. Temperature differences bend these bimetallic springs and they are usually wound in spirals or coils. The useful measuring range is from −73 to +537°C with an accuracy of ±1% of full scale [4]. Bimetallic strips are used for temperature compensation in mechanical instrument movements. They are also used in switching relays to indicate whether a limiting value is exceeded.

12.2.1.4 **Radiation Pyrometers**

High temperatures are measured with radiation pyrometers, which capture the radiation emitted by the object [4, 9]. They are important mostly for high-temperature, non-contact measurements. According to the Stefan–Boltzmann law the intensity of radiation emitted by a black body increases with the fourth power of the absolute temperature. According to Wien's law the wavelength at maximum radiation is inversely proportional to the absolute temperature. One can build thermometers based on these two laws, by measuring either the radiation at all wavelengths or the radiation at specific wavelengths [4]. In the thermocouple pyrometer (Figure 12.4) one collects the radiation emitted from the object on a blackened area whose temperature rise is measured with a thermocouple. In portable devices an eyepiece lens is behind the measuring area so that one can aim at the object.

Measuring errors result from space losses between object and measuring instrument; for well-known distances compensation can be made for the errors. Addi-

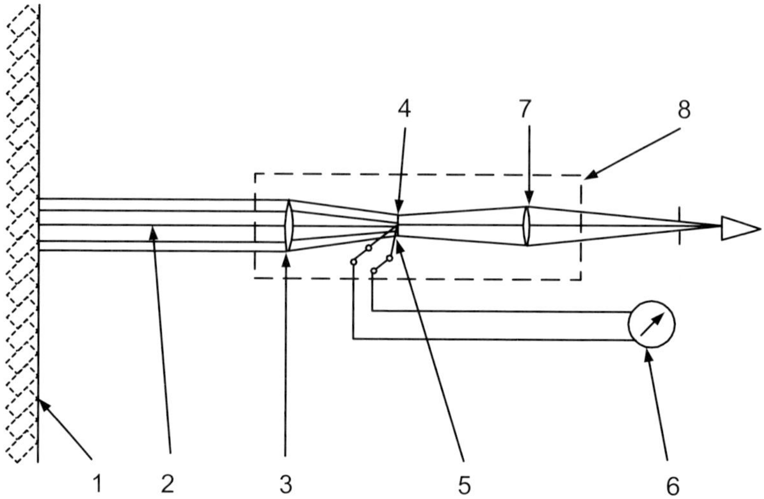

Fig. 12.4. Principle of the radiation pyrometer. 1: Radiation source; 2: radiant heat; 3: collecting lens; 4: black surface; 5: thermocouple; 6: temperature indication; 7: eyepiece; 8: measuring instrument [1].

tionally most objects are not black-body radiation emitters; this can be corrected by using the Planck radiation law to compare the radiation intensities for different wavelengths.

Optical pyrometers can be used for temperature ranges of 760–3500°C [4]. In special cases (narrow-band and total radiation pyrometers) they can be used for much lower temperatures (between −40 and +4000°C) if a non-contact measurement is desired [4].

Pyrometers are important also for qualitative observations, for example as automatic flame guards, and for protection in the combustion chambers of thermal installations so that, after the flame is extinguished, fuel oil or heating gas does not enter the combustion chamber to form an explosive mixture.

12.2.2
Pressure Measurement

Not only is the pressure a significant thermodynamic process variable, but pressure measurements are also fundamentally important for safety reasons. Besides the absolute pressure in an apparatus, one is often interested in a pressure difference, for instance in cases of flow measurements with orifices, and level measurements.

Process equipment pressures are usually measured by the elastic deformation of bellows, a Bourdon tube, a diaphragm, or a capsule [3].

Bellows pressure gauges (Figure 12.5) are very compact and permit linear pressure/path relationships. They are particularly useful for the measurement of

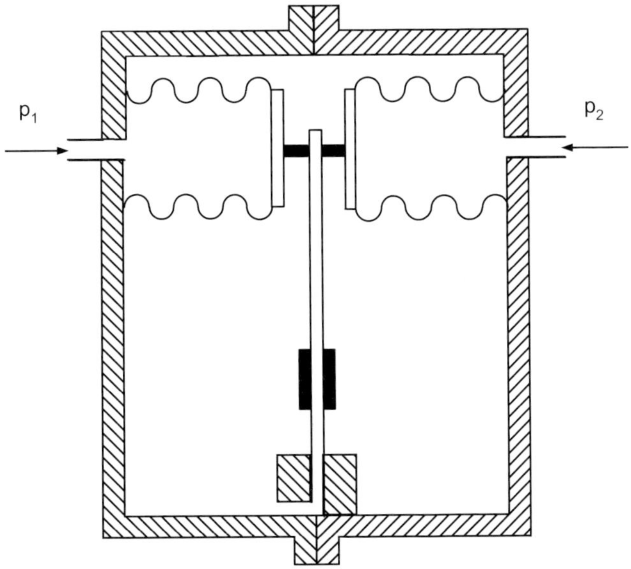

Fig. 12.5. Bellows differential pressure gauge [1].

pneumatic pressure signals and differential pressure measurement, and have been used in pneumatic controller feedback elements.

The gauges are available for pressures of 0–12 mbar (minimum range) and 0–140 bar (maximum range) [3]. For vacuum applications the range is 0–12 mbar vacuum. Accuracy is $\pm 0.1\%$ to $\pm 2\%$ of span [4].

The Bourdon pressure gauge (Figure 12.6) is a versatile instrument. In this instrument a tube with an oval cross-section (Bourdon tube) is bent or coiled inside the instrument. If the internal pressure rises, then the oval cross-section expands and the tube increases its radius of curvature. At its end a pointer is attached, which is carried by the movement of the free end. Pressure gauges are available for pressures of 0–0.3 bar (minimum range) and 0–7000 bar (maximum range) [3]. Above 100 bar the tubes are arranged in several windings. For vacuum applications the range is 0–1 bar vacuum.

In a diaphragm pressure gauge the elastic deflection of a circularly clamped metal disk is transferred to a pointer. Such devices are suitable for measuring ranges of 0–0.5 mbar and 0–70 bar [3]. For vacuum applications the range is 0–0.5 mbar and 0–1 bar vacuum. They are also suitable for differential pressure measurement if both sides of the diaphragm are connected with the respective measuring points.

Bourdon tube and diaphragm pressure gauge designs are easily calibrated and have standard designs.

A capsule is formed by joining two or more diaphragms together, with the total deflection of the assembly equal to the sum of the deflections of each capsule.

Fig. 12.6. Principle of the Bourdon pressure gauge [1].

Pressure sensor selection guidelines can be found in Ref. 3. Pressure sensors have various transducers connected to them to convert the measurement into a usable signal. Strain gauges and piezoresistive and piezoelectric pressure transducers are commonly used to produce the measured values.

Currently, smart transducers are used; they are characterized by inclusion of microprocessors and electronics to store ranges, calibrations and diagnostics, and for other tasks of data handling from the field instruments to the control room.

12.2.3
Weight

Balances compare unknown weights with one standard weight, or alternatively with a well-known force. Balances for weight measurement are very reliable and accurate instruments; their measuring accuracy is often better than ±0.1% [4]. It must be kept in mind that the gravitational attraction on the Earth's surface can vary by as much as 0.5% but this can be compensated by calibration standards [4]. The hydraulic and electrical strain gauge load cells and the electrical strain gauge are now used as primary sensors. The strain gauge type of load cell is capable of a sensitivity of 1 part in 20 000 [6]. Other more accurate sensors are available, such as the variable capacitance transducer capable of a sensitivity of 1 part in 1 000 000 [6].

Either bulk weighing of vessels or weighing platforms are generally used in the polymer industry. Types of balances used include weighing platforms, portable platform scales, and truck scales. Common methods of weighing include mechanical lever scales, spring-balance scales, and load cell weighing systems. Portable bench scales have capacities from 10 to 1000 kg. Beam-type weighing platforms have capacities from 200 to 5000 kg [4].

In electronic balances the force exercised by the load on the base is measured with a force measurement cell (for example, an electrical strain gauge) and transformed into an electrical signal. This signal is processed within the balance in a digital or analog fashion and is then displayed. These systems can be calibrated and used in fully automatic installations. As force measurement cells one can use pressure sensors with lower requirements for accuracy. These are installations in which the weight is transformed into pressure pneumatically or hydraulically; one can determine levels in tanks or reaction vessels by this method, which can also be used to meter automatically the main ingredients of a batch process, for example the water and monomer phase in an emulsion or suspension polymerization.

Strain gauges are used widely in force measurement cells. When a strip of conductive metal is stretched or compressed by a mechanical load, without reaching its elastic limit, its electrical resistance changes. In order to use this principle, one sticks strain gauges onto elastic cylinders or bending rods, whose deformation is thus transferred to the strain gauges. An elastic cylinder carries four to eight resistance strain gauges, which are arranged in a bridge circuit. The load on the cell then detunes the bridge and creates a measuring voltage proportional to the detuning. This configuration also provides compensation for resistance variations due to temperature. Modern units contain almost exclusively digital balances for signal indication and analysis. With these units one can meter in weights of ingredients in a given order.

Belt balances (Figure 12.7) allow simultaneous conveying and weighing, so they are used frequently as metering devices. In these balances, an endless conveyor belt is designed to be the load part of a balance [6]. Weight changes are measured by load cells, and then integrated over short intervals to give the rate of flow. Totalizers can give the total weight over a given time. The belt is covered continuously with the material to be weighed. The feed requires a variable metering device, for example a vibrating hopper or a screw-type feeder.

12.2.4 Liquid Level

Level measurements are used either to control level in vessels or apparatus, or to measure throughput. Level measurements are important in order to estimate inventories and their movement, or to assure the smooth operation of processes. Fur-

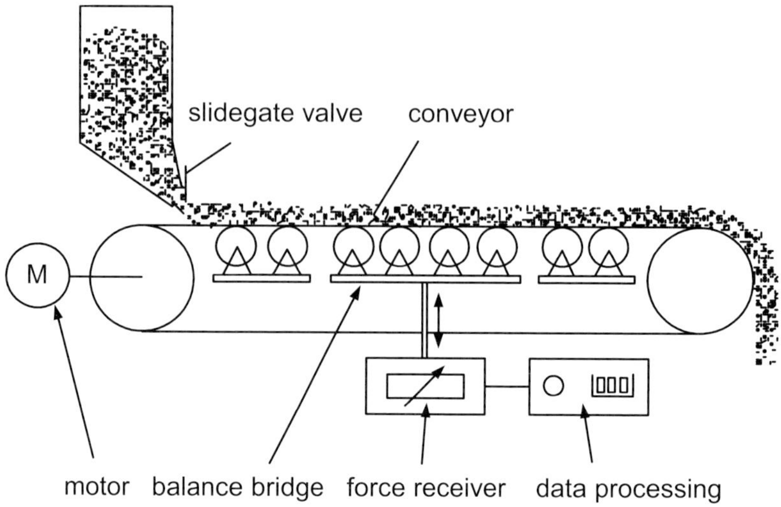

Fig. 12.7. Conveyor belt balance [1].

thermore, in batch processes, level measurements are important to limit filling and emptying steps as well as for alarming. For continuous processes, they serve to maintain a constant level in the vessels where the flow occurs (for example, the sump in a rectification column or the contents of a chemical reactor). Flow measurements are indispensable to metering and feeding functions.

A great variety of level measurement techniques are available. These involve point-contact, visual, buoyancy, float, and hydrostatic methods, and radio-frequency, ultrasonic, microwave, nuclear radiation, resistance tape, and thermal level systems [3].

Point-contact measurements of liquid level are basically length measurements. For example, one observes the wetting of a dipstick through a sight glass in the vessel, or the movement of a float that follows the surface of the liquid in the reservoir. Similarly the point-contact level in a solid hopper can be measured directly by a mechanical plumb line [4], which is attached at the end of a measuring tape that is unwound from a barrel so that the plumb line hits the solid surface; markings on it permit the measurement of the length digitally. Likewise, the length of a hanging string can monitor the position of a float on a liquid surface.

In the float method, a floating gauge is not introduced into the vessel itself but into a different container, such as a side gauge glass, which is arranged in parallel and communicates with the main vessel. In this way disturbances arising from the movement of the fluid (boiling or circulation) are avoided. Floating gauges can also be used in order to indicate the position of the separation surface between two immiscible liquids.

In the buoyancy method, displacement bodies measure levels in liquid containers. By measuring the difference in weight of a partially submerged body at various degrees of submergence, one may determine the level of the liquid in which it is submerged [3]. The device fails if the weight or the volume of the displacement body changes during the process, for example because of scaling or if the density of the medium changes.

In the hydrostatic method, the pressure at the base of the liquid, which is related to the height of the liquid above the base, is measured. The measured pressure depends directly on the liquid height according to the Bernoulli equation or mechanical energy balance [10–12] [Eq. (6), where z_2 is the level at the surface of the liquid, z_1 is the level at the base, p_2 is the pressure at the surface, p_1 is the pressure at the base, ρ is the liquid density, and g is the acceleration due to gravity]. In pressurized reservoirs one measures the pressure difference $p_1 - p_2$ between the base and the gas space.

$$z_2 = z_1 + \frac{p_1 - p_2}{\rho g} \tag{6}$$

In a variant of this method, an inert gas is bubbled through the liquid with a dip tube. If the density of the liquid is known, one can also determine the level through weighing.

It should be noted that in equations such as Eq. (6) a conversion factor may be necessary in unit systems that are termed inconsistent. In the US Customary System (USCS) the force unit is the lb-force (lb_f) and in SI units the force unit is the Newton (N). The conversion factor $g_c = 32.174\ lb_m\ ft\ lb_f^{-1}\ s^{-2}$ is used in USCS units to resolve expressions involving forces in lb_f and masses in lb_m so that Newton's law [Eq. (7), where F is force, m is the mass, and a is the acceleration] is satisfied.

$$F = \frac{ma}{g_c} \tag{7}$$

In the SI unit system, which is used in this book, $g_c = 1\ kg\ m\ s^{-2}\ N^{-1}$ and so this conversion factor is not necessary.

In polymer reactors one has to deal often with very viscous liquids or melts. Here floating gauges or displacement methods fail. An alternative is level measurement with radioactive gamma-ray beams (Figure 12.8) that can penetrate metal walls [4]. The entire reactor is traversed by gamma-radiation; its intensity is given according to the Beer–Lambert law [Eq. (8), where I, I_0 are the intensities behind and in front of the object, ε is the absorption coefficient, ρ is the density of the medium and l is the path length.

$$I = I_0 \exp(-\varepsilon \rho l) \tag{8}$$

Radiation sources include cobalt-60 and cesium-137. Receivers are scintillation counters. The specimen is usually arranged in a rod-like shape and installed on a point-shaped detector, particularly if a continuous measurement is concerned. The benefit of the method is that it operates in a non-contact fashion. However, it re-

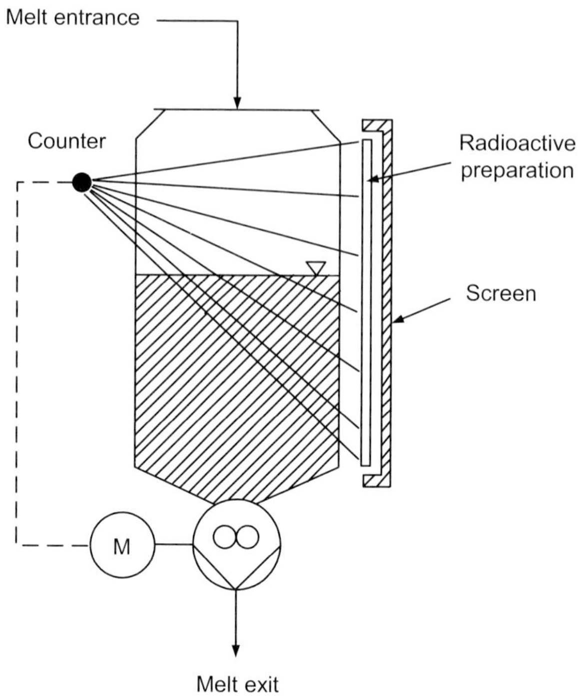

Fig. 12.8. Level measurement using radiation [1].

quires extensive safety measures [4]. The specimen must be shielded with lead against the environment, so that only the direction toward the detector is free. Mounting, dismantling, and maintenance may be undertaken by trained personnel only, and the radiation limiting values are to be checked constantly. For this reason this method should be used only when other methods fail. The accuracy is 3 mm to 1% of height span [4].

Other level-measuring methods are also possible [3]. Capacitance measurements use the change in capacitance of a capacitor, which is built in a probe through the vessel wall (Figure 12.9). Level can also be measured by the change in resistance of a resistance tape on the vessel wall.

One can also obtain levels by introducing vibrating sensors into the container at different levels until the vibrations are damped by the liquid (Figure 12.10).

Microwave radar level systems are also possible, which are not affected by density changes in the beam path as are ultrasonic beam techniques.

12.2.5 Flow

Flow measurements establish the volume or mass of a fluid per unit of time through the measurement device. The methods are similar for both liquids and gases; nevertheless it must be considered that liquids have a substantially higher

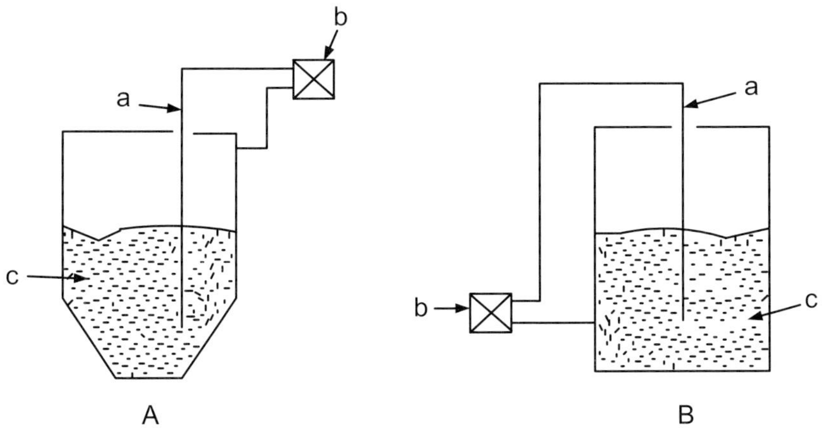

Fig. 12.9. Electrical measurement of liquid level. (A) Capacitive measurement; (B) conductivity measurement; a: electrode; b: measuring instrument; c: material [1].

density than gases. Also, the density dependence on pressure and temperature must be particularly considered with gases. Because of these considerations, different forms of implementation result [2–4]. Types of measurement systems include differential pressure, magnetic, turbine, oscillatory, mass flow, ultrasonic, and positive-displacement meters [3].

Direct flowmeters are generally counters. They couple the number of self-repeating periodic processes (for example, circulation) with time measurement. The most widely used device of this type is the oval gear flowmeter (see Figure 12.11). For cold water the flow ranges are 0.8–5 L min^{-1} to 416–2669 L min^{-1}

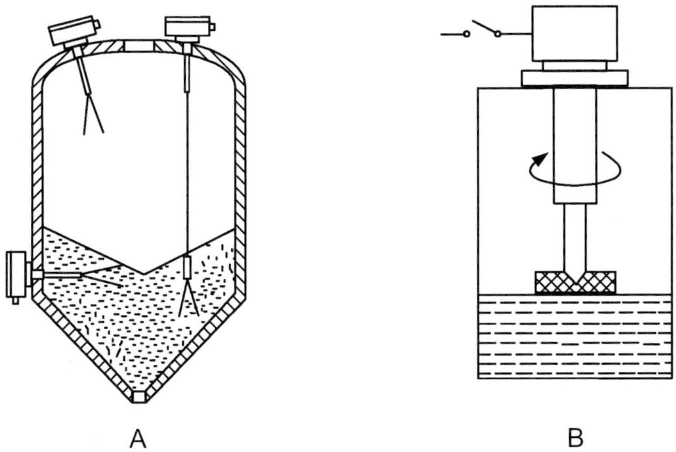

Fig. 12.10. Mechanical measurement of liquid level. (A) Vibration probes; (B) rotary wing probe [1].

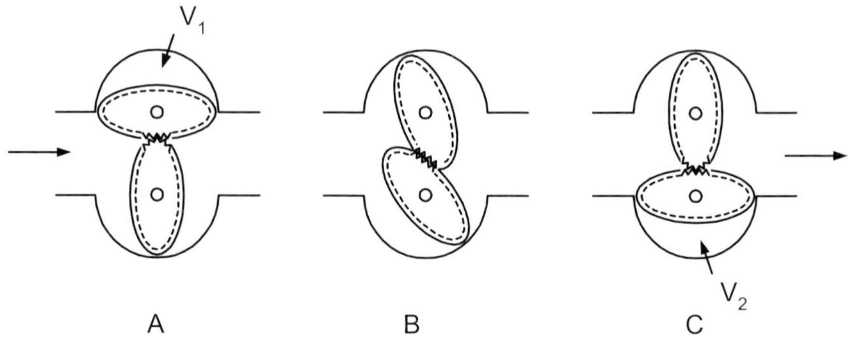

Fig. 12.11. Oval-shaped gear flowmeter [1].

and for heavy oil (5–300 cP) the flow ranges are 0.2–7.2 L min^{-1} to 167–3826 L min^{-1} [3].

In the oval wheel counter, two oval wheels run in gears moving in opposite directions; the pressure drop along the device drives them. The wheels lock successively and the volumes V_1 and V_2 convey them and release them again (Figure 12.11). With each full rotation of the pair of wheels, the volume $V = 2(V_1 + V_2)$ is conveyed through the counter [1]. The devices can be calibrated and high accuracies can be obtained (0.1% [4]); with the smallest types, inaccuracy is highest. The devices are very susceptible to contamination and must therefore be protected by fine filters of solid particles.

Gases can be measured with the rotating lobe flowmeter, which is similar to the oval gear flowmeter (see Figure 12.12). The difference is that the rotating lemniscate-like lobed impellers do not run directly on one another and are not geared together. They are coupled through a gearbox.

The devices are calibrated and can be constructed for flow stream capacities between 2 and 4800 $m^3\ h^{-1}$ [4]. They have good repeatability (0.015%) at high flow rates [4]. They can be used for temperatures up to 205°C and pressures up to 83 bar [4]. The temperature and pressure of the gas are captured and processed so that the volume is indicated in standard conditions of temperature and pressure (STP). The measured variable is deduced from the speed of the rotors, which is obtained through a magnetic clutch from the pressure chamber.

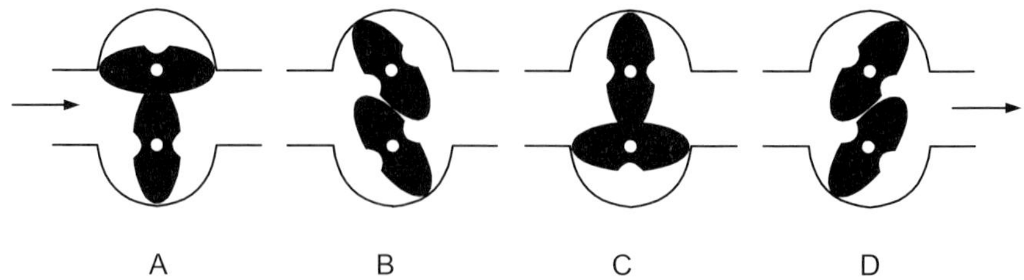

Fig. 12.12. Lobed-impeller flowmeter [1].

Among the indirect volume flowmeters, turbine flowmeters are the most important. They measure the speed of the flowing medium from the speed of an impeller, or rotor. There are devices arranged in an axial or radial direction to the flow. Turbine flowmeters can be used for liquids, gases, and vapors.

The rotating vane meter, which is arranged in the radial direction to the flow, is the most well known and is used for flows of 1–4800 $m^3\ h^{-1}$ [4]. The measuring range is 1:10 with an accuracy of about $\pm 0.1\%$ [4].

The turbine flowmeter with electrical impulse pick-up, arranged in the axial direction of the flow, is of importance for gas flows. It contains an impeller with magnetic wings which, with each passage, induce a voltage surge in a solenoid coil situated outside. The frequency of this alternating current is proportional to the flow. Reproducible measurements require turbulent flow, because only then does the slip not depend on the flow velocity. The measuring range also depends on the viscosity of the flowing medium.

There exist gas turbine flowmeters for flows up to 4500 $m^3\ h^{-1}$ [4]. Linearity is $\pm 1\%$ over a flow range of 20:1 [4]. Rangeabilities can be as much as 100:1. If one wants total volumes using flowmeters, for example for accounting purposes that require high accuracy, then the instantaneous measurement must be integrated with time. This can be accomplished with pneumatic or electrical counters.

Measuring methods which deduce the flow directly from flow characteristics are very common. There are versatile methods being used, which all depend on the velocity of the flowing medium. The most important are measurements with orifice devices, with suspended bodies, and according to a magnetic-inductive method. Beyond that there are flowmeters which use the pressure drop of a flowing liquid in a capillary, and magnetic-inductive, ultrasonic, and Coriolis flowmeters.

Restrictor devices such as venturi tubes, orifice plates, and nozzles operate on the basis of the Bernoulli equation [10, 11]. A venturi tube is depicted in Figure 12.13.

For incompressible fluids without friction or pump work at points upstream (1) and downstream (2) on a streamline in the flow, the Bernoulli equation becomes Eq. (9), where z_1, z_2 are the levels at the two points, p_1, p_2 are the pressures at the

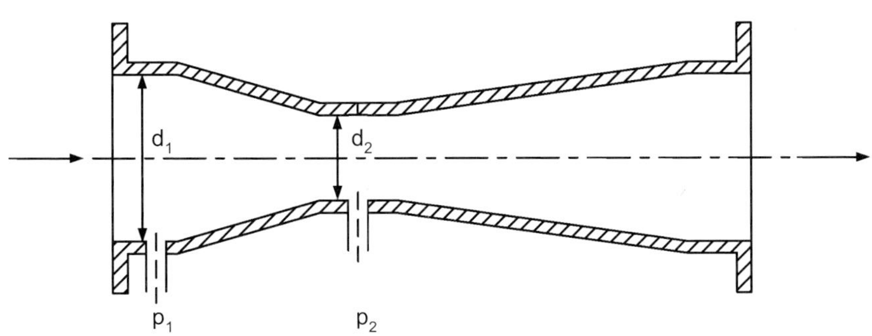

Fig. 12.13. Venturi nozzle. d_1: Diameter of the pipe; d_2: diameter of the throat.

two points, $\bar{v}_1, \bar{v}_2$ are the average velocities at the two points, α_1, α_2 are the kinetic energy correction factors at the two points, ρ is the liquid density, and g is the acceleration due to gravity.

$$\frac{p_1}{\rho} + gz_1 + \frac{\alpha_1 \bar{v}_1^2}{2} = \frac{p_2}{\rho} + gz_2 + \frac{\alpha_2 \bar{v}_2^2}{2} \tag{9}$$

The definition of the kinetic energy correction factor is the area average of the cubed velocity $\overline{v^3}$ over the average velocity cubed, $\bar{v}^3$ [Eq. (10)] [10].

$$\alpha = \frac{\frac{1}{A}\int v^3\, dA}{\left(\frac{1}{A}\int v\, dA\right)^3} = \frac{\overline{v^3}}{\bar{v}^3} \tag{10}$$

Note that $\alpha = 2$ for laminar flow and $\alpha = 1$ for plug flow. For equal heights (horizontal restrictors) the Bernoulli equation becomes Eq. (11).

$$\frac{p_1}{\rho} + \frac{\alpha_1 \bar{v}_1^2}{2} = \frac{p_2}{\rho} + \frac{\alpha_2 \bar{v}_2^2}{2} \tag{11}$$

The continuity equation can be written at two points on a streamline in the flow [Eq. (12), where A_1, A_2 are the flow areas at the two points, ρ_1, ρ_2 are the liquid densities at the two points] [10].

$$\rho_1 \bar{v}_1 A_1 = \rho_2 \bar{v}_2 A_2 \tag{12}$$

Since the fluid is incompressible, Eq. (13) holds.

$$\rho_1 = \rho_2 = \rho \tag{13}$$

Combining Eqs. (11)–(13) and eliminating $\bar{v}_1$ gives Eq. (14), where β is the ratio of diameters d_2/d_1.

$$\bar{v}_2 = \frac{1}{\sqrt{\alpha_2 - \alpha_1 \beta^4}} \sqrt{\frac{2(p_1 - p_2)}{\rho}} \tag{14}$$

To compensate for the friction loss and kinetic energy factor assumptions, the velocity equation is modified to Eq. (15), [10] where C_d is an empirical discharge coefficient.

$$\bar{v}_2 = \frac{C_d}{\sqrt{1 - \beta^4}} \sqrt{\frac{2(p_1 - p_2)}{\rho}} \tag{15}$$

Therefore, for restrictors, the volumetric flow q_v relationship is Eq. (16) [10].

$$q_v = \bar{v}_2 A_2 = \frac{C_d A_2}{\sqrt{1-\beta^4}} \sqrt{\frac{2(p_1 - p_2)}{\rho}} \tag{16}$$

The mass flow q_m relationship is Eq. (17) [11].

$$q_m = \rho \bar{v}_2 A_2 = \frac{C_d A_2}{\sqrt{1-\beta^4}} \sqrt{2(p_1 - p_2)\rho} \tag{17}$$

For compressible flow, similar equations may be used. The mass flow equation is modified to be Eq. (18) [6, 10, 12]. Y is a dimensionless expansion factor that is a function of $p_2/p_1, \beta$, and κ, where $\kappa = C_p/C_v$ is the ratio of specific heats of the gas [6, 10].

$$q_m = \frac{C_d Y A_2}{\sqrt{1-\beta^4}} \sqrt{2(p_1 - p_2)\rho_1} \tag{18}$$

Note that $Y = 1$ for an incompressible fluid.

Both factors C_d and Y can be determined for different tube nominal sizes and flow velocities with air or water, and can also be obtained from charts in Refs. 6 and 12, or tables in Ref. 2. Typically the data can also be used for other media, if the Reynolds numbers are kept constant. The most important restrictors are the venturi tube (Figure 12.13) and the orifice plate (Figure 12.14). The discharge coefficient C_d is about 0.94–0.99 for a venturi tube and about 0.6 for an orifice plate [6, 11].

The simplest implementation of a restrictor is the standard orifice plate, which also supplies the most accurate values (Figure 12.14). The orifice is a sharp-edged disk with diameter d_2 in a pipe with diameter d_1. The pressure tap positions relative to the plate vary depending on the desired design [6]. For so-called radius taps, the pressure-measuring drillings are situated inside the flanges in each case at a distance of one pipe diameter upstream and half a pipe diameter downstream of the orifice [6]. The accuracy of the orifice plates is $\pm 0.25\%$ to $\pm 0.5\%$ of actual flow [4]. The rangeability with a measurement error of less than $\pm 1\%$ of the actual flow is 3:1, which can be improved with smart transmitters [4].

The orifice plate is the restrictor most frequently used in practice. For the measuring accuracy of all standard orifices it is essential that the edge sharpness be maintained in continuous operation. If this is not the case, then the standard nozzles have advantages, and their permanent pressure loss is somewhat lower than in standard orifices. They also have advantages for very small throughputs and are used for very dirty liquids.

In a venturi tube the so-called diffuser is attached to the normal nozzle piece and serves for backpressure gain. This has the smallest permanent pressure loss among the various restrictors. The accuracy of venturi tubes is $\pm 0.75\%$ of the uncalibrated rate and $\pm 0.25\%$ of the rate calibrated in a laboratory.

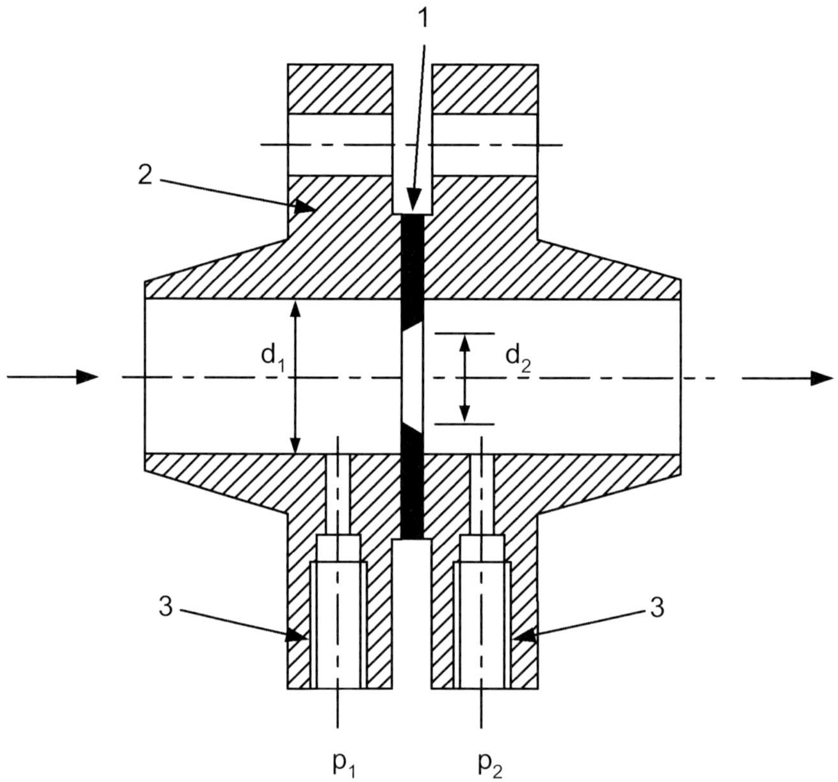

Fig. 12.14. Orifice plate and flanges. 1: Orifice plate; 2: flange supports; 3: pressure measurement drill taps [1]. Note that the pressure tap positions vary, depending on the desired design [6].

All measurements with restrictors require turbulent, evenly distributed flow. In order to smooth out disturbances in the flow stream, orifices and nozzles require a straight and smooth upstream piping length of between six and 40 pipe diameters [6] depending on the nature of the upstream fitting and between two and four pipe diameters downstream to the nearest fitting. Sources of error are clusters of gas bubbles or dirt particles (in fluids) or condensing liquids (for gases) in the differential pressure tubes. For gases the pressure ratio p_2/p_1 at the restriction must not be less than about 0.53; otherwise sonic velocity is achieved in the restriction [10].

Measuring methods with floating bodies work with vertically arranged, inverted cone-shaped or slightly tapered tubes, where the fluid flows through from bottom to top. In the rotameter, the flow holds thereby a floating body in equilibrium (see Figure 12.15), which depends on the flow velocity and the lift of the body against its weight.

The force balance is written as Eq. (19) [10], where F_W is the (downward) gravity force, F_B is the (upward) buoyancy force, and F_D is the (upward) drag force.

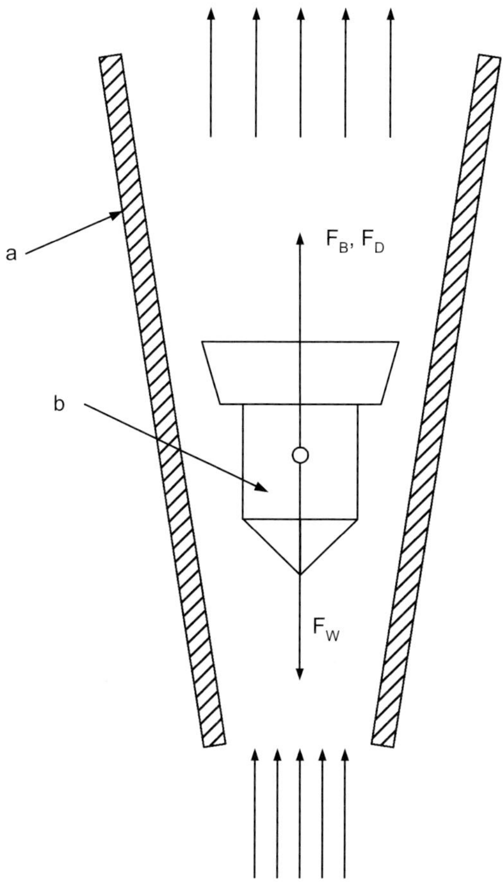

Fig. 12.15. Flow measurement with float (rotameter).
a: Conical flow pipe; b: float; F_W: gravity force; F_B: buoyancy force; F_D: drag force [1].

$$F_W = F_B + F_D \tag{19}$$

If V_f is the volume of the float, ρ_f is the density of the float, and ρ is the density of the fluid, the force balance becomes Eq. (20).

$$F_D = V_f \rho_f g - V_f \rho g \tag{20}$$

The drag force is given by Eq. (21), where C_D is the drag coefficient, A_f is the projected area of the float, and $v_{\max}$ is the maximum velocity past the float.

$$F_D = C_D A_f \frac{\rho v_{\max}^2}{2} \tag{21}$$

Substituting the drag force into the force balance we get Eq. (22).

$$C_D A_f \frac{\rho v_{\max}^2}{2} = V_f g(\rho_f - \rho) \tag{22}$$

The volumetric flow rate q_v can be related to the maximum velocity past the float by Eq. (23), where A_t is the tube area at the point of constriction and A_f is the float area.

$$q_v = v_{\max}(A_t - A_f) \tag{23}$$

To size the rotameter, $v_{\max}$ can be eliminated between these two equations and the expression simplified so that the volumetric flow rate q_v is obtained as Eq. (24), where $C_d = 1/\sqrt{C_D}$ is a discharge coefficient with values between about 0.6 and 0.8 [9].

$$q_v = C_d(A_t - A_f)\sqrt{\frac{2gV_f(\rho_f - \rho)}{\rho A_f}} \tag{24}$$

For a linearly tapered tube with the bottom diameter the same as the float diameter, the area for flow is a linear function of the height of the float h [Eq. (25)] [10].

$$d_t^2 - d_f^2 = (d_f + ah)^2 - d_f^2 = 2d_f ah + (ah)^2 \approx 2d_f ah \tag{25}$$

Substituting this into the mass flow equation gives Eq. (26), from which the flow rate depends almost linearly on the height of the floating element h and the viscosity of the medium; in each case it is calculated with the help of calibration curves, which depend on the instrument and the calibration medium (air or water).

$$q_m = \rho q_v = \rho v_{\max}\frac{\pi}{4}(d_t^2 - d_f^2) = \left(\frac{2\pi}{4}\rho v_{\max} d_f a\right)h \tag{26}$$

The floating elements are available in different forms; for simple gas measurements the floating element is a sphere; for fluids one mostly finds a cylindrical floating element (Figure 12.15), which is shaped downward like a cone and at the upper edge like a flat truncated cone-like disk. This disk is sharply grooved in the side, in order to induce a rotation in the equilibrium position. This movement stabilizes its position in the center of the conical pipe.

The position of the floating element can be observed directly, if the conical guide tube consists of glass or quartz. If that is not possible, then its position can also be measured magnetically, if it contains a magnet and the pipe consists of a non-magnetic material.

Rotameters are useful for flows from 0.01 $cm^3\ min^{-1}$ to 920 $m^3\ h^{-1}$ of liquid, or 0.3 $cm^3\ min^{-1}$ to 2210 $m^3\ h^{-1}$ of gas [4]. The measurement accuracy for industrial rotameters is approximately $\pm 1\%$ to $\pm 2\%$ of full scale over a 10:1 range [4].

Magnetic-inductive flowmeters consist of one nonmagnetic, insulated pipe, which is traversed by a magnetic field. If a conductive fluid moves through the pipe, the ions contained in it are deflected crosswise to the flow direction and to the magnetic field. Two electrodes, which are attached at the same height inside the pipe, then indicate a voltage difference, which is proportional to the flow velocity. The method is independent of the flow profile, density, viscosity, and contamination. Moreover it does not build up an additional pressure drop in the line. The measuring accuracy is within $\pm 1\%$ [1]. For the time being, there are still difficulties for nominal pressures over 16 bar.

When a fluid moves in a tube that is rotated, it experiences a Coriolis force proportional to its mass and velocity and to the angular velocity of the tube. In the Coriolis flowmeter, the tube is not rotated but vibrated. Vibrations in the tube are set up and a small elastic deformation in the tube results from the Coriolis force. The deformation magnitude is related to the mass flow rate. These flowmeters have the advantage that the measurement is independent of the temperature, density, pressure, viscosity, or flow profile [3]. The accuracy is high, $\pm 0.15\%$ within the range of 10:1 of full scale rate [4], but the initial cost is high. The flow range is 0–28300 kg min^{-1} [4]. The high accuracy can be very useful for ingredients that are precisely metered to polymerizers, despite the high cost. Because of their operating principle, Corilis flowmeters can simultaneously obtain the density of the fluid.

12.2.6
Densitometry, Dilatometery, and Gravimetry

Density is the mass per unit volume of a substance. Specific gravity is the density of the substance relative to a standard such as liquid water at 4°C. The density of liquids is monitored by hydrometers, weighing a fixed volume (density balance), the Coriolis method (described in Section 12.2.5), and vibrating methods. Hydrostatic, displacement float, sonic, and radiometric methods are among others also used [4].

A hydrometer consists of a tube closed at both ends, with one end enlarged into a bulb that contains fine lead shot or mercury to cause the instrument to float upright in a liquid. In the glass tube is a scale calibrated so that the reading on it, level with the surface of the liquid in which the hydrometer is floating, indicates the number of times the liquid is heavier or lighter than water, that is, the specific gravity of the liquid. Hydrometers have been connected photometrically or mechanically to produce a useable electrical signal [4].

The density balance operates on the principle of weighing a fixed volume. In it the liquid flows through a U-tube, which forms the load part of a balance (Figure 12.16). The deflection of the balance can be indicated directly, and usually one uses electrical or pneumatic transducers so that the measured values can be transferred and processed further. Density balances must be installed so as to avoid vibrations.

In the tuning fork densitometer one utilizes the dependency of the frequency of a tuning fork or a vibrating plate, which is excited to produce an under-damped oscillation, on the density of the surrounding medium. The resonant frequency of

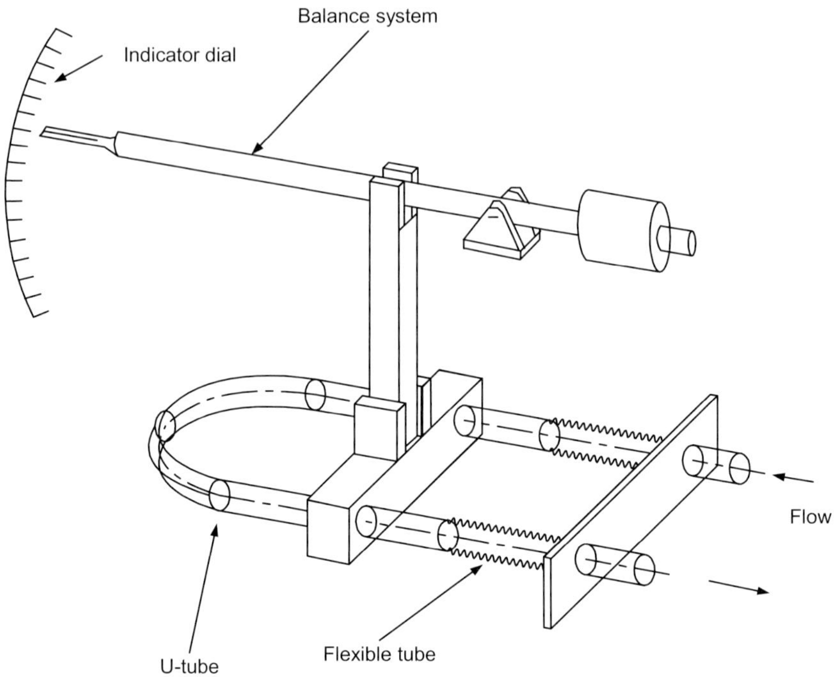

Fig. 12.16. Density balance [1].

an oscillating system is approximately proportional to the square root of the oscillating mass [4]. A vibrating flow U-tube may also be used for density measurement [8].

Vibrating and density balance densitometers can also be used for compressed gases ($\rho > 2$ kg m^{-3}). Gases at standard pressure are measured with gas density balances, with which one determines the buoyancy of a closed hollow sphere filled with a comparison gas.

A sonic densitometer consists of a device for measuring the speed of sound in the liquid [4, 8]. The speed of sound c is given by Eq. (27), where E is the bulk modulus and ρ is the mass density [4]. Unfortunately the liquid must be clear of particles that might scatter the sound waves and limit the received signal.

$$c = \sqrt{\frac{E}{\rho}} \tag{27}$$

The density of liquids can also be measured radiometrically, by the absorption of gamma-rays in the medium. The intensity of the exiting radiation depends on the density of the medium and is given according to the Beer–Lambert law [Eq. (28), where I, I_0 are the intensities behind and in front of the object, ε the absorption coefficient, ρ the density of the medium and l the path length].

$$I = I_0 \exp(-\varepsilon \rho l) \tag{28}$$

Unlike level measurements the radiometric measurement of the density requires high outlet intensity and thus higher safety precautions. However, it has a broad range of application and is particularly interesting for measurements in polymer technology; one can measure very viscous melts, emulsions, and suspensions.

Dilatometers measure the volume shrinkage during the course of liquid polymerization reactions and are mainly used for laboratory measurement of monomer conversion. They are based on the principle that polymers are denser than their monomers. Volume changes are monitored as monomer is converted to polymer by following the change in height of the solution inside a graduated capillary tube. Conversion is monitored with a computer-linked photodetector that tracks the meniscus in the capillary and records the height changes [13].

The percentage of total solids in a polymer sample can be determined by the gravimetric method through moisture weight loss. The sample is loaded onto a pan and the weight determined. Then it is put into an oven at high temperature for a time to remove all volatiles. It is then reweighed and the percentage of solids is determined.

12.2.7 Viscosity

Viscosities are of interest in polymer technology in order to follow the course of a polymerization reaction or to monitor continuously the quality of a product. Viscosity μ is the constant of proportionality between applied shear stress τ and the resulting shear rate $\dot{\gamma}$ according to Newton's Law of viscosity [Eq. (29)] [11, 13].

$$\tau = \mu \dot{\gamma} \tag{29}$$

Viscosity may be constant (Newtonian), shear thickening (dilatant), or shear thinning (pseudoplastic) with shear rate. For polymer systems, solution or melt, the viscosity can be related to the molecular weight of the polymer, as discussed in Refs. 4, 8, and 13.

In most cases viscosity is measured by capillary viscometers or rotating viscometers. In a capillary viscometer one measures the pressure drop by means of constant laminar flow in a capillary; the constant flow can be achieved by a pump and the pressure drop is obtained by a differential pressure transmitter whose "plus" and "minus" sides are connected to the capillary. The pressure drop is then directly proportional to the viscosity according to the Hagen–Poiseuille law [4, 11] [Eq. (30), where μ is the viscosity, r is the capillary radius, l is the capillary length, Δp is the pressure drop, and q_m is the mass flow rate]. The capillary viscometer may also be employed in-line for monitoring of molecular weight in polymerizations, as described in Ref. 14.

$$\mu = \frac{\pi r^4 \Delta p \rho}{8 q_m l} \tag{30}$$

A method to obtain a measure of molecular size that is quick and cheap is the melt indexer [13]. The melt index is defined as the number of grams of polymer extruded in 10 min through a capillary 2.1 mm in diameter and 8 mm long at a certain temperature and pressure (ASTM D1238).

Among the different possible ways to measure viscosities in rotating viscometers, the coaxial cylinder apparatus is the most commonly used in practice. The measured liquid intersperses the annular gap between the stationary inner cylinder (bob) and the rotating outer cylinder (cup). Therefore a velocity gradient builds between the inner and outer cylinders (Couette flow). The momentum, which is transferred by this downward gradient to the inner cylinder, is directly proportional to the viscosity. Deflection is compensated by a torsion bar and the equilibrium deflection is measured electrically. The measurement of the angular velocity of the cup and the angular deflection of the bob makes it possible to determine the viscosity [4, 11].

Besides the coaxial device, the cone-and-plate viscometer is also used. In this device an inverted cone faces a solid plate and the apex of the cone just touches the plate. The measured liquid is in the free gap. The viscosity of the measured fluid is computed from the torque on the cylinder-driving shaft [4, 11].

The Mooney viscometer, used particularly in the rubber industry, is a variant of the cone-and-plate viscometer; it restricts the sample to a disk-shaped cavity (ASTM D1646) [4].

Vibrating-reed viscometers can also be used for continuous in-line polymer viscosity [4]. The amplitude of the probe vibration depends on the viscosity of the fluid. As the viscosity of the fluid increases, the resistance to probe vibration increases.

12.2.8 Composition

The composition of raw materials, finished products, and samples of the various steps of a reaction is normally measured at the laboratory using the appropriate physical and chemical analytical methods. However, sampling and analysis are time-consuming and in many cases the result of the analysis is only of current interest and too late for control decisions to be made. In order to monitor compositions continuously, one needs automatically functioning analytical instruments that can continuously obtain and show the composition of a mixture. Some devices are fast and precise enough to be able to generate signals for control loops. The controllers would then adjust the desired values of other input variables such as flow, temperature, or pressure in a cascade control scheme.

Optical methods are common [4, 8, 13]. Infrared (IR) spectrographic analysis makes it possible in many cases to follow the appearance or the disappearance of one or more characteristic absorption frequency bands. These frequency bands correspond to frequencies of vibrations of the bonds in the molecules. One must first analyze the spectrum of the IR radiation and then measure the corresponding frequencies. More recently the Fourier transform infrared technique (FTIR) has been used for faster data acquisition and handling than traditional IR spectrographic analysis. IR and FTIR can be applied to polymer solutions or solid films for composition analysis and are particularly useful for copolymer composition determination.

Optical analytical devices are also built for measuring radiation in the ultraviolet (UV) and the visible spectral region, but the spectral absorption bands obtained here are usually so broad that these devices are used only for special tasks.

The refractive index (RI) of a mixture is a function of the composition of the mixture and the respective refractive indices of the constituents [8]. The mixture refractive index follows mixture laws such as the Lorentz–Lorenz law. Operational measuring instruments are usually differential refractometers or critical-angle refractometers [4]. A large disadvantage in the method is that it only provides meaningful results when a two-component system is considered. However, a differential refractometer is commonly used as a concentration detector in the effluent of a gel permeation chromatography (GPC) column for molecular weight determination.

Raman spectroscopy is dependent on the collision of incident light quanta with the molecule, inducing the molecule to undergo a change [13]. It is now being used to provide a means of studying pure rotational and vibrational transitions in molecules. Raman scattering of light by molecules may be used to provide their chemical composition and molecular structure and is currently being applied to polymers, as shown in Refs. 8, 15, and 16.

Apart from optical methods, one uses magnetic (for paramagnetic materials, for example oxygen in diamagnetic gases) and electrical methods. Examples of the latter are conductivity measurements (of ionic liquids, for example purity of boiler feeding water), ionization methods (for example, the flame ionization detector in gas chromatographs or the photo-ionization of gases with UV light to measure tracking of hydrocarbons in air), electro-chemical potential methods (for example, pH measurements), and occasionally polarographic methods.

Nuclear magnetic resonance (NMR) is based on the principle that when a hydrogen-containing compound is in a strong magnetic field and exposed to radio-frequency signals the compound absorbs energy at discrete frequencies [13]. This technique can be used to measure chain molecular structure, copolymer composition, and copolymer sequence lengths. It can also deduce isotactic/atactic ratios and other structure variations, as shown, for example, in Ref. 17. Mass spectrometry and NMR are currently not in routine on-line process use, but can be used to calibrate other on-line methods.

Many methods depend on the separation of a fluid mixture. Among these, gas chromatography (GC) stands out [8]. Suitable devices for on-line control were developed from laboratory gas chromatographs and operate very reliably. However, they can be expensive because of the associated program controls. The principle of gas chromatography is that a carrier gas (helium) is passed over a tubular column of a fine solid. A sample is injected into the carrier gas stream and the gas effluent from the column is run past a detector such as a flame ionization detector. Calibration is based on the fact that, all conditions being equal, a given hydrocarbon will require the same length of time to pass through the column to the detector (elution time) [4].

A mass spectrometer source produces ions, and information about a sample may be obtained by analyzing the dispersion of ions when they interact with the sample using the mass-to-charge ratio. Sometimes mass spectrometers are used after a separation step such as gas chromatography or liquid chromatography for fraction identification.

12.2.9 Surface Tension

In emulsion polymerizations particularly, it may be of some interest to measure the surface tension of the polymerization. The surface tension can give an indication of whether or not micelles are present, which is important in particle nucleation above the critical micelle concentration (CMC) [18, 19].

The on-line method used is usually the bubble pressure method. A dip tube is inserted below the liquid surface and bubbles are formed by compressed gas. Bubbles formed within a liquid are compressed by surface tension. The resulting pressure rises with decreasing bubble radius. This increased pressure, in comparison to the outside of the bubble, is used to measure surface tension. During the process of bubble formation and breakage, the pressure can be measured in the bubble. From the pressure oscillation the surface tension can be calculated.

12.2.10 Molecular Weight Distribution (MWD)

It is widely recognized that a reliable method of monitoring molecular weight distribution (MWD), and the various molecular weight averages (M_n, M_w, and M_z) during the polymerization process is of importance to final polymer quality. The polydispersity index is defined as the ratio of the weight to the number average molecular weight [Eq. (31)], which is a measure of the spread of the MWD.

$$PD = \frac{M_w}{M_n} \tag{31}$$

Monitoring the molecular weight distribution or its averages from a batch or continuous polymer reactor in real time would be desirable. Monitoring and feedback control of polymerizations can provide fundamental benefits for improved quality.

Traditionally, gel permeation chromatography (GPC) or size-exclusion chromatography (SEC) has been used to determine MWD [8, 13]. In GPC/SEC, polymer solutions are injected into one or more columns in series, packed with porous particles. The packing has small pores and during elution the polymer molecules may or may not, depending on their size, penetrate into the pores. Therefore, smaller molecules have access to a larger fraction of pores than the larger ones, and the chains elute in a decreasing order of molecular weights. For each type of polymer an empirical correlation exists between molecular weights and elution volumes. This can be used to calibrate the GPC/SEC, which allows the evaluation of average molecular weights and molecular weight distributions.

Direct column calibration for a given polymer requires the use of narrow MWD samples of that polymer. The chromatograms of such standards give narrow peaks and with each standard is associated the retention volume of the peak maximum. There are a number of polymers for which narrow MWD standards are commercially available. More recently triple-detector instruments have been designed which include a differential viscometer, a light-scattering instrument, and a differential refractometer that monitors the column effluent. A calibration curve can be obtained from this arrangement as long as all signals are calibrated [13].

For on-line purposes, the viscosity measures previously mentioned (see Section 12.2.7) have been used as a proxy for molecular weight averages in on-line control. Some vendors are commercializing more rapid GPC/SEC instruments for on-line control, with certain instruments already available.

12.2.11 Particle Size Distribution (PSD)

The particle size distribution (PSD) can have a fundamental effect on the physical properties of dispersions that are common polymer products. The measurement of just the average particle size may not be sufficient. For example, the presence of populations of different sizes resulting in a multimodal distribution could have a strong influence on final properties and may need to be controlled.

There are several particle size measurement techniques in use, such as optical imaging, electron imaging, optical diffraction and scattering, electrical resistance changes, sieving, sedimentation, and ultrasonic attenuation [4].

Optical (larger than 1 μm) and scanning electron microscopy (SEM) techniques literally give the clearest picture of a PSD. However, analyzing the images may be tedious without image analyzers. Nevertheless, this method can be used as a check or calibration on the indirect methods.

The Coulter-counter particle size analyzer (larger than 0.5 μm) is used for measuring volumes of individual particles. Particles are suspended in a conductive fluid, into which electrodes are placed. As a particle passes through an aperture between the electrodes, it displaces its own volume of electrolyte, and there is a measurable change in the electrical resistance of the system. The change becomes a precise measure of particle volume. These volumes can then be put into size bins and the PSD can be constructed.

Today there are two principal light scattering technologies that are commercially available: light scattering intensity measurement (also known as static, or Rayleigh, scattering) and dynamic light scattering measurement [also known as quasi-elastic light scattering (QELS) or photon correlation spectroscopy (PCS)]. It should be noted that light scattering techniques are applied not only to particle size measurement but also to macromolecule size measurement.

In static light scattering measurements, the light intensity scattered in solution by a particle, which is small compared with the wavelength of the incident laser beam, is proportional to the concentration multiplied by the molecular weight. When the particles are very small compared with the wavelength of the light, the intensity of the scattered light is uniform in all directions (Rayleigh scattering); for larger particles (above approximately 250 nm diameter), the intensity is angle-dependent (Mie scattering). If the concentration of the particles in solution is known or is measured during the analysis process, the particle size averages and distributions can be determined.

Dynamic light scattering provides a relatively fast and simple method for submicron particle sizing [8]. When a beam of light passes through a colloidal dispersion, the particles or droplets scatter some of the light in all directions. Random intensity fluctuations in scattered laser light arising from the Brownian motion of colloidal particles are analyzed to give either a simple mean size and polydispersity index or complete distribution data, even for multimodal distributions.

Turbidimetry has been used traditionally in industry to obtain a measure of average particle size and even the entire PSD, and is a measure of the attenuation of a beam of light passing through a suspended particle sample [8].

Acoustic attenuation spectroscopy measurements can be made without the need for sample dilution and can be used in the particle size range of 10 nm to 100 μm. As sound travels through a slurry or colloid, it is attenuated. The level of attenuation is related to the particle size distribution as discussed in Refs. 20 and 21. Acoustic attenuation measurements can be made on high-concentration and/or opaque samples [8].

Packed column hydrodynamic chromatography (HDC), a technique for separating particles based on their size by eluting in the order largest to smallest, operates on a principle similar to that of GPC/SEC [8]. The sample under investigation and a small-molecule marker solution are introduced so that the effluent flow is not interrupted. HDC has a dynamic operating range from about 20 nm to 1.2 μm. An ultraviolet (UV) detector response is used to calculate the concentration of particles of different sizes present in the sample. The subsequent computation of particle size distribution requires a calibration procedure employing a particle size standard.

Capillary hydrodynamic fractionation (CHDF) is a hydrodynamic method for measurement of nanometer-sized particles. In this method, slurry containing the particles is forced through a capillary. Flow rate through the capillary is highest in the center of the capillary due to the laminar flow velocity profile. Larger particles extend into the high-flow region while smaller particles travel closer to the wall of the capillary. Larger particles are swept through the capillary ahead of the smaller particles, producing particle size fractionation. The particles are detected at the end of the capillary by a UV detector. The measurable particle size range for CHDF is about 15 nm to 2 μm.

For on-line control these techniques must be evaluated for speed, reliability, and sample dead time [4]. As in the MWD techniques, manufacturers are moving more toward on-line implementation of the more recent methods, but many of these techniques still remain in the control laboratory.

12.3
Sensor Signal Processing

12.3.1
Sensors and Transmitters

Sensor signal processing involves operations such as amplification, compensation, filtering, and normalization. As discussed (see Section 12.1), transmitters transform the signals leaving the sensor into a standardized signal, and the sensor plus the transmitter are considered to be the transducer or measurement instrument [7]. Transmitters should be attached operationally as closely as possible to the sensor location in order to minimize transmission error. To protect instruments against outside effects from contamination, damage, and corrosion, they should be placed in sturdy housings. A measurement amplifier is connected to the signal source of an electrical sensor and converts the signal coming from the sensor into the standardized direct current signal.

The pneumatic or electrical signals depict only the analog measured variable. The direct measured variable is not always the desired result, however. Thus, for example, one obtains from flow measurements with diaphragms a differential pressure, not the mass flow, q_m, which for compressible flow in an orifice is given by Eq. (32), where $\Delta p = p_1 - p_2$ is the pressure drop between the taps upstream and downstream respectively, ρ_1 is the upstream density, C_d is an empirical discharge coefficient, A_2 is the area of the orifice, Y is a dimensionless expansion factor, and $\beta = d_2/d_1$ is the ratio of the orifice to the upstream pipe diameter (see Section 12.2.5) [10].

$$q_m = \frac{C_d Y A_2}{\sqrt{1-\beta^4}} \sqrt{2\rho_1 \Delta p} \qquad (32)$$

As can be seen from Eq. (32), the mass flow is proportional to the product $\sqrt{\rho_1 \Delta p}$. As density and flow can be subject to variations during the process, the downstream pressure will change and so will the density, and the two measured values are coupled together. Additionally, density is also a function of temperature so this must also be measured and a correction added. A smart transducer provides functions beyond those necessary for generating a correct representation of a sensed or controlled quantity. A smart transducer can be used in this situation to correct the flow for pressure and temperature.

12.3.2
Converters

If one wants to use the analog signals for more complicated calculations or in a process control system, one must digitize them and after processing transform them back to analog signals, for example to drive the final control elements or actuators. Converters are used for this purpose. Measurement converters transform the analog signal into a digital one (A/D converter) or a digital signal into an analog one (D/A converter). Any decimal number can be converted into the binary system by summing the appropriate multiples of the different powers of 2. Equation (33) demonstrates the conversion of binary into decimal numbers as well as that of decimal into binary numbers; thus 10101101 (binary) is equivalent to 173 (decimal).

$$(1 \times 2^7) + (0 \times 2^6) + (1 \times 2^5) + (0 \times 2^4) + (1 \times 2^3) + (1 \times 2^2) + (0 \times 2^1) + (1 \times 2^0) = 173 \tag{33}$$

The raw binary sensor data can be encoded according to several binary formats, such as: pure binary, two's complement binary, signed binary, binary coded decimal (BCD), ASCII, or Gray codes. In order to convert the digits of the decimal system to binary digits, which are required for digital technology, one can directly code these digits. Very frequently BCD is used. Every decimal digit is assigned to the corresponding binary digit. In the original binary code the number 35 would be shown as 100011, whereas in the BCD code it would be shown as 0011;0101.

It must be cautioned that during this conversion a quantization error results, because of round-off errors in the decimal places; for example, 1.354 V (analog) might become 1.35 = 0001;0011;0101 (BCD). Many process control applications use A/D and D/A conversions of at least 12-bit unsigned binary representation to obtain a resolution of 1 part in $2^{12} = 0.024\%$ [7].

12.3.3
Indicators

The numerous measurement signals of a process-engineering unit must be indicated so that they can be used and also recorded in some fashion. They are therefore combined in a central control room; there they provide a current overview of the state of the entire plant.

Indicators show the measured value in an analog or digital way. The analog representation consists of a pointer before a calibrated scale, and recently also in LED displays and analog representations on monitors. Digital outputs are indicated as number displays or on counters. Apart from the measured value, indicator instrument panels also contain alarm functionality (optical and/or acoustical), which indicates if a critical measured value is exceeded in order to automatically shut down the installation. Once the signal is processed it is sent to the controller, the output of which, the controlled variable, is then sent to the actuator.

In the past these functions were performed by panel boards consisting of indicators, alarms, strip-chart recorders, and single-loop controllers. Today, distributed control systems (DCS) and data historians perform these functions (see Section 12.4.6).

12.3.4 Filtering Techniques

As mentioned above (see Section 12.1.2), during a measurement stochastic errors can be introduced. The effects of process and measurement noise can be minimized by signal conditioning or filtering [7]. Analog filters have been commonly used for many years. A high-frequency noise (low-pass) filter can be represented by Eq. (34), where $x(t)$ is the filter input, $y(t)$ is the filter output, and τ_F is the filter time constant. This equation can be discretized to produce a digital filter with sampling interval Δt [Eq. (35)].

$$\tau_F \frac{dy(t)}{dt} + y(t) = x(t) \tag{34}$$

$$\tau_F \frac{y_n - y_{n-1}}{\Delta t} + y_n = x_n \tag{35}$$

When rearranged this gives the exponential smoothing filter in Eq. (36), where the filter factor, $0 < \alpha \leq 1$, is defined by Eq. (37).

$$y_n = \alpha x_n + (1 - \alpha) y_{n-1} \tag{36}$$

$$\alpha \equiv \frac{1}{1 + \frac{\tau_F}{\Delta t}} \tag{37}$$

This filter can be implemented in a digital computer. Other filters, such as a moving-average filter, are also possible [7].

12.4 Regulatory Control Engineering

12.4.1 General

Measurement instruments supply information on the current operating conditions of a plant. These form the basis to control the process and to keep conditions constant so that the optimal-quality results are obtained. For this purpose one uses controllers which determine automatically the sequence of control events. For example, a process could be that of the tank shown in Figure 12.17. The tank has a

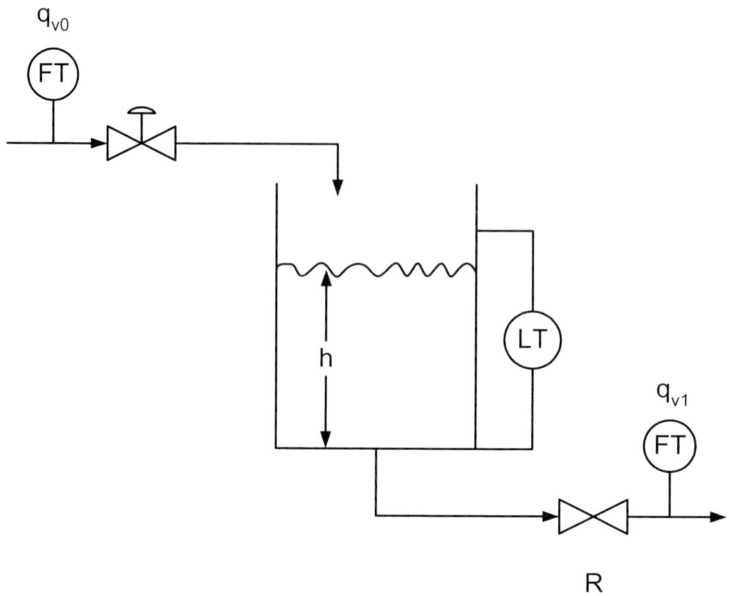

Fig. 12.17. Tank with liquid level height h, inlet valve with volumetric flow q_{v0}, and outlet volumetric flow q_{v1} through a restrictor with resistance R.

stream flowing into it through a valve and a stream leaving it through a restriction. A more comprehensive example of a polymer process as shown in Figure 12.33 will be discussed later (see Section 12.5).

In order to be able to keep process or measured variables constant, or to change them systematically, they must first be measured continuously and recorded. Flow rates into and out of the tank (denoted by the flow transducer, or FT) and the level (denoted by the level transducer, or LT) are measured variables. One therefore obtains information about the instantaneous state of the desired measured variable to be controlled, the controlled variable, and about its change with time. In the case of the tank, this variable would be the level in the tank as measured by the level transducer. The controlled variable value must then be compared with the desired value, called the "set point". Any deviations from the set point, called the "error", can then be corrected through suitable actions. Devices that execute this comparison and provide output signals, which can be used for the correction sent to the manipulated variable, are called feedback controllers or regulators. In the case of the tank, the manipulated variable could be the flow into the tank measured by the flow transducer. If one wants to have a process follow certain time-varying conditions (as is the case often with batch processes), then one can use controllers to send the suitable programmed sequence signals. Often measured or unmeasured variables that affect the process and cause deviations from the set point are called "disturbance variables". In the case of the tank, this variable could be an uncontrolled leak from the tank or a secondary flow into the tank.

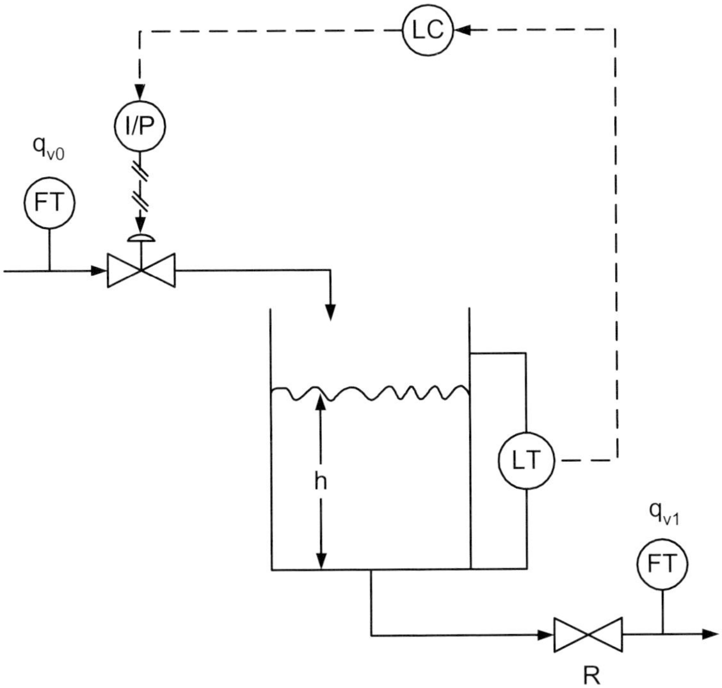

Fig. 12.18. Feedback level control on a tank.

The signals leaving the controller are used to affect temperatures, pressures, and flow rates by means of actuators. Actuators are usually valves whose flow area changes with the signal coming from the controller. Metering machines, or positive displacement pumps, can also be used as actuators.

Measurement, comparison, and adjustment constitute an entity called a control loop. Very schematically one may describe the control loop in terms of sensors, controllers, actuators, and the process to be controlled. Shown in Figure 12.18 is a level control system for a tank. Broken lines denote control signals. The controller (denoted by LC) receives the level signal from the level transducer (LT) that sends a signal to the current-to-pressure transducer (I/P) that then applies a pneumatic pressure to the valve diaphragm.

Shown in Figure 12.19 is the control loop pictured as a block diagram. In the control loop the controlled variable y is continuously measured and compared with the set point y_{sp}; it is desirable to have the deviation or error $e = y_{sp} - y$ made as small as possible. The controller then decides what control action to make on the manipulated variable u. Disturbances in the process denoted by d influence the controlled variable y and are compensated for by the manipulated variable u.

In contrast to the controlled system, the tank in Figure 12.17 is not under feedback control. It could be operated in a manual semi-batch mode, where the inlet

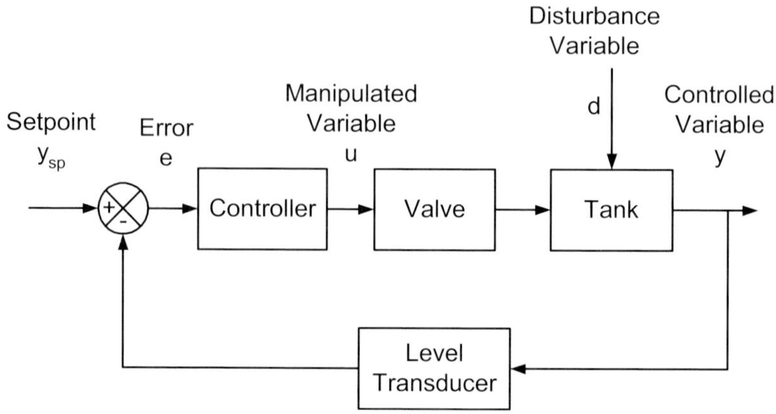

Fig. 12.19. Block diagram of the tank level control loop.

valve is used to fill the tank to a certain level. The control consists of a sequence of operating conditions, which must be carried out under a well-defined plan. It is used around well-defined processes, for example to automate the loading step of a batch emulsion.

By analogy with the control loop hierarchy, one can split the control chain into the process to be controlled and the control system, as shown (for example) in Figure 12.18. In the given example, the process consists of the tank, which can be filled, the control valve with the appropriate piping, the restrictor, the level and flow transducers and perhaps a recorder; the control system covers the programmable controller, the signal transmission, and the I/P converter.

Expenditure for measurement, control, and actuator systems in the process industries is substantial. The proportional cost is higher the smaller the process is, because the absolute expenditure depends only on the number of measurement points and control loops, not on the size of the installation.

12.4.2 Process Dynamics

It is the task of the control system of a plant to drive the controlled variable (temperature, pressure, flow, and so on) to the desired values, by either constant or prescribed trajectories. In order to accomplish this, the controlled variable must be continuously measured and the dynamics of the process must be known. In control technology one distinguishes processes with self-regulation and without self-regulation. In the first case the output variable y assumes a new equilibrium state after the input variable u is changed; an example is the flow through a pipe after a change at the control valve. Another example is the tank mentioned above: when the inlet flow is changed a new level is reached, as long as the outlet flow has a restrictor whose flow depends on the height of liquid in the tank. One calls such systems "self-regulating", because the output reaches a new steady state when subjected to step-changes in the input. In the second case a change in the input does not produce a new steady state. For instance, in a tank that is continuously fed and drained by manipulation of an outlet valve, the level remains constant only if the difference between output flow and input flow is zero. One calls such systems "non-self-regulating", because the output does not reach a new steady state when subjected to step-changes in the input.

12.4.2.1 First-order System

In many cases the relationships between plant inputs and outputs may be derived in the form of a mathematical model, which can then be used to understand and control the plant. For the tank example in Figure 12.17 the dynamic mass balance becomes Eq. (38), where ρ is the liquid density, A is the cross-sectional area of the tank, h is the height of liquid in the tank, and q_{v0}, q_{v1} are the inlet and outlet volumetric flow rates [7, 22]. If we assume constant density this becomes Eq. (39).

$$\frac{d}{dt}(\rho A h) = \rho q_{v0} - \rho q_{v1} \tag{38}$$

$$A\frac{dh}{dt} = q_{v0} - q_{v1} \tag{39}$$

The flow restrictor may be modeled, to a first approximation, as a linear relationship between flow and height of liquid in the tank [Eq. (40), where R is the valve resistance]. The material balance becomes Eq. (41).

$$q_{v1} = \frac{h}{R} \tag{40}$$

$$RA\frac{dh}{dt} = Rq_{v0} - h \tag{41}$$

Since this is a self-regulating process, a steady state exists, governed by Eq. (42), where the subscript s denotes the steady-state value. Subtracting Eq. (42) from Eq. (41) yields Eq. (43).

$$0 = Rq_{v0s} - h_s \tag{42}$$

$$RA\frac{dh}{dt} = R(q_{v0} - q_{v0s}) - (h - h_s) \tag{43}$$

Since we are mainly only interested in deviations from the steady state, in this case we may define deviation variables by Eqs. (43), where u is the manipulated flow deviation variable and y is the controlled level deviation variable.

$$\begin{aligned} u &= q_{v0} - q_{v0s} \\ y &= h - h_s \end{aligned} \tag{44}$$

In terms of the deviation variables, our mass balance becomes Eqs. (45). This is in the form of a first-order system [Eq. (46), where the time constant is given by Eq. (47) and the process gain by Eq. (48).

$$RA\frac{dy}{dt} = Ru - y \tag{45}$$

$$\tau\frac{dy}{dt} = K_p u - y \tag{46}$$

$$\tau = RA \tag{47}$$

$$K_p = R \tag{48}$$

The dynamic behavior of a dynamic system can be well represented through the so-called "step response". The dynamic evolution of the output variable can be monitored in response to a step-change of the input. We may ask how this system responds to a step-change of magnitude M in the input flow rate. The first-order system becomes that described by Eqs. (49), where $H(t)$ is the Heaviside step function defined by Eq. (50).

$$\begin{aligned}\tau\frac{dy}{dt} &= K_p M H(t) - y \\ y(0) &= 0\end{aligned} \tag{49}$$

$$H(t) = \begin{cases} 0 & \text{if } t < 0 \\ 1 & \text{if } t \geq 0 \end{cases} \tag{50}$$

The solution to this differential equation, which may be found with the assistance of Laplace transforms [7, 22] is Eq. (51).

$$y(t) = K_p M(1 - e^{-t/\tau}) \tag{51}$$

This result is plotted in Figure 12.20 (denoted by one tank) for the case of $M = 1$ m^3 min^{-1}, $K_p = 1$ min m^{-2}, $\tau = 1$ min. Note that the output is over-damped and approaches a steady value after long times.

12.4.2.2 Second-order System

In a similar fashion one may analyze the system of two non-interacting tanks in series, as shown in Figure 12.21.

The system mass balances are Eqs. (52), where the subscripts 1 and 2 denote tanks 1 and 2 respectively [22].

$$\begin{aligned}A_1\frac{dh_1}{dt} &= q_{v0} - q_{v1} \\ A_2\frac{dh_2}{dt} &= q_{v1} - q_{v2}\end{aligned} \tag{52}$$

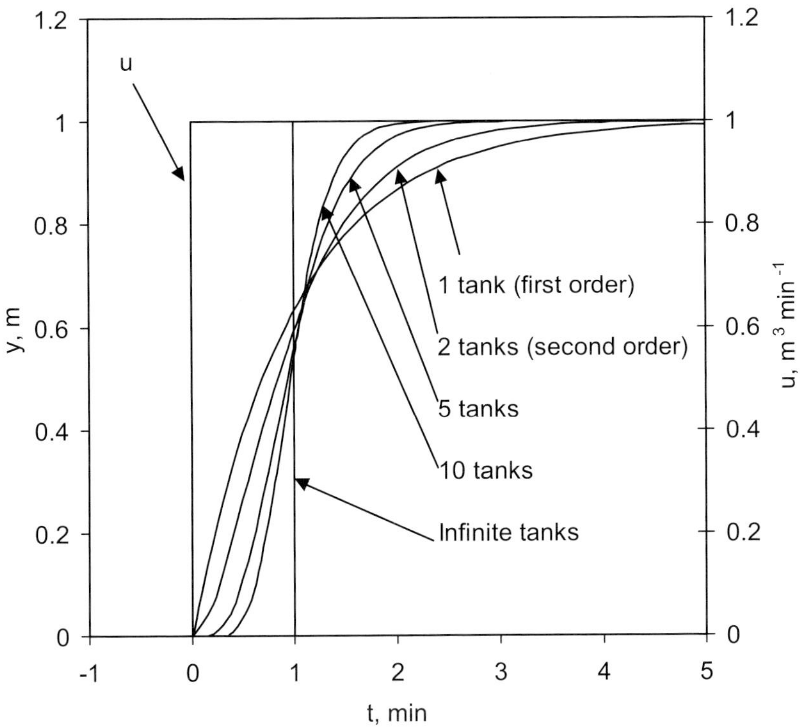

Fig. 12.20. Step responses of $n = 1, 2, 5, 10$, and an infinite number of tanks in series.

The outlet restrictors may again be considered linear as a first approximation [Eqs. (53)].

$$
\begin{aligned}
q_{v1} &= \frac{h_1}{R_1} \\
q_{v2} &= \frac{h_2}{R_2}
\end{aligned}
\tag{53}
$$

The material balances become those given by Eqs. (54), and at steady state they are given by Eqs. (55).

$$
\begin{aligned}
R_1 A_1 \frac{dh_1}{dt} &= R_1 q_{v0} - h_1 \\
R_2 A_2 \frac{dh_2}{dt} &= \left(\frac{R_2}{R_1}\right) h_1 - h_2
\end{aligned}
\tag{54}
$$

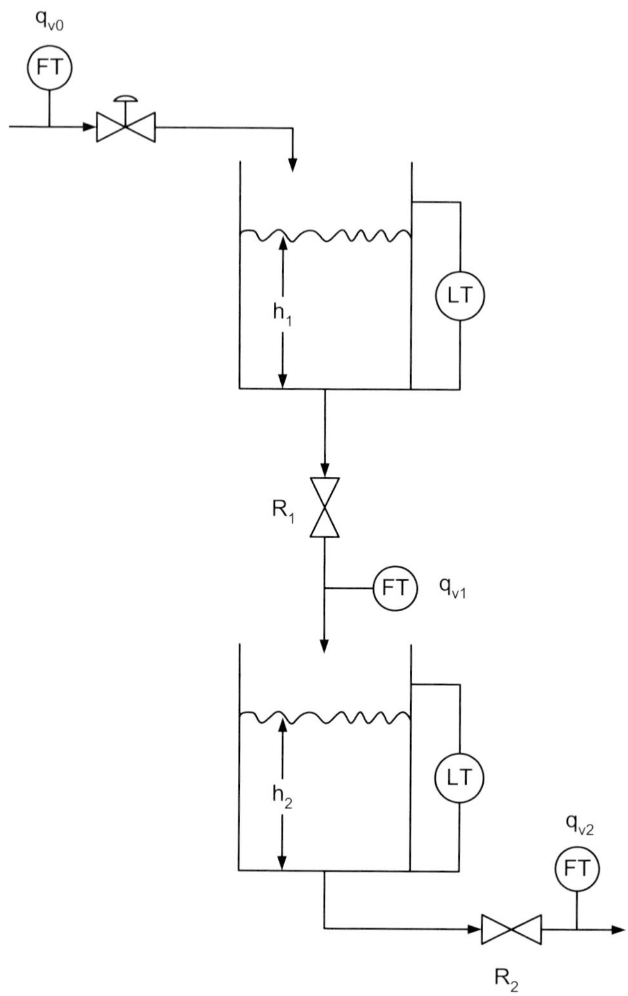

Fig. 12.21. Two tanks in series without feedback control.

$$\begin{aligned} 0 &= R_1 q_{v0s} - h_{1s} \\ 0 &= \left(\frac{R_2}{R_1}\right) h_{1s} - h_{2s} \end{aligned} \tag{55}$$

Subtracting Eqs. (55) from (54) yields Eqs. (56).

$$\begin{aligned} R_1 A_1 \frac{dh_1}{dt} &= R_1(q_{v0} - q_{v0s}) - (h_1 - h_{1s}) \\ R_2 A_2 \frac{dh_2}{dt} &= \left(\frac{R_2}{R_1}\right)(h_1 - h_{1s}) - (h_2 - h_{2s}) \end{aligned} \tag{56}$$

Introducing deviation variables given by Eqs. (57), the mass balances become Eqs. (58), which are in the form of two first-order systems [Eqs. (59)] where the time constants and process gains are given by Eqs. (60).

$$\begin{aligned} u_0 &= q_{v0} - q_{v0s} \\ y_1 &= h_1 - h_{1s} \\ y_2 &= h_2 - h_{2s} \end{aligned} \tag{57}$$

$$\begin{aligned} R_1 A_1 \frac{dy_1}{dt} &= R_1 u_0 - y_1 \\ R_2 A_2 \frac{dy_2}{dt} &= \left(\frac{R_2}{R_1}\right) y_1 - y_2 \end{aligned} \tag{58}$$

$$\begin{aligned} \tau_1 \frac{dy_1}{dt} &= K_{p1} u_0 - y_1 \\ \tau_2 \frac{dy_2}{dt} &= K_{p2} y_1 - y_2 \end{aligned} \tag{59}$$

$$\begin{aligned} \tau_1 &= R_1 A_1, \quad K_{p1} = R_1 \\ \tau_2 &= R_2 A_2, \quad K_{p2} = \frac{R_2}{R_1} \end{aligned} \tag{60}$$

These mass balances [Eqs. (58)] may be combined into one equation in y_2 by differentiating the second and then eliminating y_1 by successive substitutions [22] to afford Eq. (61).

$$\tau_1 \tau_2 \frac{d^2 y_2}{dt^2} + (\tau_1 + \tau_2) \frac{dy_2}{dt} + y_2 = K_{p1} K_{p2} u_0 \tag{61}$$

This is the equation of a second-order system. Again we may ask how this system responds to a step-change of magnitude M in the input flow rate. The second-order system becomes that described by Eqs. (62).

$$\begin{aligned} \tau_1 \tau_2 \frac{d^2 y_2}{dt^2} + (\tau_1 + \tau_2) \frac{dy_2}{dt} + y_2 &= K_{p1} K_{p2} M H(t) \\ y_2(0) &= 0 \end{aligned} \tag{62}$$

The solution to this differential equation, which again may be found with the assistance of Laplace transforms, is Eq. (63) [22].

$$y_2(t) = K_{p1} K_{p2} M \left(1 - \frac{\tau_1}{\tau_1 - \tau_2} e^{-t/\tau_1} - \frac{\tau_2}{\tau_2 - \tau_1} e^{-t/\tau_2}\right) \tag{63}$$

For the special case of equal time constants ($\tau_1 = \tau_2 = \tau/2$), the solution is Eq. (64).

$$y_2(t) = K_{p1} K_{p2} M \left(1 - e^{-2t/\tau} - \frac{2t}{\tau} t e^{-2t/\tau}\right) \tag{64}$$

This result is plotted in Figure 12.20 (denoted by two tanks) for the case of $M = 1$ m^3 min^{-1}, $K_{p1} K_{p2} = 1$ min m^{-2}, $\tau = 1$ min. Note that the response of the two-tank process is over-damped and slightly lagged compared with the single tank, producing an S-shaped curve.

12.4.2.3 High-order and Dead Time Systems

One may proceed in a similar fashion to obtain solutions for more and more tanks in series with equal time constants, with the constraint that the total time constant is equal to the sum of the identical individual time constants so that $\tau_1 = \tau_2 = \cdots = \tau_n = \tau/n$, creating systems of higher and higher order. These results are plotted in Figure 12.20 for the cases with the numbers of tanks in series $n = 1, 2, 5, 10, \infty$ with $M = 1$ m^3 min^{-1}, $K_{p1} K_{p2} \ldots K_{pn} = 1$ min m^{-2}, $\tau = 1$ min. As the number of tanks increases, the response approaches that of a pure delay or dead time, θ, equal to the total time constant of the infinite tanks, $\theta = \tau$.

12.4.2.4 First-order Plus Dead Time System

Since dead time is a frequently occurring phenomenon, many systems can be represented by a combination of a first-order system plus dead time (FOPDT) system. In the time domain the FOPDT system equation is Eq. (65).

$$\tau \frac{dy(t)}{dt} = K_p u(t - \theta) - y(t) \tag{65}$$

For a step-change in input flow rate of magnitude M the system becomes Eqs. (66), the solution to which is Eq. (67).

$$\begin{aligned} \tau \frac{dy(t)}{dt} &= K_p M H(t - \theta) - y(t) \\ y(0) &= 0 \end{aligned} \tag{66}$$

$$y(t) = K_p(1 - e^{-(t-\theta)/\tau}) M H(t - \theta) \tag{67}$$

This result is plotted in Figure 12.22 in original height $h = y + h_s$ variable for the case of the FOPDT tank with parameters $M = 1$ m^3 min^{-1}, $K_p = 10$ min m^{-2}, $\tau = 20$ min, $\theta = 2$ min, $h_s = 4$ m and $t_0 = 10$ min is the initial time of the step-change in flow rate.

If at time t_0 in a system the input variable q_{v0} is increased by the amount Δq_{v0} and the output variable instantaneously changes by Δh, then we are dealing with a self-regulating system without time delay. If the output variable follows with de-

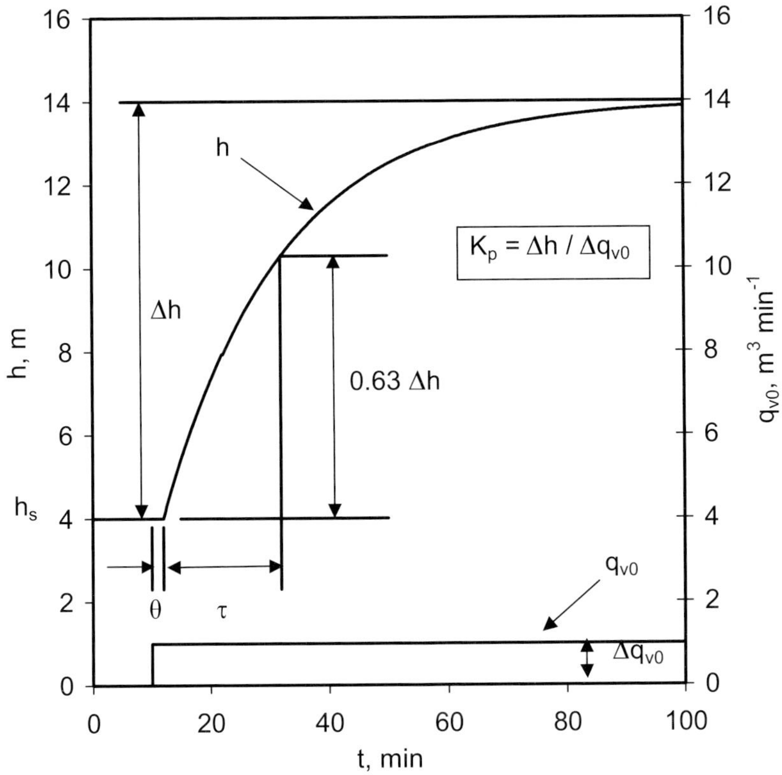

Fig. 12.22. Step response of a first-order plus dead time process.

lays, then delay elements are present such as storage/accumulation or flow lags. If the response of the output variable begins also after a certain time θ, then there exists a system with dead time.

Figure 12.22 suggests a method to determine these parameters from a step test experiment [6, 7, 22]. The parameters can be geometrically obtained from Figure 12.22 and this forms the basis for model identification. At time t_0 the steady-state input to the process is stepped by an amount Δq_{v0} and the steady-state output change Δh measured. Then the process gain is given by Eq. (68).

$$K_p = \frac{\Delta h}{\Delta q_{v0}} \tag{68}$$

The dead time θ is the delay in the response after the input step at time t_0. The time constant τ is the time after the output has risen $0.63\Delta h$ above its original steady value h_s after the input step at time t_0.

Dead time is always to be expected if there are transport processes in the system. A pure dead time is observed during the metering of compounds with a weigh belt, because for every adjustment at the entrance slide, the new quantity is only observed at the output after the residence time on the belt (Figure 12.7). Dead time and time lags are to be expected in mixing control in a pipe, because in addition to the transport time longitudinal mixing also plays a role. Higher-order lags are often hard to distinguish from dead time, as shown in Section 12.4.2.3.

12.4.2.5 **Integrating System**

Suppose the situation of the tank in Figure 12.17 is that the effluent flow restrictor is replaced by a fixed flow rate. This is a non-self-regulating or integrating process. The mass balance is Eq. (69), as before [22].

$$A\frac{dh}{dt} = q_{v0} - q_{v1} \tag{69}$$

However, this time the effluent flow is flow controlled by a valve or pump, not the height of liquid in the tank. At steady state for a fixed value of q_{v1} Eq. (70) applies; that is, at steady state, the inlet and outlet flows are exactly matched.

$$0 = q_{v0s} - q_{v1} \tag{70}$$

Subtracting Eq. (70) from Eq. (69) gives Eq. (71).

$$A\frac{dh}{dt} = q_{v0} - q_{v0s} \tag{71}$$

Introducing deviation variables [Eqs. (72)], the mass balance becomes Eq. (73), where the process gain is given by Eq. (74).

$$\begin{aligned} u &= q_{v0} - q_{v0s} \\ y &= h - h_s \end{aligned} \tag{72}$$

$$\frac{dy}{dt} = K_p u \tag{73}$$

$$K_p = \frac{1}{A} \tag{74}$$

For a step-change in input flow rate of magnitude M the system becomes that described by Eqs. (75), the solution to which is Eq. (76), which is a ramp function with slope K_p.

$$\begin{aligned} \frac{dy}{dt} &= K_p M H(t) \\ y(0) &= 0 \end{aligned} \tag{75}$$

$$y(t) = K_p M t \tag{76}$$

Thus the tank will overflow or run dry if the inlet and outlet flows are not exactly matched.

12.4.2.6 Integrator plus Dead Time System

With dynamic systems that are non-self-regulating one finds, likewise, step responses with and without dead time. An integrator plus dead time process would be modeled like Eq. (77).

$$\frac{dy(t)}{dt} = K_p u(t - \theta) \tag{77}$$

For a step-change in input flow rate of magnitude M the system becomes as in Eqs. (78), the solution to which is Eq. (79).

$$\begin{aligned} \frac{dy(t)}{dt} &= K_p M H(t - \theta) \\ y(0) &= 0 \end{aligned} \tag{78}$$

$$y(t) = K_p M(t - \theta) H(t - \theta) \tag{79}$$

This result is plotted in Figure 12.23 in original height variable $h = y + h_s$ with parameters $M = 1\ \text{m}^3\ \text{min}^{-1}$, $K_p = 2\ \text{m}^{-2}$, $\theta = 2$ min, $h_s = 4$ m and $t_0 = 10$ min.

The parameters for this process may be obtained from a step test. At time t_0 the steady-state input to the process is stepped by an amount Δq_{v0} and the gain K_p is found from the measured slope [Eq. (80)].

$$\text{Slope} = \frac{\Delta h}{\Delta t} = K_p \Delta q_{v0} \tag{80}$$

The dead time θ is the delay in the response after the input step at time t_0.

12.4.3 Controllers

Controllers are devices that are meant to keep a specific controlled variable constant despite outside disturbances. One differentiates between continuous, digital, and On–Off controllers, depending on whether the output is continuous, discrete, or on–off. The great majority of controllers have been continuous, but the digital controller is now commonplace due to the widespread use of computers and distributed control systems.

Referring again to Figure 12.19, the controller compares the measured value of the controlled variable with the desired value, the set point, computes the manipulated variable, and influences the actuator so that the set point and the controlled

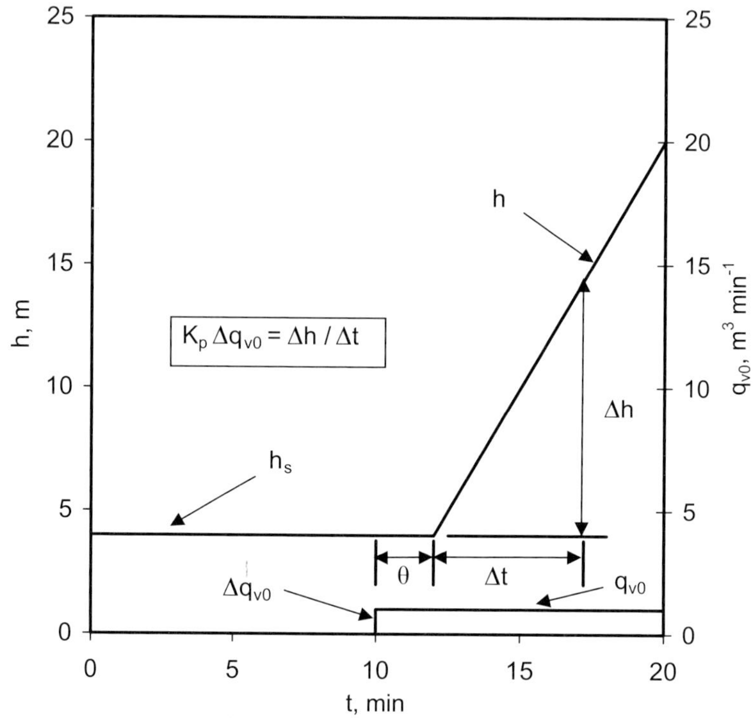

Fig. 12.23. Step response of an integrator plus dead time process.

variable can be as close to each other as possible. Independently of design and application, one finds three basic functions, which can be simply represented mathematically: namely the proportional, integral, and derivative functions.

12.4.3.1 Proportional Control

The proportional controller (P controller) assigns to each value of the deviation a specific value of the manipulated variable. For our tank example, the P controller mode takes the form of Eq. (81), where q_{v0}, q_{vs} are the new output and the steady-state values of the flow rate respectively, K_c is the controller gain, and h_{sp}, h are the current level input height set point and measured value respectively.

$$q_{v0}(t) = q_{vs} + K_c[h_{sp} - h(t)] \tag{81}$$

The steady-state value of the flow rate q_{vs}, or bias, is the flow rate when the deviation from the set point is zero. The sign on the gain term, or controller action, is adjusted to be either positive or negative to create negative feedback control in which the controlled variable approaches a stable steady state. We may write this controller in terms of deviation variables, according to Eqs. (82), where e is called the error and u is the controller output, to obtain the deviation form [Eq. (83)].

$$e = h_{sp} - h(t)$$
$$u(t) = q_{v0}(t) - q_{vs} \tag{82}$$

$$u(t) = K_c e(t) \tag{83}$$

Pure proportional controllers have the disadvantage that in order to change the actuator, the controller needs a certain deviation; therefore this deviation can never be entirely eliminated and is termed "offset".

As given here the controller gain has units of flow/height or $m^2 \, min^{-1}$. In commercial controllers the manipulated and controlled variables are often made dimensionless with the range and standard signal of the transmitter and actuator so that the units of the gain would be, for example, %/% or mA/mA. Also, in older controllers the term "proportional band" (PB) is used instead of "controller gain" and is defined by Eq. (84).

$$PB = \frac{100\%}{K_c} \tag{84}$$

12.4.3.2 **Integral Control**

The integral controller (I controller) assigns the integral over time of the error to a specific change in the output variable. In the case of large errors over time, the action is large; in the case of small errors over time, the action is small.

The I controller obeys Eq. (85), where τ_I is the reset time, in deviation variables.

$$u(t) = \frac{1}{\tau_I} \int_0^t e(t)\, dt \tag{85}$$

The output changes until the error has disappeared. Pure integral-only controllers have the disadvantage that they are sluggish for large τ_I and that they may be prone to oscillation for small τ_I.

A disadvantage of integral control is reset windup. This results from the situation where, if the controller output saturates and the error remains large for an extended period of time, the integral gets large and does not recover immediately even if the error then goes to zero or reverses sign. The integral has to be manually reset in this situation in actual implementation.

12.4.3.3 **Derivative Control**

Derivative controllers (D controllers) are practical only in combination with other controller modes. They provide a fast engagement at the beginning of the control action. Here the derivative controller output is proportional to the rate of change of the error [Eq. (86), where τ_D is the derivative time].

$$u(t) = \tau_D \frac{de(t)}{dt} \tag{86}$$

The derivative mode of a D controller causes theoretically a large change in output of extremely short duration for a step-change in error. In practice, the effect of a derivative controller is to cause a short increase in the output variable, which then returns to the previous value.

12.4.3.4 PI, PD, and PID Control

These basic controllers are used in combinations; P, PI, PID, and also PD controllers are common. The appropriate equations are linear combinations of the selected components. The PI controller obeys the relationship of Eq. (87). It reacts to a deviation with fast correcting action, and corrects the rest of the deviation slowly. It kicks in immediately and operates very accurately in the long term.

$$u(t) = K_c\left(e + \frac{1}{\tau_I}\int_0^t e(t)\,dt\right) \tag{87}$$

The PD controller obeys Eq. (88). It kicks in first very strongly; the output then reverts back to a value that corresponds to the output of a P controller.

$$u(t) = K_c\left(e + \tau_D\frac{de(t)}{dt}\right) \tag{88}$$

If one combines all three controller actions, then a PID controller results. It obeys Eq. (89). This is the universal type of controller, which acts fast and avoids lasting offset.

$$u(t) = K_c\left(e + \frac{1}{\tau_I}\int_0^t e(t)\,dt + \tau_D\frac{de(t)}{dt}\right) \tag{89}$$

The PID form implemented usually includes a derivative mode filter such as a first-order filter to eliminate noise, which would be written in the time domain as Eqs. (90), where $e_F(t)$ is the filtered error and $0.05 < \alpha < 0.2$ is the dimensionless filter constant [7].

$$\begin{aligned} \alpha\tau_D\frac{de_F(t)}{dt} &= -e_F(t) + e(t) \\ u(t) &= K_c\left(e + \frac{1}{\tau_I}\int_0^t e(t)\,dt + \tau_D\frac{de_F(t)}{dt}\right) \end{aligned} \tag{90}$$

The PID controller may also use the rate of change of the measured variable (for example, $h(t)$) instead of the error $e(t)$ to eliminate set point change derivative kick. Derivative kick is mitigated if a derivative mode filter is used.

12.4.3.5 Digital Controllers

Modern implementations of controllers are not analog but digital due to the widespread use of computers and digital control systems. Digital control algorithms can be derived from the continuous versions using standard numerical approximations of the analog controllers. Digital signals are discrete in nature and arise from sampling continuous measurements at equal time intervals of width Δt, or they may arise from naturally discrete signals such as, for example, from analyzers.

We may illustrate the nature of digital PID algorithms by starting with the ideal PID controller [7, 22, 23] according to Eq. (91), where $q_v(t), q_{vs}$ are the new output and the steady-state bias values of the flow rate respectively, and $e(t) = h_{sp} - h(t)$ is the current error in the level height set point and measured value.

$$q_v(t) = q_{vs} + K_c\left(e(t) + \frac{1}{\tau_I}\int_0^t e(t)\,dt + \tau_D\frac{de(t)}{dt}\right) \tag{91}$$

A simple digital form of this equation may be written for the nth time interval using the rectangular rule of integration approximation to the integral and a first-order backward finite difference approximation of the derivative to yield the positional form, Eq. (92), where $e_n, q_{v,n}$ are the error and controller output respectively at the nth sampling instant, and Δt is the sampling period.

$$q_{v,n} = q_{vs} + K_c\left[e_n + \frac{1}{\tau_I}\sum_{i=1}^{n} e_i\Delta t + \tau_D\left(\frac{e_n - e_{n-1}}{\Delta t}\right)\right] \tag{92}$$

We may write this equation for the $(n-1)$st sampling instant [Eq. (93)].

$$q_{v,n-1} = q_{vs} + K_c\left[e_{n-1} + \frac{1}{\tau_I}\sum_{i=1}^{n-1} e_i\Delta t + \tau_D\left(\frac{e_{n-1} - e_{n-2}}{\Delta t}\right)\right] \tag{93}$$

If we subtract this Eq. (93) from Eq. (92), we obtain the velocity form [Eq. (94)], since it calculates the change in output.

$$q_{v,n} - q_{v,n-1} = K_c\left[(e_n - e_{n-1}) + \frac{1}{\tau_I}e_n\Delta t + \tau_D\left(\frac{e_n - 2e_{n-1} + e_{n-2}}{\Delta t}\right)\right] \tag{94}$$

Solving for the output at the nth instant, we have a form [Eq. (95)] that calculates the output directly.

$$q_{v,n} = q_{v,n-1} + K_c\left[(e_n - e_{n-1}) + \frac{1}{\tau_I}e_n\Delta t + \tau_D\left(\frac{e_n - 2e_{n-1} + e_{n-2}}{\Delta t}\right)\right] \tag{95}$$

One major advantage of the velocity form Eq. (95) over the positional form Eq. (92) is that, as the summation is lacking, it has inherent anti-reset windup. Also, an initial value of the output bias is not required [7].

If the sampling period is small compared with the process time constant

($\Delta\tau \ll \tau$) these integral and derivative approximations are accurate and the conventional tuning techniques used for continuous controllers discussed in Section 12.4.3.6 may also be used for digital controllers [7].

12.4.3.6 Controller Tuning

Controller parameters are tuned to provide both performance and stability and there are many different rules for the tuning parameters of a control loop [6, 7, 22]. After deciding on the controller structure, one decides on the desired closed-loop response criteria. Then one must distinguish between processes where there is at least an approximate model with known parameters and the case when the process model is unknown.

Process model is known An approximate model of the process may be obtained by the step response test noted earlier (see Section 12.4.2.4) (or from first principles). When an approximate model of the process is known we may obtain the tuning parameters directly. We use here the example of the first-order plus dead time process, since its dynamics are so representative of the polymer equipment dynamics. Here we have chosen for the tuning criteria to minimize the integral of the time-weighted absolute error (ITAE) [Eq. (96)] [7].

$$ITAE = \int_0^\infty t|e(t)|\,dt \tag{96}$$

Tab. 12.1. Minimum ITAE model controller tuning rules based on a FOPDT process [22].[a]

Controller type	*Type of response*	K_c	τ_I	τ_D
P	disturbance	$\frac{0.49}{K_p}\left(\frac{\tau}{\theta}\right)^{1.084}$	–	–
PI	set point	$\frac{0.586}{K_p}\left(\frac{\tau}{\theta}\right)^{0.916}$	$\frac{\tau}{\left[1.03 - 0.165\left(\frac{\theta}{\tau}\right)\right]}$	–
PI	disturbance	$\frac{0.859}{K_p}\left(\frac{\tau}{\theta}\right)^{0.977}$	$\frac{\tau}{0.674}\left(\frac{\theta}{\tau}\right)^{0.680}$	–
PID	set point	$\frac{0.965}{K_p}\left(\frac{\tau}{\theta}\right)^{0.855}$	$\frac{\tau}{\left[0.796 - 0.147\left(\frac{\theta}{\tau}\right)\right]}$	$0.308\tau\left(\frac{\theta}{\tau}\right)^{0.929}$
PID	disturbance	$\frac{1.357}{K_p}\left(\frac{\tau}{\theta}\right)^{0.947}$	$\frac{\tau}{0.842}\left(\frac{\theta}{\tau}\right)^{0.738}$	$0.381\tau\left(\frac{\theta}{\tau}\right)^{0.995}$

[a] Tunings should only be applied in the range $0.1 < \theta/\tau < 1.0$.
B. A. Ogunnaike, W. H. Ray, *Process Dynamics, Modeling, and Control*,

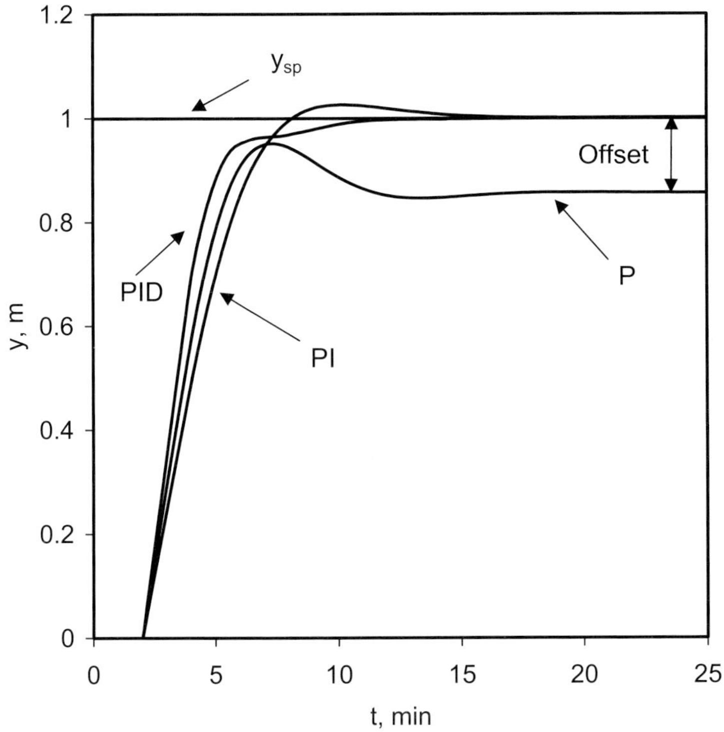

Fig. 12.24. Responses for step-change in set point with ITAE tuned P, PI, and PID controllers on a first-order plus dead time process.

The minimum ITAE tuning rules based on a FOPDT process are given in Table 12.1 [22]. Note that the tunings in Table 12.1 should only be applied in the range $0.1 < \frac{\theta}{\tau} < 1.0$.

Figure 12.24 shows the dynamic response of P, PI, and PID controller types to a step-change in the input of the first-order plus dead time (FOPDT) process of Figure 12.22 with parameters $K_p = 10$ min m^{-2}, $\tau = 20$ min, $\theta = 2$ min. For the FOPDT example the tuning for the P controller is $K_c = 0.595$ min m^{-2}, for the PI controller it is $K_c = 10$ min m^{-2}, $\tau_I = 19.7$ min, and for the PID controller it is $K_p = 0.691$ min m^{-2}, $\tau_I = 25.6$ min, $\tau_D = 0.725$ min. The derivative mode filter was used for the PID controller with a filter constant of $\alpha = 0.1$. The control loop was simulated numerically for Figure 12.24. It can be seen that the P controller produces a long-term offset, which the PI controller eliminates, but with some overshoot of the set point. The addition of the derivative action for the PID controller eliminates the overshoot and produces the best controller performance.

Process model is unknown If the process model is unknown, the continuous cycling method may be used [7]. The rules of Ziegler and Nichols [24] have proven

Tab. 12.2. Zeigler–Nichols controller tuning parameters based on the continuous cycling method [7, 22, 24].

Controller type	K_c	τ_I	τ_D
P	$0.5K_{cu}$	–	–
PI	$0.45K_{cu}$	$\frac{P_u}{1.2}$	–
PID	$0.6K_{cu}$	$\frac{P_u}{2}$	$\frac{P_u}{8}$

themselves useful in practice. For this test the controller is first brought into automatic as a pure P controller with small gain, the reset time τ_I is set to the largest value, and the derivative time τ_D is set to the lowest value. Then the gain K_c is slowly increased. When the controlled variable has an oscillatory response, the ultimate gain K_{cu} has been reached and the oscillation has the ultimate period P_u. The rules of Ziegler and Nichols then give the controller settings in Table 12.2. From these starting values one can then find the optimal tunings very quickly by small systematic modifications.

Figure 12.25 shows responses for a step-change in set point with P and PID controllers on the first-order plus dead time process with parameters. The ultimate gain for the P-only controller, found by trial and error, is $K_{cu} = 1.68\ \text{m}^2\ \text{min}^{-1}$, and the ultimate period from Figure 12.25 is found to be $P_u = 7.4$ min. The PID controller is tuned with Ziegler–Nichols parameters from Table 12.2 of $K_c = 1.008\ \text{m}^2\ \text{min}^{-1}$, $\tau_I = 3.7$ min, $\tau_D = 0.925$ min. Note that the PID response in Figure 12.25 has a much greater overshoot and much longer settling time than the ITAE response given in Figure 12.24. Other alternative tuning rules have been developed since the Ziegler–Nichols rules [7].

12.4.3.7 On–Off Controllers

On–Off controllers output the manipulated variable in discrete values (on/off, in/out) at discrete time intervals. The ideal On–Off controller is given by Eqs. (97), where $\mathrm{u_{max}}$ and $\mathrm{u_{min}}$ are the on and off values of the output and $e(t)$ is the error [7].

$$u(t) = \begin{cases} \mathrm{u_{max}} & \text{if } e(t) \geq 0 \\ \mathrm{u_{min}} & \text{if } e(t) < 0 \end{cases} \tag{97}$$

Figure 12.26 illustrates an On–Off controller applied to the FOPDT process with parameters $K_p = 10\ \text{min m}^{-2}$, $\tau = 20$ min, $\theta = 2$ min. Controller outputs Off and On are $u_{\text{min}} = 0\ \text{m}^3\ \text{min}^{-1}$ and $u_{\text{max}} = 0.2\ \text{m}^3\ \text{min}^{-1}$ respectively. Although the controller is very simple and there are no tuning parameters other than the on and off values, it can be seen that continuous cycling results, which may or may not be desirable.

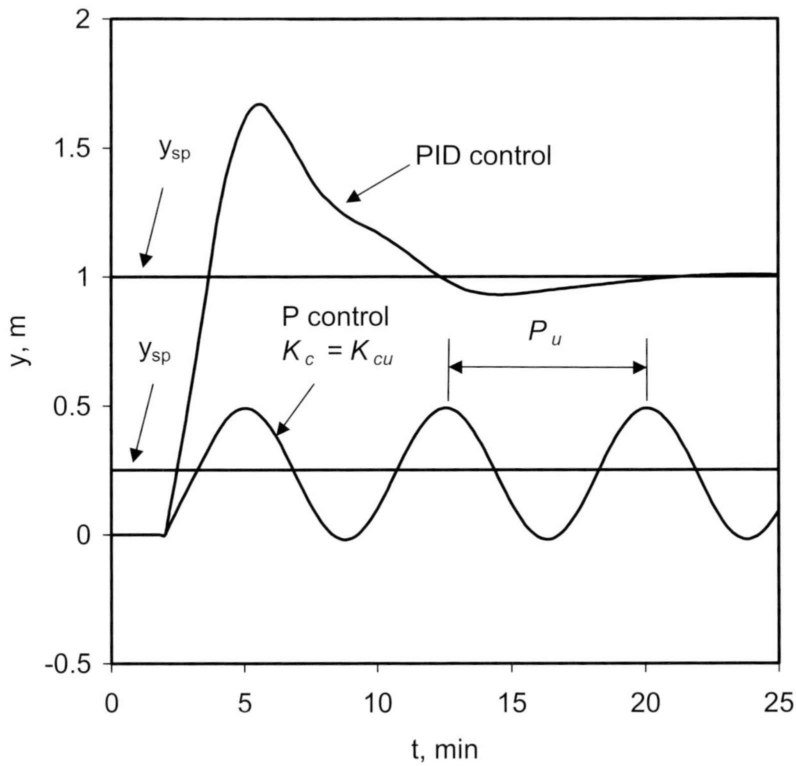

Fig. 12.25. Responses for a step-change in set point with P and PID controllers on a first-order plus dead time process. The gain for the P controller is the ultimate gain K_{cu} and cycles at the ultimate period P_u. The PID controller is tuned with Ziegler–Nichols tunings.

A modification of the ideal On–Off controller is the two-point controller, which switches the output when an upper or lower limit value is exceeded, the difference between these two points being the deadband [6]. Limit signal transducers can be used as two-point controllers (such as contact thermometers).

The benefits of the two-point controller as compared with the continuous controller are that it is simple, durable, reliable, and cheap, requires no auxiliary power, and does not lead to instability in the control loop. The disadvantages are the wear on the actuator working parts, the cyclical offset, and the fact that large power circuits must be switched. Two-point controllers are therefore hardly ever used in an industrial environment.

12.4.3.8 **Self-operated Regulators**

Controllers which acquire sufficient force from the measurement to move the actuator do not require auxiliary power. Such controllers have a simple and durable construction, and many combine measuring instrument, controller, and actuator

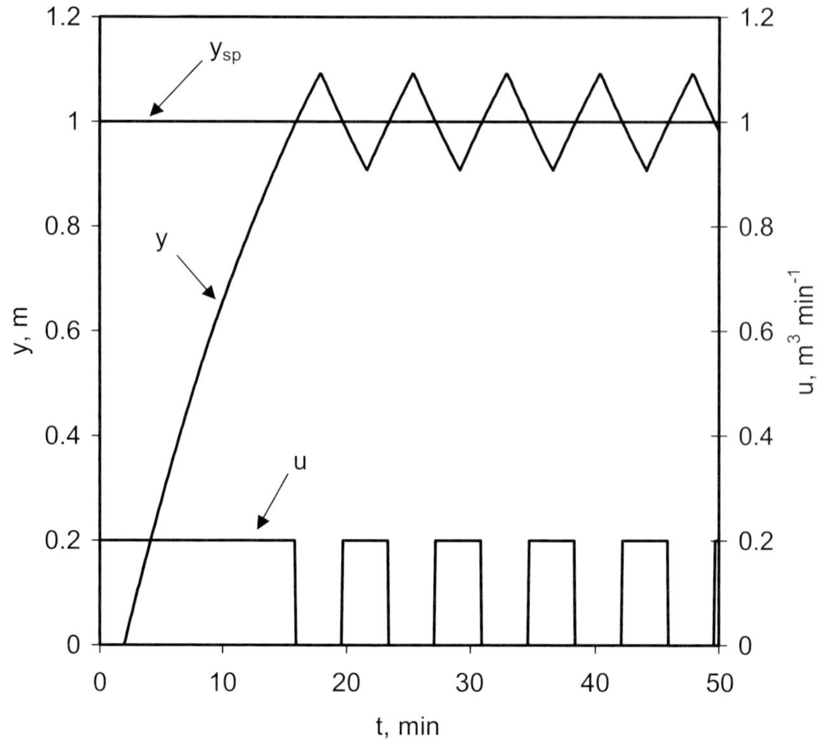

Fig. 12.26. Responses for step-change in set point with an On–Off controller on a first order plus dead time process.

functions. They have the advantage that they keep functioning if the auxiliary power fails. Therefore simple pressure control valves of this type are suitable as safety valves, in which a spring holds the plug of a valve in the seat, until an adjustable overpressure overcomes the force of the spring (Figure 12.27).

Here the exit pressure affects a diaphragm and holds a valve in an equilibrium position against the force of a spring. If the pressure rises, this closes the cross-section of the plug against the seat, and if the pressure lowers it opens again. This is also the method of operation of the pressure controllers in pneumatic transmission lines. Supply pressure-reducing valves are controllers without auxiliary power.

Simple flow controllers use the differential pressure at an adjustable restrictor: the plus and minus sides are on the two sides of a spring-supported diaphragm. These controllers are used as backflow (check) valves or as isolation valves, which shut off if the flow is reversed or too high. Level controllers with floats are in condensate separators often coupled with simple temperature controllers as in Figure 12.28. These devices contain an electrical jump switch and operate therefore as two-point On–Off controllers. Controllers without auxiliary energy are used, where it is not essential to have high control accuracy and when one must do without the auxiliary power for safety reasons.

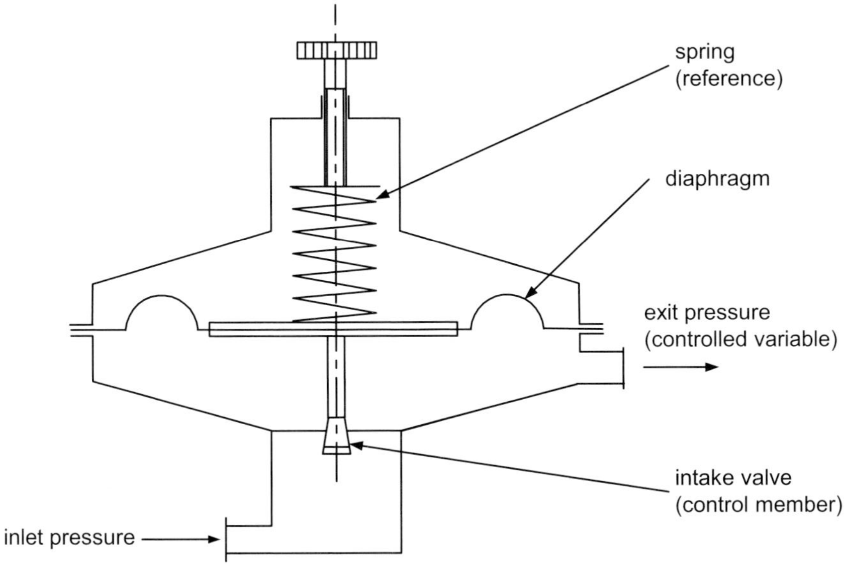

Fig. 12.27. Pressure controller with no auxiliary power [1].

Single-loop controllers with auxiliary power use compressed air or, more commonly now, electrical power to receive and transmit signals. These devices are much more accurate than those without the auxiliary power; also they are not integrated in the measurement apparatus or the actuator. Distributed control systems are rapidly replacing the single-loop controller panel boards unless only a small number of control loops are involved.

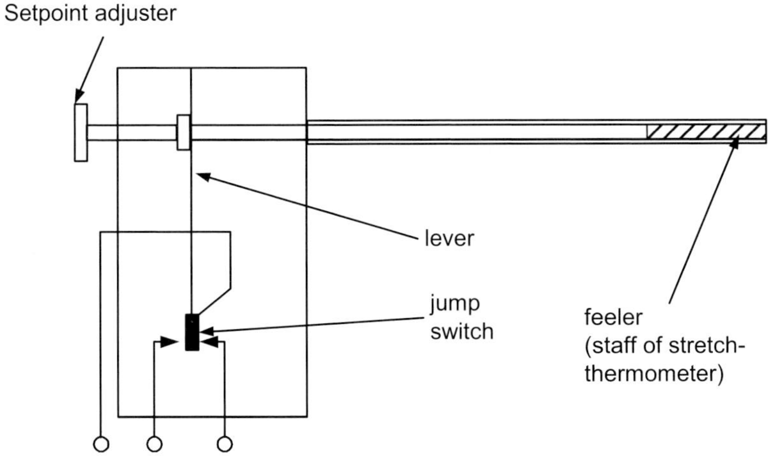

Fig. 12.28. Electrical discrete temperature controller [1].

12.4.4
Valve Position Controllers

The output signal of a controller is sent electrically or pneumatically to a valve positioner that modulates the supply pressure to the control valve actuator and must make the stem position proportional to the input signal from the controller despite loads on the valve [6, 25]. There the electrical signal must be converted to a pneumatic positioning pressure or the pneumatic signal must be brought to a higher pressure in order to move the stem of the actuator reliably and to adjust the position of the valve accurately.

In the pneumatic position controller, which is attached directly to the valve actuator, the pneumatic adjusting signal output from the controller works on a membrane and summing beam [6]. If the pressure signal rises, then a connection to the plant compressed-air network is opened to a pneumatic spool valve then to the actuator that moves the valve stem downward until a compensating spring on the summing beam holds the signal line pressure in equilibrium. Then the valve stem lifts until a new equilibrium position that is sensed by a cam lever is achieved. Thus a complete feedback loop exists where in this case the actuator is the process, the stem position is the controlled variable, the input signal is the set point, and the pneumatic amplifier network is the controller.

Electro-pneumatic valve positioners are used with diaphragm-actuated, sliding-stem control valves. The electro-pneumatic valve positioner receives an electronic input signal from a control device and modulates the supply pressure to the control valve actuator, providing an accurate valve stem position [25].

12.4.5
Single-loop Controllers

Single-loop controllers, which used to be more prevalent, are still used in some applications, such as laboratory or pilot plant. They are a rectangular package that has a faceplate for displaying the set point of a controller, to compare the set point with the actual value, and to enable the switch from manual control to automatic control or to cascade control. Single-loop controllers can be combined in a compact unit, which is built directly into a panel board. The faceplates of pneumatic and electrical control devices are very similar to each other. The transition to process control systems with newer display technologies is resulting in the disappearance of single-loop controllers.

12.4.6
Digital Control Systems

Since the 1970s the hardware used in the implementation of process controls has evolved from pneumatic analog technology to electronic analog technology to microprocessor-based controls. Although pneumatic and electronic analog process controls may still be in service in older facilities, almost all of the newer plants now utilize microprocessor technology in implementing process control. Because of rapid advances in computer technology and the increasing computation power available on a single microprocessor, the hardware is becoming more powerful in terms of available functionality. Excellent descriptions of control system hardware is available in Refs. 4, 6, and 7 as well as in the technical literature provided by the major vendors such as Honeywell, ABB–Bailey, Fisher Controls, and Siemens and also in the specialized trade shows, publications, and books sponsored by the Instrument Society of America. From the perspective of polymer reaction engineering, there are four types of digital control hardware, which are listed in order of increasing functionality, complexity, and cost:

1. *Single-loop controllers* are a stand-alone microprocessor-based version of the older analog single-loop controller (see Section 12.4.5). They can operate independently or be part of a distributed control system and can execute the standard PID algorithm but also increasingly cascade control and gain scheduling.
2. *Programmable logic controllers* (PLCs) were the first digital technology to successfully compete with conventional technology in industrial control applications. They were originally developed to replace hard-wired relay logic. PLCs are extremely important in batch control applications, because they are ideally suited to program large and complex batch sequencing cycles using ladder diagrams and a higher-level programming language. Because of their logic handling capability PLCs can also be used to implement process interlocks, which are designed to prevent process conditions that would unduly stress equipment (perhaps resulting in minor damage) or lead to off-specification product.
3. *Personal computer (PC) controllers* are appropriate for the control of a laboratory or small- to medium-size pilot plant installations. The typical PC is augmented by the appropriate input/output (I/O) interface software. Software tools are available for the implementation of the previously discussed PID controllers using the PC. Advanced control algorithms can also be executed at the PC level. The PC also provides graphic capability for the monitoring of the performance of the control loop and can also provide data historian capability, which can be very important for polymer process development.
4. The *distributed control system*, the process control architecture referred to as a DCS, was introduced in the mid-1970s and has since become the standard for large installations. The main concept behind the DCS is that microprocessor-based nodes are interconnected by a digital communications network often referred to as the "data highway". A typical DCS installation handles seamlessly process I/O from the field, including interfaces with process analyzers (that is, gas chromatographs), which are important in advanced control applications. The DCS can be interfaced with PLCs, with operator stations for process monitoring and control including alarm management, with engineering workstations for the system configuration activities, and also with a set of host computers which can provide product quality information, data compression, reporting, and archival capabilities, as well as advanced control calculations.

12.4.7 Actuators

Actuators govern the flow of energy and mass streams in the process [25]. Through the actuators the controller output results in changes to the controlled variable (temperature, pressure, flow, level, and so on) and thus the control loop is closed. Actuators consist of the control positioner (for example, a valve positioner) and the actuator element. The control positioner transfers the signal output of the controller in a controlled movement to bring the actuator element into a new position.

Examples of actuators are rheostats to set and control electrical current, valves, flaps, and metering pumps for liquids. The actuator, which is most frequently encountered in chemical process engineering, is the diaphragm valve (Figure 12.29).

The input actuating pressure lies on one side of a diaphragm and exercises a force on it; the spring works against this force. The actuating pressure shifts the position of the diaphragm so as to create equilibrium of forces with the spring. The diaphragm is connected with the valve stem; the stem (and the plug) therefore make a stroke proportional to the actuating pressure. Between the plug and the valve seat a variable annular gap opens or closes, thus resulting in variable flow. The seat and the plug are in a housing, which is solidly anchored and also holds the stem. For high-temperature service the housing is cooled with fins in order to protect the bushing packing. With dangerous material one uses valves with sealed bellows and therefore no bushing is needed. Additional equipment, for example position controllers or limit switches, can be attached to the valve as necessary.

In the pressure-free state the valve stem goes naturally to a terminal position. In Figure 12.30(a), as the pneumatic controller output signal increases, increased pressure on the diaphragm compresses the spring, thus pushing the stem in and closing the valve further. Such a valve is termed "air-to-close" (A–C). By reversing either the plug/seat or the spring/air inlet orientation, the valve becomes "air-to-open" (A–O) as in Figure 12.30(b). Normally the choice of A–C or A–O valve is based on safety considerations. One chooses the way the valve should operate (full flow or no flow) safely in case of a transmitter failure.

Standard diaphragm valves have a diaphragm area from 200 to 1000 cm^2 and a stroke of 15–30 mm [1]. Very large valves are equipped with diaphragms of up to 3000 cm^2 and have a 100 mm stroke. These large valves have large power requirements and operate with 3 bar pressure. In addition a valve positioner is needed. To achieve higher strokes, specialized cylinders are used.

The valve stem, plug, and seat (called the "trim") have different forms: ball, plug, disk, or gate valves [7, 25]. Plugs are normally fabricated as rotationally symmetric parabolic cones. There are numerous other plug shapes, which can be selected for each individual case. The flow rate as a function of stem position, or lift, ℓ, is called the valve characteristic, $f(\ell)$. A valve characteristic may be linear, quick opening, or equal percentage. The plug shape, seat, and cage primarily determine the valve characteristic, the sensitivity to contamination, and flutter avoidance.

To size a valve for liquids we may return to a modified form [Eq. (98)] of the equation given for restrictors [7, 12, 25].

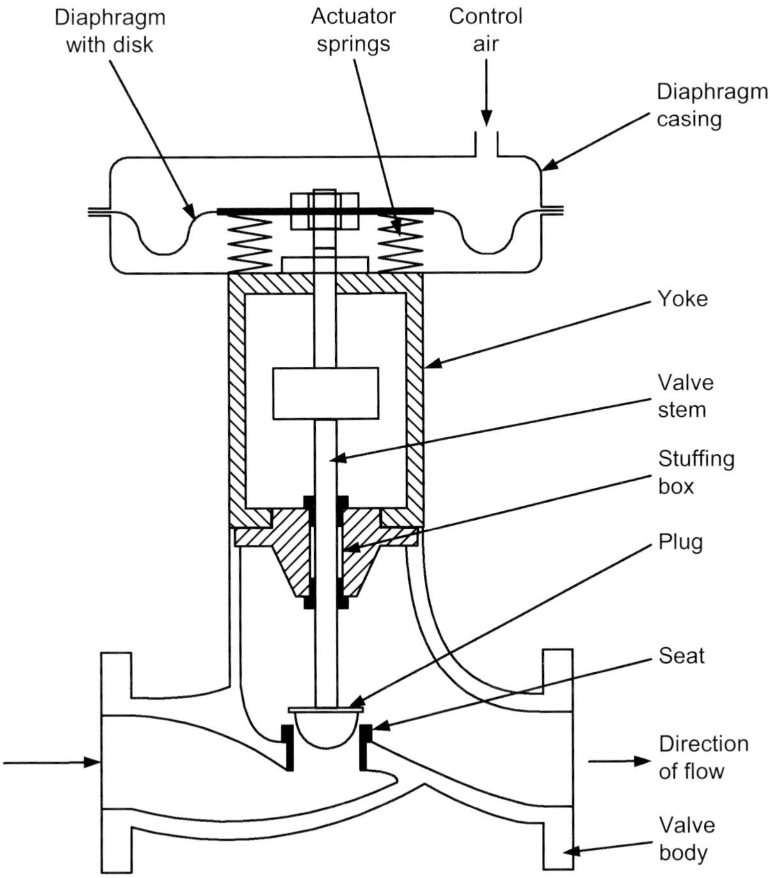

Fig. 12.29. Diaphragm valve [1].

$$q_v = C_V f(\ell)\sqrt{\frac{\Delta p}{d}} \tag{98}$$

Here q_v is the volumetric flow rate, C_V is the valve coefficient, $\Delta p = p_1 - p_2$ is the pressure drop across the valve, and d is the relative liquid density (specific gravity) [Eq. (99), where ρ is the fluid density at operating temperature and ρ_{w0} is the water density at a standard temperature and pressure (usually 4°C and 1 bar)].

$$d = \frac{\rho}{\rho_{w0}} \tag{99}$$

The valve characteristic $0 \leq f(\ell) \leq 1$ is given by Eq. (100), where $0 \leq \ell \leq 1$ is the lift and $20 \leq R \leq 50$ is a valve design parameter [7].

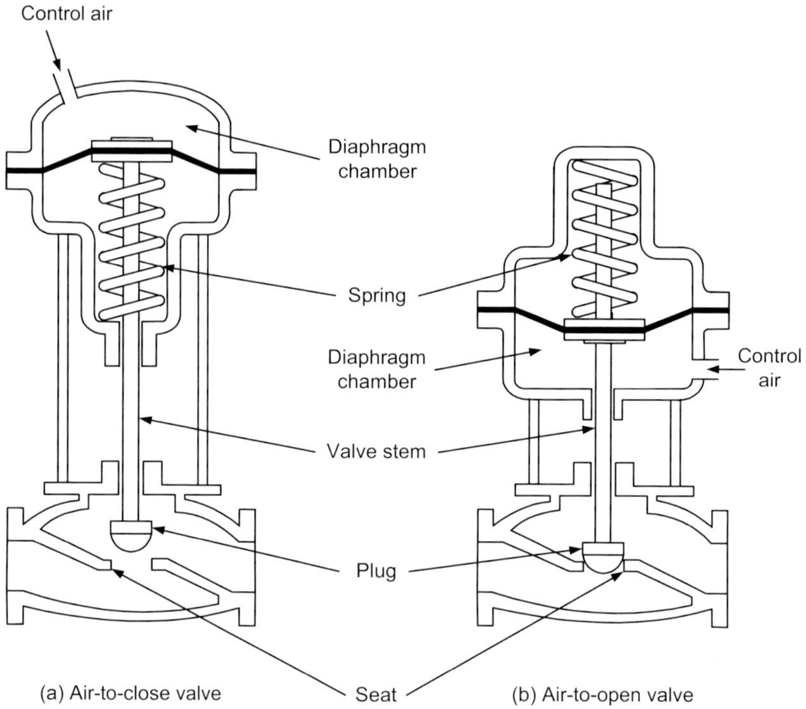

Fig. 12.30. Diaphragm valves with different rest states: (a) air-to-close; (b) air-to-open [1].

$$f(\ell) = \begin{cases} \ell, & \text{for a linear valve} \\ \sqrt{\ell}, & \text{for a quick opening valve} \\ R^{\ell-1}, & \text{for an equal percentage valve} \end{cases} \tag{100}$$

In USCS units the valve coefficient, C_V, is the full open flow rate in gal min^{-1} of water at 60°F (15.6°C) that will flow through the valve with a 1 psi (6.89 kPa) pressure drop. In SI units the valve coefficient (called K_V) is the full open flow rate in m^3 h^{-1} of water at 20°C that will flow through the valve with a 1 bar pressure drop. International control valve standards are listed in Ref. 25.

For gases the pressure ratio p_2/p_1 at the valve must not be less than about 0.53; otherwise sonic velocity, c, is achieved in the restriction [10]. At the sonic velocity the throughput cannot be increased as the valve cross-section continues to open [10]. The pressure inside the valve must be lower than the liquid vapor pressure; otherwise flashing or cavitation occurs.

Besides diaphragm valves, one frequently uses butterfly valves (particularly for gases in lines with large diameter), which are constructed with a stop (Figure 12.31) or without (Figure 12.32). Butterfly valves without a stop (so-called regulating flaps; see Figure 12.32) do not seal completely; their leakage flow is within 5–10% of the expected value when the valve is fully open.

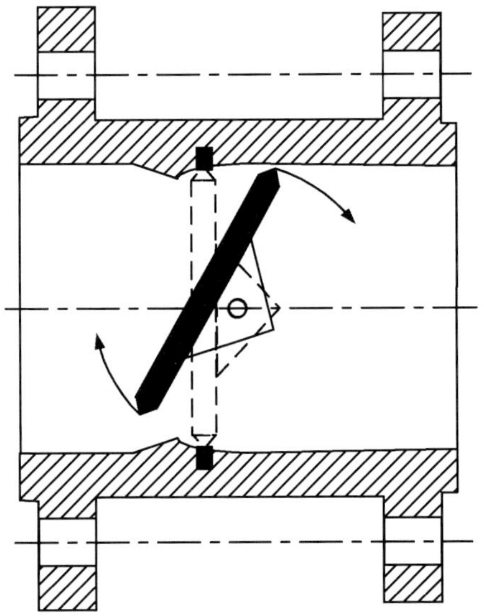

Fig. 12.31. Butterfly valve with a stop [1].

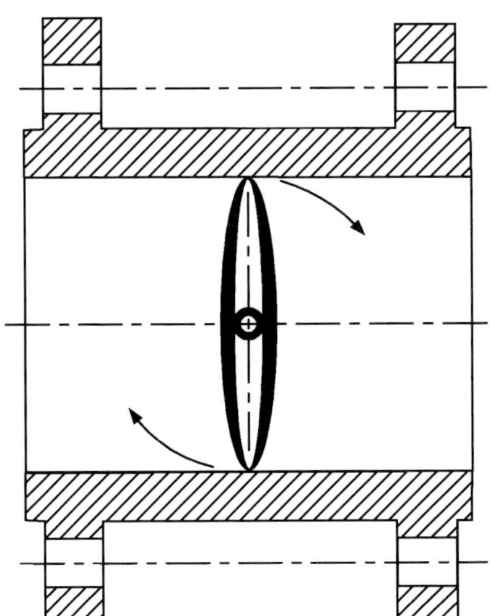

Fig. 12.32. Butterfly valve without a stop [1].

With large nominal sizes and higher pressure, gate valves are used. It is advantageous that they have only a low-pressure drop and are insensitive against contamination. Gate valves have one large stroke; one cannot equip them with a diaphragm control drive, but the stem moves with gear racks, piston drive units, or electric motors.

12.5 Advanced Control Engineering

The polymerization reactor is usually at the heart of the manufacturing process, impacting both downstream processing and final customer-related polymer properties. The following factors have contributed to the industrial significance of polymer reactor control:

- the need to improve fixed-asset productivity by optimizing reactor yield and uptime;
- the trend toward shorter manufacturing campaigns for the different polymer grades manufactured in the same reactor, or toward more frequent on-line product transitions to reduce product inventories and hence working capital;
- global competition, which imposes tough requirements for polymer grade uniformity;
- safety and environmental considerations regarding the stable operation of a potentially thermally unstable reaction.

The measurement and control techniques discussed in the previous sections, many of which are generic in nature, are not always adequate to meet the following practical considerations specific to a polymerization reactor, which pose additional challenges to the successful application of a control strategy:

- Polymerization reactors are well known both theoretically and experimentally to exhibit multiple steady states [26, 27] and in some cases they may also exhibit oscillations in terms of monomer conversion and polymer particle diameter [28]. In other cases it may be necessary to choose an open-loop unstable state as the reactor operating point. Furthermore, polymerization reactors can be highly exothermic and result in reactor thermal runaway.
- On-line measurements of polymer architecture such as composition, molecular weight, and degree of branching which were discussed in Sections 12.2.8 and 12.2.10, are not always available and for some polymer systems may be simply unavailable. In many cases the control engineer may have to rely on polymer properties inferred from infrequent laboratory analysis of reactor samples or from laboratory analysis of the final polymer after it has experienced significant post-reactor processing, thus introducing large dead times in the control loop.
- The relationship between reactor operating conditions, such as monomer conver-

sion, temperature, residence time, polymer composition, and viscosity, and the customer-related final properties, such as tensile strength, elongation at break, and processability in an injection molding machine, may not always be well defined. Variability in the polymer isolation process and long-term polymer structure changes such as "aging" may result in the fact that, although in some cases the reactor may be operating "on-aim" within well-defined manufacturing specifications, the final polymer may not process satisfactorily when delivered to the customer.

- The control system for a polymerization reactor must be sufficiently robust to handle unmeasured disturbances, which impact polymer reactor operation. These disturbances typically result either from trace amounts of polymerization inhibitors left over after monomer purification before the polymerization reaction or from trace amounts of other compounds which may be present in a typical polymerization recipe and which may be affecting the reaction.
- A polymerization reactor often produces several grades (in terms of composition and viscosity) of the same polymer and therefore the control strategy must be easily adapted to a multi-product plant and in some cases to on-line grade transitions. In the case of a multi-product plant it may be necessary to operate the reactor in terms of rather short campaigns in order to minimize the finished-product inventory and thus the working capital. In these cases the reactor control system must be designed in such a way as to achieve fast startups while minimizing off-specification polymer formation.

The purpose of this section is to highlight several control engineering techniques, some of which are known in the literature as "advanced control", which can be deployed to meet some of the challenging aspects of polymer reactor control. These techniques have been extensively discussed in literature reviews such as Refs. 29–32 and in books [18, 19].

The discussion will be facilitated by focusing on the polymerization process shown in Figure 12.33 [33]. Although the process is generic, the flowsheet captures many of the elements of actual free-radical polymerization reactor installations. Monomers A and B are continuously added with initiator, solvent, and chain-transfer agent. In addition, an inhibitor may enter with the fresh feeds as an impurity. These feed streams are combined (stream 1) with the recycle stream (stream 2) and flow to the reactor (stream 3), which is assumed to be a jacketed well-mixed tank. A coolant flows through the jacket to remove the heat of polymerization. Polymer, solvent, unreacted monomers, initiator, and chain-transfer agent flow out of the reactor to the separator (stream 4), where polymer, residual initiator, and chain-transfer agent are removed. Unreacted monomers and solvent (stream 7) then continue on to a purge point (stream 8), which represents venting and other losses and is required to prevent accumulation of inerts in the system. After the purge, the monomers and solvent (stream 9) are stored in the recycle hold tank, which acts as a surge capacity to smooth out variations in the recycle flow and composition. The effluent (stream 2) recycle is then added to the fresh feeds.

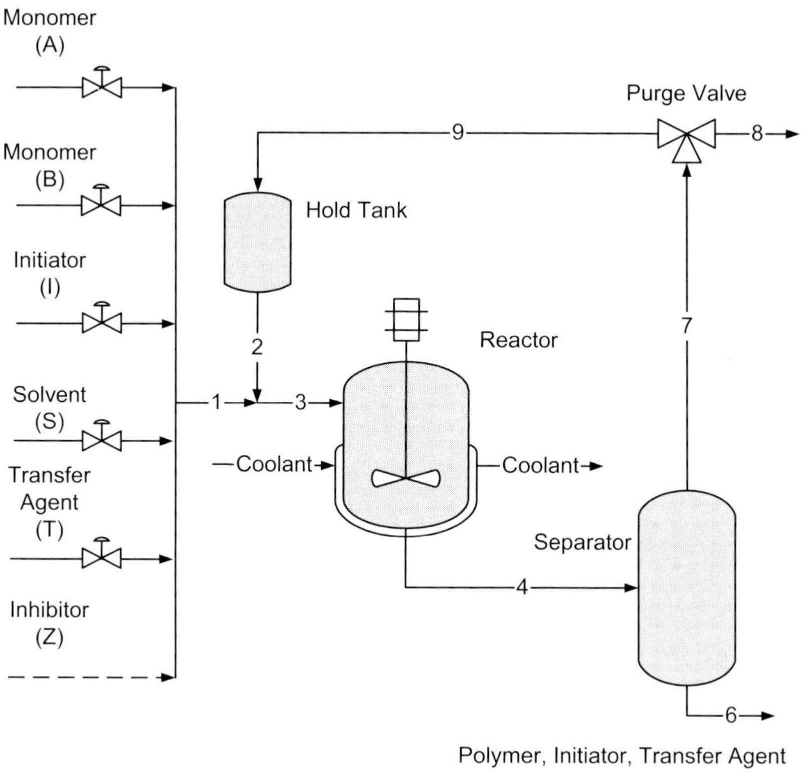

Fig. 12.33. Copolymerization with recycle loop [33].

It has been found that the hierarchical approach summarized in Figure 12.34 is very useful in the successful application of process control in a complex industrial environment. Process knowledge, which is usually captured in an experimentally validated mathematical model, is the cornerstone of a successful control strategy. This is particularly true for polymerization reactors, where the in-depth knowledge of process operation in terms of the effect of operating variables on polymer properties can be used to great advantage in the design of the control system and can result in a much more straightforward (and therefore easy to maintain) strategy than would have been possible otherwise (see, for example, the control strategy discussed in Ref. 30). Process knowledge together with the appropriate sensors, transmitters, and analyzers are the prerequisites for the design of the basic control system to regulate pressure, temperature, level, and flow (PTLF). Only when the elements of the regulatory control system are in place and are properly designed and maintained can the control engineer attempt, in increasing order of complexity, the implementation of more advanced regulatory control strategies, multivariable model-based control algorithms, and on-line scheduling and optimization strategies to compute set points for the regulatory controls. In many instances advanced control applications have failed in an industrial environment not because

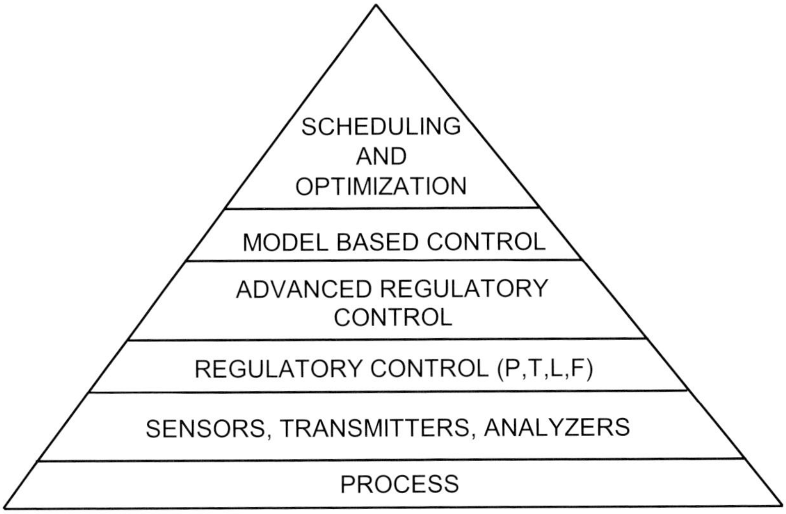

Fig. 12.34. The process control hierarchy.

the algorithms were necessarily faulty but because the basic regulatory control system performed poorly, either because of inadequate design or because one of the critical measurements (that is, a process analyzer) was poorly maintained.

12.5.1 Feedforward Control

The traditional PID feedback controller, which was discussed previously (see Section 12.4.3.4), is very widely used, because it requires minimal process knowledge for its design. In particular, a mathematical model of the process is not required although it can be quite useful for appropriate tuning. Furthermore, if process conditions change, the PID controller can be retuned to maintain satisfactory performance. A properly tuned PID controller can be quite robust in maintaining good steady-state operation in the face of unmeasured disturbances.

However, since control action can only occur if a deviation occurs between the set point and the measured variable, perfect control is not possible. Therefore feedback control fails to provide predictive control action to compensate for the effects of known disturbances. A more serious limitation, which is particularly important for polymer reactor control, is that the controlled variable cannot always be measured on-line.

Feedforward control was developed to counter some of these limitations. Its basic premise is to measure the important disturbance variables and then take corrective compensatory action based on a process model. The quality of control is directly related to the fidelity and accuracy of the process model. Two implementations of feedforward control which are widely used in polymer reactor control will be discussed, namely feedforward control design based on steady-state models, and ratio control. Furthermore, the powerful combination of feedforward and feedback control will be considered, because it utilizes the best of both approaches: feedforward control works by reducing the effects of measured disturbances, and feedback control provides the necessary "trim" to compensate for the effect of model and measurement inaccuracies as well as measurement error.

12.5.1.1 Steady-state Model Feedforward Control

To illustrate this approach the polymerization process previously described in Figure 12.33 is considered. The presence of the recycle stream introduces disturbances in the reactor feed which perturb the polymer properties. The objective of the feedforward control is to compensate for these disturbances by manipulating the fresh feeds in order to maintain constant feed composition and flow to the reactor. Feedforward control of the recycle allows the designer to separate the control of the reactor from the rest of the process.

As shown in Ref. 33, the feedforward control equations were obtained by writing component material balances around the recycle addition point. For example, for monomer A this balance is given by Eq. (101).

$$q_{n3} = q_{na1} + y_{a2}q_{n2} \tag{101}$$

This equation is then solved in Eq. (102) for the fresh feed of monomer A since it is desired to keep the goal flow of monomer A to the reactor (q_{na3}) constant.

$$q_{na1} = q_{na3} - y_{a2}q_{n2} \tag{102}$$

The recycle composition (y_{a2}) is typically measured by online gas chromatographs, which may have significant time delays. If a faster response time of the analyzer is required, an infrared or Raman spectroscopy probe may be used. As discussed in Section 12.2.8 flow q_{n2} is typically measured and controlled by manipulating the recycle valve to maintain the desired inventory in the feed tank. Any disturbances in the recycle composition or flow will cause variations in the fresh feed in order to keep the reactor feed constant. Similar feedforward controllers are implemented for monomer B and solvent, which are also present in the recycle stream.

As shown in Refs. 22 and 33, the performance of the feedforward control allows perfect compensation of disturbances that can arise, for example, from a step-change in the purge ratio, so that the reactor polymer characteristics (composition, molecular weight) are unaffected. Without the presence of the feedforward controllers the reactor dynamics and hence its control can be affected directly by the presence of three lags in series (reactor, separator, hold tank) and thus become unnecessarily more complex.

12.5.1.2 Ratio Control

Ratio control is a form of feedforward control which is widely used in the chemical industry and has proven very useful in polymerization reactor control. As is evident from its name, its purpose is to keep the ratio of two process variables at a

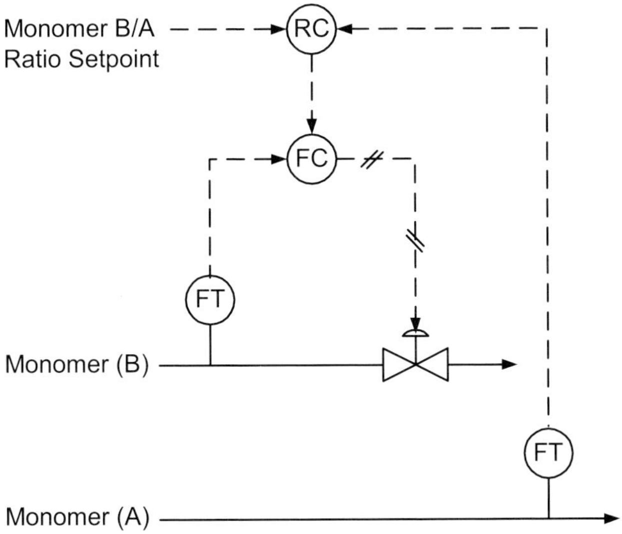

Fig. 12.35. Ratio control.

given value; hence it can be deployed when the objective is to maintain a certain proportion (or stoichiometry) of reactants in the reactor.

The implementation of ratio control can be described by referring back to the polymerization process in Figure 12.33 and to the discussion in Ref. 7.

Typically, flow controllers are designed for each of the reactor feed streams (for example, monomer, initiator, chain transfer) and each one of these controllers has a set point, which is dependent on the particular polymer being made. However, when ratio control is implemented as shown in Figure 12.35, one of the reactor feed streams (monomer A in this case) is chosen as the reference stream. The measured flow rate of monomer A is then transmitted to the ratio station RC, which multiplies the signal by the desired ratio K_R (typically determined by the polymer chemist) to calculate the set point y_{sp} for the flow controller of monomer B. Dynamic simulations for the polymerization reactor shown in Figure 12.33 and reported in Ref. 33 showed that the selection of flow ratios as manipulated variables reduced the interactions among the output variables such as composition and molecular weight. It is important that the flow controller for monomer B be tightly tuned with short settling times so that when there are disturbances in the flow of monomer A, any transient mismatches between monomer A and B flows are minimized.

12.5.2
Cascade Control

Cascade control is also widely used in the chemical process industries and especially in cases where there may be nonlinear behavior in the dynamics of the con-

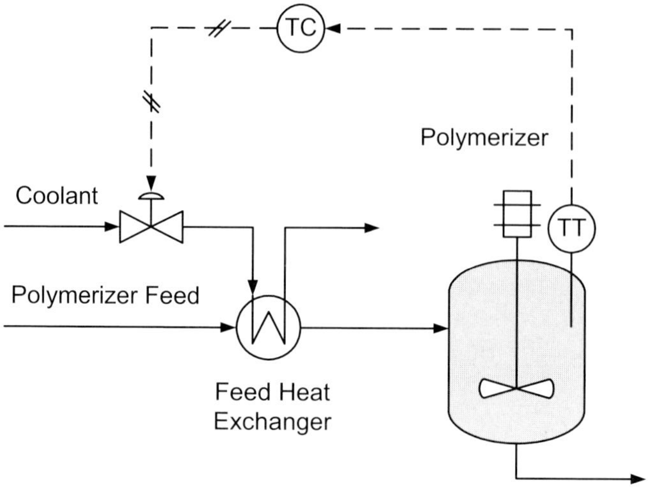

Fig. 12.36. Conventional temperature control of an adiabatic polymerizer.

trol loop. It also addresses the main drawback of conventional feedback control, namely the fact that control action only occurs where the controlled variable deviates from the set point. Unlike feedforward control, which requires that disturbances be explicitly measured and a model be available to calculate controller output, cascade control introduces an additional measurement and an additional feedback controller. The secondary measurement is typically located so that it recognizes the upset conditions sooner than the controlled variable.

The concept of cascade control has been traditionally very much used for effective reactor temperature control. An example of the implementation of cascade control is shown in Figures 12.36 and 12.37.

In many instances polymerization reactors are operated adiabatically. In the case shown in Figure 12.36, in which only traditional feedback control is used, measurement of the reactor temperature is used to manipulate the heat exchanger flow to cool the reactor feed so that the reactor adiabatic temperature rise is adequate to remove the heat of polymerization. This conventional scheme may regulate reactor temperature satisfactorily but disturbances that occur in the feed line may result in a rather sluggish response of the temperature controller. Polymer properties are very sensitive to temperature excursions and in many cases this sluggish response of the temperature control loop may not be acceptable. Cascade control as shown in Figure 12.37 resolves the problem by introducing an additional measurement, namely the temperature of the reactor feed, and an additional controller. The cascade control structure has the following characteristics:

- The output signal of the primary (frequently referred to as the "master") reactor temperature control loop serves as the set point of the secondary (frequently referred to as the "slave") reactor feed temperature control loop.

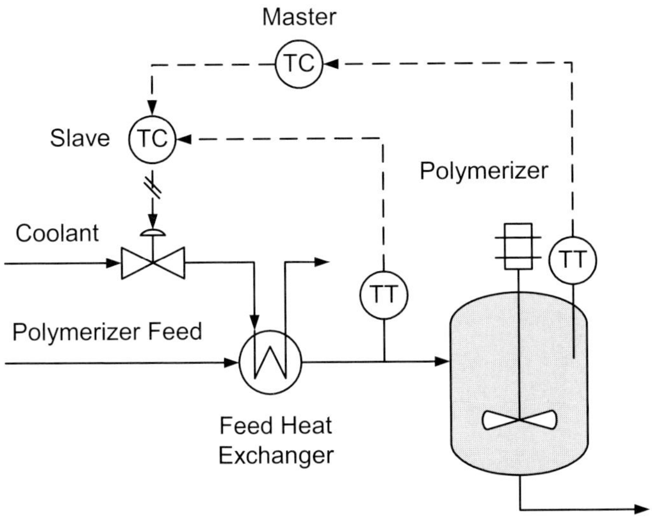

Fig. 12.37. Cascade temperature control of an adiabatic polymerizer.

- The two feedback controllers are nested with the secondary control loop located inside the primary control loop.
- There is one manipulated variable but two sensors and two controlled variables.

The secondary controller responds rapidly to any temperature disturbances in the reactor feed line and provides improved reactor temperature control. As discussed in Ref. 7, a similar cascade control scheme can be implemented in the case where the reactor is jacketed and the reactor temperature is controlled by manipulating the cooling medium (typically water) inlet stream. In this case the additional measurement is the temperature of the jacket, which is compared with a set point provided by the master reactor temperature controller. The resulting error signal is the input to the controller for the cooling water makeup.

It is obvious from the preceding discussion that for a cascade control system to function effectively, the secondary control loop must be selected and tuned so as to have a faster response than the primary controller. According to Ref. 7, the secondary controller is normally a P or PI controller with derivative action hardly used. The primary controller is typically a PI or PID controller.

12.5.3 Feedforward–Feedback Control

The combination of feedforward and feedback control provides a very powerful practical strategy for the control of polymer properties such as composition and molecular weight. Typically it is still fairly difficult to have on-line direct measurements of polymer composition, so the control design has to incorporate the available off-line reactor sample composition measurements obtained at the laboratory, typically using IR or NMR techniques. Similarly, despite advances in size exclusion chromatography/gel permeation chromatography technology (SEC/GPC), on-line SEC/GPC is not routinely available for most industrial polymer reactor control applications. In many cases the control engineer has therefore to rely on off-line measurements of molecular weight. Typically measures of molecular weight used in control applications are the inherent viscosity and/or the melt index. The latter is very common in polyolefin production and recent advances in capillary rheometry, as reported in the vendor literature, provide the capability of continuous measurement of melt index and viscosity during polymer production using on-line or at-line instrumentation. It is also very important to establish appropriate set points and specifications for the inherent viscosity and melt index by relating them to the underlying molecular weight distribution, as shown for example in Ref. 34.

Returning to the polymerization process shown in Figure 12.33 and following the discussion in Ref. 30 (which dealt with an emulsion polymerization reactor) as well as the previously discussed concept of the process control hierarchy, it is very important to use process understanding (typically captured in dynamic simulations using experimentally validated models) in the design of the control system for polymer composition and molecular weight. For example, it was shown for a specific case, which is however typical of a wide class of continuous polymer reactors in industrial practice, that the reactor temperature control is crucial because inherent viscosity is extremely sensitive to temperature variations. As discussed above (see Section 12.5.2), cascade control can be effectively deployed to control reactor temperature within very tight specification limits. Achieving good reactor temperature control can become particularly challenging in multi-product semi-batch polymerization reactors, because physical properties of the reactor contents vary from run to run and within a run and the standard PID controllers used in a cascade design may not be able to perform satisfactorily over the entire range of operation required. In these cases more advanced temperature control strategies based on adaptive control [7] or model-based control can be effectively used, as shown, for example, in Ref. 35 for an industrial reactor and in Ref. 36 for a laboratory-scale one. Moreover, adaptive cascade control strategies can provide better performance without the need for retuning than a traditional PI cascade control system in the case of jacketed stirred tank reactors in which multiple products are produced and the overall heat-transfer coefficient is unknown and can vary significantly as a result of fouling, as shown, for example, in Ref. 37.

An additional difficulty in the control of polymer properties is that in some cases the control problem is multivariable, in the sense that there are interactions between the molecular weight and composition loops and therefore when a manipulated variable is chosen to control molecular weight it may also affect composition. It is important to use process knowledge to validate the selection of manipulated variables. For example, for the polymerization reactor shown in Figure 12.33, process simulations showed that one way to decouple polymer quality control is to take advantage of the fact that polymer composition is naturally very sensitive to changes in reactor feed composition but inherent viscosity is relatively insensitive to reactor feed composition changes. As discussed in Ref. 33, there exists a much more formal way for feedback controller design. It consists in creating an approximate linear multivariable model from the nonlinear polymer reactor model using step test data and then using the techniques of relative gain array (RGA) and singular value decomposition analysis (SVD) (as described, for example, in Ref. 7) to determine the best pairings of controlled and manipulated variables for robust multi-loop control.

A typical control strategy for polymer composition, which can be implemented for the reactor shown in Figure 12.33, is illustrated in Figure 12.38. The measurements used are the flows of the fresh monomer feeds, the recycle flow, the monomer composition of the recycle feed and the total monomer reactor feed provided by two on-line gas chromatographs, and the polymer composition provided by laboratory analysis of reactor samples. The supervisory control consists of three levels implemented in a cascade fashion:

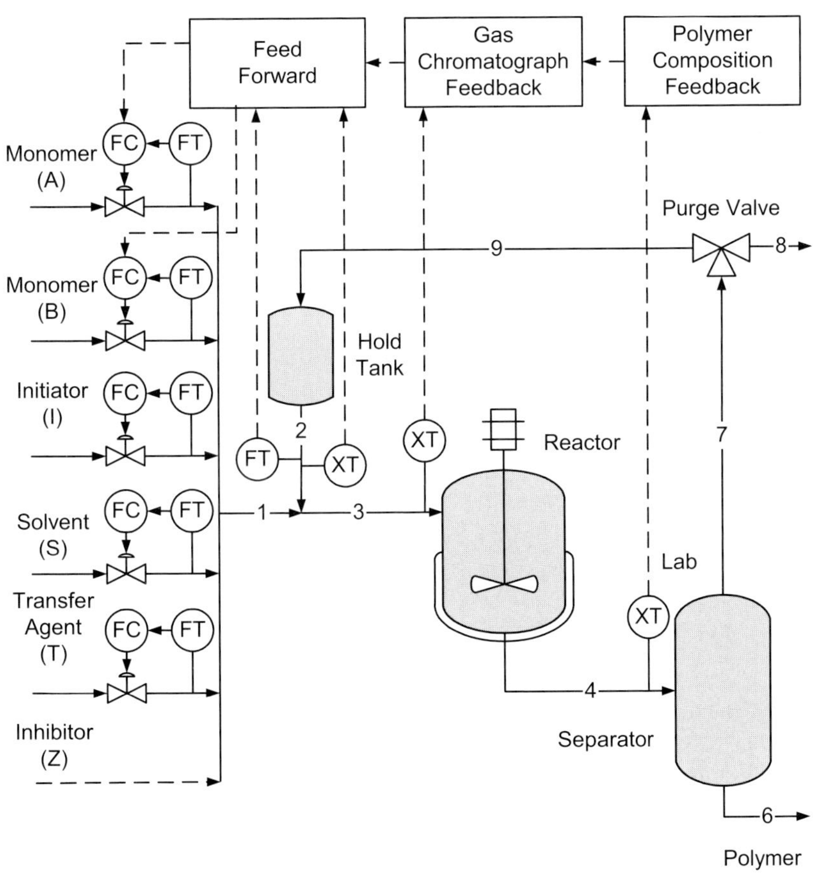

Fig. 12.38. Copolymer composition control strategy.

1. *Feedforward controller*, previously discussed, which maintains the total monomer feed rate and monomer A feed composition at the appropriate set points for the specific polymer grade being produced. The feedforward controller can be easily extended to n monomers by specifying the total monomer reactor feed rate and the feed composition for $n - 1$ monomers.
2. *Gas chromatograph feedback controller*, which uses the velocity algorithm for digital PID control, discussed in Section 12.4.3.4 to calculate flow correction factors for the monomers from the gas chromatograph measurement of the actual monomer feed composition. This controller provides the necessary integral action so that the offset between the actual reactor monomer feed composition and its set point, which may be caused by flowmeter inaccuracies or other unmeasured disturbances, is minimized.
3. *Polymer composition feedback controller*, which updates the set points for the reactor monomer feed composition based on the laboratory analysis of a reactor sample. This controller thus provides the necessary integral action so that the offset between the measured composition of the reactor sample and the polymer grade composition is minimized.

It is important to note that these feedforward and feedback controllers have been designed hierarchically, in the sense that each level in the structure will not activate unless the levels below it are functioning properly. Furthermore, in practice extensive data validation checks must be incorporated so that robust performance can be assured even when the gas chromatograph or laboratory analysis measurements may be unavailable or faulty.

The control of inherent viscosity shown in Figure 12.39 uses the same approach as the composition control. Depending on the polymer grade being manufactured, the initiator or the chain-transfer agent may be used to control emulsion viscosity. An inherent viscosity feedback controller adjusts automatically the set point of the monomer to transfer agent ratio controller based on the measured viscosity value, (Section 12.2.7) and provides the necessary integral action so that the difference between the reactor inherent viscosity and the polymer grade inherent viscosity specification is minimized. As in the case of composition control, extensive data and controller output checks must be incorporated in any practical implementation to provide robust performance.

12.5.4
State Estimation Techniques

In the design of the composition and viscosity feedback controllers it is very important to establish whether the polymer reactor dynamics need to be taken explicitly into account. The choice of sampling frequency balances the requirements for good quality control versus the need to minimize analytical costs. Usually, when the reactor residence time is much shorter than the sampling frequency, integral control is appropriate, because the time between measurements is usually sufficient for the effect of an adjustment to a process variable set point to be com-

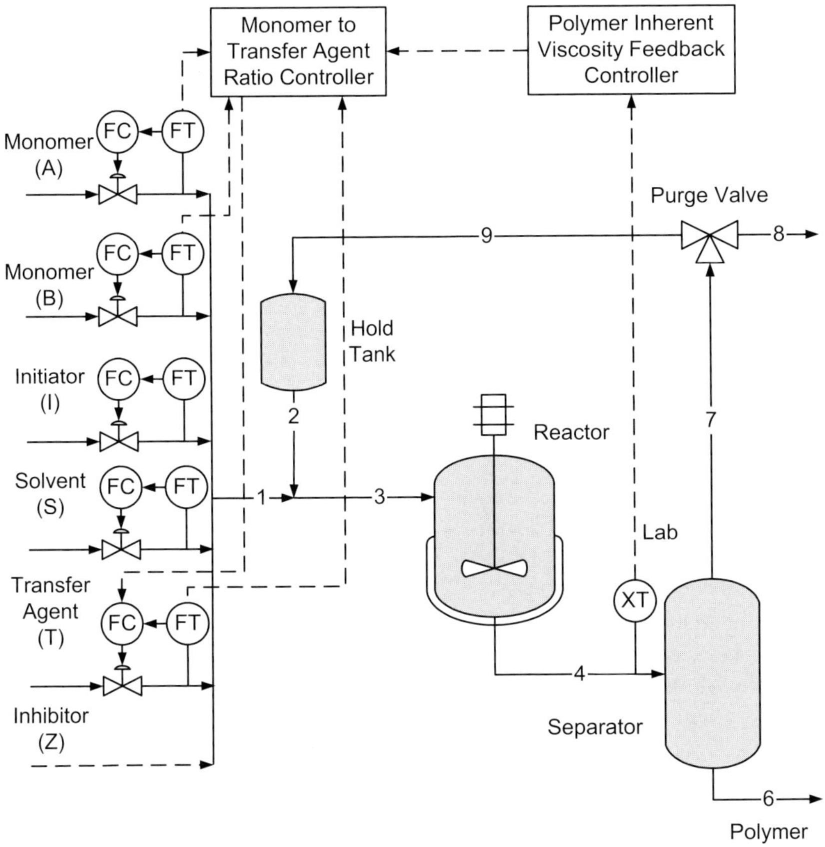

Fig. 12.39. Inherent viscosity control strategy.

plete within this interval. In other cases the sampling dead time introduced by the periodic analysis of polymer concentration, polymer composition, and molecular weight may necessitate the incorporation of on-line state estimators of polymer properties.

Reaction calorimetry aims to measure heat released from a polymerization in order to infer monomer conversion and polymerization rate (as reviewed, for example, in Refs. 8, 38, and 39). Careful measurement and balancing of mass and energy flows are necessary for success of this technique. For example, the commercial Mettler–Toledo RC1 jacketed reactor acts as a calorimeter supplying mass balance, polymerization heat generation, and transport data.

On-line estimation may also be accomplished using first-principles polymerization kinetic models implemented on-line in the form of an extended Kalman filter (EKF) (as illustrated for example in Refs. 8 and 40–42). It should be pointed out that the choice of techniques for on-line estimation of polymer properties is still an active area of research and is very much dependent on the specifics of the polymer chemistry and available on-line instrumentation.

12.5.5
Model Predictive Control

The previously discussed single-loop and appropriately chosen multi-loop feedforward and PID feedback control strategies may not be adequate for the effective control of polymer properties, particularly in the case when the polymerization reactor exhibits strongly nonlinear dynamic behavior, or when there are strong interactions between the controlled variables, or when there are constraints on the manipulated variables. From the advanced process control techniques such as internal model control (IMC), inferential control, and adaptive control that have been developed by the academic process control community for these tough multivariable control problems, model predictive control (MPC) has reached the stage where it is having a significant impact on industrial practice. MPC algorithms are rapidly becoming imbedded in the distributed control system (DCS) software libraries, which facilitates their use. As reported in a 2003 survey [43], by the end of 1999 there were at least 4500 industrial MPC applications worldwide, mainly in oil refineries and petrochemical plants.

The structure of MPC is shown in the block diagram of Figure 12.40 [7]. A mathematical model of the process is used to predict the current values of the output (controlled) variables. The model is usually implemented in the form of a multivariable linear or nonlinear difference equation. It is typically developed from data collected during special plant tests consisting of changing an input variable or a disturbance variable from one value to another using a series of step-changes with different durations, or more advanced protocols such as the pseudo random-binary sequence described in Ref. 7. The residuals (that is, the difference between the pre-

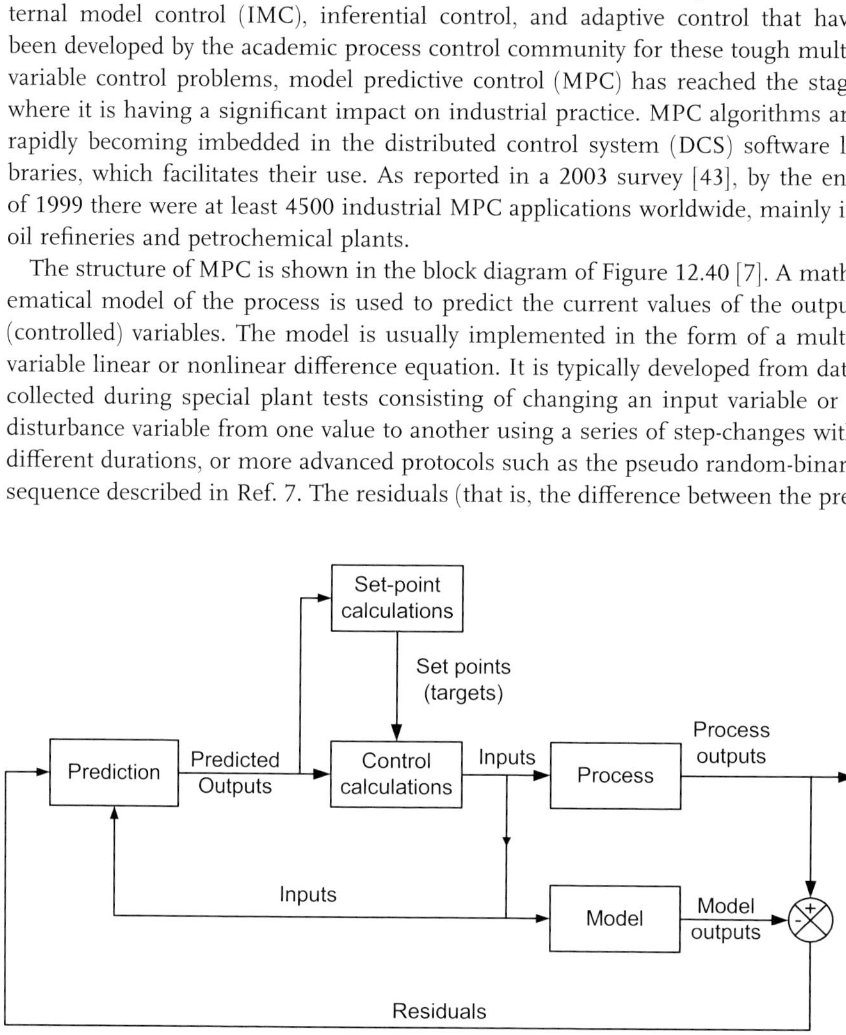

Fig. 12.40. Block diagram for model predictive control [7]. D. E. Seborg, T. F. Edgar, and D. A. Mellichamp, *Process Dynamics and Control*, Copyright © 2003 John Wiley & Sons, Inc. This material is used by permission of John Wiley & Sons, Inc.

dicted and actual output variables) serve as a feedback signal to the prediction block and are used in two types of control calculations that are performed at each sampling instant, namely set point calculations and control calculations.

A unique feature of MPC is that inequality constraints can be incorporated in both the set point and control calculations. In practice inequality constraints arise as a result of physical limitations on plant equipment such as pumps, control valves, and heat exchangers. The set points for the control calculations are typically calculated from an economic optimization of the process based on a steady-state model. Typical optimization objectives can include maximizing a process function, minimizing a cost function, or maximizing a production rate. The objective of the control calculations in the control block is to determine a sequence of control moves (changes in the manipulated variables) so that the predicted response moves to the set point in an optimal manner, for example by following a reference trajectory. The calculated control actions are implemented as set points to regulatory control loops. A detailed explanation of the different design choices that are necessary for the effective implementation of design of the MPC controller is available in Refs. 7 and 22. It is also essential to point out that the quality of MPC is very strongly dependent on the availability of a reasonably accurate process model that can capture the interactions between input, output, and disturbance variables.

Although MPC control has become an established technology for tough control problems in petrochemical plants, its application in polymer reactor control is currently transitioning from purely academic studies using simulated examples to applications in industrial reactors. Referring to the polymerization process previously described in Figure 12.33 [33], several academic researchers (for example, in Refs. 44–46) have designed and implemented MPC controllers. These controllers were used successfully for plant startup, minimization of off-grade product during grade transitions, and regulation around a set point. Recently reported industrial applications of MPC in polymer reactors, for example in Refs. 47 and 48, have focused on the transition control requirements, which typically consist in completing the transition in either a minimum time period or with minimum amount of off-specification product. MPC is well suited to the grade transition problem [49], which often translates into aggressive manipulated variable moves, often against manipulated variable limits or plant constraints. Depending on the specific process reported, benefits have also been achieved in product consistency or increased production rate.

12.5.6 Batch and Semi-batch Control

12.5.6.1 Operation and Variability

Batch reactors are the most common reactor used in polymerization engineering. They may vary in size from a five-gallon pilot unit to a 30 000-gallon (or greater) production size [19]. Removal of the heat of polymerization is accomplished by circulating coolant through a jacket or by refluxing monomer and solvent. The main advantage of batch reactors is the flexibility to accommodate multiple products.

They are well suited for low-volume products and for products for which there are numerous grades (as in specialty polymers), because each batch can be made according to its own recipe and operating conditions without the costs incurred when a continuous reactor is shut down and restarted. Process control of batch reactors must address the main disadvantage of batch reactors versus continuous ones, namely variability within a batch and/or variability from batch to batch. This variability is particularly important in free-radical batch polymerization, where the time of formation of a single chain is only a very small fraction of the batch time and therefore inhomogeneity results from the fact that polymer chains are formed under very different conditions during the course of the batch. This is especially significant for composition control in a free-radical batch copolymerization reactor where, unless special control strategies are deployed, polymer chains formed early in the reaction may contain a higher fraction of the more reactive monomer than the chains formed later in the reaction (compositional drift). On the contrary, in step growth polymerization (polyamides, polyesters), where the growth time of an individual chain is approximately the batch time, the effects of the changing reaction environment, and hence within-batch inhomogeneities, are much less of an issue, since all chains will see the same changing environment [19].

Operation in semi-batch reactor mode is very common in polymer reaction engineering practice. Typically, one way to address the issue of compositional drift in free-radical batch copolymerization is to operate the reactor under the so-called "starved feed" policy. In this case the monomer feed rate is automatically adjusted to maintain a constant rate of reaction, as inferred for example by reactor pressure (Section 12.2.2). In this starved feed operation the reaction environment is maintained constant during the batch and therefore the monomer composition in the reactor feed is equal to the desired polymer composition. It is also possible to implement more sophisticated control strategies during the batch by establishing operating trajectories for initiator addition, monomer addition, and/or reactor temperature to achieve desired polymer properties in minimum time, to maximize productivity, or to tailor the polymer molecular weight distribution. This is typically accomplished by solving off-line an optimization problem using a kinetic model of the process as shown, for example, in Refs. 50–54. These essentially open-loop trajectories constitute a form of feedforward control and are then implemented as part of the batch sequential logic and recipe management system using ladder logic and binary logic diagrams as shown in Ref. 7.

If monomer conversion and molecular weight information is available during the batch (for example, through on-line densitometry, energy balance estimation, or on-line gel permeation chromatography) it can be incorporated as part of a feedback adaptive predictive control strategy, as for example in Ref. 55, to maintain the molecular weight at a desired value, while bringing the reaction to a specified monomer conversion in minimum time by manipulating initiator feed rate and coolant jacket temperature. In addition, if a fundamental model of the process is available including reaction kinetics and an energy balance, control moves during the batch can be calculated by solving a nonlinear dynamic optimization problem within the context of the previously discussed model predictive control to account for the wide variety of constraints typically encountered in batch systems. This approach has been described [56] and applied recently [57] to a set of commercial polymerization reactors exhibiting challenging dynamic behavior that prevented conventional control from delivering optimum manufacturing performance.

12.5.6.2 Statistical Process Control

In many cases of batch and semi-batch polymerization control there are no on-line measurements of polymer quality (for example, polymer composition, molecular weight) during the batch and these measures of end-use properties are only available at the end of the batch. In this case recipe modifications from one run to the next are common. The minimal information needed to carry out this type of run-to-run control is a static model relating the manipulated variable to the quality variables at the end of the batch. As pointed out in Ref. 7, this model can be as simple as a steady-state (constant) gain relationship or a nonlinear model that includes the effects of different initial conditions and the batch time. The philosophy of statistical process control can be very useful in this case, since the polymer quality variable (for example the Mooney viscosity in elastomer manufacture) can be plotted for each successive batch on a Shewhart (x-bar) chart with the upper and lower control limits placed at three standard deviations above and below the target. A point outside the control limits means that the batch is out of control and the batch recipe and possibly the sequence logic must be adjusted for the next batch. If the quality variable for the batch is within the control limits, no control action is taken to prevent manipulations of the batch process based on stochastic variations within it.

Very often in DCS-operated batch polymer reactors the primary process variables such as pressure, temperature, level, and flow (Section 12.2.1–12.2.4) are recorded during the batch as well as the quality variables at the end of the batch. However, it may be very difficult to obtain a kinetic model of the polymerization process due to the complexity of the reaction mechanism, which is frequently encountered in the batch manufacture of specialty polymers. In this case it is possible to use advanced statistical techniques such as multi-way principal component analysis (PCA) and multi-way partial least squares (PLS), along with an historical database of past successful batches to construct an empirical model of the batch [8, 58, 59]. This empirical model is used to monitor the evolution of future batch runs. Subsequent unusual events in the future can be detected during the course of the batch by referencing the measured process behavior against this "in-control" model and its statistical properties. It may therefore be possible to detect a potentially bad batch before the run is over and to take corrective action during the batch in order to bring it on aim.

12.5.7 Future Trends

Trends in industry, which reinforce the importance of measurement of polymer reactor process control, include the increasing emphasis by customers on receiving a more uniform product with desired property specifications from their polymer suppliers; the increasing computing power provided by DCS manufacturers; and the availability inside the DCS of advanced control modules such as the MPC so that the implementation of nonlinear multivariable control is easier and faster. As new polymer reactors are designed (for example, to manufacture polymers with tailor-made properties using novel metallocene catalysts or living radical polymerization technology), it would be beneficial to incorporate process control considerations in the process design phase, so that the controllability of the plant can be established before it is actually constructed. Furthermore, it is expected that the on-line control not only of average polymer properties but also of polymer distributions such as the particle size distribution, as shown for example in Ref. 60, and the branching distribution will become important, together with recipe management, production scheduling, and production optimization. The instrumentation and control methodologies that will be needed to be deployed to meet these needs are a challenging and vibrant area of investigation for academic researchers and industrial practitioners alike.

Notation [61, 62]

Symbols

Symbol		Units
A	area	m^2
a	acceleration	$m\ s^{-2}$
a	resistance coefficient based on calibration	$°C^{-1}$
b	resistance coefficient based on calibration	$°C^{-2}$
c	resistance coefficient based on calibration	$°C^{-3}$
c	speed of sound	$m\ s^{-1}$
C_D	drag coefficient	–
C_p	heat capacity at constant pressure	$J\ mol^{-1}\ K^{-1}$
C_v	heat capacity at constant volume	$J\ mol^{-1}\ K^{-1}$
C_d	discharge coefficient	–
C_V	valve coefficient in USCS units	$gal\ min^{-1}\ psi^{-1/2}$
d	diameter	m
d	disturbance variable	various
d	relative liquid density (specific gravity)	–
d	differential operator	–
d_i	discrete deviation or error	various
E	modulus of elasticity	Pa (= $N\ m^{-2}$)
e	controller error, $y_{sp} - y$	various
e_F	filtered error	various
f	valve characteristic	–
F	force	N (= $kg\ m\ s^{-2}$)

g	gravitational acceleration	$m\ s^{-2}$
g_c	gravitational conversion factor in SI units	$1\ (= kg\ m\ s^{-2}\ N^{-1})$
g_c	gravitational conversion factor in USCS units	$lb_m\ ft\ s^{-2}\ lb_f$
h	height	m
$H(t)$	Heaviside step function	–
I	radiant intensity	$W\ m^{-2}$
K_c	controller gain	various
K_{cu}	ultimate controller gain	various
K_p	process gain	various
K_V	valve coefficient in SI units	$m^3\ h^{-1}\ bar^{-1/2}$
l	length	m
ℓ	valve lift	–
m	mass	kg
M	magnitude of step-change in input variable	various
M_n	number average molecular weight	$kg\ mol^{-1}$
M_w	weight average molecular weight	$kg\ mol^{-1}$
M_z	z-average molecular weight	$kg\ mol^{-1}$
n	number of measurements	–
p	pressure	$Pa\ (= N\ m^{-2})$
P_u	ultimate period	s
PB	proportional band	%
PD	polydispersity index	–
q_m	mass flow rate	$kg\ s^{-1}$
q_n	molar flow rate	$mol\ s^{-1}$
q_v	volume flow rate	$m^3\ s^{-1}$
r	radius	m
R	resistance	Ω
R	valve resistance coefficient	$s\ m^{-2}$
R	equal percentage valve parameter	–
s	sample standard deviation	various
t	Celsius temperature	°C
T	thermodynamic temperature	K
t	time	s
u	manipulated deviation variable	various
v	velocity	$m\ s^{-1}$
$\bar{v}$	velocity average	$m\ s^{-1}$
V	volume	m^3
x	input deviation variable	various
$x(t)$	continuous input variable	various
x_i	discrete measured value	various
x_n	discrete input variable	various
$\bar{x}$	sample mean	various
y	controlled deviation variable	various
$y(t)$	continuous output variable	various
y	mole fraction	–

Y	expansion factor	–
y_n	discrete output variable	various
y_{sp}	set point	various
z	space coordinate	m

Greek

Symbol		SI unit
α	filter factor	–
α	kinetic energy correction factor	–
β	diameter ratio, d_2/d_1	–
$\dot{\gamma}$	shear rate or rate of deformation	s^{-1}
Δp	pressure drop	Pa (= N m^{-2})
Δt	time interval	s
Δu	change in input deviation variable	various
Δy	change in output deviation variable	various
ε	absorption coefficient	$m^2\ kg^{-1}$
θ	dead time	s
κ	ratio of specific heats, C_p/C_v	–
μ	viscosity	Pa s
μ	population mean	various
ρ	mass density	kg m^{-3}
ρ_{w0}	mass density of water at reference temperature	kg m^{-3}
σ	population standard deviation	various
σ	surface tension	N m^{-1}
$\sum$	summation operator	–
τ_D	controller derivative time	s
τ_F	filter time constant	s
τ_I	controller reset time	s
τ	shear stress	Pa (= N m^{-2})
τ	time constant	s

Indices

0	refers to inlet, reference, standard, or in front of
1	refers to datum, upstream, or outlet of tank 1
2	refers to surface, downstream, in throat, or outlet of tank 2
B	refers to buoyancy
c	refers to conversion factor or controller
D	refers to drag or derivative control
F	refers to filter
f	refers to float
I	refers to integral control
i	refers to *i*th sample

m refers to mass
max refers to maximum
min refers to minimum
n refers to nth sample, tank number, or moles
p refers to pressure or process
s refers to steady state value
sp refers to set point
t refers to temperature or tube
u refers to ultimate
v refers to volume
w refers to water
W refers to weight
$\bar{x}$ bar above symbol refers to average
$\dot{y}$ dot above symbol refers to time rate of change

References

1 A. ECHTE, Chapter 4: "Meß-, Steuer- und Regelungstechnik", *Handbuch der Technischen Polymerchemie*, Wiley-VCH, Weinheim, **1993**.

2 R. P. BENEDICT, *Fundamentals of Temperature, Pressure, and Flow Measurements*, 3rd ed., Wiley-Interscience, New York, **1984**.

3 D. M. CONSIDINE, *Process/Industrial Instruments and Controls Handbook*, 5th ed., McGraw-Hill, New York, **1999**.

4 B. G. LIPTÁK, *Instrument Engineers' Handbook*, 4th ed., *Process Measurement and Analysis, Volume I*, CRC Press, New York, **2003**.

5 D. M. LEVINE, P. P. RAMSEY, R. K. SMIDT, *Applied Statistics For Engineers and Scientists Using Microsoft Excel and MINITAB*, Prentice Hall, **2001**.

6 R. H. PERRY (Editor), D. W. GREEN (Editor), J. O. MALONEY (Editor), *Chemical Engineers' Handbook*, 7th ed., McGraw-Hill, New York, **1997**.

7 D. E. SEBORG, T. F. EDGAR, D. A. MELLICHAMP, *Process Dynamics and Control*, John Wiley & Sons, New York, **2003**.

8 O. KAMMONA, E. G. CHATZI, C. KIPARISSIDES, "Recent developments in hardware sensors for the on-line monitoring of polymerization reactions", *J. Macromol. Sci., Rev. Macromol. Chem. Phys.*, **1999**, *C39*, 57.

9 E. A. AVALLONE (Editor), T. BAUMEISTER (Editor), *Marks' Standard Handbook for Mechanical Engineers*, 10th ed., McGraw-Hill, New York, **1996**.

10 W. L. MCCABE, J. C. SMITH, P. HARRIOTT, *Unit Operations of Chemical Engineering*, 5th ed., McGraw-Hill, New York, **1993**.

11 R. B. BIRD, W. E. STEWART, E. N. LIGHTFOOT, *Transport Phenomena*, 2nd ed., Wiley & Sons, New York, **2002**.

12 *Flow of Fluids through Valves, Fittings, and Pipe*, Technical Paper 410, Crane Co., 300 Park Ave., New York, NY, **1988**.

13 F. RODRIGUEZ, C. COHEN, C. OBER, L. A. ARCHER, *Principles of Polymer Systems*, 5th ed., Taylor and Francis, New York, **2003**.

14 M. P. VEGA, E. L. LIMA, J. C. PINTO, "In-line monitoring of weight average molecular weight in solution polymerization using intrinsic viscosity measurements", *Polymer*, **2001**, *42*, 3909.

15 O. ELIZALDE, J. R. LEIZA, J. M. ASUA, "On-line monitoring of all acrylic emulsion polymerization reactors by Raman spectroscopy", *Macromolecular Symposia*, **2004**, *206*, 135.

16 M. M. Reis, P. H. H. Araújo, C. Sayer, R. Giudici, "Comparing near infrared and Raman spectroscopy for on-line monitoring of emulsion co-polymerization reactions", *Macromolecular Symposia*, **2004**, *206*, 135.
17 F. J. Wyzgoski, P. L. Rinaldi, E. F. McCord, M. A. Stewart, D. R. Marshall, "Poly(*n*-butyl acrylate-*co*-carbon monoxide-*co*-ethylene) characterization by high-temperature two-dimensional NMR at 750 MHz", *Macromolecules*, **2004**, *37*, 846.
18 F. J. Schork, P. B. Deshpande, K. W. Leffew, *Control of Polymerization Reactors*, Marcel Dekker, New York, **1993**.
19 F. J. Schork, "Reactor operation and control", in *Polymer Reaction Engineering*, C. McGreavy (Editor), Kluwer Academic Publishers, Dordrecht, **1993**.
20 A. K. Hipp, G. Storti, M. Morbidelli, "Acoustic characterizations of concentrated suspensions and emulsions. 1. Model Analysis", *Langmuir*, **2002**, *18*, 391.
21 A. K. Hipp, G. Storti, M. Morbidelli, "Acoustic characterizations of concentrated suspensions and emulsions. 2. Experimental validation", *Langmuir*, **2002**, *18*, 405.
22 B. A. Ogunnaike, W. H. Ray, *Process Dynamics, Modeling, and Control*, Oxford University Press, New York, **1994**.
23 C. L. Smith, *Digital Computer Process Control*, International Textbook Company, Scranton, PA, **1972**.
24 J. G. Ziegler, N. B. Nichols, "Optimum settings for automatic controllers", *Trans. ASME*, **1942**, *64*, 759.
25 *Control Valve Handbook*, 3rd ed., Fisher Controls International, **2001**.
26 A. K. Adebekun, K. M. Kwalik, F. J. Schork, "Steady-state multiplicity during solution polymerization of methyl methacrylate in a CSTR", *Chem. Eng. Sci.*, **1989**, *44*, 2269.
27 W. H. Ray, C. M. Villa, "Nonlinear dynamics found in polymerization processes – a review", *Chem. Eng. Sci.*, **2000**, *55*, 275.
28 G. R. Meira, "Forced oscillations in continuous polymerization reactors and molecular weight distribution control. A survey", *J. Macromol. Sci – Rev. Macromol. Chem.*, **1981**, *C20*, 207.
29 M. Embirucu, E. L. Lima, J. C. Pinto, "A survey of advanced control of polymerization reactors", *Polymer Eng. Sci.*, **1996**, *36*, 433.
30 J. P. Congalidis, J. R. Richards, "Process control of polymerization reactors: an industrial perspective", *Polym. Reaction Eng.*, **1998**, *6*, 71.
31 M. A. Dubé, J. B. P. Soares, A. Penlidis, A. E. Hamielec, "Mathematical modeling of multicomponent chain growth polymerizations in batch, semi-batch, and continuous reactors. A review", *Ind. Eng. Chem. Res.*, **1997**, *36*, 966.
32 W. H. Ray, "Process modeling for polymerization process operation and control", presented at the *Tutorial on Polymerization Reactor Control, AIChE Annual Meeting*, November, **2003**.
33 J. P. Congalidis, J. R. Richards, W. H. Ray, "Feedforward and feedback control of a solution copolymerization reactor", *AIChE J.*, **1989**, *35*, 907.
34 K. C. Seavey, Y. A. Liu, N. P. Khare, T. Bremner, C.-C. Chen, "Quantifying relationships among the molecular weight distribution, non Newtonian shear viscosity, and melt index for linear polymers", *Ind. End. Chem. Res.*, **2003**, *42*, 5354.
35 R. W. Chylla, D. R. Haase, "Temperature control of semi-batch polymerization reactors", *Computers Chem. Eng.*, **1993**, *17*, 257.
36 G. Defaye, N. Regnier, J. Chabanon, L. Caralp, C. Vidal, "Adaptive-predictive temperature control of semi-batch reactors", *Chemical Engineering Science*, **1993**, *48*, 3373.
37 D. Tyner, M. Soroush, M. C. Grady, "Adaptive temperature control of multi product jacketed reactors", *Ind. End. Chem. Res.*, **1999**, *38*, 4337.
38 H. U. Moritz, "Polymerization calorimetry: a powerful tool for reactor control", in *Polymer Reaction Engineering, Vol. 248*, VCH, New York, **1989**.

39 T. F. McKenna, S. Othman, G. Fevotte, A. M. Santos, H. Hammouri, "An integrated approach to polymer reaction engineering: a review of calorimetry and state estimation", *Polym. Reaction Eng.*, **2000**, *8*, 1.

40 K. B. McAuley, J. F. MacGregor, "On-line inference of polymer properties in an industrial polyethylene reactor", *AIChEJ.*, **1991**, *37*, 825.

41 M. F. Ellis, T. W. Taylor, K. F. Jensen, "On-line molecular weight distribution estimation and control in batch polymerization", *AIChEJ.*, **1994**, *40*, 445.

42 M.-J. Park, S.-M. Hur, H.-K. Rhee, "Online estimation and control of polymer quality in a copolymerization reactor", *AIChEJ.*, **2002**, *48*, 1013.

43 S. J. Qin, T. A. Badgwell, "A survey of industrial model predictive control technology", *Control Eng. Practice*, **2003**, *11*, 733.

44 B. R. Maner, F. J. Doyle III, "Polymerization reactor control using autoregressive-plus Voltera-based MPC", *AIChEJ.*, **1997**, *43*, 1763.

45 L. Özkan, M. V. Kothare, C. Georgakis, "Control of a solution copolymerization reactor using multi-model predictive control", *Chem. Eng. Sci.*, **2003**, *58*, 1207.

46 R. Bindlish, J. B. Rawlings, "Target linearization and model predictive control of polymerization processes", *AIChE J.*, **2003**, *49*, 2885.

47 H. Seki, M. Ogawa, S. Ooyama, K. Ahamatsu, M. Ohshima, W. Yang, "Industrial application of nonlinear Model Predictive Control to polymerization reactors", *Control Eng. Prac.*, **2001**, *9*, 819.

48 D. Bartusiak, R. W. Fontaine, C. O. Schwanke, A. R. Gomatam, "Nonlinear Model Predictive Control of a polymerization reactor", presented at the *Tutorial on Polymerization Reactor Control, AIChE Annual Meeting*, November, **2003**.

49 M. Takeda, W. H. Ray, "Optimal-grade transition strategies for multi-stage polyolefin reactors", *AIChE Journal*, **1999**, *45*, 1776.

50 I. Saenz de Buruaga, P. D. Armitage, J. R. Leiza, J. M. Asua, "Nonlinear control for maximum production rate of latexes of well-defined polymer composition", *Ind. End. Chem. Res.*, **1997**, *36*, 4243.

51 G. Maschio, T. Bello, C. Scali, "Optimization of batch polymerization reactors: Modelling and experimental results for suspension polymerization of methyl methacrylate", *Chem. Eng. Sci.*, **1992**, *47*, 2609.

52 G. Maschio, T. Bello, C. Scali, "Optimal operation strategies to control the molecular weight distribution of polymer products", *Chem. Eng. Sci.*, **1994**, *49*, 5071.

53 I. M. Thomas, C. Kiparissides, "Computation of the near optimal temperature and initiator policies for a batch polymerization reactor", *Can. J. Chem. Eng.*, **1984**, *62*, 284.

54 D. Tyner, M. Soroush, M. C. Grady, J. R. Richards, J. P. Congalidis, "Mathematical modeling and optimization of a semi-batch polymerization reactor", *Proceedings IFAC Symposium on Advanced Control of Chemical Processes, Pisa. Italy*, **2001**, *3*, 983.

55 W. E. Houston, F. J. Schork, "Adaptive predictive control of a semi batch polymerization reactor", *Polym. Process Eng.* **1987**, *5*, 119.

56 T. Peterson, E. Hernandez, Y. Arkun, F. J. Schork, "A nonlinear DMC algorithm and its application to a semi-batch polymerization reactor", *Chem. Eng. Sci.*, **1992**, *42*, 737.

57 J. W. Wassick, D. Coffey, B. Callihan, "Nonlinear Model Predictive Control of a commercial polymerization semi-batch reactor", presented at the *Tutorial on Polymerization Reactor Control, AIChE Annual Meeting*, November, **2003**.

58 P. Nomikos, J. F. MacGregor, "Multi-way partial least squares in monitoring batch processes", *Chemometrics and Intelligent Laboratory Systems*, **1995**, *30*, 97.

59 T. Kourti, P. Nomikos, J. F. MacGregor, "Analysis, monitoring, and fault diagnosis of batch processes using multiblock and multiway PCA", *J. Proc. Cont.*, **1995**, *5*, 277.

60 F. J. Doyle, C. A. Harrison, T. J. Crowley, "Hybrid model based approach to batch-to-batch control of particle size distribution in emulsion polymerization", *Computers Chem. Eng.*, **2003**, *27*, 1153.

61 I. Mills, T. Cvitas, K. Homann, N. Kallay, K. Kuchitsu, *Quantities, Units and Symbols in Physical Chemistry (The Green Book)*, Blackwell Science, **1993**.

62 J. Inczedy, T. Lengyel, A. M. Ure, *Compendium of Analytical Nomenclature (definitive rules 1997)*, 3rd ed., Blackwell Science, **1998**.

13
Polymer Properties through Structure[1]

Uday Shankar Agarwal

Since polymers display an incredible range of properties, their potential for applications is ever increasing. The link between polymer structure and the resulting properties has long been realized, and emerged as the concept of tailor making of polymers. In earlier times, developments were largely targeted on achieving the desired properties through new monomers, copolymer and blend compositions, physical and chemical additives, and control of polymer molecular weights, as well as through structural changes derived from processing. In recent years, the increased expertise in anionic polymerization and the emergence of controlled radical polymerization techniques have led to relatively easy access to several precisely controlled topologies such as those of telechelic, block, graft, star-shaped, and several other branched homoploymers and copolymers with controlled molecular weights. For example, phenomenal growth has been recorded in the area of synthesis of block copolymers, as well as in examination of their self-assembling characteristics leading to unique control of nanoscale morphology. This is providing tools for development of properties suitable for applications extending from those requiring mechanical toughness to electronic and biological uses. Another rapidly developing area is nanocomposites, where the attainable range of properties is further expanded by incorporating inorganic and organic nanoparticles in polymeric matrices.

In this chapter we aim to demonstrate the relationship of the structure of polymers with their thermal, solution, and rheological behavior. Besides providing a general review of such behavior, we will emphasize some of the recent developments in these areas.

13.1
Thermal Properties of Polymers

The thermal behavior of polymers is different from that of simple compounds in that, on heating, the transition of polymers from solids to liquids occurs not at

1) The symbols used in this chapter are listed at the end of the text, under "Notation".

Handbook of Polymer Reaction Engineering. Edited by T. Meyer, J. Keurentjes

ISBN: 3-527-31014-2

a sharp temperature, but over a range of temperatures. Such behavior is at least partly attributable to the mixture of chain lengths of molecules that make up a polymeric material. In addition to a crystalline solid phase that gains liquid-like mobility at the corresponding crystalline melting point, polymeric materials display an amorphous phase that undergoes a gradual transformation to a liquid-like state over a temperature range called the glass transition temperature.

13.1.1 Crystalline and Amorphous Polymers

Crystallization of polymers is important because it imparts several highly useful properties such as strength, toughness, stiffness, and solvent and chemical resistance. A polymer can crystallize when it is cooled from a melt, or concentrated from a diluted solution. This is because the approaching chains may fold and pack into a regular, long-range, three-dimensional, positional and orientational order characterized by a certain unit cell of that polymer. Figure 13.1 shows the

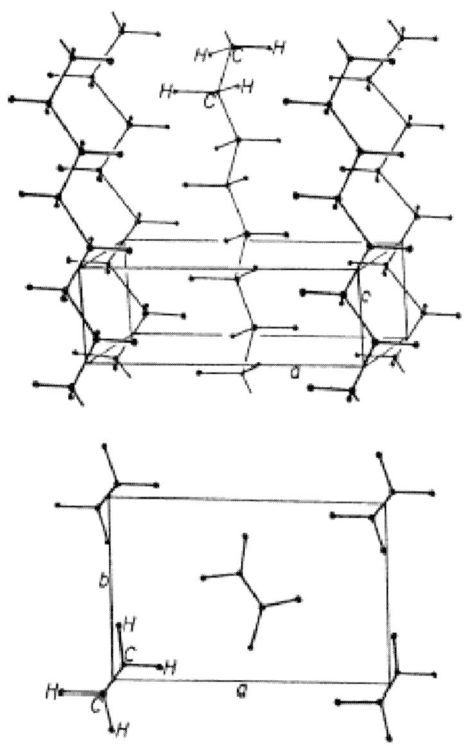

Fig. 13.1. Arrangement of polyethylene molecules in the orthorhombic unit cell of dimensions $a = 7.41$ Å, $b = 4.94$ Å, $c = 2.55$ Å (from C. W. Bunn, in *Fibers from Synthetic Polymers*, ed. R. Hill, Elsevier, Amsterdam, **1953**).

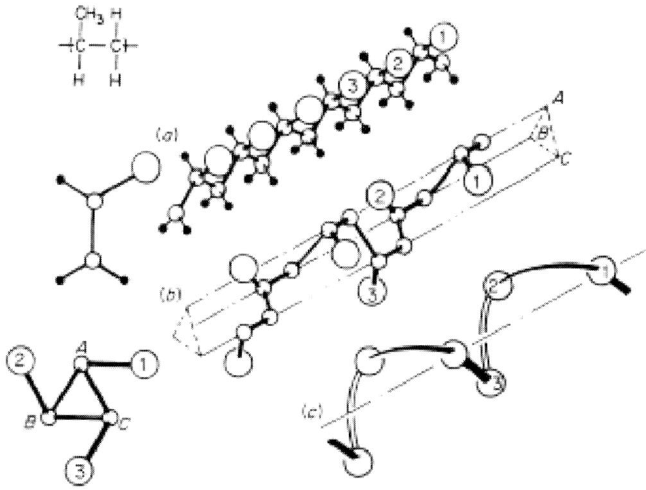

Fig. 13.2. Helical conformation of an isotactic vinyl polymer crystalline chain (from F. Rodriguez, *Principles of Polymer Systems*, 2nd Ed., McGraw Hill Intl., Auckland, **1982**).

unit cell in a crystal of a commercially important crystalline polymer: polyethylene. Here, the chains are extended in planar zigzag (*trans*) conformation, which corresponds to low energy. However, steric hindrances can lead to other chain conformations in crystals. For example, in the presence of bulky substituents, the vinyl polymer chains rotate to take up a rotated helical conformation (Figure 13.2).

If a given polymer lacks a tendency to get ordered, the chains stay randomly coiled and entangled during solidification (Figure 13.3). Such polymers (for example, poly(methyl methacrylate)) are called amorphous. Crystallinity is best detected by wide-angle X-ray diffraction (WAXD, Figure 13.4), where the peak positions provide information about the repeat distances in the crystal lattice. Unlike small molecules that display sharp peaks during WAXD, even crystallizable polymers display broad WAXD peaks (Figure 13.4). This is because polymers are never

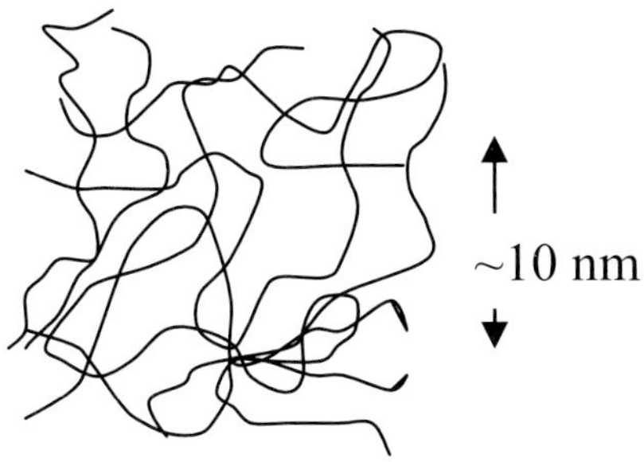

Fig. 13.3. Lack of long-range order in an amorphous polymer.

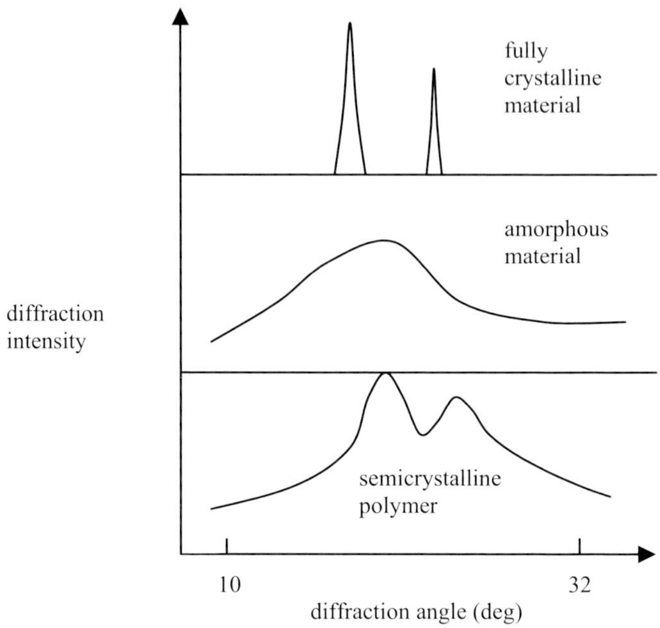

Fig. 13.4. WAXD patterns of fully crystalline solids, amorphous solid/melts, and semicrystalline polymers.

100% crystalline due to entanglements between chains, and are therefore called semicrystalline. The extent of crystallization (x_c) can vary from nearly 100% for polytetrafluoroethylene or linear polyethylene to nearly zero for noncrystallizable polymers such poly(methyl methacrylate). The extent of crystallization is determined by considering the variation of properties (ρ, such as density, specific heat, and electrical resistivity) between amorphous phase polymer and crystalline phase polymer to be linearly additive [Eq. (1)].

$$\rho = (1 - x_c)\rho_a + x_c\rho_c \tag{1}$$

13.1.2
Influence of Polymer Structure on Crystallizability of Polymers

Since crystallization requires that the participating chains to come close together, the spatial regularity and the packing possibilities of the polymers are the most important factors in determining their crystallizability. For example, isotactic and syndiotactic polypropylene are stereoregular, and show a higher tendency to crystallize than atactic polypropylene (Figure 13.5). Randomly arranged or bulky pendent groups reduce this tendency because they hinder the close alignment of the polymer molecules that is necessary for intermolecular interactions. For example,

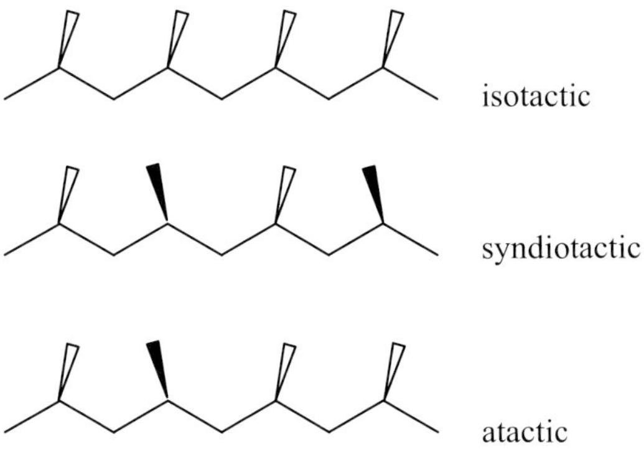

Fig. 13.5. The regularity of a polypropylene chain structure decides its crystallizability. The isotactic and syndiotactic polypropylene crystallize, but atactic polypropylene does not.

poly(vinyl acetate) with large pendent groups (–O–CO–CH_3) is amorphous, but crystallizes easily when hydrolyzed to poly(vinyl alcohol), which has small pendent groups (–OH). Crystallizability is also induced by existence of functional groups capable of intermolecular hydrogen bonding (such as amides, urethanes, urea, and so on) or functional groups capable of polar interactions (such as –Cl and –CN). Block copolymers have large segments of regular structure capable of participating in crystallization, and hence show a higher tendency to crystallize than random copolymers.

13.1.3 The Glass Transition Temperature

An amorphous polymer can exist in the glassy (hard, brittle, and stiff) or the rubbery (soft) state, depending on the temperature being lower or higher than the so-called glass transition temperature (T_g). The glass-to-rubber transition at T_g is associated with an increase in free volume resulting in an increase in cooperative rotational and translational motion of the chain segments (20–50 atoms), as the thermal energy overcomes the intermolecular restraints imposed by secondary bonding forces between polymer chains. This results in marked changes, such as an increase in the number of possible chain conformations and an increase in the free volume. This is manifested as jumps in the thermal expansion coefficient (Figure 13.6) and specific heat (Figure 13.7), and a drop of a few orders of magnitude in the mechanical modulus (Figure 13.8). Even as the chain segments gain mobility above T_g, the chain entanglements persist, imparting rubberlike elasticity (Figure 13.8). With further increases in temperature and the associated energy and motion of the molecules, translation/slip of the entire molecules becomes possible. A consequence of Eq. (1) is that the changes at T_g are less drastic in semicrystalline polymers as the crystallites impart rigidity, even at temperatures higher than T_g.

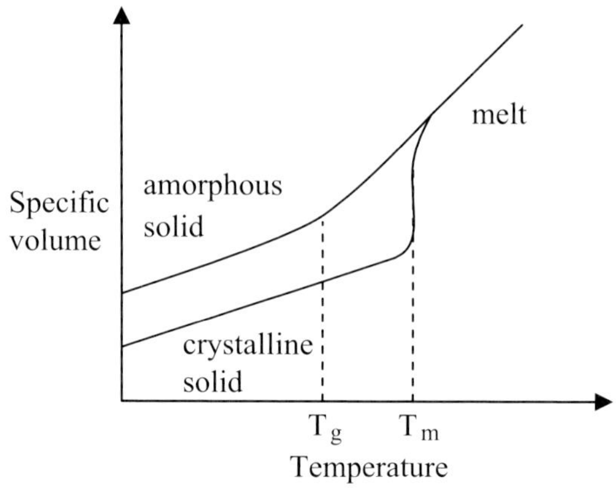

Fig. 13.6. Volume expansion (dilatometry) in amorphous and crystalline polymers with increasing temperature.

13.1.4
Influence of Polymer Structure on T_g of Polymers

The T_g of a polymer is important as it defines the maximum temperature at which an amorphous solid polymer still displays structural rigidity. For a given polymer structure, the T_g increases with an increase in molecular weight (M) [Eq. (2), where $T_{g,\infty}$ is the T_g of a polymer with infinite molecular weight, and K is a constant]. T_g is affected by structural features of the polymer chains, such as chain stiffness and intermolecular forces.

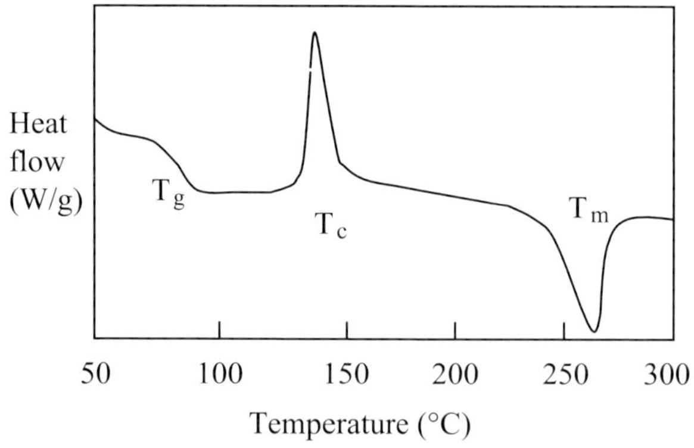

Fig. 13.7. Heat flux during heating of poly(ethylene terephthalate) (PET) at a constant rate in a differential scanning calorimeter (DSC), showing changes in specific heat at T_g, crystallization exotherm at T_c, and melting endotherm at T_m.

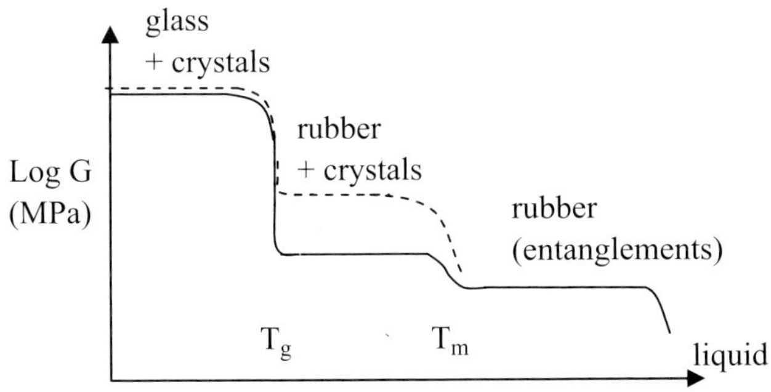

Fig. 13.8. Influence of temperature on mechanical modulus of a semicrystalline polymer. The broken line indicates the corresponding curve at higher crystallinity.

$$T_g = T_{g,\infty} - \frac{K}{M} \tag{2}$$

For example, while low T_g values are exhibited by flexible chains such as polyethylene with easy rotation about the backbone, a higher T_g results from chain stiffness such as that resulting from cyclic structures in the backbone (as in BPA polycarbonate). This concept is used in enhancing the T_g of PET by partial replacement of the aliphatic glycol in it with an alicyclic glycol, thereby making the polymer more suitable for high-temperature applications [1]. While bulky side groups (for example, in poly(α-methylstyrene)) enhance the T_g because of the resulting steric hindrance, long and flexible side groups decrease the T_g as they enhance the separation between polymer backbones and introduce additional chain-ends. Secondary attractive forces (for example, polarity) enhance the T_g. The extent of syndiotacticity in polymers such as PMMA can be influenced by the method of polymerization, and in-turn influence the T_g [2, 3].

When a random or alternating copolymer is made up of non-interacting monomers, its T_g can be calculated from the weight fractions (w_i) and glass transition temperatures (T_{gi}) of its components, by means of Eq. (3).

$$\frac{1}{T_g} = \frac{w_1}{T_{g1}} + \frac{w_2}{T_{g2}} \tag{3}$$

When a polymer is desired to be flexible at the temperature of use, this effect (Eq. 3) can be used to plasticize a polymer *internally* (that is, reduce its T_g) by incorporating comonomers such as vinyl acetate and vinyl chloride. In the case of block or graft copolymers, microphase separation of the blocks often leads to separate glass transitions of the corresponding blocks [4, 5]. The case is similar for immiscible blends. Plasticization can be achieved *externally* by addition of low molecular weight, high-boiling organic compounds (plasticizers) that can enhance inter-backbone distance and drastically reduce T_g. For example, dioctyl phthalate is often added to turn the otherwise rigid plastic poly(vinyl chloride) into a flexible material [6].

13.1.5 The Crystallization Temperature and the Melting Point

When heated beyond its T_g, a crystallizable polymer undergoes crystallization that is detected calorimetrically as an exotherm (Figure 13.7). The rate of crystallization first increases with temperature, reaching a maximum at the so-called crystallization temperature, T_c. With a further increase in temperature, the solid undergoes a first-order transition into a melt at T_m (Figures 13.7, 13.8), corresponding to zero free energy change on melting [Eq. (4)].

$$T_m = \frac{\Delta H_m}{\Delta S_m} \tag{4}$$

The T_m is high for polymers with a lower ΔS_m; this is due, for example, to chain stiffness in PEEK (poly(ether ether ketone), $T_m = 395°C$) or to specific interactions that persist into melts (nylon-6, $T_m = 228°C$). T_m can also be high for polymers exhibiting high ΔH_m values, for example through regular and frequent polar interactions. Random addition of a comonomer leads to reductions in both the T_m and the extent of crystallization, while long homopolymer segments in block and graft copolymers or in immiscible blends may crystallize into their respective crystalline forms and display two melting points.

Heating a polymer above T_m to a melt allows its fabrication into the desired shape in processing operations such as injection molding, extrusion, fiber spinning and blow-molding. During cooling from melt, a semicrystalline polymer undergoes crystallization at temperatures below T_m but above T_g. Since crystallization involves translation from a highly disordered melt to a highly ordered state under viscous conditions, the extent, size, and perfection of crystallization depends on crystallization conditions such as the rate of cooling and the temperature of crystallization. For example, rapid cooling often results in a large number of crystals, as well as reductions in the extent and perfection of crystallization. On the other hand, crystallization at temperatures just below T_m for long periods results in larger and more perfect crystals.

13.1.6 Tuning Polymer Crystallization for Properties

Crystallization plays a major role in determining the thermal and mechanical behavior of polymers as it permits polymers to retain rigidity at temperatures exceeding the T_g. Crystallization also enhances the barrier properties and solvent resistance, as small molecules cannot diffuse through crystalline domains. Crystallization of polymers can be tuned, and thus can provide very useful control over the subsequent processing and final properties. For example, crystallization from dilute solutions can lead to disentanglement and high drawability in polyethylene, producing fibers with a very high modulus and strength [7]. Crystallization and morphology development in polyethylene during its formation on heterogeneous catalysts leads to a "nascent" state with special characteristics such as a high T_m compared to molded polyethylene. One hypothesis attributes this to the influence of strain resulting from the temperature gradient at the polymerization site, in a manner somewhat similar to strain-induced extended chain crystallization from stirred solutions [8, 9]. Polymer molecules being formed at the heterogeneous catalyst fragment surface experience strong elongational flow, which could influence the nascent crystallization process [10, 11] Innovations in polyethylene reactor technology are directed toward controlling the morphology and properties of the polymerization product [12]. One expectation is that polyethylene as a nascent reactor product can be directly drawn into high-modulus, high-strength fibers similar to those produced via the solution route [13].

Crystallization in step-growth polymers such as polyesters and nylons is known to assist their subsequent solid-state polymerization because exclusion of reactive end-groups from crystalline domains enhances their effective concentration in the amorphous domains [14, 15]. However, the condensation reaction between the last fraction of end-groups may be hindered by crystallization [16, 17]. The possibility and rate of crystallization can also be enhanced by processes that enhance orientation, such as shearing and fiber drawing [18]. For example, partial replacement of terephthalic units with isophthalic units in PET reduces crystallinity, so that no crystallization in seen in 70:30 random poly(ethylene terephthalate-*co*-ethylene isophthalate) under quiescent conditions. However, heating its amorphous fiber above its T_g under a moderate tensile force results in rapid stress-induced crystallization [19]. The reduction in crystallization by copolymerization has been employed to enhance drawability of melt-spun polyester and polyamide fibers [20].

On the other hand, a high rate of crystallization during cooling from melts is desired, as it determines the cycle time and thus the suitability of a polymer for very important shape-forming processes such as injection molding. The crystallization rate is easily enhanced by reducing the molecular weight of polymers, but that is also detrimental to the mechanical properties. Therefore, nucleating agents are often used to enhance the crystallization rate. The activity of several heterogeneous nucleating agents is related to oriented deposition of polymer chains on the surfaces of inorganic particles [21]. Nucleating agents such as talc, by virtue of their highly oriented crystalline surfaces, induce epitaxial crystallization of polymers such as PE and PET [22]. The activity of organic compounds (salts) such as sodium benzoate for PET crystallization is related to formation of ionic chain-ends that aggregate in polymer melts to form nucleating clusters [23]. The crystallization enhancement that occurs on replacing a small fraction ($< 1\%$) of the ester segments in PET with amide segments was attributed to self-assembly of the amide segments [24]. These homogenous nucleators may offer an advantage in impact properties. Their incorporation in PET can be carried out by simple solid-state modification reactions, thus avoiding the degradation reactions accompanying melt-mixing

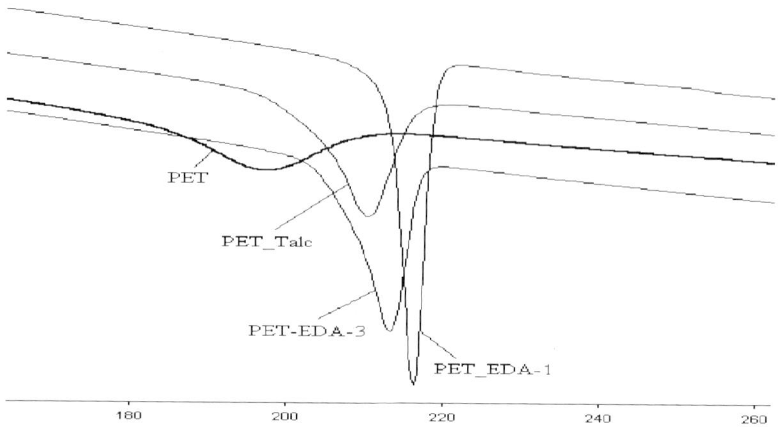

Fig. 13.9. DSC cooling scans of (i) PET homopolymer ($[\eta] = 0.62$ dL g^{-1}); (ii) PET chemically modified in the solid-state with ethylene diamine (EDA) vapor to replace 2.2 mol% of ester functionality with amide functionality (nitrogen content = 0.31%, PET–EDA-1: $[\eta] = 0.5$ dL g^{-1}, PET–EDA-3: $[\eta] = 0.61$ dL g^{-1}); and (iii) PET melt mixed with 0.3% talc. The modified samples undergo crystallization at a higher temperature (that is, sooner during the cooling process). (Reprinted from *Polymer*, Volume 43, page 5709, Copyright **2002**, U. S. Agarwal, G. de Wit, P. J. Lemstra, with permission from Elsevier).

methods [25]. Figure 13.9 shows the enhanced tendency to crystallize of the thus-modified PET during cooling from melt.

The optical clarity of polymers generally decreases with the appearance of crystallinity. Semicrystalline polymers are generally opaque because of the scattering due to the difference in the refractive indices of the crystalline and the amorphous domains. However, enhancement of the nucleation density with suitable nucleating agents can lead to reduction of spherulite sizes (below the wavelength of light) and hence to transparent semicrystalline polymers. For example, the recently developed derivatives based on D-sorbitol dissolve in a polypropylene melt and are thus easily dispersed. On cooling, these additive molecules aggregate, and provide nucleation sites for fine PP spherulites [26]. More recently, nanoparticles (such as clay) and carbon nanotubes are being explored as potential nucleating agents [27].

13.1.7
Morphology of Crystalline Polymers

Early X-ray diffraction work in the 1920s indicated the polymers to be semicrystalline, with crystallites dispersed in an amorphous matrix. The longest dimension of the crystallites in polymers is 5–50 nm, which is a fraction of the length of the extended polymers (up to 5 μm). Based on this, the crystalline morphology was conceived to be of the "fringed micelle" type, where the long polymer chains pass successively through the lengths of several crystalline and amorphous domains (Figure 13.10).

Later work showed that crystalline domains are most often made up of flat rib-

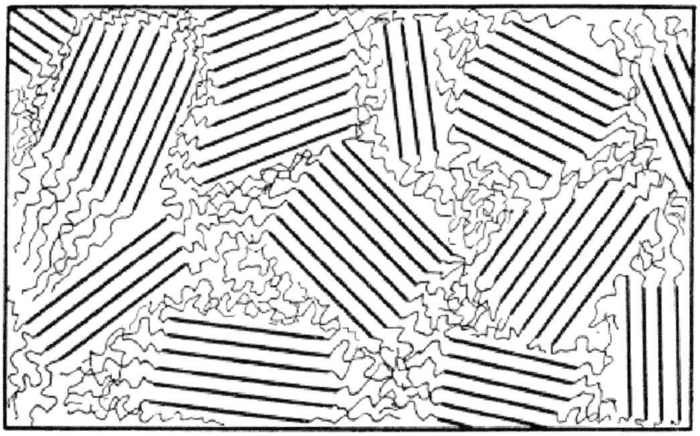

Fig. 13.10. Fringed micelle model of semicrystallinity in polymers. (Reprinted from Paul J. Flory, *Principles of Polymer Chemistry*. Copyright **1953**, Cornell University and Copyright **1981** Paul J. Flory. Used by permission from the publisher, Cornell University Press).

bons like "lamellae", within which the chains are largely folded (insets *a* and *b* of Figure 13.11) and oriented along the smaller-thickness direction (~10 nm). The lamellae grow to dimensions of approximately 1 μm by addition of segments to a "fast-growing" direction (inset *a* of Figure 13.11). The growth in the thickness direction is by a not-so-regular stacking of multiple lamellae, with some interlamellar links (inset *b* of Figure 13.11). The crystal thickness increases with the temperature of crystallization, or with subsequent annealing at a higher temperature.

During cooling from melt or concentration from solution, polymers form spherulites which are spherical aggregates of lamellae (Figure 13.11). These spherulites

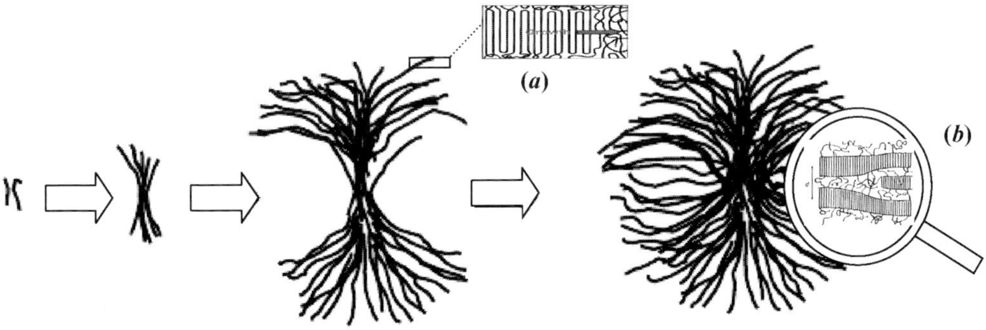

Fig. 13.11. Growth of a spherulite with lamellar growth. (a) Lamellae grow as chain segments (parallel to lamellar thickness direction) are added successively in the "fast growth" direction. (b) Lamellae are stacked somewhat irregularly, with some interlamellar links (adapted from G. Strobl, *The Physics of Polymers*, Springer, Berlin, 1996 and J. M. Schultz, *Makromol Chem. Makromol Symp.* **1988**, *15*, 339).

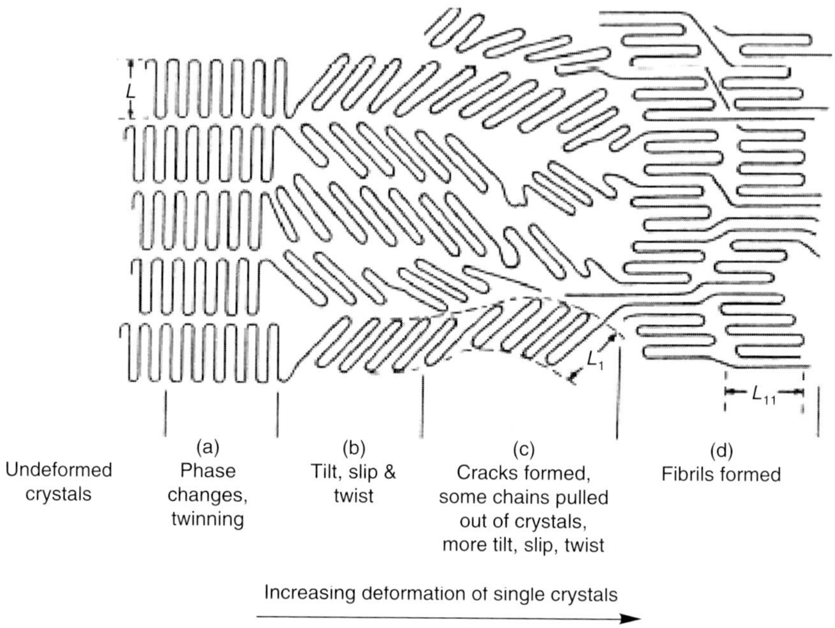

Fig. 13.12. Mechanical deformation (drawing) of spherulitic structures breaks the constituent lamellae into blocks and tie molecules. The blocks tilt, with the chains aligning along the load axis, and make up highly oriented microfibrils. (From A. Peterlin, in *Macromolecular Reviews*, Volume 1 (ISBN 0470-68245-0). Copyright **1967**, John Wiley. Used by permission from the publisher, John Wiley & Sons, Inc.).

grow radially from a central nucleus, to dimensions ranging from about 0.1 μm to a few millimeters, until they meet the neighboring spherulites. Polymer chains are oriented tangentially around each nucleus. On deformation of such material under stress (for example, during fiber drawing), the spherulitic structure is destroyed and crystalline rearrangements lead to polymer chain orientation coinciding with the stress direction (Figure 13.12). This imparts a high modulus and strength to the fibers in the axial direction.

13.1.8
Tailoring Polymer Properties through Modification, Additives, and Reinforcement

The properties of polymers are determined by the nature and composition of the structural units, as well as the molecular architecture. Polymers are most often made from the corresponding monomers. In comparison with developing new monomers and polymerization methods, a less expensive route to tailor-making of polymers is through blending, copolymerization, architecture control during polymerization, post-polymerization chemical modification, and additives. Detailed discussions on such possibilities are available in several reviews [37–42]. Here we will briefly highlight two recent developments in the areas of block copolymers and polymeric nanocomposites.

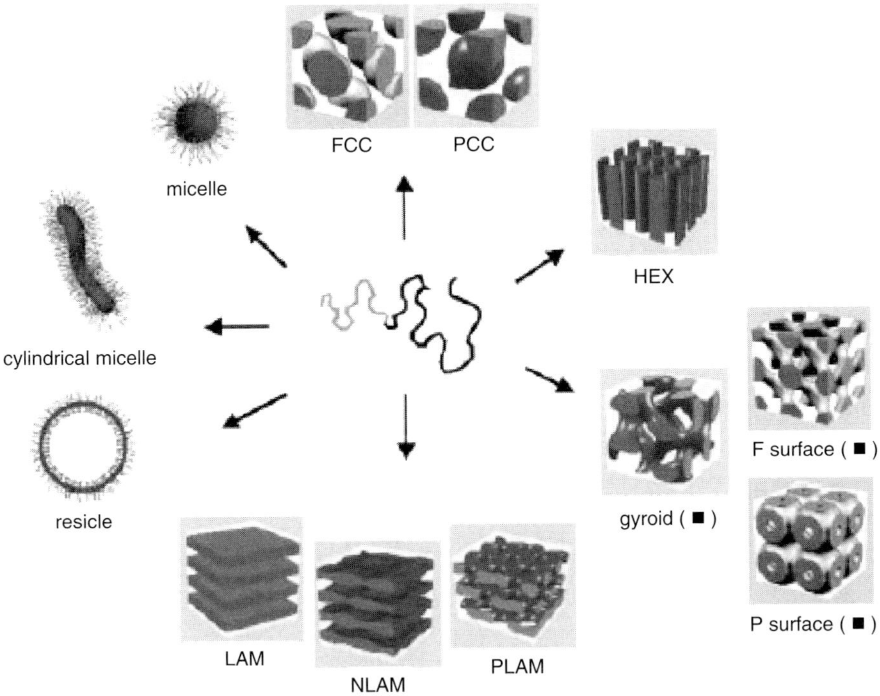

Fig. 13.13. Self-organization of block copolymers into various possible morphologies (from S. Forster, T. Plantenberg, *Angew. Chem. Int. Ed.* **2002**, *41*, 688).

13.1.8.1 New Morphologies through Block Copolymers

Traditionally, well controlled block copolymers were prepared by living anionic polymerization. However, the advent of controlled radical polymerization techniques [28, 29] has led to more facile synthesis of block copolymers in large quantities. The sizes of the blocks are controlled by the ratio of monomer to initiator, which thus also provides control over their relative sizes. Since the blocks are covalently bonded, their phase separation on the macroscopic scale is prevented. However, a repulsive interaction between the constituent blocks can lead to chain segregation, and thus self-organization into a variety of periodicities in the range of 10–20 nm [4, 5] (Figure 13.13). Self-assembling behavior of polymers leads to interesting characteristics that can be gainfully employed in improvement of mechanical properties [30]. Shell long ago commercialized PI–PS–PI triblock copolymers in which the glassy PS domains provided physical crosslinks, and thus elastomeric behavior, even in the absence of chemical crosslinks [31]. Other morphologies may find applications in catalysis, membranes, electro-optics, production of nanoparticles, absorption/release of drugs, and so on. For example, selective removal of one block from a bicontinuous morphology (gyroid, Figure 13.13) leads to a nanoporous structure with a network of interconnected nanochannels (diameter 20–30 nm). Metal plating of these channels can provide a very high internal surface area for catalytic activity [32]. Homopolymers and block copolymers have also been used to design novel materials through their capacity to organize organic and inorganic nanomaterials [33, 36].

13.1.8.2 Polymeric Nanocomposites

Conventionally, high-modulus, high-strength fibers are used as reinforcing materials in polymer composites, and fillers such as $CaCO_3$, talc, silica, glass and carbon black are used to reduce resin cost or modify electrical conductivity, dimensional stability, abrasion resistance and other properties. In recent years, enhancement of polymer properties by composite formation with nanoparticles has developed as an area of intensive research. Nanostructured materials such as nanoclay, graphite nanosheets, carbon nanotubes, and polyhedral silsesquioxane (POSS) are being explored as potential nanoscale reinforcing agents in polymeric matrices [43, 44]. In particular, single-wall carbon nanotubes (SWNTs) are unique among nanoparticles due to their large aspect ratio and their mechanical as well as electronic and conductive properties. For example, the modulus and strength of nanocomposites with 1 wt.% SWNT compare with those of conventional fiber composites with 10 wt.% carbon fibers [45]. Incorporation of 1 wt.% SWNT in PMMA is shown to result in a seven-fold increase in the strain to fracture [46]. In addition, several electrical, thermal, barrier, and crystallization properties are enhanced. For example, threshold concentrations for electrical conductivity of SWNTs in polymers have been reported to be as low as 0.1% [47]. Crystallization in PP, nylon-66, and PPT, and thermal stability in PET, PPT, PBT, and nylon-66 have been shown to be enhanced by incorporation of nanoparticles. One critical issue is how to achieve a good dispersion of the nanoparticles in the polymeric matrix. This is often difficult by melt-mixing in the high-viscosity polymers. Better dispersion of nanoparticles often requires either solvent processing, or mixing with monomers followed by in-situ polymerization [48–50].

13.2 Polymer Conformation and Related Properties

A large variety of polymer properties are attributable directly not to their chemical nature, but to their macromolecular constitution – that is, the long chains. In particular, several properties of polymers are related to, and can thus be estimated from, the conformational characteristics of the long chains.

13.2.1 The Chain Conformation

Polymer molecules mostly exist as random coils in solutions and melts. Their largest dimension is much smaller than the fully extended chains, but several times

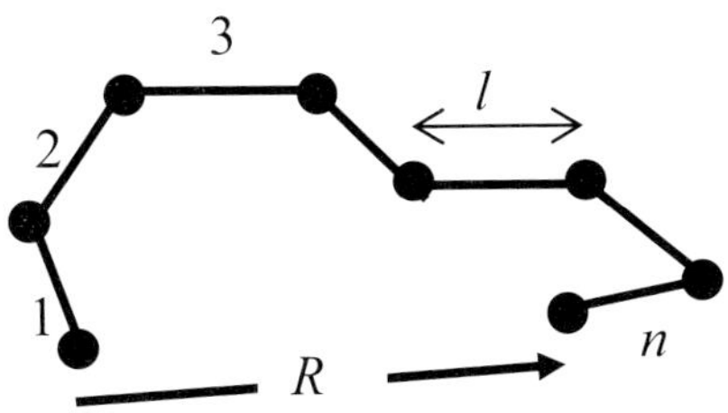

Fig. 13.14. End-to-end distance of a polymer coil made up of *n* segments, each of length *l*.

larger than the dimensions based on polymer density. The size of a polymer molecule is often described by its characteristic dimension: the root mean square end-to-end distance, $(\overline{r^2})^{1/2}$. Representing a polymer molecule as a chain made up of n bonds of length l (Figure 13.14), it can be shown that Eq. (5) applies, where $f(\theta) = \left(\dfrac{1 - \cos\theta}{1 + \cos\theta}\right)$ and $g(\varphi) = \left(\dfrac{1 + \cos\varphi}{1 - \cos\varphi}\right)$ are 1 for a hypothetical freely jointed chain, and are greater than 1 for real chains where the bond angle is fixed at θ and rotation about the bonds is restricted unequally at some value of φ between 0 and 2π (Figure 13.15).

$$\overline{r^2} = \langle R \cdot R \rangle = f(\theta)g(\varphi)\alpha' n l^2 \tag{5}$$

The factor α' is 1 for the unperturbed coil defined in this way, and is larger (or smaller) when a polymer molecule is expanded (or compressed) due to polymer–solvent interactions. Thus, for an idealized freely jointed chain in theta solvent, we have $\overline{r^2} = nl^2$, and the corresponding mean square radius of gyration $\overline{S^2} = nl^2/6$. When one accounts for the steric effects that prevent distant chain segments from overlapping (excluded volume effect), the dependence of $(\overline{r^2})^{1/2}$ is predicted to be $n^{3/5}$ (closer to experimental observation), rather than $n^{1/2}$ [Eq. (5)]. This polymer coil size is much larger than that based on polymer density, and hence markedly influences the viscosity behavior of polymers.

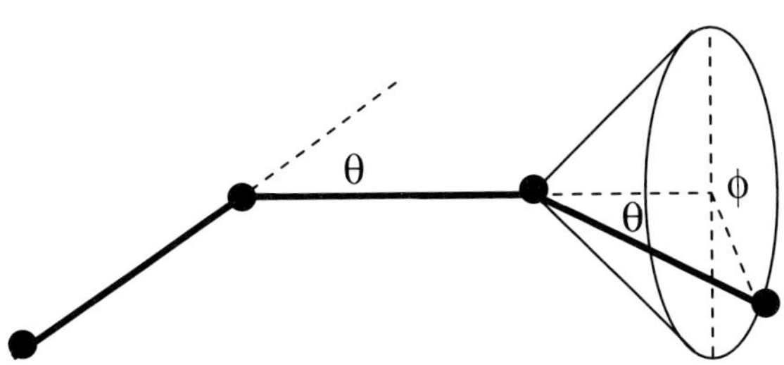

Fig. 13.15. Possible restriction on bond angles (θ) and on rotation about the bonds at some values of ϕ between 0 and 2π.

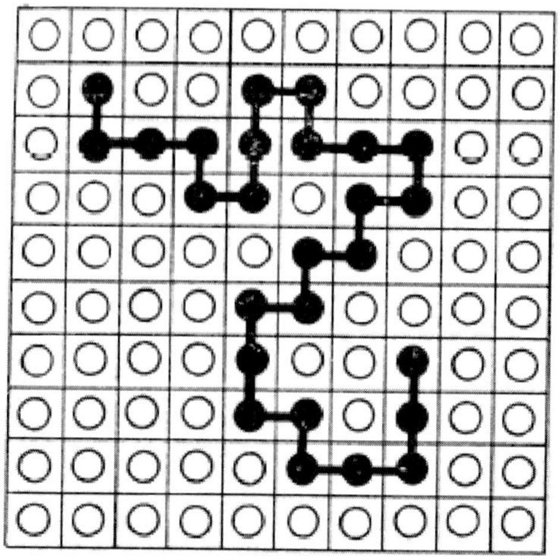

Fig. 13.16. Segments of a polymer chain in a liquid lattice. (Reprinted from Paul J. Flory, *Principles of Polymer Chemistry*. Copyright **1953** Cornell University and Copyright **1981** Paul J. Flory. Used by permission from the publisher, Cornell University Press).

13.2.2
Solubility of Polymers

Flory [51] and Huggins [52] were the first to examine the case of overlapping chains in solutions. They represented the chains of n segments by placing n beads on the interconnected cells of a lattice (Figure 13.16), and the other cells were occupied by solvent molecules. Statistical analysis of the possible conformations allows calculation of the entropy, while the enthalpic interactions between the polymer segments and solvent molecules are represented by the so-called χ parameter, which is temperature-dependent: $\chi < 0$ represents good solvents, and $\chi = \frac{1}{2}$ represents the so-called θ-condition where Gaussian statistics prevails. The corresponding free energy change for mixing per site is given by Eq. (6), where ϕ is the fraction of polymers made up of n segments each.

$$\frac{\Delta G_m}{k_B T} = \frac{\phi}{n} \ln \phi + (1 - \phi) \ln(1 - \phi) + \chi\phi(1 - \phi) \tag{6}$$

This allows the calculation of osmotic pressure by Eq. (7), where c is the mass concentration of the polymer of molecular weight M.

$$\pi = \frac{c R_g T}{M} \left[\ln(1 - \phi) + \left(1 - \frac{1}{n}\right)\phi + \chi\phi^2 \right] \tag{7}$$

Equating the derivative of the mixing free energy [Eq. (6)] with respect to ϕ allows

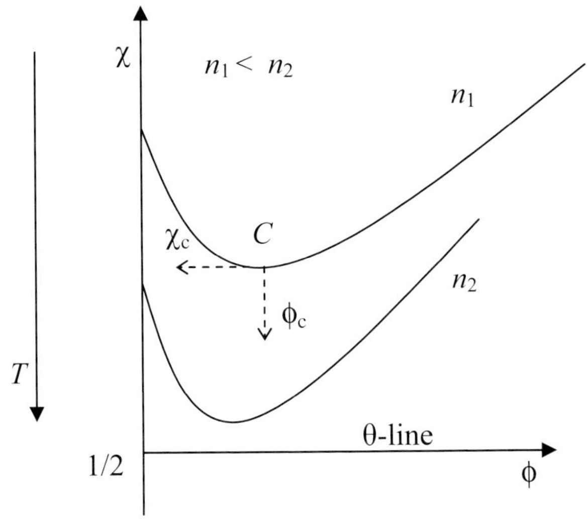

Fig. 13.17. Phase separation in polymer–solvent systems predicted by Flory–Huggins theory. The system exhibits UCST behavior: that is, phase separation at temperatures below a critical point *C*. Coexistence curves for two different molecular weights are shown.

the calculation of two ϕ values (for χ values larger than a critical value χ_c) corresponding to minimum energy, and thus the coexistence curve (Figure 13.17) that describes separation of polymer solutions into two phases at low temperatures ($\chi = A + (B/T) > \chi_c$) [53]. The molecular weight dependence of $\chi_c(= 0.5 + n^{-1/2})$ yields a downward shift of the curve in Figure 13.17, indicating an increasing tendency toward phase separation with increasing molecular weight of the polymers. Extension of the calculations to the case of polymer–polymer blends [51, 52] suggests incompatibility at χ values greater than $\chi_c = 2/n$ for two polymers with an equal number of segments. This strong tendency for phase separation of polymers when no specific polymer–polymer enthalpic interactions (for example, favorable hydrogen bonding) are present is attributed to the only small increase in entropy on mixing of high molecular weight polymers. A detailed discussion on blending of polymers and the resulting morphologies and properties is beyond the scope of this review; the interested reader is referred to more specialized texts [54–56].

13.2.3 Dilute Solution Zero-shear Viscosity

Rheology is the science of deformation and flow. Viscosity (η) is a measure of the stress (τ) required to make a fluid deform (flow) at a desired strain rate [Eq. (8), where $\dot{\gamma} = dv_x/dy$; Figure 13.18].

$$\tau_{yx} = \eta\dot{\gamma}_{yx} \tag{8}$$

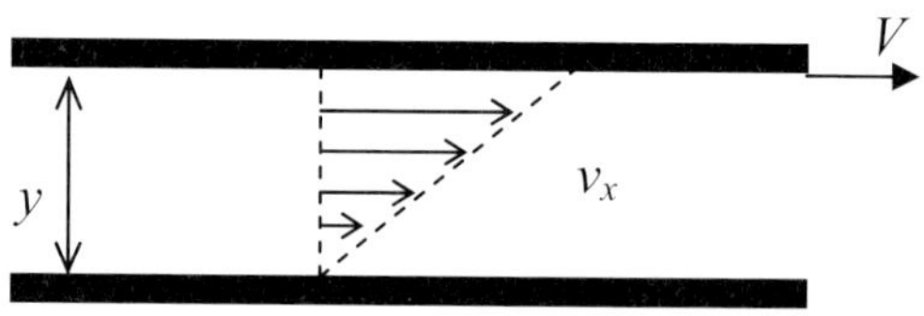

Fig. 13.18. Simple shear flow between two parallel plates separated by a distance y, and moving with relative velocity $V = y\dot{\gamma}_{yx}$.

A unique characteristic of polymers is that even a small concentration (c) of a high molecular weight polymer can significantly enhance the viscosity of a solution (η) as compared to the viscosity of the solvent (η_s). This is because the expanded polymer coils in slowly deforming dilute solutions behave as rigid spheres with a radius of the order of $(\overline{r^2})^{1/2}$, resulting in a large *polymer* volume fraction ϕ [Eq. (9)] with the corresponding solution viscosity (Einstein, for $\phi \ll 1$) given by Eq. (10).

$$\phi = \frac{4}{3}\pi(\overline{r^2})^{3/2}(cN_A/M) \tag{9}$$

$$\eta = \eta_s(1 + 2.5\phi) \tag{10}$$

A polymer's capacity to enhance solution viscosity is described in terms of its intrinsic viscosity ($[\eta]$):

$$[\eta] = \lim_{c\to 0} \frac{\eta - \eta_s}{c\eta_s} \tag{11}$$

Combining Eqs. (9)–(11) with Eq. 2.1, and realizing that $n \sim M$, one finds that $[\eta]$ has an $M^{1/2}$ dependence on polymer molecular weight. In reality, the factor α' [Eq. (5)] depends to a small extent on M, resulting in the Mark–Houwink relation [Eq. (12)].

$$[\eta] = KM^a \tag{12}$$

The constants K and a are obtained by fitting experimental data for a given polymer–solvent system. The intrinsic viscosity $[\eta]$ is conveniently measured by observing the time for flow of a certain volume of solution (and solvent) through a capillary, and thus provides an easy indication of the molecular weight from the tabulated Mark–Houwink constants [57].

13.2.3.1 Polymers as Dumbbells

More details of the rheological behavior of polymer solutions can be evaluated by representing a polymer molecule as a dumbbell: that is, two beads connected by spring (connecting vector $\underline{Q}$, Figure 13.19). Their contribution to the fluid

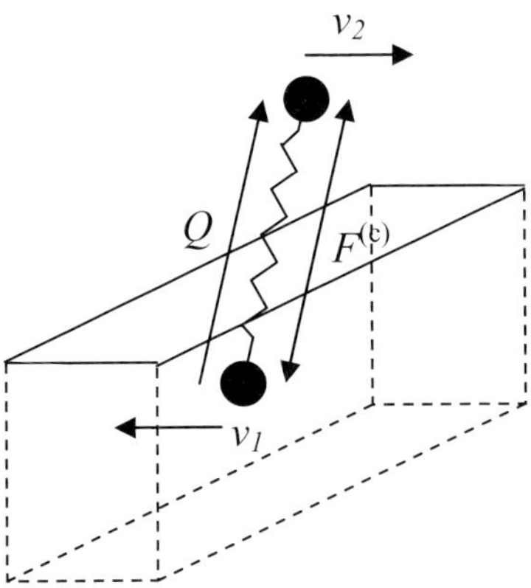

Fig. 13.19. Microscopic definition of polymer contribution to stress on a plane, as the stress exerted by the dumbbell bead above the plane on the bead below the plane.

stress [Eq. (10)] on any plane comes from the tensile or compressive spring force ($\underline{F}^{(c)} = H\underline{Q}$, H being the spring constant) transmitted through the springs [58] [Eq. (13), where $\bar{n}$ is the number concentration of the springs].

$$\tau - \eta_s \dot{\gamma} = (\bar{n} k_B T \delta + \bar{n} \langle Q F^{(c)} \rangle) \tag{13}$$

For example, a linearly elastic dumbbell model representation (Figure 13.19) of polymer molecules predicts Eq. (14) to apply, where $\lambda = \xi/4H$ is a time constant and ξ is the bead friction factor.

$$\eta = \eta_s + \bar{n} k_B T \lambda \tag{14}$$

When it is desired to represent a real polymer as a dumbbell, λ can be estimated from the intrinsic viscosity measurement by means of Eq. (15).

$$\lambda = \frac{[\eta] \eta_s M}{N_A k_B T} \tag{15}$$

13.2.3.2 **Polymers as Chains of Beads and Springs**

A more realistic representation of polymer chains is the Rouse model [59], which considers a polymer molecule to be a linear chain of N *free-draining* beads interconnected by springs (each of time constant $\lambda_H = \xi/4H$), and predicts Eq. (16) to apply.

$$[\eta] = \frac{N_A k_B T \lambda_H}{\eta_s M} \frac{N^2 - 1}{3} \tag{16}$$

This implies that $[\eta]$ has a linear dependence on molecular weight. This is not consistent with the experimental observations [Eq. (12)]. When hydrodynamic interaction and excluded volume effects between chain segments are considered, the corresponding Zimm model [60] correctly predicts the $\sim M^{1/2}$ dependence [Eq. (17)].

$$[\eta] \sim \frac{N_A}{M} N^{3/2} l^3 \tag{17}$$

13.2.4 Viscosity of Concentrated Solutions and Melts

The models discussed in Section 13.2.3 for dilute polymer solutions all predict $[\eta]$ to be independent of concentration; that is, they predict linear dependence of viscosity on concentration. When the concentration of polymer in solution is higher than the overlap concentration ($c^* \sim 1/[\eta] \sim M^{-1/2}$), the random coils get entangled with each other ($\phi \geq 1$). This is responsible for a faster than linear increase in viscosity with concentration (Figure 13.20), which often follows the scaling relation of Eq. (18) [61]. That is, if η/η_s is plotted against c/c^* or $[\eta]c$, then the curves of various molecular weights can be superimposed.

$$\frac{\eta}{\eta_s} = F\left(\frac{c}{c^*}\right) = G([\eta]c) \tag{18}$$

While accounting for the concentration effects by analyzing a single chain placed in an "effective medium" made up of the other chains, the dependence expressed

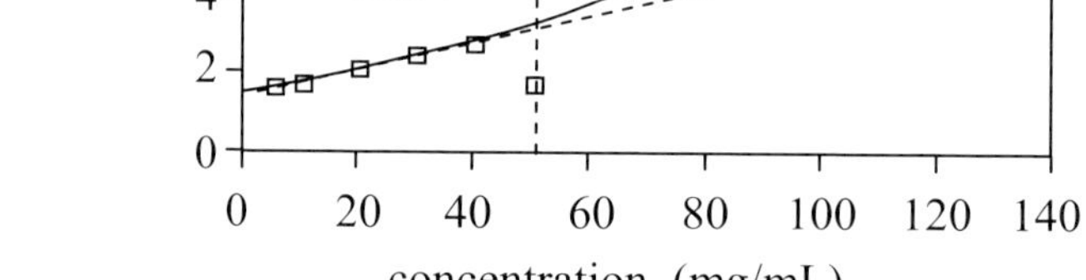

Fig. 13.20. Concentration dependence of viscosity of synapse (molecular weight 18.55 kDa) solution is linear at $c < c^*$, $c^* \sim 50$ mg mL^{-1}. Redrawn from Malvern application note no. MRK511-01 (http://www.particular.ie/m%20wt%20protein%20nano%20zs.pdf).

in Eq. (19) was obtained [62], indicating Rouse-like [Eq. (16)] linear dependence on N at intermediate concentrations (Figure 13.20).

$$\eta \sim \eta_s \left(1 + \frac{c^2 l^6}{24} N\right) \tag{19}$$

The transition from the Zimm regime [Eq. (17), $[\eta] \sim N^{1/2}$] at dilute concentrations to Rouse-like behavior [Eq. (16), $[\eta] \sim N$] at the intermediate concentrations is related to the hydrodynamic screening at high concentrations: the hydrodynamic interactions between segments become negligible because the local velocity around the beads falls quickly.

At even higher molecular weights and concentrations, the viscosity in concentrated solutions and melts goes from a linear to a 3.4 power dependence on M [Eq. (20); Figure 13.21].

$$\eta \sim c^{4-5} M^{3.4} \tag{20}$$

This behavior is described by confinement of a polymer molecule to its own tube whose walls are defined by the network of surrounding entangled chains (Figure 13.22) [63, 65]. Hence, chain motion (including stretching and relaxation) is permitted only curvilinearly to the tube (reptation), while motion of the chain perpendicular to the tube is hindered. During flow, the relaxation of chain stretching along the tube axis is instantaneous, while a nonrandom orientation of tube segments contributes to the stress.

13.2.5 Nonlinear Polymers

Modern synthesis methods have allowed preparation of polymers with a great variety of branching architectures, such as random, comb, uniform and nonuniform star, brush, and dendritic [66, 67]. It is understandable that the conformations and hence several properties of these structures can vary remarkably among linear polymers of similar compositions and molecular weights. The compression of the branched molecule compared to linear molecules of same molecular weight can be quantified in terms of the factors defined in Eqs. (21) and (22), where the subscripts b and l represent the branched and the linear polymers.

$$g = (\overline{S^2})_b / (\overline{S^2})_l \tag{21}$$

$$g' = [\eta]_b / [\eta]_l \tag{22}$$

The factors g and g' can often be related, depending on the branching architecture. For example, for star polymers with p equal arms [68, 69], the relationship is given by Eq. (23).

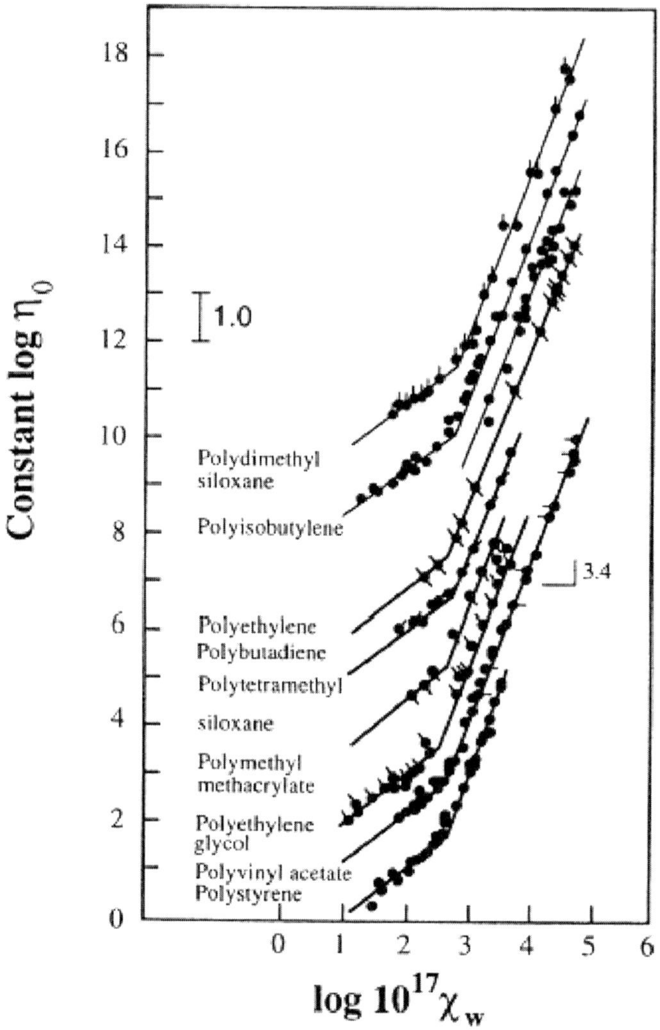

Fig. 13.21. Variation of polymer melt viscosity with molecular weight. (from C. W. Macosko, *Rheology: Principles, Measurements and Applications*, Copyright **1994** VCH. Used by permission from the publisher, John Wiley & Sons, Inc.).

$$g = \frac{3}{p} - \frac{2}{p^2} = [0.2624 + (1 - 0.2624)(g')^{1.088}](g')^{0.6087} \tag{23}$$

Such characteristics of several other branched architectures have been examined [70]. The molecular compression behavior by branching allows high loading of the polymeric additives without the excessive viscosity enhancement associated with polymers. As we shall discuss in Section 13.3.6, the rheological effects of

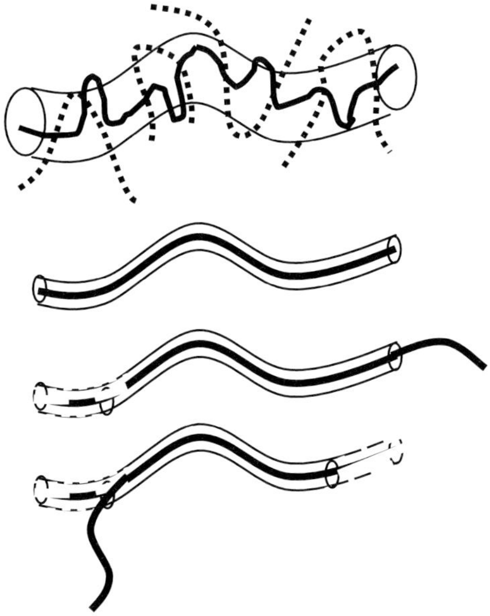

Fig. 13.22. Confinement of a polymer chain's motion in its own tube defined by a fixed network of surrounding chains. Back and forth motion in the tube leads to tube renewal and thus to the loss of the original tube orientation and stress (adapted from M. Doi, S. F. Edwards, *The Theory of Polymer Dynamics*, Clarendon, Oxford, **1986**).

branching are more substantial and are put to use in strongly flowing polymer melt systems.

13.2.6
Rigid Rod-like Polymers

Some polypeptides and para-linked synthetic aromatic polymers take up elongated forms, forming rodlike structures (rather than flexible coils). Their solution viscosity displays a strong dependence on the polymer molecular weight according to Eq. (24), where L and R are the length ($\sim M$) and the radius, and ρ_b is the bulk density of the rodlike molecules [64].

$$[\eta] = \frac{2(L/R)^2}{45\rho_b(\ln(L/2R) - 0.8)} \tag{24}$$

At higher concentrations (number concentration $> 1/L^3$), the motion of rodlike polymers is highly hindered in the directions perpendicular to the rod axis, and the viscosity increases spectacularly with mass concentration (ρ) and molecular weight [Eq. (25)] [64].

$$\eta \propto \rho^3 M^6/\ln(N) \tag{25}$$

Due to this effect, the increase in molecular weight during their step-growth polymerization leads to a reduction in the reaction rate of such polymers [71], and provides the possibility of controlling the kinetics and molecular weight distribution by orienting flow [72]. Spontaneous liquid-crystalline ordering of these polymers at still higher concentrations allows their processing into very strong fibers [73].

13.3
Polymer Rheology

In Section 13.2 we discussed the equilibrium conformations of polymers and their influence on the polymeric behavior. We shall now discuss the rheological properties of polymers associated with their very special characteristic that their conformation changes under imposed flows.

13.3.1
The Viscous Response: Shear Thinning

One important characteristic of polymers is their shear thinning behavior: the viscosity decreases with increasing rate of deformation (Figure 13.23). This behavior at $\dot{\gamma} \sim \lambda^{-1}$ is attributed to the partial disentanglement of the chains as they get extended in the flow direction. Since the relaxation time increases with molecular weight [for example, $\lambda \sim M^{1.5}$ at low concentrations; Eqs. (12), (15)] and with polymer concentration, shear thinning begins at lower shear rates for polymers with

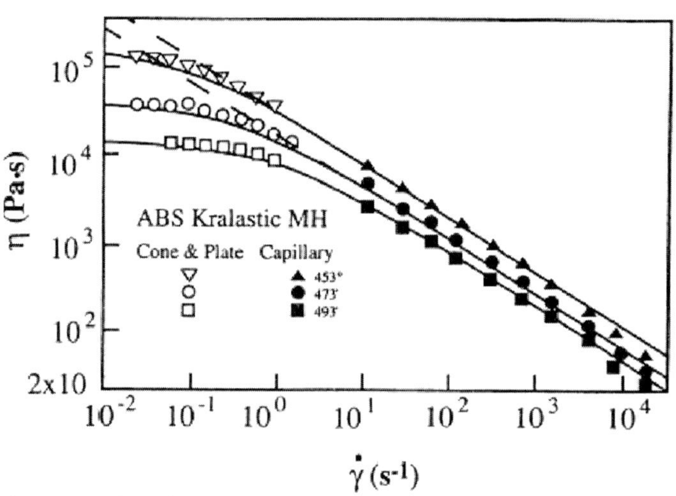

Fig. 13.23. Shear thinning behavior of an ABS polymer melt at three different temperatures. Broken lines are power law fits and solid lines represent the Cross model. (from C. W. Macosko, *Rheology: Principles, Measurements and Applications*, Copyright **1994** VCH. Used by permission from the publisher, John Wiley & Sons, Inc.).

higher molecular weights and at higher concentrations. The shear thinning behavior is responsible for the processibility of many polymer melts. For example, high strain rates of extrusion are employed lest they should undergo stress and thermal degradation, and the circumferential shear (due to screw rotation) during extrusion reduces the viscosity sufficiently to enable axial flow (due to the pressure developed) [74].

Empirically, the shear thinning behavior is most conveniently represented by the power law relationship expressed by Eq. (26) [75] (Figure 13.23) or, more appropriately by a four parameter Cross model [Eq. (27), where η_o and η_∞ correspond to the low (zero) and the high shear plateaus, and λ and n (< 1) correspond to the location of transition and the slope on a log–log plot (Figure 13.23)] [76].

$$\eta = K(\dot{\gamma})^{n-1} \tag{26}$$

$$\frac{\eta - \eta_\infty}{\eta_o - \eta_\infty} = [1 + (\lambda\dot{\gamma})^2]^{(n-1)/2} \tag{27}$$

It is interesting to examine the ability of the simple molecular models (Section 13.2.2) to represent the shear thinning phenomena. The linearly elastic dumbbell and the Rouse models represent chain extension (unlimited at increasing $\dot{\gamma}$), yet these predict the viscosity to be independent of shear rate. The finitely extensible nonlinear elastic (FENE) dumbbell model considers the connecting spring to stiffen infinitely as the extension approaches the maximum possible value Q_o. This permits prediction of the shear thinning behavior starting at ($\lambda\dot{\gamma} \sim 1$) [58]. Similarly, when one accounts for hydrodynamic interaction, the linearly elastic dumbbell model also predicts shear thinning according to Eq. (28).

$$\eta - \eta_s = \frac{nk_B T\lambda(1 - \xi)}{1 + 2\xi(1 - \xi/2)(\lambda\dot{\gamma})^2} \tag{28}$$

13.3.2
Normal Stresses during Shear Flow

In contrast to simple fluids under shear that generate stresses parallel to the direction of shear, polymeric fluids also generate *normal* stresses perpendicular to the flow direction [Eqs. (29), (30), where N_1 and N_2 are called the first and the second normal stress difference, respectively, and are dependent on the shear rate].

N_1 is positive, corresponding to a tendency of the polymeric fluid to push apart the two plates between which the fluid is sheared (Figure 13.24). Such a tendency of polymeric fluids is manifested in interesting phenomena such as die-swell and rod-climbing (Figure 13.25).

$$\tau_{xx} - \tau_{yy} = N_1 \tag{29}$$

$$\tau_{xx} - \tau_{zz} = N_2 \tag{30}$$

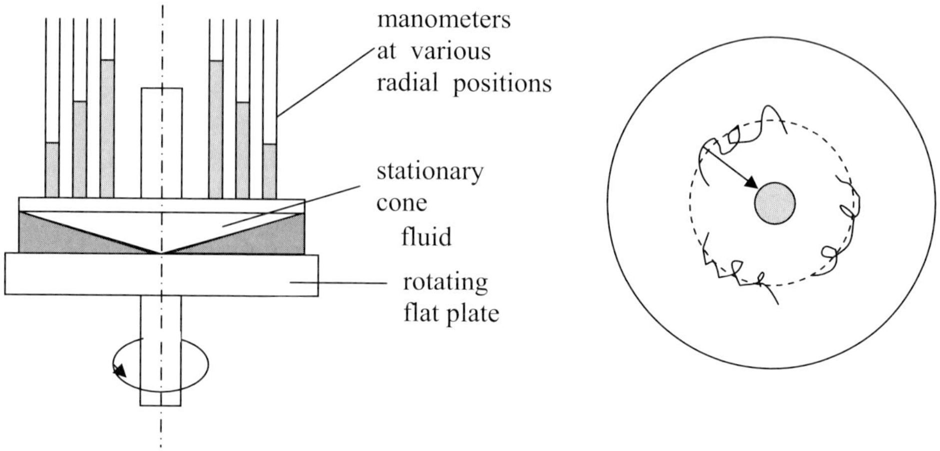

Fig. 13.24. When a polymeric fluid is sheared between a cone and a plate, stresses are generated that are also normal to the flow direction, as is evident from the monomeric fluid levels.

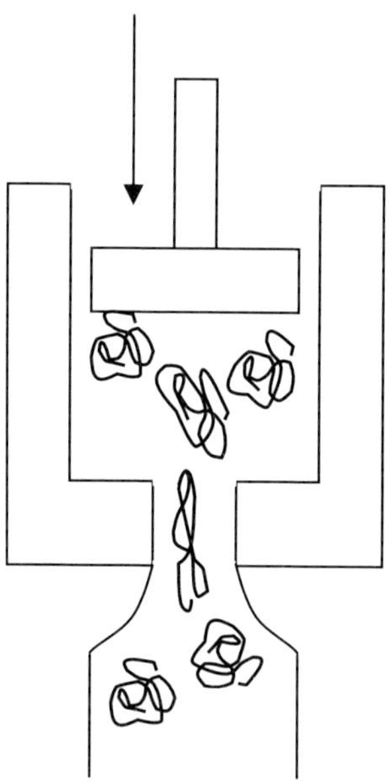

Fig. 13.25. Normal stresses in shearing polymeric fluids are responsible for their swelling on exit from a die.

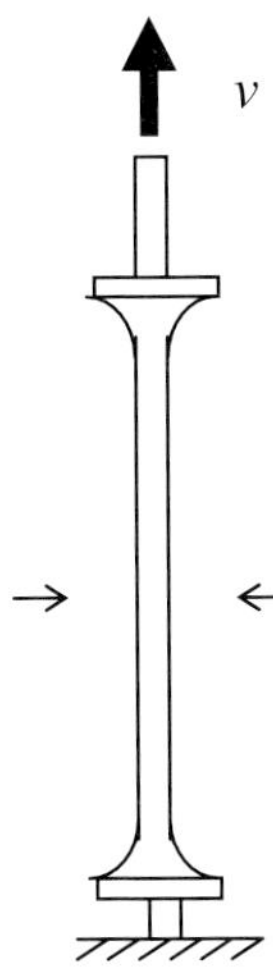

Fig. 13.26. Uniaxial extensional flow generated between a fixed plate and a plate moving at desired velocity (v), while the entire geometry is held in a constant-temperature bath. Measurement of the sample dimension and the required tension force allows calculation of the elongational viscosity.

13.3.3
Extensional Thickening

A flow that converges or diverges is called extensional or compressional (Figure 13.26). For example, stretching of a filament creates extensional flow with the corresponding velocity gradient ($\dot{\gamma}_{xx} = dv_x/dx$) resulting in tensile stress [Eq. (31), where η_e is called the extensional viscosity].

$$\tau_{xx} - \tau_{yy} = \eta_e \dot{\gamma}_{xx} \tag{31}$$

With an increase in flow time or strain at low $\dot{\gamma}_{xx}$, η_e increases to its asymptotic "steady-state" value given by Eq. (32).

$$\eta_e = 3\eta \tag{32}$$

However, many polymers show strain hardening, that is, an increase in η_e beyond this asymptotic value at high strain rates (Figure 13.27). Such behavior of high molecular weight polymers is responsible for stabilizing stretching processes involved in the production of films and fibers.

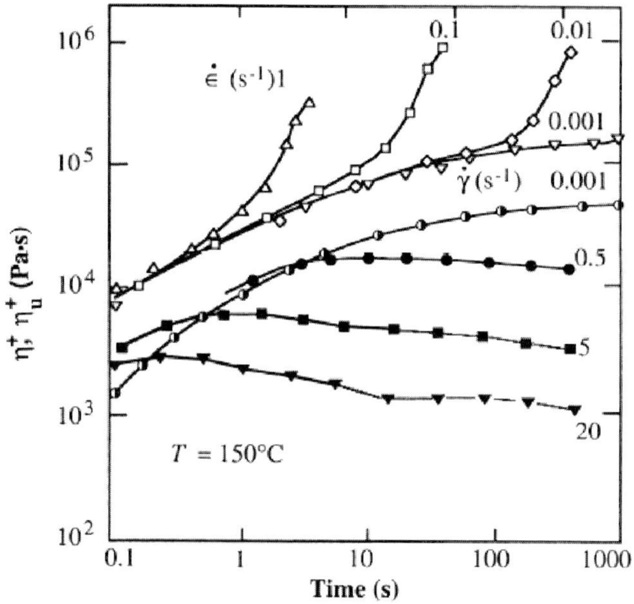

Fig. 13.27. Uniextensional extensional viscosity (η_u^+, open symbols) and shear viscosity (η^+, closed or half-filled symbols) of low-density polyethylene (LDPE) as functions of flow time after inception of the steady extensional rate ($\dot{\varepsilon}$) or shear rate ($\dot{\gamma}$), respectively. (from C. W. Macosko, *Rheology: Principles, Measurements and Applications*, Copyright **1994** VCH. Used by permission from the publisher, John Wiley & Sons, Inc.).

13.3.4 The Elastic Response

13.3.4.1 Ideal Elastic Response

Polymeric materials often tend to behave as springs (that is, they have a tendency to retract on stretching), thus displaying some degree of elasticity. An ideal elastic material responds instantaneously to application or removal of stress, with the strain (γ) being proportional to the stress (Hookean), independently of the strain rate [Eq. (33)]. Here, the constant G is the modulus of the elastic material.

$$\tau = G\gamma \tag{33}$$

13.3.4.2 Rubberlike Elasticity

Elastic polymers contain a crosslinked network to restrain the gross mobility of their chains. When the chain segments between crosslinks are deformed affinely in a macroscopic sample due to application of an external force, the resulting decrease in chain entropy brings in a springlike retractive force (with spring constant $3k_BT/\overline{r^2}$) [77]. Generalization to a three-dimensional deformation to account for cross-sectional area changes corresponding to an axial strain of $(\alpha - 1)$ results in

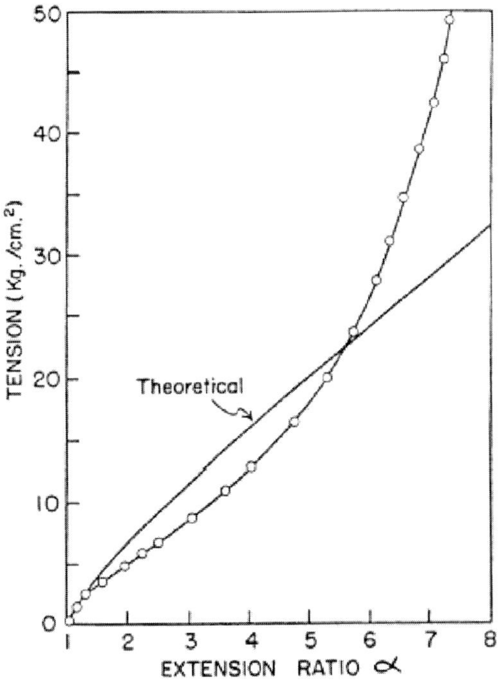

Fig. 13.28. Stress–extension curve for a sample of natural rubber compared with the ideal rubber prediction [Eq. (35)] (Reprinted from Paul J. Flory, *Principles of Polymer Chemistry*. Copyright 1953 Cornell University and Copyright 1981 Paul J. Flory. Used by permission from the publisher, Cornell University Press).

the retractive force per unit *deformed* cross-sectional area according to Eq. (34), where $G = N_e k_B T$, and N_e is the number of crosslinked chains per unit volume.

$$\tau_{xx} = G\left(\alpha^2 - \frac{1}{\alpha}\right) \tag{34}$$

The α^2 dependence indicates that the chains become progressively harder to stretch. As seen in Figure 13.28, this simple model based on (infinitely extensible) Gaussian chains fails to explain the extreme stiffening behavior at very high strains near the maximum possible chain extension.

13.3.5
The Viscoelastic Response

Elastic behavior in polymers is rubberlike and thus not Hookean, and viscous behavior is shear thinning and thus not Newtonian. Further, real polymers display characteristics that are a combination of such elastic and viscous responses. This is best illustrated by the Maxwell model [75, 78], which considers the viscoelastic

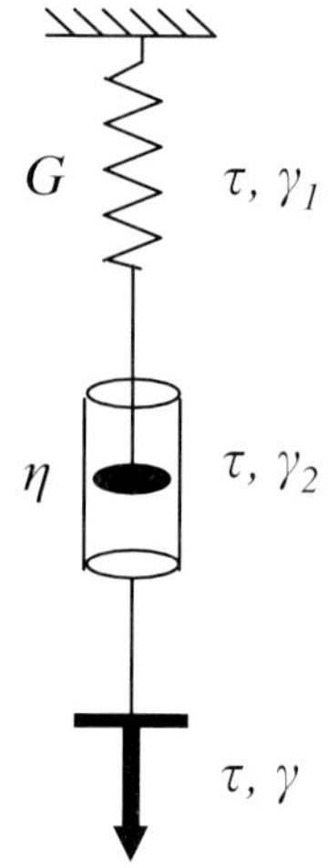

Fig. 13.29. The Maxwell model for a viscoelastic material considers an elastic spring and a viscous dashpot connected in series.

behavior to be a series combination of the Hookean elasticity as represented by a linear spring and the Newtonian viscous behavior as represented by a dashpot (Figure 13.29). The stress (τ) being the same for the spring and the dashpot, and the total strain (γ) being the sum of the strains of the spring and the dashpot, one obtains Eq. (35) or, substituting $\lambda = \eta/G$, Eq. (36).

$$\frac{d\gamma}{dt} = \frac{d\gamma_1}{dt} + \frac{d\gamma_2}{dt} = \frac{1}{G}\frac{d\tau}{dt} + \frac{\tau}{\eta} \tag{35}$$

$$\tau + \lambda\frac{d\tau}{dt} = \eta\frac{d\gamma}{dt} \tag{36}$$

In Figure 13.30, we compare the creep and stress relaxation responses of such a material with the responses of an ideal elastic solid and a viscous Newtonian fluid. It is easily realized from the creep behavior that a small value of η and large value of G (and thus a small value of λ) correspond to the viscoelastic response approximating to the viscous response. Further, the creep response of the viscoelastic material at short times (for example, instantaneously) closely approximates to the elastic behavior, because viscous flow occurs only over time. At long times, the accumulated viscous response represents the overall deformation. Similarly, the stress relaxation behavior of the viscoelastic material suggests that λ is a measure of time taken for stress to relax to $1/e$ of the original stress, and hence this is called the relaxation time of the material. The effect of material characteristics λ and process time (t) are thus easily put together in the dimensionless parameter called the Deborah number De, as in Eq. (37).

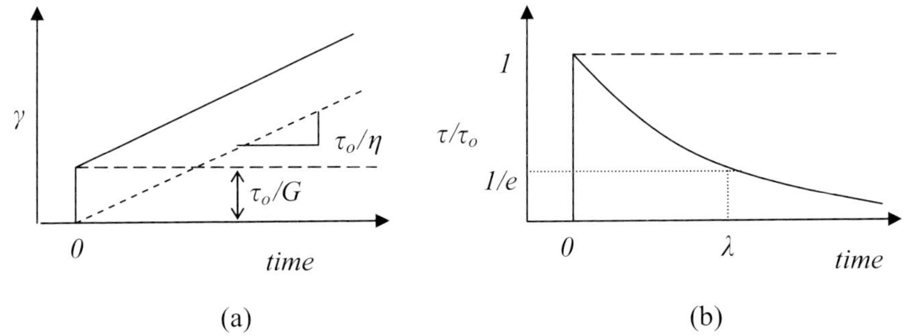

Fig. 13.30. (a) Creep behavior (strain response to an applied step-stress τ_o) and (b) stress relaxation behavior (stress response to step-strain γ_o). The behavior of an elastic solid (broken lines with long dashes), a viscous liquid (broken line with short dashes, and dotted line), and a Maxwell viscoelastic material (solid line) are compared.

$$De = \lambda/t \tag{37}$$

$De > 1$ and $De < 1$ correspond to predominantly elastic and viscous responses respectively. De determines the level of shear stress that can relax during the melt processing, and hence the residual chain orientation, morphology, and properties in the solid-state article fabricated by melt processing.

The Maxwell model [Eq. (36)] can also be expressed in the integral form of Eq. (38) [75]. That is, the stress can be looked upon as the sum of every incremental strain $\dot{\gamma}(t')\Delta t'$ multiplied by the corresponding exponentially decreasing modulus $Ge^{-(t-t')/\lambda}$.

$$\tau = \int_{-\infty}^{t} Ge^{-(t-t')/\lambda}\dot{\gamma}(t')\,dt' \tag{38}$$

13.3.5.1 Linear Viscoelasticity in Dynamic Oscillatory Flow

When a sinusoidally varying strain (Figure 13.31) of frequency ω and a small amplitude γ_{max} is imposed on a fluid according to Eqs. (39) and (40), the corresponding stress also oscillates [Eq. (41)], albeit with a phase delay (δ, Figure 13.31) after the imposed strain rate that can be decomposed into two strain-dependent terms [Eq. (42)], of which the first is the elastic contribution in phase with the strain [Eq. (39)], and the second is the viscous contribution in phase with the strain rate [Eq. (40)].

$$\gamma_{yx} = \gamma_{max}\sin(\omega t) \tag{39}$$

$$\dot{\gamma}_{yx} = \dot{\gamma}_{max}\cos(\omega t) \tag{40}$$

$$\tau_{yx} = \tau_{max}\cos(\omega t - \delta) \tag{41}$$

$$\tau_{yx} = G'\gamma_{max}\sin(\omega t) + G''\gamma_{max}\cos(\omega t) \tag{42}$$

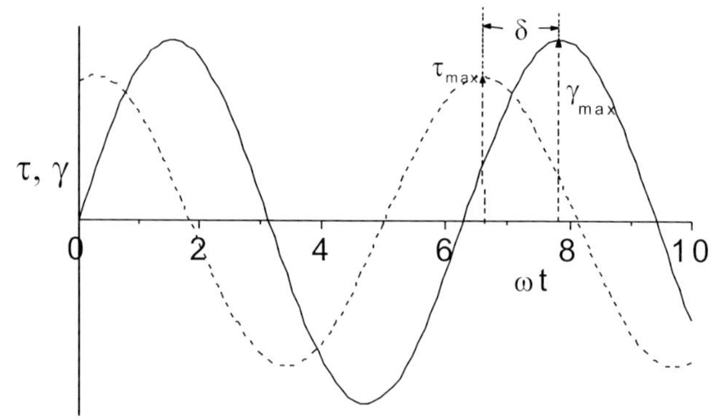

Fig. 13.31. Imposed sinusoidal strain (solid line), and the corresponding sinusoidal stress (broken line) response of a viscoelastic material.

$G' = \tau_{\max} \cos \delta / \gamma_{\max}$ and $G'' = \tau_{\max} \sin \delta / \gamma_{\max}$ are called the storage modulus and the loss modulus, corresponding to the elastic and viscous contributions, respectively. For the Maxwell model [Eq. (36)], their dependence on frequency is given by $G' = G\dfrac{(\omega\lambda)^2}{1+(\omega\lambda)^2}$ and $G'' = G\dfrac{\omega\lambda}{1+(\omega\lambda)^2}$. With an increase in frequency, a polymer behaves increasingly like a solid (storage/elastic), $G' \to G$: that is, the storage modulus equates to the elastic modulus. This single-mode Maxwell model correctly represents the observed $G' \sim \omega^2$ and $G'' \sim \omega$ dependence in the terminal zone (long time, small ω), and a crossover at $\omega = 1/\lambda$ (Figure 13.32). However, real polymers display not a single time constant and a single modulus, but a spectrum

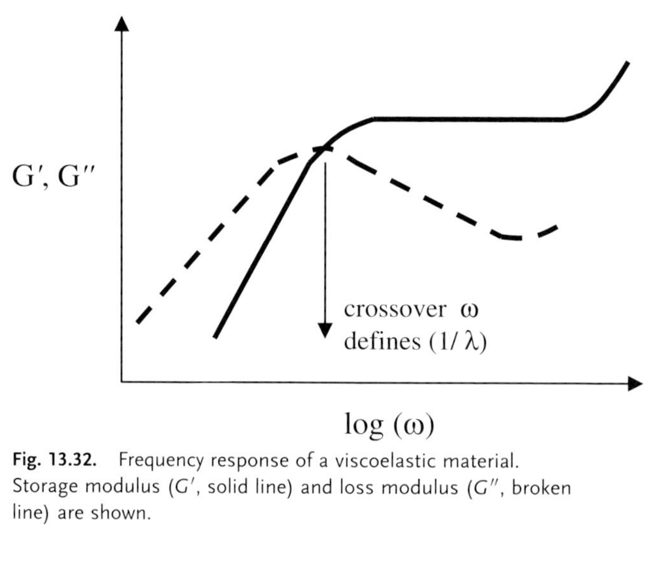

Fig. 13.32. Frequency response of a viscoelastic material. Storage modulus (G', solid line) and loss modulus (G'', broken line) are shown.

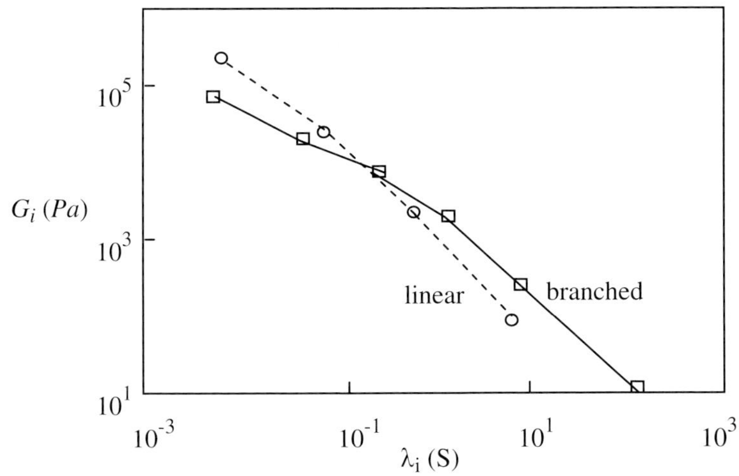

Fig. 13.33. Relaxation spectra for linear and branched polyethylene (PE). The spectrum of the branched PE includes a relaxation time that is 10 times longer than the longest relaxation time of linear PE. This is consistent with the long relaxation times predicted for branched PE. (redrawn from S. H. Wasserman in *Metallocene Catalyzed Polymers*, Eds. G. M. Benedikt, B. L. Goodall, BF Goodrich, Brecksville, **1998**).

of relaxation times (λ_i) and modulus parameters (G_i) corresponding to different modes of relaxation. The Maxwell model can then be generalized by considering the stress to be a sum (superposition) of the contributions of each mode of relaxation, as represented by Eq. (43), where G_i and λ_i can be found by carrying out a multiparameter fit of the stress response to step-strain or dynamic strain.

$$\tau = \int_{-\infty}^{t} \sum_i (G_i e^{-(t-t')/\lambda_i}) \dot{\gamma}(t')\, dt' \qquad (43)$$

These values of G_i and λ_i constitute the relaxation spectrum of a polymer, and are useful in correlating the viscoelastic behavior of polymers with their molecular characteristics, such as long-chain branching in polyethylene (Figure 13.33) [79].

13.3.6 Influence of Polymer Branching Architecture in Bulk Polymers

Polymer chemistry has played a very important role in providing polymeric material with pre-designed molecular architectures. In particular, the branching architecture has been seen to greatly influence the rheological behavior of polymer melts as the reptation of the arms is severely suppressed. Relaxation of the arms can take place only by arm withdrawal, which is exponentially slowed with increasing arm length [63]. It is only after sufficient movement of the arms that the motion of branching points (otherwise hindered by the network around them) can slowly take place (Figure 13.34) [80]. Some spectacular effects of these entanglements in long-chain branched industrial polyethylene (LDPE) are the following: very long relaxation times (Figure 13.33), greater extensional thickening (Figure 13.27), and greater shear thinning (Figure 13.35) [81, 82]. The extensional thickening is critical for imparting the characteristics necessary for film blowing, blow-molding and foaming applications of polyolefins, and the increased shear thinning allows easier extrusion at high shear rates. Efforts have been made to enhance processibility of polyesters by introducing branching, and lower melt and solution viscosities are obtained. However, the solid-state properties were found to be poor [83].

Hyperbranched polymers are compact, highly branched, three-dimensional macromolecules with a high density of end-groups [84]. Their compact structure results in inherently low viscosity, as evidenced by the value of the Mark–Houwink constant α being less than 0.5. Hyperbranched aliphatic polyesters have been applied as toughners for epoxy thermosets without a significant increase in viscosity [85]. Hyperbranched polymers can be used as rheology control agents during the processing of polymers, for example to obtain a dramatic decrease in viscosity [86], as lubricants [87], or for sharkskin elimination [88].

Dendrimers are a relatively new class of highly and regularly branched polymers with well defined numbers of branching layers and a low molecular weight between branches. Dendritically branched polymers may have somewhat higher molecular weights, yet lower intrinsic viscosity, than corresponding regular star-shaped polymers, due to their dense internal structure [89]. They can display high elasticity [90], and this makes them potentially interesting rheology modifiers. The large number of peripheral terminal groups contribute to their solubilities, and offer potential for loading high functionalities when utilized as additives, for example for solubility enhancement, fiber dyeability, crystallization suppression, and suchlike [91, 92].

Dense grafting of side chains onto linear backbones, and homopolymerization of macromonomers, are both used to synthesize macromolecular brushes. Steric repulsion of the side chains forces the main chain into an extended wormlike conformation, resulting in liquid-crystalline phases, and lower dynamic shear moduli than linear flexible coils in concentrated solutions [93, 94]. Densely grafted polymeric brushes on sliding surfaces have been found to reduce friction, and therefore have potential for providing biolubrication for artificial implants [95].

13.3.7
Polymers as Rheology Modifiers

Due to their large coil size and hence their great influence on the shear and elongational behavior of solvents, polymers are used as rheology-modifying additives in applications ranging from fuels, lubricants, coatings, and sprays to enhanced oil recovery and turbulent drag reduction [96]. For example, macromolecular coils of poly(alkyl methacrylate) and ethylene–propylene copolymers that expand with temperature are used to offset the solvent viscosity decrease upon heating (Figure

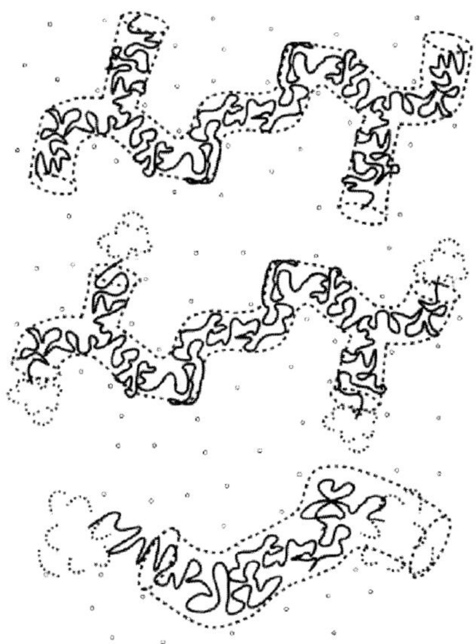

Fig. 13.34. Relaxation in a branched H-polymer melt involves retraction of arms by fluctuations, followed by crossbar reptation as branch points move along the dilated tube (Reprinted with permission from T. C. B. McLeish et al., *Macromolecules*, Volume 32, Page 6734. Copyright **1999**).

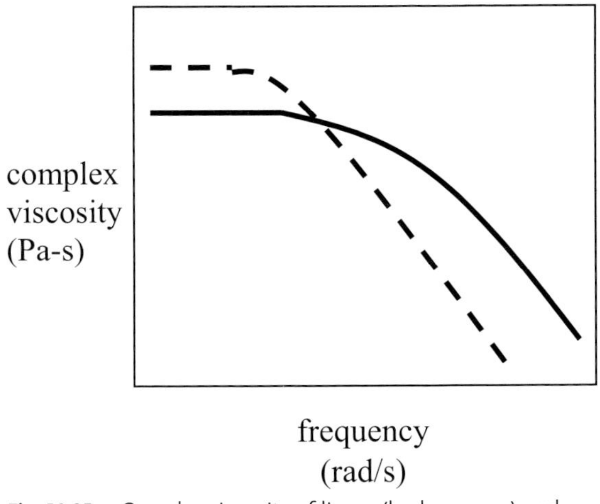

Fig. 13.35. Complex viscosity of linear (broken curve) and long-chain branched (solid curve) polymers. Long-chain branching results in stronger shear thinning properties.

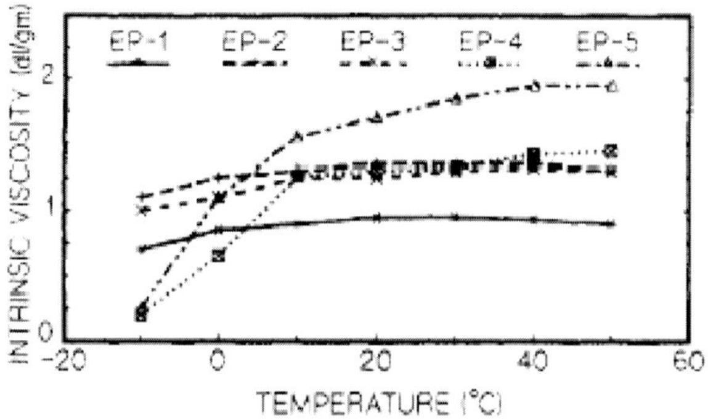

Fig. 13.36. Temperature effect on intrinsic viscosity of ethylene–propylene copolymers in toluene. The calculated hydrodynamic volume of EP-5 ($M_w = 322\,000$, ethylene 80 mol%) increases from 5.6×10^{-18} to 43.7×10^{-18} cm^3/molecule as the temperature increases from -10°C to 50°C. (Reprinted with permission from A. Sen, I. D. Rubin, *Macromolecules*, Volume 23, Page 2519. Copyright **1990**, American Chemical Society).

13.36), resulting in flatter viscosity–temperature curves [97]. This kind of application in automotive lubricating oils is termed viscosity index improvement [96]. Extensional thickening by polymers is responsible for the use of polymeric additives as anti-misting agents to resist fluid breakup into minute droplets [98, 99]. Another application of polymers is as drag-reducing agents where a few parts per million of polymeric additives can locally enhance the elongational viscosity in order to dampen turbulence, thereby enhancing fluid flow by as much as 100%. At the high deformation rates during such applications, polymers can undergo chain scission and thus lose their efficiency in rheology modification [100]. Enhancement of shear stability against such chain scission by way of control of chain architecture is of much interest [101]. Once again, developments in living polymerization techniques in recent decades have provided access to new polymeric topologies. Some of these have now been optimized sufficiently to yield polymers of high enough molecular weight to be of interest for rheology modification [102].

13.3.8 Rheological Control with Block Copolymers

Nanophase separation of block copolymer melts introduces interesting rheological behavior. For example, the transient elongational viscosity behavior of a triblock copolymer of styrene and olefins was found to be strongly dependent on the initial orientation of the cylindrical domains [103].

While conventional processing of polymers involves melting, block copolymers such as PS–*b*–PnBA display melt-like behavior simply on application of pressure at room temperature; this is due to pressure-induced miscibility of the blocks (baroplastics [104]).

Precipitation and gelation of long-chain paraffins in waxy crude oils hinders their flow and hence their recovery and processing. Flow characteristics can be improved with addition of a diblock copolymer with a crystalline polyethylene block and an amorphous poly(ethylene–propylene) block which self-assembles in the oil. The former block provides nucleation sites for crystallization of the paraffins, and the latter provides steric stabilization for the wax crystals [105].

13.3.9
Polymer-like Structures through Noncovalent Associations

As compared with the all-covalently linked monomers forming polymeric structures, weak noncovalent associating interactions between macromolecules/ oligomers also provide a possibility of forming chainlike structures. Examples of such interactions are van der Waals and Coulomb interactions, hydrophobic interactions, hydrogen bonds, and ionic bonds. Use of reversible associations like these have been successful in tailoring rheological behavior. For example, polymers with ionically associating groups have been described as useful drag-reducing agents, as their shear stability is enhanced due to sacrificial breaking of the reversible associations at high shear, and reassociation in appropriate low-shear conditions [106, 107]. Hydrophobically associating polymers (HEURs) are used for controlling the rheological behavior of industrial coatings, since they provide sufficient shear viscosity through network formation [108]. As compared to solutions of high molecular weight polymers that show divergent trends with solvent evaporation, linear polymers and networks can be built by strongly associating 2u-ureido-4-pyrimidone end-groups [109]. High sensitivity of the associating behavior to temperature provides opportunities for on-line tuning of rheological behavior.

Introduction of ionic groups or proteins into polymers (forming ionomers) leads to physical associations at the temperature of use [110, 111]. For example, Surlyn (DuPont) is a copolymer of ethylene and methacrylic acid that shows enhanced zero-shear viscosity and elastomeric green strength. Viscoelastic characteristics are also enhanced due to loss of associations at the appropriate dissociation temperature. Ion pair associations are exploited to obtain miscibility in otherwise immiscible polymers [112].

Molecules that are capable of spatial complementarity and weak reversible noncovalent interactions with each other can self-assemble from a less ordered state (such as a solution) to an ordered structure (such as a solid crystal) without human intervention [113]. In this manner, precise tuning of oligopeptide associating structures provides an completely new route to a hierarchy of structures such as tapes, ribbons, fibrils, fibers, and fiber networks, which can be useful as materials [114–116].

13.4
Summary

In this chapter, we have discussed the interrelationships of the molecular structures of polymers with their thermal, solution, and rheological behavior. The thermal response is largely determined by interactions at short distances, such as symmetry and steric effects. On the other hand, the solution and rheological behavior is dominated by topological considerations such as the long-chain structure and branching. Thus, a large variety of polymer properties can be tuned by characteristics such as (co)polymer composition, average molecular weight, molecular weight distribution, branching architectures, reversible associations, and (nano)additives. In the context of the present handbook, it seems appropriate to emphasize once again the role of the emerging polymerization techniques in achieving these characteristics in polymers.

Notation

a	Mark–Houwink constant
c	mass concentration of the polymer [kg m^{-3}]
c^*	critical overlap concentration of the polymer [kg m^{-3}]
De	Deborah number
$f(\theta)$	factor for coiled chain dimension corresponding to bond angle
$\underline{F}^{(c)}$	spring force in the connector of a bead–spring model [N]
g, g'	factors describing compression of branched polymer versus linear polymer
$g(\varphi)$	factor for coiled chain dimension corresponding to rotation about bonds
G	elastic modulus [Pa]
G', G''	the storage modulus and the loss modulus [Pa]
ΔG_m	free energy change on mixing [J]
H	spring constant in a bead–spring model
ΔH_m	heat of melting [J]
k_B	Boltzmann's constant [J K^{-1}]
K	Mark–Houwink constant
l	length of bonds (segments) in polymer chain [m]
L	length of the rodlike molecules [m]
M	molecular weight [kg kmol^{-1}]
n	number of bonds (segments) in the polymer chain
$\bar{n}$	number concentration of the springs
N	number of beads in the bead–spring model
N_A	Avogadro's number
N_e	number of crosslinked chains per unit volume [m^{-3}]
N_1, N_2	first and second normal stress difference [Pa]
p	number of arms in a star polymer
$\underline{Q}$	connector (spring) vector in a bead–spring model
R	polymer chain's end-to-end distance vector [m]
R_g	universal gas constant [J K^{-1} kmol^{-1}]
$\overline{S^2}$	mean square radius of gyration [m^2]
ΔS_m	entropy of melting [J K^{-1}]
t	time [s]

T	temperature [K]
T_g	glass transition temperature [K]
$T_{g,\infty}$	glass transition of a polymer with infinite molecular weight [K]
T_m	melting temperature [K]
v_x	x component of velocity [m s^{-1}]
w_i	weight fraction of species i
x_c	extent of crystallization

Greek

α	extension ratio of an elastic material
α'	factor for polymer coil expansion due to polymer–solvent interactions
$\dot{\gamma}$	shear rate [s^{-1}]
γ	strain
δ	phase delay in dynamic analysis
η	viscosity of a polymer solution [Pa s]
η_s	shear viscosity of the solvent [Pa s]
η_e	extensional viscosity [Pa s]
$[\eta]$	intrinsic viscosity of a polymer in solution [m^3 kg^{-1}]
θ	angle between consecutive bonds in a polymer chain [rad]
λ	time constant of a polymer [s]
λ_H	time constant of a polymer segment [s]
ξ	bead friction factor
π	osmotic pressure [Pa]
ρ	polymer property such as mass concentration, density, specific heat
ρ_b	bulk density of a polymer
τ_{ij}	stress tensor [Pa]
ϕ	polymer volume fraction in solution
φ	angle describing rotation of bonds at fixed θ [rad]
χ	Flory's interaction parameter for mixing
χ_c	critical value of χ for phase separation
ω	angular frequency [s^{-1}]

Acronyms

BPA	bisphenol-A
EDA	ethylene diamine
HEUR	hydrophobically associating polymer (hydrophobically modified, ethoxylated urethane resin)
LDPE	low-density polyethylene
PBT	poly(butene terephthalate)
PE	polyethylene
PEEK	poly(ether ether ketone)
PET	poly(ethylene terephthalate)
PI	polyimide

PMMA	poly(methyl methacrylate)
PnBA	poly(*n*-butyl acrylate)
POSS	polyhedral silsesquioxane
PP	polypropylene
PPT	poly(propylene terephthalate)
PS	polystyrene
SWNT	single-wall carbon nanotube
WAXD	wide-angle X-ray diffraction

References

1 D. P. R. Kint, S. Munoz-Guerra, *Polym. Int.*, **2003**, *52*, 321.

2 K. Fuchs, C. Friedrich, J. Weese, *Macromolecules*, **1996**, *29*, 5893.

3 E. S. Sang et al., *Macromolecules*, **2003**, *36*, 7994.

4 I. W. Hamley, *The Physics of Block Copolymers*, Oxford University Press, Oxford, **1999**.

5 N. Hadjichristidis, S. Pispas, G. A. Floudas, *Block Copolymers: Synthetic Strategies, Physical Properties, and Applications*, John Wiley & Sons, New York, **2003**.

6 J. K. Sears, J. R. Darby, *Technology of Plasticizers*, John Wiley & Sons, New York, **1982**.

7 P. J. Lemstra, R. Krischbaum, *Polymer*, **1985**, *26*, 1372.

8 H. D. Chanzy, E. Bonjour, R. H. Marchessault, *Coll. Polym. Sci.*, **1974**, *252*, 8.

9 J. T. E. Cook et al., *Polymer* **2000**, *41*, 8615.

10 U. S. Agarwal, P. J. Lemstra, *Chem. Engng. Sci.*, **2001**, *56*, 4007.

11 U. S. Agarwal, *Ind. Engng. Chem. Res.* **2004**, *143*, 7275.

12 G. Cecchin, G. Morini, A. Pelliconi, *Macromol. Symp.*, **2001**, *173*, 195.

13 K. Tsobkallo et al., *Polymer*, **2003**, *44*, 1613.

14 S. Fakirov, in *Solid State Behavior of Linear Polyesters and Polyamides*, Eds. J. M. Schultz, S. Fakirov, Prentice Hall, New Jersey, **1990**.

15 F. J. Medellin-Rodriguez, R. Lopez-Guillen, M. A. Waldo-Mendoza, *J. Appl. Polym. Sci.*, **2000**, *75*, 78.

16 B. Duh, *J. Appl. Polym. Sci.*, **2001**, *81*, 1748.

17 Y. Ma , U. S. Agarwal, D. J. Sikkema, P. J. Lemstra, *Polymer*, **2003**, *44*, 4085.

18 G. Guerra, G. Titomanlio, *Macromol. Symp. Ser. 185 – Flow-Induced Crystallization of Polymers*, Wiley-VCH, Weinheim, **2002**.

19 U. S. Agarwal et al., Chapter 5 in *Polymer Science and Fiber Science – Recent Advances* pp. 43–66, Ed. R. E. Fornes, R. D. Gilbert, VCH, New York, **1992**.

20 Y. Funatsu, N. Suzuki, M. Fukuhara, JP 2001011733 A2 20010116, **2001**, Chem. Abstr. 134: 72858.

21 F. L. Binsbergen, *J. Polym. Sci., Polym. Symp.*, **1977**, *59*, 11–19.

22 H. G. Haubruge et al., *Macromolecules* **2003**, *36*, 4452.

23 R. Legras et al., *Polymer* **1984**, *25*, 835.

24 K. Bouma, R. J. Gaymans, *Polym. Engng. Sci.*, **2001**, *41*, 466.

25 U. S. Agarwal, G. de Wit, P. J. Lemstra, *Polymer*, **2002**, *43*, 5709.

26 T. A. Shepard et al., *J. Polym. Sci., Polym. Phys.* **1997**, *35*, 2617.

27 A. R. Bhattacharya et al., *Polymer*, **2003**, *44*, 2373.

28 K. Matyjaszewski, J. Xia, *Chem. Rev.*, **2001**, *101*, 2921.

29 C. J. Hawker, A. W. Bosman, E. Harth, *Chem. Rev.*, **2001**, *101*, 3661.

30 R. Adhikari et al., *Macromol Rapid Commun.*, **2004**, *25*, 653.

31 G. Holden, E. T. Bishop, N. R. Legge, in *Proc. Intl. Rubber Conference*, MacLaren, London, **1968**.

32 T. Hashimoto, K. Tsutsumi, Y. Funaki, *Langmuir*, **1997**, *13*, 6869.

33 S. I. Stupp, P. V. Braun, *Science*, **1997**, *277*, 1242.

34 R. B. Thompson, V. V. Ginzburg, M. W. Matsen, A. C. Balazs, *Science*, **2001**, *292*, 2469.

35 Z. Zhang, M. A. Horsch, M. H. Lamm, S. C. Glotzer, *Nano Lett.*, **2003**, *3*, 1341.

36 A. Jain et al., *Macromolecules*, **2002**, *35*, 4862.

37 G. G. Swift, C. E. Carraher, C. Bowman, *Polymer Modification*, Plenum Press, New York, **1997**.

38 R. O. Ebewele, Chapters 5 and 9 in *Polymer Science and Technology*, CRC Press, Boca Raton, **2000**.

39 D. R. Paul, C. B. Bucknall, *Polymer Blends: Formulation and Performance*, John Wiley & Sons, New York, **2000**.

40 L. A. Utracki, *Polymer Blends Handbook*, Kluwer Ac. Pub. Dordrecht, **2003**.

41 C. Vasile, A. K. Kulshreshtha, *Handbook of Polymer Blends and Composites*, Rapra Technology, Shrophire, UK, **2002**.

42 S. Al-Malaika, *Reactive Modifiers for Polymers*, Blackie, London, **1997**.

43 E. T. Thostenson, Z. Ren, T.-W. Chou, *Composites Sci. Technol.* **2001**, *61*, 1899.

44 H. C. Zeng, in *Handbook of Organic-Inorganic Hybrid Materials and Nanocomposites*, Ed. H. S. Nalwa, American Scientific, Stevenson Ranch, Vol. 2, pp. 151–180, **2003**.

45 D. Qian, E. C. Dickey, R. Andrews, T. Rantell, *Appl. Phys. Lett.*, **2000**, *76*, 2868.

46 Y. Sabba, E. L. Thomas, *Macromolecules*, **2004**, *37*, 4815.

47 S. Barrau et al., *Macromolecules*, **2003**, *36*, 5187; J. Sandler et al., *Polymer* **1999**, *40*, 5967; M. J. Biercuk et al., *Appl. Phys. Lett.* **2002**, *80*, 2767; Z. Ounaies et al., *Composites Sci. Technol.* **2003**, *63*, 1637; O. Reghev et al., *Adv. Materials* **2004**, *16*, 248.

48 M. Q. Zhang, M. Z. Rong, K. Friedrich, in *Encyclopedia of Nanoscience and Nanotechnology*, Ed. H. S. Nalwa, American Scientific, Stevenson Ranch, **2004**.

49 L. Chazeau, C, Gauthier, G. Vigier, J. Y. Cavaille, in *Handbook of Organic-Inorganic Hybrid Materials and Nanocomposites*, Ed. H. S. Nalwa, American Scientific, Stevenson Ranch, **2003**.

50 G. Schmidt, M. M. Malwitz, *Curr. Opin. Coll. Interf. Sci.*, **2003**, *8*, 103.

51 P. J. Flory, *J. Chem. Phys.* **1942**, *10*, 51.

52 M. L. Huggins, *J. Am. Chem. Soc.*, **1942**, *64*, 1712.

53 M. Muthukumar, S. F. Edwards, in *Comprehensive Polymer Science*, Vol. 2, Eds. G. Allen, J. C. Bevington, Pergamon, Oxford, **1989**.

54 D. R. Paul, C. B. Bucknall, *Polymer Blends*, Vols. 1–2, John Wiley & Sons, New York, **2000**.

55 L. A. Utracki, *Polymer Blends Handbook*, Kluwer, Dordrecht, **2002**.

56 O. Olabisi, L. M. Robeson, M. T. Shaw, *Polymer Polymer Miscibility*, Academic Press, London, **1979**.

57 J. Brandrup, E. H. Immergut, Eds., *Polymer Handbook*, 2nd ed., Wiley, New York, **1989**.

58 R. B. Bird, C. F. Curtiss, R. C. Armstrong, *Dynamics of Polymeric Liquids*, Vol. 2, *Kinetic Theory*, Wiley, New York, **1987**.

59 P. E. Rouse, *J. Chem. Phys.*, **1953**, *21*, 1272.

60 B. H. Zimm, *J. Chem. Phys.*, **1956**, *24*, 269.

61 S. Onogi et al., *J. Polym. Sci. Part C*, **1966**, *15*, 381.

62 S. F. Edwards, M. Muthukumar, *Macromolecules*, **1984**, *17*, 586.

63 P. G. de Gennes, *Scaling Concepts in Polymer Physics*, Cornell University Press, Ithaca, **1979**.

64 M. Doi, S. F. Edwards, *The Theory of Polymer Dynamics*, Clarendon, Oxford, **1986**.

65 T. Perkins, D. E. Smith, S. Chu, *Science*, **1994**, *264*, 819.

66 J. E. MacGrath, *Anionic Polymerization: Kinetics, Mechanism and Synthesis*, ACS Symp. Ser. No. 166, Washington DC, 1981.

67 K. A. Davis, K. Matyjaszewski, *Adv. Polym. Sci.*, **2002**, *159*, 1.

68 B. H. Zimm, W. H. Stockmayer, *J. Chem. Phys.*, **1949**, *17*, 1301.

69 G. S. Grest, L. J. Fetters, J. S. Huang, D. Ritcher, *Adv. Chem. Phys.*, **1996**, *94*, 67.

70 J. J. Freire, *Adv. Polym. Sci.*, **1999**, *143*, 35.

71 D. B. Cotts, G. C. Berry, *Macromolecules*, **1981**, *14*, 930.
72 U. S. Agarwal, D. V. Khakhar, *Nature*, **1992**, *360*, 53.
73 L. Vollbracht, in *Comprehensive Polymer Science*, Vol. 5, Eds. G. Allen, J. C. Bevington, Pergamon Press, Oxford, **1989**.
74 A. C. Dierckes, W. R. Schowalter, *Ind. Eng. Chem. Fund.*, **1966**, *5*, 263.
75 R. B. Bird, R. C. Armstrong, O. Hassager, *Dynamics of Polymeric Liquids*, Vol. 1, *Fluid Mechanics*, Wiley, New York, **1987**.
76 M. M. Cross, *J. Colloid. Sci.*, **1965**, *20*, 417.
77 J.-P. Queslel, J. E. Mark, Chapter 9 in *Comprehensive Polymer Science*, Vol. 2, Ed. G. Allen, J. C. Bevington, Pergamon, Oxford, **1989**.
78 J. C. Maxwell, *Phil. Trans. Roy. Soc.*, **1867**, *A157*, 49.
79 S. H. Wasserman, in *Metallocene Catalyzed Polymers*, Ed. G. M. Benedikt, B. L. Goodall, BF Goodrich, Brecksville, **1998**.
80 T. C. B. McLeish et al., *Macromolecules*, **1999**, *32*, 6734.
81 H. A. Malmberg et al., *Macromolecules*, **2002**, *35*, 1038.
82 M. Gahleitner, *Prog. Polym. Sci.*, **2001**, *26*, 895.
83 N. Hudson et al., *Macromolecules*, **2000**, *33*, 9255.
84 C. Gao, D. Yan, *Prog. Polymer Sci.*, **2004**, *29*, 183.
85 R. Mezzenga, L. Boogh, J. A. E. Manson, *Composites Sci. Technol.* **2001**, *61*, 787.
86 Y. H. Kim, O. W. Webster, *Macomolecules*, **1992**, *25*, 5561.
87 T. J. Mulkern, N. C. B. Tan, *Polymer*, **2000**, *41*, 3193.
88 Y. Hong et al., *Polymer*, **2000**, *41*, 7705.
89 D. M. Knauss, T. Huang, *Macromolecules*, **2003**, *36*, 6036.
90 J. R. Dorgan et al., *Macromolecules*, **2003**, *36*, 380.
91 P. Froehling, *J. Polym. Sci., Polym. Chem.*, **2004**, *42*, 3110.
92 C. R. Yates, W. Hayes, *European Polym. J.*, **2004**, *40*, 1257.
93 Y. Tsukahara et al., *Polymer*, **1995**, *36*, 3413; *Adv. Polym. Technol.* **2000**, *11*, 210.
94 B. Zhang et al., *Polymer*, **2004**, *45*, 4009.
95 J. Klein et al., *Nature*, **1994**, *370*, 634.
96 D. N. Schulz, J. E. Glass, *Polymers as Rheology Modifiers*, ACS Symp. Ser. No. 62, Washington DC, **1991**.
97 A. Sen, I. Rubin, *Macromolecules*, **1990**, *23*, 2519.
98 S. Nigen, K. Walters, *J. Non-Newt. Fl. Mech.* **2001**, *87*, 233.
99 V. Bargeron et al., *Nature*, **2000**, *405*, 772.
100 J. A. Odell, A. Keller, Y. Rabin, *J. Chem. Phys.*, **1988**, *88*, 4022.
101 U. S. Agarwal, R. A. Mashelkar, *J. Chem. Phys.*, **1994**, *100*, 6055.
102 L. Xue, U. S. Agarwal, P. J. Lemstra, *Macromolecules*, **2002**, *35*, 8650.
103 Y. Kobori et al., *Macromolecules*, **2003**, *36*, 1656.
104 B. Zhang et al., *Polymer*, **2004**, *45*, 4009.
105 H. S. Ashbaugh, L. J. Fetters, D. H. Adamson, R. K. Prid'homme, *J. Rheology*, **2002**, *46*, 763.
106 S. Malik, S. N. Shintre, R. A. Mashelkar, *Macromolecules*, **1993**, *26*, 55.
107 R. M. Kowalik et al., *J. Non-Newt. Fl. Mech.* **1987**, *24*, 1.
108 A. J. Reuvers, *Prog. Organic Coatings*, **1999**, *35*, 171.
109 R. P. Sijbesma et al., *Science*, **1997**, *278*, 1601.
110 W. A. Petka, J. L. Harden, K. P. McGrath, D. Wirtz, D. A. Tirrell, *Science*, **1998**, *281*, 389.
111 C. W. Lantman, W. J. MacKnight, R. D. Lundberg, in *Comprehensive Polymer Science*, Vol. 2, Eds. G. Allen, J. C. Bevington, Pergamon, Oxford, **1989**.
112 M. Rutkowski, A. Eisenberg, *J. Appl. Polym. Sci.*, **1985**, *30*, 3317.
113 G. M. Whitesides, M. Boncheva, *Proc. National Acad. Sci.* **2002**, *99*, 4769.
114 S. Zhang, D. M. Marini, W. Hwang, S. Santoso, *Curr. Opin. Biol.* **2002**, *6*, 865.
115 A. Aggeli et al., *Proc. Nat. Acad. Sci.* **2001**, *98*, 11 857.
116 H. A. Lashuel et al., *J. Am. Chem. Soc.* **2000**, *122*, 5262.

14
Polymer Mechanical Properties[1]

Christopher J. G. Plummer

14.1
Introduction

14.1.1
Long-chain Molecules

A polymer is made up of macromolecules, that is, long chains of covalently linked chemical units, called monomer units. In many synthetic macromolecules, these units are identical, and the term "polymer" implies their number per chain exceeds about 10^2, although it may reach over 10^5 in many instances. Branched and highly crosslinked architectures are also important in practice, but it is above all the presence of long linear monomer sequences that leads to the unique properties of polymers. For conciseness, therefore, the emphasis in the present overview of mechanical properties will be on isotropic, chemically homogeneous polymers composed of flexible linear or lightly crosslinked macromolecules.

Long chains of atoms or molecules linked by single covalent bonds are usually "flexible" in that rotation about the bonds is relatively unhindered, so that they have many internal degrees of freedom. The forces that oppose any attempt to constrain the macromolecule, such as maintaining its ends at fixed positions, are therefore essentially entropic in origin. This entropic contribution to their mechanical properties is one of the key features of polymers; it gives rise to the phenomenon of rubber elasticity, for example, as will be discussed in Section 14.2. Another characteristic of polymers is the important role of their glass transition temperature, T_g. Polymers with regular configurations, such as polyethylene (PE), may crystallize, but generally remain partly amorphous, even at temperatures, T, well below their melting point, T_m. Moreover, polymers with random configurations, such as atactic polystyrene (aPS), often show little crystallinity at any T and are consequently referred to as "amorphous". In either case, a significant proportion of a polymer becomes glassy at $T < T_g$. The glass transition therefore has a pro-

1) The symbols used in this chapter are listed at the end of the text, under "Notation".

Handbook of Polymer Reaction Engineering. Edited by T. Meyer, J. Keurentjes

ISBN: 3-527-31014-2

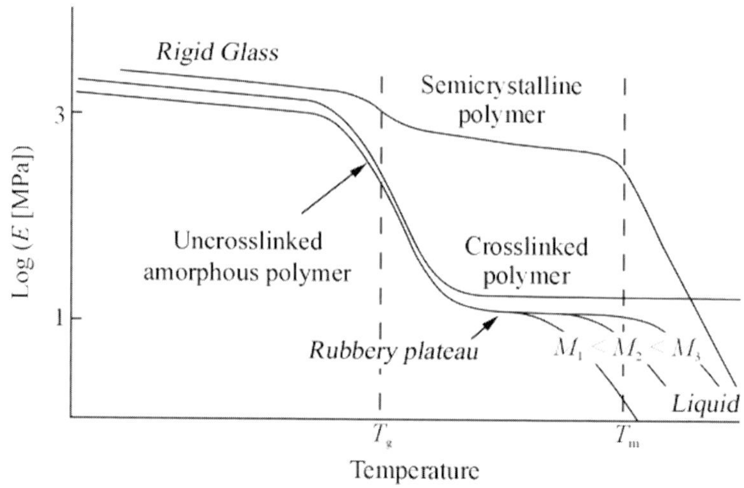

Fig. 14.1. Sketch of the low strain mechanical response of various types of polymer.

found effect on polymer properties, whereas in many low molar mass materials it only has observable physical consequences if solidification is associated with extremely rapid cooling rates.

Figure 14.1 shows schematically the effect of T on the low strain mechanical behavior of various types of polymer, along with the principal regimes of behavior. Below T_g, unoriented polymers are rigid solids with a Young's modulus, E, of between about 1 and 4 GPa, whose low strain response is conditioned by the cohesive forces between nonbonded atoms. In organic polymers, these are mainly due to weak Van der Waals interactions so that, even below T_g, the moduli are much lower than in ionic and metallic crystals as well as in three-dimensional covalently bonded structures such as diamond. As T is increased above T_g, E drops by a factor of more than 10^3 in amorphous polymers such as aPS, although the decrease is less marked in semicrystalline polymers, whose T_m is typically very much greater than T_g, and the reinforcing effect of the relatively rigid crystalline regions persists. Remarkably, even in uncrosslinked amorphous polymers, the modulus remains nearly constant as T is increased further. This gives rise to a "rubbery plateau", whose extent depends strongly on the molar mass, M, as indicated by Figure 14.1. Thus, the plateau becomes effectively infinite for an infinite, crosslinked network of chains, but disappears altogether when M is less than some critical molar mass, M_c. Beyond the plateau there is a transition to viscous liquid behavior, a transition that is defined by T_m in semicrystalline polymers.

14.1.2
Simple Statistical Descriptions of Long-chain Molecules

A real macromolecule contains bonds with fixed valence angles, as shown in Figure 14.2(a) for PE. However, as one moves along an instantaneous snapshot of a

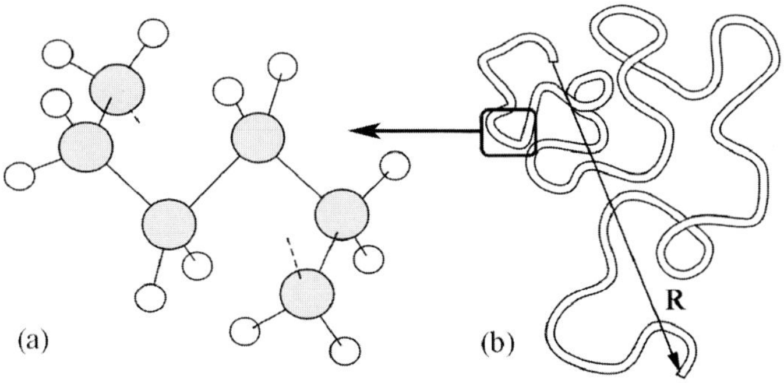

Fig. 14.2. Schematic representations of an isolated polyethylene chain at two length scales.

typical macromolecular chain, taking any bond as a starting point, the correlation between the direction of the skeletal bonds and that of the starting bond decreases rapidly. Thus, taken as a whole, a linear molecule resembles a long, flexible piece of string, as shown in Figure 14.2(b). It may therefore be modeled as a sequence of N freely jointed bonds of length b, in which there is no correlation between the vector corresponding to the nth bond, $\mathbf{r}_n$, and $\mathbf{r}_{n+1}$. The end-to-end vector $\mathbf{R}$ that links one extremity of the chain to the other is then given by Eq. (1).

$$\mathbf{R} = \sum_{n=1}^{N} \mathbf{r}_n \tag{1}$$

The magnitude of $\mathbf{R}$ in an unconstrained chain is uncorrelated with its direction, so that its value averaged over a large number of arbitrarily chosen conformations must be zero. Because thermal motion causes a real chain to change its local conformation every few picoseconds, for more familiar time scales it would therefore be perceived as a "fuzzy ball" with $\langle \mathbf{R} \rangle = 0$ and a time-averaged mean square end-to-end distance given by Eq. (2) ($\mathbf{r}_n$ and $\mathbf{r}_m$ are uncorrelated so that $\langle \mathbf{r}_n \cdot \mathbf{r}_m \rangle$ must vanish for $n \neq m$).

$$\langle \mathbf{R}^2 \rangle = \sum_{n=1}^{N} \sum_{m=1}^{N} \langle \mathbf{r}_n \cdot \mathbf{r}_m \rangle = \sum_{n=1}^{N} \langle \mathbf{r}_n^2 \rangle = Nb^2 \tag{2}$$

For a real macromolecule, the value of b characteristic of the equivalent freely jointed chain is greater than the mean length of the skeletal bonds, l_o, owing not only to fixed valence angles but also to the presence of rigid units and steric hindrance to rotation about single bonds. If N_l is the number of skeletal bonds in the chain, $\langle \mathbf{R}^2 \rangle = C_\infty N_l l_o^2$, where the constant C_∞ is the Flory "characteristic ratio", also called the "chain stiffness" [1, 2]. It is of the order of 10 in flexible polymers (7.3 in PE and 9.6 in aPS, for example [3]), but may be much higher for more rigid architectures such as para-linked benzene rings. C_∞ provides the link between molecular architecture and general models for macroscopic physical properties governed by chain statistics, often expressed in terms of $b = C_\infty N_l l_o^2 / R_{\max}$, where R_{max} is the maximum end-to-end distance of the chain (which depends on both l_o and the bond angles).

In the limit of large N, the freely jointed chain model implies the probability density $P(\mathbf{R}, N)$ for a given $\mathbf{R}$ to be a Gaussian distribution [Eq. (3)], whence the frequent use of the term "Gaussian chain" to refer to idealized flexible linear macromolecules [4].

$$P(\mathbf{R}, N) = \left(\frac{3}{2\pi N b^2}\right)^{3/2} \exp\left(-\frac{3\mathbf{R}^2}{2Nb^2}\right) \tag{3}$$

This result is a simplification in that it does not take into account long-range interactions and solvent effects, for example. It nevertheless provides a working description of conformations in solid and molten polymers, and hence forms the basis for a wide range of theoretical models in polymer physics and mechanics, some of which will be discussed in what follows.

14.2 Elasticity

14.2.1 Deformation of an Elastic Solid

The stress state associated with an elemental cube may be defined in terms of the six independent quantities in Eq. (4), where the first subscript refers to the direction of the normal to the plane on which the stress acts and the second subscript refers to the direction of the stress (at equilibrium the net torque is zero so that $\sigma_{ij} = \sigma_{ji}$).

$$\boldsymbol{\sigma}_{ij} = \begin{bmatrix} \sigma_{xx} & \sigma_{xy} & \sigma_{xz} \\ \sigma_{xy} & \sigma_{yy} & \sigma_{yz} \\ \sigma_{xz} & \sigma_{yz} & \sigma_{zz} \end{bmatrix} \tag{4}$$

The deformation may likewise be defined in terms of the components of engineering strain [Eq. (5)].

$$\varepsilon_{ij} = \begin{bmatrix} \varepsilon_{xx} & \varepsilon_{xy} & \varepsilon_{xz} \\ \varepsilon_{xy} & \varepsilon_{yy} & \varepsilon_{yz} \\ \varepsilon_{xz} & \varepsilon_{yz} & \varepsilon_{zz} \end{bmatrix} \tag{5}$$

The generalized form of Hooke's law, which proposes a linear relationship between stress and strain at vanishingly small strains, is then given by Eq. (6).

$$\varepsilon_{ij} = \mathbf{E}_{ijkl}\boldsymbol{\sigma}_{kl} \tag{6}$$

In an isotropic solid, there is no coupling between tensile and shear stresses, and Hooke's law takes the familiar form of Eq. (7), where ν is Poisson's ratio and G is the shear modulus.

$$\varepsilon_{xx} = \frac{1}{E}\sigma_{xx} - \frac{\nu}{E}(\sigma_{yy} + \sigma_{zz}), \quad 2\varepsilon_{xy} \equiv \gamma = \frac{1}{G}\sigma_{xy} \tag{7}$$

Moreover, it is easily shown that Eq. (8) holds, so that a complete description of the small strain response requires only two independent parameters.

$$G = \frac{E}{2(1 + \nu)} \tag{8}$$

Similarly, the bulk (compression) modulus, K, is given by Eq. (9).

$$K = \frac{E}{3(1 - 2\nu)} \tag{9}$$

It is implicit in such descriptions of small-strain elasticity that the stress is unaffected by the deformation, which justifies the assumption of linearity. For the large deformations associated with rubber elasticity and shear yielding in glassy and semicrystalline polymers, however, this is no longer valid, and it is more usual to express deformations in terms of the deformation ratios, λ, such that the point (x, y, z) is transformed to $(\lambda_1 x, \lambda_2 y, \lambda_3 z)$. If the coordinate system is defined with respect to the deformed system, it is always possible to choose the axes so that they are not rotated by the deformation. The shear components of the deformation are then equal to zero, which permits considerable simplification. For example, the incremental work per unit volume of *undeformed* material is given by Eq. (10), where the f_i are the applied forces along the axes of a unit cube of undeformed material [5].

$$dw = f_1 d\lambda_1 + f_2 d\lambda_2 + f_3 d\lambda_3 \tag{10}$$

14.2.2 Thermodynamics of Rubber Elasticity

Perhaps the most striking difference between rubbers and other materials is their capacity for large reversible or nearly reversible deformations at their service temperature, and it is this aspect of their behavior that will be described here. The commercial exploitation of natural rubber developed rapidly with the discovery that crosslinking greatly improves its mechanical properties, giving the first of what is now a broad class of materials often referred to as "elastomers". Elastomers are characterized by flexible chains, T_g well below room temperature, little or no crystallinity in the undeformed state, and light chemical or physical crosslinking (through the presence of rigid glassy or crystalline microdomains, for example).

Although E drops significantly as T is raised above T_g, K changes relatively little, so that $K \gg E$ and, from Eq. (9), $\nu \approx 0.5$. Volume changes may hence be considered negligible compared with other types of deformation. This justifies the use of the Helmholtz free energy in the thermodynamic analysis of rubber elasticity, defined by Eq. (11).

$$A = U - TS \tag{11}$$

where U is the internal energy and S is the entropy. For a reversible change of state, the first law of thermodynamics gives Eq. (12), where dW is the work carried out *on* the system and dQ is the heat transfer *into* the system, leading to Eq. (13).

$$dU = dQ + dW = TdS + dW \tag{12}$$

$$dA = dU - TdS - SdT = dW - SdT \tag{13}$$

The work done when a specimen subject to a force f undergoes an incremental elongation dl is $dW = fdl$. The corresponding change in free energy is given by Eq. (14), from which Eq. (15) follow.

$$dA = fdl - SdT \tag{14}$$

$$f = \left(\frac{\partial A}{\partial l}\right)_T, \quad S = -\left(\frac{\partial A}{\partial T}\right)_l \tag{15}$$

From the identity of Eq. (16), one obtains Eq. (17), and hence Eq. (18).

$$\frac{\partial}{\partial T}\left(\frac{\partial A}{\partial l}\right)_T = \frac{\partial}{\partial l}\left(\frac{\partial A}{\partial T}\right)_l \tag{16}$$

$$\left(\frac{\partial f}{\partial T}\right)_l = -\left(\frac{\partial S}{\partial l}\right)_T \tag{17}$$

$$f = \left(\frac{\partial A}{\partial l}\right)_T = \left(\frac{\partial U}{\partial l}\right)_T - T\left(\frac{\partial S}{\partial l}\right)_T = \left(\frac{\partial U}{\partial l}\right)_T + T\left(\frac{\partial f}{\partial T}\right)_l \tag{18}$$

Measurements of $f(T)$ at large fixed deformations, such as shown in Figure 14.3, indicate that $f(T) \approx CT$ for $T > T_g$, where C is a constant. The second term in Eq. (18) therefore dominates under these conditions and the response is almost entirely entropic [4, 6]. (The increase in f with decreasing T below T_g is due to thermal contraction.)

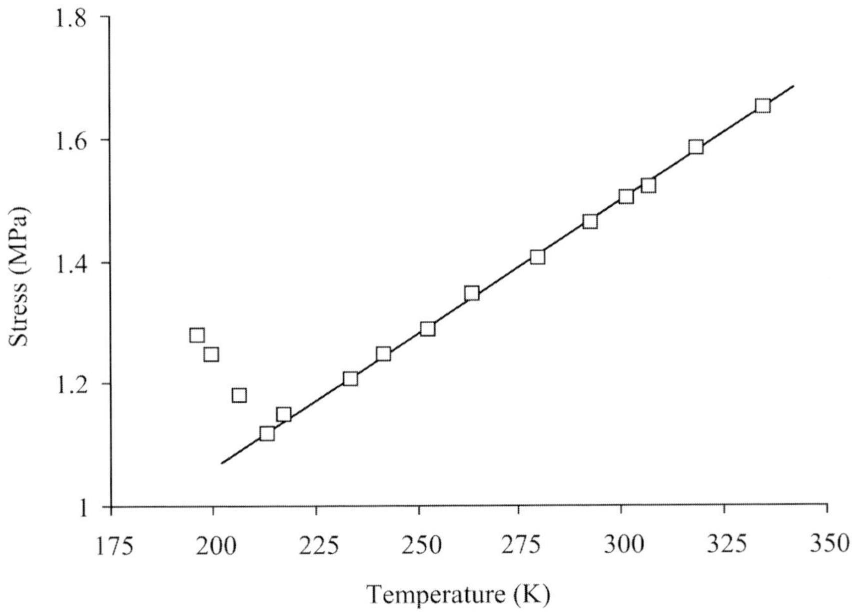

Fig. 14.3. Tensile stress at constant extension (350%) as a function of T in a crosslinked rubber (after Ref. 2).

Another characteristic of elastomers is that their temperature increases during rapid deformation, for which $dQ \approx 0$. In this case, one can show that Eq. (19) applies, where C_l is the specific heat at fixed l.

$$\left(\frac{\partial T}{\partial l}\right)_S = \frac{T}{C_l}\left(\frac{\partial f}{\partial T}\right)_l \tag{19}$$

Given that $(\partial f/\partial T)_l > 0$ (Figure 14.3), T must indeed increase with l. It may also be shown that the length of an elastomer subject to a constant force should *decrease* as T is increased, which is again consistent with observation, albeit somewhat counterintuitive, given that crystalline and glassy solids tend to become less rigid on heating [4].

14.2.3 Statistical Mechanical Approach to Rubber Elasticity

The entropy changes that give rise to rubber elasticity may be modeled in terms of the chain statistics introduced in Section 14.1.2. For a chain whose end-to-end vector is fixed and equal to $\mathbf{R}$, the number of conformations, $\Omega(\mathbf{R})$, that the chain can adopt is proportional to $P(\mathbf{R}, N)$. From Eq. (3), one thus obtains Eq. (20), where k is Boltzmann's constant and S_o is a constant.

$$S = S_o + k \ln \Omega = S_o - \frac{3k\mathbf{R}^2}{2Nb^2} \tag{20}$$

Hence, for constant U, the free energy is obtained from Eq. (21).

$$A = -TS = A_o - Tk \ln \Omega = A_o + \frac{3kT\mathbf{R}^2}{2Nb^2} \tag{21}$$

The magnitude of the restoring force on a chain with end-to-end vector $\mathbf{R}$ is therefore given by Eq. (22).

$$f = \frac{\partial A}{\partial \mathbf{R}} = \frac{3kT\mathbf{R}}{Nb^2} \tag{22}$$

The total free energy change when a large deformation is applied to the chain, so that $\mathbf{R} = (x, y, z)$ changes to $\mathbf{R}' = (\lambda_1 x, \lambda_2 y, \lambda_3 z)$, is given by Eq. (23).

$$\Delta A = -T\Delta S = \frac{3kT(\mathbf{R}'^2 - \mathbf{R}^2)}{2Nb^2} = \frac{3kT((\lambda_1^2 - 1)x^2 + (\lambda_2^2 - 1)y^2 + (\lambda_3^2 - 1)z^2)}{2Nb^2} \tag{23}$$

In an elastomer the forces are transferred to individual chains through the crosslinks, which constitute the nodes of a continuous network. If the chains linking these nodes are assumed to be freely jointed chains composed of N links of length b, Eq. (2) implies the spatial separation of the nodes, $\mathbf{R}$, to be given by Eq. (24) in the undeformed state.

$$\langle \mathbf{R}^2 \rangle = \langle x^2 \rangle + \langle y^2 \rangle + \langle z^2 \rangle = Nb^2 \tag{24}$$

Because $\mathbf{R}$ has no preferred direction, Eq. (25) holds.

$$\langle x^2 \rangle = \langle y^2 \rangle = \langle z^2 \rangle = \frac{Nb^2}{3}. \tag{25}$$

If there are n_x chains per unit volume, then assuming affine deformation of the nodes (so that they move as if fixed to a uniform elastic background), it follows from Eq. (23) that the total change in the free energy per unit volume, a, for a deformation $\lambda_1, \lambda_2, \lambda_3$ is given by Eq. (26).

$$\Delta a = \frac{n_x kT(\lambda_1^2 + \lambda_2^2 + \lambda_3^2 - 3)}{2} \tag{26}$$

From the incompressibility criterion, $\lambda_1\lambda_2\lambda_3 = 1$. Hence, for a uniaxial extension along x, that is $\lambda_1 = \lambda$, $\lambda_2 = \lambda_3 = \lambda^{-1/2}$, Eq. (27) applies and, from Eq. (10), the force f_1 acting on a unit cube of undeformed material (equivalent to the nominal or "engineering" stress) is given by Eq. (28).

$$\Delta a = \frac{n_x kT}{2}\left(\lambda^2 + \frac{2}{\lambda} - 3\right) \tag{27}$$

$$f_1 = \frac{\partial w}{\partial \lambda} = \frac{\partial \Delta a}{\partial \lambda} = n_x kT\left(\lambda - \frac{1}{\lambda^2}\right) \tag{28}$$

This remarkable result implies that the deformation behavior depends only on the number of network chains per unit volume and on T. The detailed chemistry of the chain is therefore unimportant, with the proviso that the chains must be much longer than the statistical segment length b. For very small deformations, $\lambda = 1 + \varepsilon_{xx}$, and σ_{xx} is given by Eq. (29).

$$\sigma_{xx} \approx f_1 \approx 3n_x kT\varepsilon_{xx} \tag{29}$$

Given that $\nu \approx 0.5$ and hence that $E = 3G$ [Eq. (9)], Eq. (30) follows.

$$G = n_x kT \tag{30}$$

There are many shortcomings with this simple treatment, and, as illustrated in Figure 14.4, the behavior of real elastomers deviates markedly from Eq. (28), particularly at large λ. A particular difficulty with Eq. (3) is that it predicts finite probabilities for $|\mathbf{R}| > R_{max}$, which is clearly unphysical. To account for the finite extensibility of the chains, and hence for the observed work hardening, it is necessary to chose more realistic expressions for $P(\mathbf{R}, N)$, such as the inverse Langevin function or the associated analytical approximations [4]. Moreover, in the case of relatively stiff chains such as DNA or certain polypeptides, the worm-like chain model of Porod and Krattky may be more appropriate than the Gaussian chain. Even if the chain statistics are correctly described, non-affine deformation of the network junctions, long-range inter- and intra-chain interactions (entanglement), specific interactions such as hydrogen bonding, strain-induced crystallization (observed in natural rubber, for example), the presence of fillers, and so forth may need to be taken into account in real rubbers and elastomers [7–9].

14.3 Viscoelasticity

14.3.1 Linear Viscoelasticity

The discussion has so far implicitly been limited to behavior in the rubbery plateau regime in Figure 14.1, where the conditions for rubber elasticity in amorphous

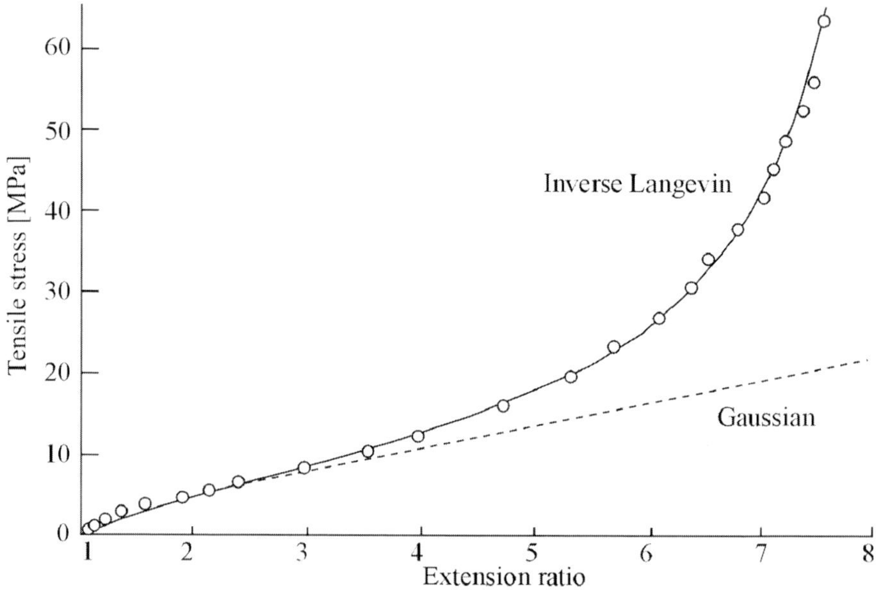

Fig. 14.4. Experimental stress–extension ratio curve for natural rubber along with a best fit of the predicted behavior for network described by the inverse Langevin function, and of Gaussian network [Eq. (28)] with the same network parameters (after Ref. 7).

polymers are met, namely long flexible chains, weak intermolecular forces ($T > T_g$) and anchoring of the chains to form a network (the nature of this anchoring in uncrosslinked polymers will be discussed in Section 14.3.3). The corresponding constitutive expressions [Eq. (28), for example] predict the response to deformation to depend weakly on T and to be independent of the deformation rate. Although this is reasonably consistent with experiment, the situation changes drastically near T_g, for example, where time, t, becomes a very important factor, and the mechanical behavior is said to be strongly "viscoelastic".

In a *linear* viscoelastic material, the moduli and compliances $E(t)$, $G(t)$, $D(t)$, and $J(t)$ (tensile and shear modulus and tensile and shear compliance respectively) are functions of t, although they remain independent of stress or strain. Development of constitutive equations for such materials is generally based on the idea that the effects of small increases in stress or strain are additive, which is known as the Boltzmann superposition principle [4, 10, 11]. A strain $\varepsilon(t)$ may be considered to result from a sum of step strains applied at time u and maintained for a time $t - u$ [Eq. (31)].

$$d\varepsilon(u) = \frac{d\varepsilon(u)}{du} du \tag{31}$$

In uniaxial tension, the Boltzmann superposition principle implies Eq. (32).

$$\sigma(t) = \int_{-\infty}^{t} E(t-u)\frac{d\varepsilon}{du}du \tag{32}$$

Similarly, for a given stress history, Eq. (33) results.

$$\varepsilon(t) = \int_{-\infty}^{t} D(t-u)\frac{d\sigma}{du}du \tag{33}$$

Linear viscoelastic materials thus retain a memory of their entire mechanical history although, conveniently for the experimentalist, the memory fades with $t-u$, and for practical purposes the lower limit of integration in Eqs. (32) and (33) is often set to zero. Equations (32) and (33) may be used to calculate the response to an arbitrary deformation using $E(t)$ or $D(t)$ determined from simple static experiments (creep or relaxation).

Another important class of experiments involves periodic tensile or shear strains of the type $\varepsilon(t) = \varepsilon_o \cos \omega t = \mathrm{Re}(\varepsilon_o e^{i\omega t})$. In a linear elastic solid deformed in tension, σ oscillates in phase with ε with amplitude $E\varepsilon_o$. In a linear viscous liquid, $\sigma = \eta d\varepsilon/dt$ and hence varies as $-\eta\varepsilon_o\omega \sin \omega t = \eta\varepsilon_o\omega \cos(\omega t + \pi/2)$. In a linear viscoelastic material, however, intermediate behavior is observed [Eqs. (34), where $0 < \delta < \pi/2$].

$$\begin{aligned} \varepsilon(t) &= \varepsilon_o \cos \omega t \\ \sigma(t) &= \sigma_o \cos(\omega t + \delta) \end{aligned} \tag{34}$$

It is hence possible to define two dynamic moduli, the storage modulus, E' (or G', in the case of a shear experiment), which gives the stress component in phase with the strain, and the loss modulus, E'' (or G''), which gives the stress component out of phase with the strain. Thus, in tension, Eqs. (35) is obtained, where E', E'', and $\tan \delta$ are given by Eq. (36).

$$\sigma(t) = \sigma_o \cos(\omega t + \delta) = E'\varepsilon_o \cos \omega t + E''\varepsilon_o \sin \omega t \tag{35}$$

$$E' = \frac{\sigma_o}{\varepsilon_o}\cos \delta, \quad E'' = \frac{\sigma_o}{\varepsilon_o}\sin \delta, \quad \tan \delta = \frac{E''}{E'} \tag{36}$$

Equation (35) is sometimes more conveniently expressed in complex notation as Eq. (37), where the complex modulus E^* is defined by Eq. (38).

$$\sigma(t) = \mathrm{Re}(\sigma_o e^{i(\omega t+\delta)}) = \mathrm{Re}(E^*\varepsilon_o e^{i\omega t}) \tag{37}$$

$$E^* = \frac{\sigma_o}{\varepsilon_o}(\cos \delta + i \sin \delta) \tag{38}$$

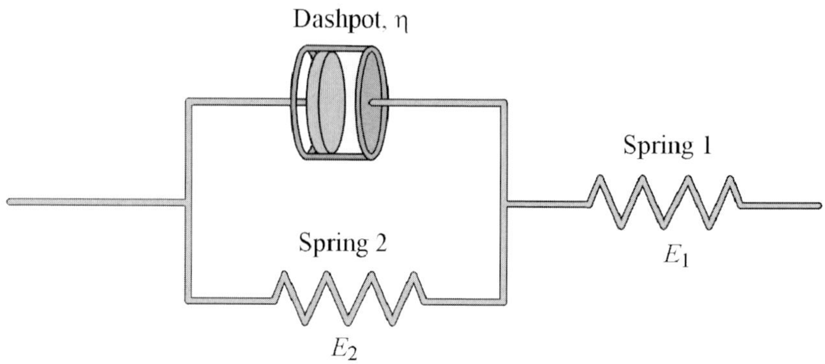

Fig. 14.5. The Zener model (or "standard linear solid").

Viscoelastic materials are often modeled using linear combinations of fixed Hookean and Newtonian elements (springs and dashpots). Although such models may not fully account for the complexity of real materials, generalizations of this approach have been extensively developed and methods and approximations for transforming between the different viscoelastic functions are widely available in the literature [10, 11]. For the sake of illustration, one simple example will be considered here, namely the Zener element, a possible representation of which is shown in Figure 14.5. At very high frequencies most of the deformation in a Zener element is taken up by spring 1 and the limiting modulus E_o is equal to E_1. At very low frequencies, where the stress in the dashpot becomes negligible, the deformation is taken up by spring 1 and spring 2 in series, and the limiting modulus $E_\infty = E_1E_2/(E_1 + E_2)$. However, at intermediate frequencies, the behavior becomes strongly influenced by η [Eqs. (39), where $\Delta E = E_o - E_\infty$ and the relaxation time is given by Eq. (40)].

$$E^* = E_o - \Delta E\frac{1 - i\omega\tau}{1 + \omega^2\tau^2}; \quad \tan\delta = \frac{\Delta E\omega\tau}{E_\infty + E_o\omega^2\tau^2} \tag{39}$$

$$\tau = \eta\frac{\Delta E}{E_o^2} \tag{40}$$

The behavior predicted by Eqs. (39) for values of E_1 and E_2 appropriate to the glass transition in an amorphous polymer (compare Figure 14.1) is shown in Figure 14.6. The model accounts for the qualitative features of experimentally observed transitions, namely a step-like drop in modulus as ω decreases below τ^{-1}, and a characteristic peak in tan δ. However, more complex models involving many relaxation times, that is, a discrete or continuous relaxation time spectrum, are necessary if more quantitative agreement with experiment is to be obtained (for an example of a discrete relaxation time spectrum derived from a molecular model, see Section 14.3.3) [10, 11].

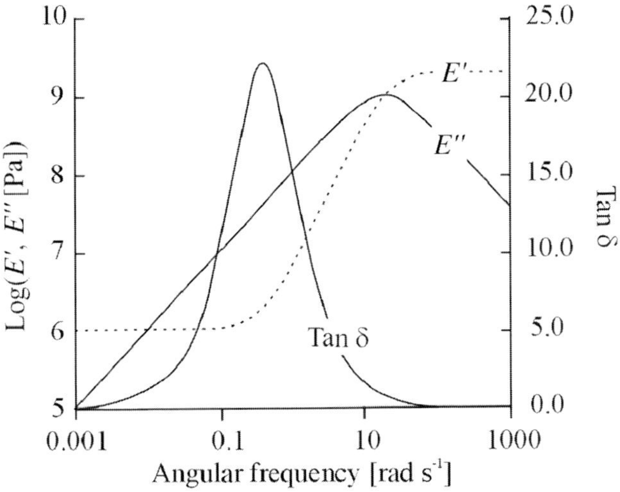

Fig. 14.6. Predictions of the Zener model for the dynamic behavior in the vicinity of a transition ($E_1 = 2 \times 10^9$ Pa, $E_2 = 10^6$ Pa, $\tau = 0.05$ s).

Similar behavior to that in Figure 14.6 is observed if the viscoelastic functions are measured as a function of T at fixed ω, because the viscous response, which tends to dominate the relaxation times, is a strongly decreasing function of increasing T, particularly close to T_g. The effect of increasing T is therefore equivalent to that of decreasing ω (or increasing t in a static experiment). Thus, in dynamic torsion temperature sweeps on poly(methyl methacrylate) (PMMA), the glass transition at about 120°C is associated with a clear peak in tan $\delta(T)$. As shown in Figure 14.7, other transitions may also be observed at T well below T_g in this type of experiment. These "secondary" relaxations are important features of both glassy and semicrystalline polymers because they show that molecular motions are not entirely suppressed at $T < T_g$. Indeed, such motions play an important role in plasticity (see Section 14.4.1). Conventionally, the first relaxation encountered as T decreases is labeled α, and may correspond to either the melting/crystallization transition (for semicrystalline polymers) or the glass transition. The secondary relaxations are designated β, γ, and so on, in order of their appearance as T decreases further (see Figure 14.7). They may often be assigned to specific types of molecular motion, such as side-group motions or cooperative motion of several main-chain units, and often show an Arrhenius T dependence. Activation of side-group motions typically requires more energy than main-chain motions. Thus the main-chain relaxation in bisphenol-A polycarbonate (PC) occurs at about −80°C, which is about 230 K below T_g (150°C), whereas relaxation of the aromatic side groups in aPS is observed at 50°C, which is only 50 K below T_g (100°C) [5].

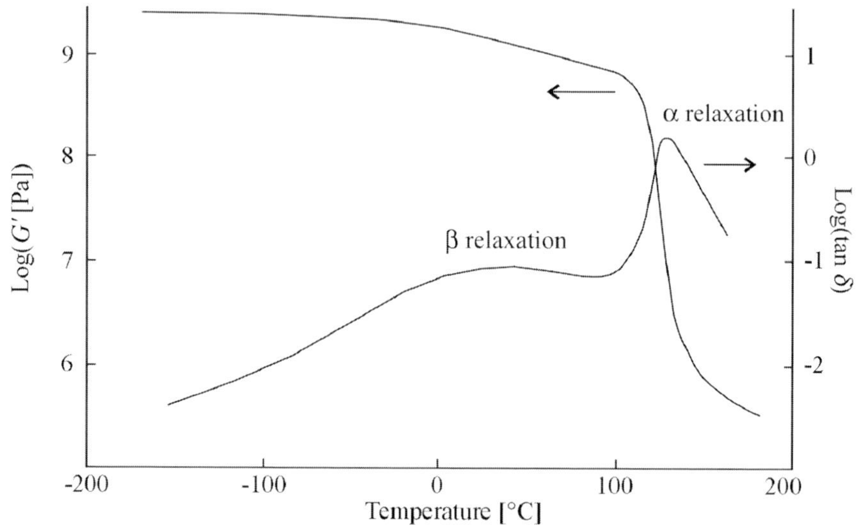

Fig. 14.7. Dynamic behavior as a function of T of poly(methyl methacrylate) (PMMA) tested in torsion at 10 Hz.

14.3.2 Time–Temperature Superposition

The idea of time–temperature equivalence introduced in Section 14.3.1 is of considerable practical importance because one would often like to predict the long-term response of materials on the basis of experiments carried out on a laboratory time scale. This is to some extent possible in polymers, for which it has been widely verified that viscoelastic functions determined at different T over a fixed range of ω or t, slightly adjusted to take into account the effect of T and density ρ on the elastic response [through Eq. (30)], superpose if the ω or t scale is multiplied by a shift factor, $a_T(T, T_r)$, where T_r is some convenient reference temperature. A typical master curve obtained in this way is shown in Figure 14.8 for stress relaxation data from polyisobutene, taking $T_r = -66.5°C$. The effect is to expand the t or ω scale of the measurement carried out at T_r, revealing the whole of the α transition in this case.

Williams, Landel, and Ferry (WLF) observed that if T_r is set to T_g, the variation of $\log a_T$ with $T - T_r$ is similar for a wide variety of polymers [10]. They rationalized this in terms of the molecular response, starting with Doolittle's equation [Eq. (41)] for the viscosity, where A and B are constants. f_v is the fractional free volume, equivalent to the "unoccupied" volume divided by the total volume of the polymer (the "occupied" volume includes that necessary to accommodate thermal vibrations).

$$\ln \eta = \ln A + \frac{B}{f_v} \tag{41}$$

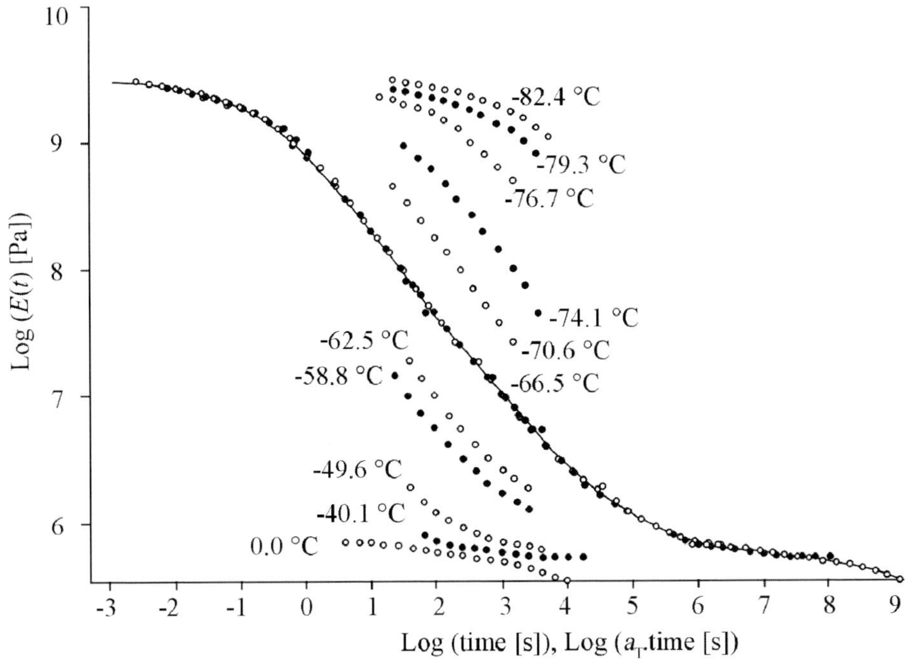

Fig. 14.8. Stress relaxation data for polyisobutene at different temperatures T, as indicated and superposed onto the data for $-66.5°C$ by shifting along the log(t) axis (after Ref. 7).

Equation (41) is based on the idea that the greater f_v, the greater the molecular mobility (owing to reduced crowding), and the lower η. For $T > T_g$, f_v is given by Eq. (42), where α_f is the coefficient of thermal expansion of the fractional free volume and f_g is the fractional free volume associated with the glass transition.

$$f_v \approx f_g + \alpha_f (T - T_g) \tag{42}$$

If the T dependence of all the relaxation times is assumed to be that of η in Eq. (41), the shift factor for the scaled viscoelastic functions is given by Eq. (43), and hence Eq. (44) follows.

$$a_T = \frac{\eta(T)}{\eta(T_g)} \tag{43}$$

$$\log a_T = -\frac{\dfrac{B}{2.303 f_g}(T - T_g)}{\dfrac{f_g}{\alpha_f} + T - T_g} \equiv \frac{C_1^g (T - T_g)}{C_2^g + T - T_g} \tag{44}$$

Equation (44) is the well-known WLF equation. "Universal" values of the various physical parameters in Eq. (44) lead to $C_1^g = 17.44$ and $C_2^g = 51.26$ K [10]. These are of the same order of magnitude as C_1^g and C_2^g obtained empirically (34 and 80 K for PMMA, for example), and indeed, time–temperature superposition has been found to work well for a wide range of single-phase polymers, with the proviso that it begins to break down for the relatively fast vibrational modes characteristic of the glassy state [12]. Moreover, although superposition may work for $T \gg T_g$, at temperatures above about $T_g + 50$ K the shift factors tend to show an Arrhenius dependence rather than following the WLF equation.

14.3.3 Molecular Models for Polymer Dynamics

The starting point for molecular models for polymer dynamics based on the ideas introduced in Section 14.2.3 is the Rouse model for an isolated chain in a viscous medium, in which the chain is taken to behave as a sequence of m "beads" linked by Gaussian springs [Figure 14.9(a)] [13–16]. The chain interacts with the solvent via the beads, and the solvent is assumed to drain freely as the chain moves. Hence, Eq. (22) leads to Eqs. (45), where N' is the number of links between adjacent beads, ζ is a friction coefficient per bead and $\mathbf{r}_i$ is the position of the ith bead.

$$\zeta \frac{d\mathbf{r}_i}{dt} + \frac{3kT}{N'b^2}(2\mathbf{r}_i - \mathbf{r}_{i-x} - \mathbf{r}_{i+1}) = 0 \tag{45}$$

Rouse solved the $m - 1$ simultaneous equations [Eqs. (45)] by transforming them into a set of uncoupled equations for the normal modes of motion of the chain.

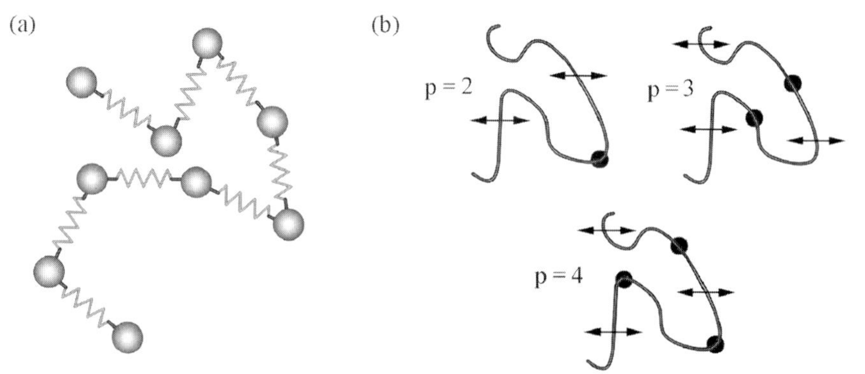

Fig. 14.9. (a) The bead–spring model for the dynamics of an isolated chain in solution; (b) sketch of the second to fourth normal modes of motion (the circles represent fixed nodes and the arrows indicate coherent motion of the sub-chains defined by the nodes).

The moduli of a dilute solution containing n chains per unit volume may then be expressed in terms of a discrete relaxation time spectrum, where each relaxation time corresponds to one of the normal modes of motion [Figure 14.9(b)]. For simple shear, this leads to Eqs. (46) and (47), with τ_p given by Eq. (48) [13–16].

$$G(t) = nkT\sum_{p=1}^{m} e^{-t/\tau_p} \tag{46}$$

$$G^*(\omega) = nkT\sum_{p=1}^{m}\frac{\omega^2\tau_p^2}{1+\omega^2\tau_p^2} + inkT\sum_{p=1}^{m}\frac{\omega\tau_p}{1+\omega^2\tau_p^2} \tag{47}$$

$$\tau_p = N'b^2\zeta\left[24kT\,\sin^2\left\{\frac{p\pi}{2(m+1)}\right\}\right]^{-1}, \quad p = 1, 2 \ldots, m \tag{48}$$

Because the length of chain corresponding to each Gaussian spring is the shortest unit that can relax in the model, these expressions are only physically meaningful for $m \gg 1$. In this limit, Eq. (48) may be replaced by Eq. (49), where $\zeta_o = \zeta/N'$ is a "monomeric friction coefficient" and $N = mN'$ is the total number of statistical segments per chain.

$$\tau_p \approx \frac{N'^2m^2b^2\zeta_o}{6\pi^2p^2kT} = \frac{N^2b^2\zeta_o}{6\pi^2p^2kT} \tag{49}$$

Hence, the longest "Rouse" relaxation time, τ_R, is proportional to N^2. For $t > \tau_R$, the motion of the chain becomes essentially diffusive, and the diffusion constant of the center of gravity of the chains is given by the Einstein relation, Eq. (50).

$$D_R = \frac{kT}{N\zeta_o} \tag{50}$$

Rouse-like behavior is not in fact observed in dilute solutions, for which it is necessary to take into account the influence of the chain on the motion of the solvent, and deviations from Gaussian statistics arising from polymer–solvent interactions [17, 18]. These factors are incorporated in the Zimm model, which predicts the diffusion constant to be proportional to $N^{-\nu}$, for example, where ν depends on the solvent quality, in better agreement with experimental data [4, 14]. Indeed, although it was first proposed for isolated chains, the Rouse model turns out to be more appropriate to polymer melts, where flexible linear chain conformations are approximately Gaussian and hydrodynamic interactions are relatively unimportant [4, 14–16].

The Rouse model is nevertheless inadequate to describe the high-frequency response associated with bond rotations and local cooperative motions, important for the glassy state (see, for example, Ref. 4). Moreover, as the chain length increases (or the concentration increases in a solution of long chains), the fact that

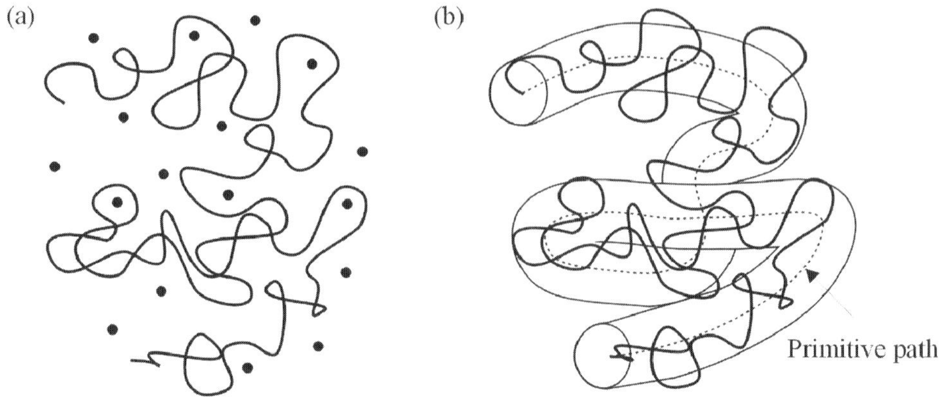

Fig. 14.10. (a) Schematic representation of entanglement constraints; (b) the "tube" model.

chains cannot cross one another leads to additional constraints on chain motion, known as entanglement, which remain effective at $T > T_g$. Figure 14.10(a) shows the origin of these constraints schematically for an arbitrary chain in a polymer melt. The dots correspond to the nearest neighbors of the chain, which are assumed for the sake of illustration to intersect the plane of the drawing. Because the chain cannot cross its neighbors, it cannot move very far in directions perpendicular to its local trajectory. Diffusion is therefore limited to snake-like motion of the chain along its own contour, called "reptation" [14, 18]. A convenient way of representing this effect in three dimensions is to assume the chain to be trapped in a tube with a primitive path length L, as shown in Figure 14.10(b), so that only motion within or along the tube is possible. For an undeformed Gaussian chain, the tube is also Gaussian, with an end-to-end vector $\mathbf{R}$. The primitive path may thus be represented by N_L statistical segments of length a, such that $L = N_L a$ and $\langle \mathbf{R}^2 \rangle = N_L a^2 = La$. Because $\langle \mathbf{R}^2 \rangle$ must also equal Nb^2, Eq. (51) follows.

$$L = Nb^2/a \tag{51}$$

The tube model forms the basis for detailed theoretical approaches to the dynamics of an entangled polymer melt pioneered by Doi and Edwards [13, 14]. The present discussion will be restricted to a simple description of some of its basic features. Consider, for example, the relaxation behavior of a chain subjected to a step shear strain at $t = 0$. As shown in Figure 14.11, the applied strain results in a deformation of the tube, and hence of the chain trapped inside it. The first relaxation occurs within the tube at times $t < \tau_e$, where τ_e is the Rouse relaxation time for a chain with $N_e = (a/b)^2$ statistical segments, and hence a mean square end-to-end distance equal to a^2. At $t > \tau_e$, the constraints due to the tube begin to dominate and the only way the stress can relax further relax is for the chain to escape the deformed tube and re-establish a random (unperturbed) conformation. As sketched in Figure 14.11, it achieves this through Brownian motion backward and

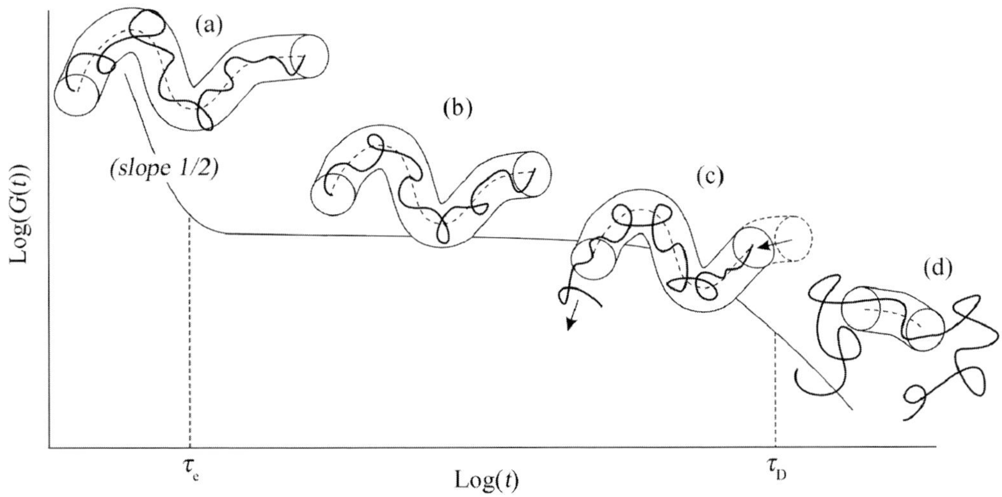

Fig. 14.11. Relaxation according to the tube model (small strain limit): (a) Rouse relaxation; (b) rubbery plateau; (c) to (d) reptation.

forward along the tube. The time necessary for the chain to diffuse a distance L and hence relax fully is $\tau_D = L^2/D_R$, where D_R is given by Eq. (50). Hence, Eq. (52) is obtained from Eq. (51).

$$\tau_D = \frac{N^3 b^4 \zeta_o}{a^2 kT} = \frac{6\pi^2 N^3}{N_e^3}\tau_e \tag{52}$$

Because τ_D, which is known as the "reptation time", is significantly longer than τ_e for large N, in Figure 14.11 there is an intermediate regime of roughly time-independent behavior, which is identified with the rubbery plateau. Thus, entanglement may be considered to play an equivalent role in this regime to the chemical crosslinks in an elastomer. The greater N, that is, the higher M, the greater the extent of the rubbery plateau (compare Figure 14.1).

By analogy with an elastomer [Eq. (30)], the small strain rubbery plateau modulus may be set equal to $G_e = n_e kT$, where n_e is the "entanglement density", assumed to play an equivalent role to n_x in an elastomer. G_e may therefore be expressed in terms of an entanglement molar mass, M_e, according to Eq. (53), where N_A is Avogadro's number.

$$G_e = \frac{\rho N_A kT}{M_e} \tag{53}$$

In terms of the tube model, M_e is interpreted as the mass that corresponds to N_e statistical segments. Values of n_e and M_e deduced from G_e are found to vary widely according to chemical structure; $n_e \approx 4 \times 10^{25}\ \mathrm{m}^{-3}$ in aPS, which has relatively bulky chains, but is nearly an order of magnitude higher in PC, which has compact, relatively rigid chains [19].

The network and tube models both imply entanglement to be ineffective for $M < 2M_e$, so that M_c, the critical mass introduced in Section 14.1.1, may be taken to equal $2M_e$. As well as defining the threshold of rubber-like behavior above T_g, M_c marks the transition from a regime in which the melt viscosity $\eta \propto M$ (Rouse dynamics), to a regime in which $\eta \propto M^{3.4}$, characteristic of an entangled melt. The original theory of Doi and Edwards predicts $\eta \propto M^3$, but better agreement with experiment has since been obtained by incorporating the effects of primitive path fluctuations and constraint release (if all the chains reptate simultaneously, the constraints in Figure 14.10(a) cannot be considered strictly permanent for times of the order of τ_D) [4, 5, 14]. Even so, in the limit of very long monodisperse chains, for which the tube effectively remains fixed, or free chains trapped in a crosslinked network, the M^3 dependence is recovered. A strong dependence of the viscosity on M is a general feature of entangled polymers and is of great practical importance, because the choice of M in applications is often dictated by the need to reach a compromise between fracture resistance, which tends to improve with M (see Section 14.4.4) and melt processing, which is facilitated by low viscosities.

14.3.4
Nonlinear Viscoelasticity

Although it is a powerful means of investigating molecular structure and of basic characterization, and provides a general indication of the influence of M on flow behavior, the restrictions imposed by linear viscoelasticity make it inapplicable to a wide variety of practical problems. Nonlinearity is often associated with the phenomenon of "shear thinning", that is, a reduction of the viscosity with shear rate in steady flow, characteristic of many polymer melts at intermediate shear rates [15]. This contrasts with the Newtonian behavior implied by the Boltzman superposition principle for steady flow [Eq. (54), in a liquid, $G(s)$ must vanish as $s \to \infty$].

$$\sigma(t) = \int_{-\infty}^{t} G(t-u)\dot{\gamma}\, du = \dot{\gamma} \int_{0}^{\infty} G(s)\, ds = \dot{\gamma}\eta \tag{54}$$

Newtonian regimes are nevertheless widely observed in polymer melts in the high and low shear rate limits, where the viscosities are designated by η_∞ and η_o respectively. This is reflected by the empirical expressions widely used in engineering practice to describe the response to steady shear flow, an example being the Cross model [Eq. (55)], which reduces to the well known "power law" of Eq. (56) when $\eta_\infty \gg \eta \gg \eta_o$, with $n = 1/(m+1)$ between 10 and 20 for most polymer melts.

$$\frac{\eta_o - \eta}{\eta - \eta_\infty} = K\dot{\gamma}^m \tag{55}$$

$$\frac{\sigma}{\sigma_o} = \left(\frac{\dot{\gamma}}{\dot{\gamma}_o}\right)^{1/n} \tag{56}$$

However, these simple empirical expressions are far from universal, and fail to account for effects specific to nonlinear behavior, such as the appearance of finite first and second normal stress differences $\sigma_{xx} - \sigma_{yy} = N_1(\dot{\gamma})$ and $\sigma_{yy} - \sigma_{zz} = N_2(\dot{\gamma})$ in steady shear flow. (For a linear viscoelastic material in shear, σ_{xx}, σ_{yy} and σ_{zz} are equal to the applied pressure, usually atmospheric pressure.) These may be linked to the development of molecular anisotropy in polymer melts subject to flow, and are responsible for the Weissenberg effect, which refers to the tendency for a nonlinear viscoelastic fluid to "climb" a rotating rod inserted into it, as well as practically important phenomena such as die swell [20].

There has been considerable effort to develop phenomenological models for nonlinear viscoelasticity that provide a comprehensive description of such phenomena by extending the formalism of linear viscoelasticity to include situations where the viscoelastic functions can no longer be considered independent of stress or strain [21]. These are exemplified by the K-BKZ (Kaye-Bernstein, Kearsley, and Zapas) formalism, which includes a "damping function" in order to account for the acceleration of the relaxation rate at large deformations. The popularity of the K-BLZ approach is partly linked to the fact that the Doi–Edwards tube model, with the so-called "independent alignment" approximation, leads to expressions of the same form when generalized to large deformations, providing a molecular basis for its interpretation (in terms of the tube model, a large deformation will significantly increase L, giving rise to a supplementary relaxation process called "chain retraction" on a time scale τ_R, during which L returns to its equilibrium value [13, 14]). Phenomenological models, as well as expressions derived from the generalized Doi–Edwards approach are able to account for many of the observed features of polymer melts, and, although they are cumbersome, numerical techniques are now powerful enough to permit their incorporation into complex flow simulations of direct relevance to industrial practice.

14.4 Yield and Fracture

14.4.1 Yield in Polymers

The stresses and strains of practical importance for polymers at $T < T_g$ or T_m are often relatively large, and, unlike the small strains referred to in the preliminary discussion of the glassy state in Section 14.1.1, they may result in significant plastic deformation, in that the associated changes in specimen morphology are not recovered after stress release over laboratory time scales. A plastically deformed polymer with sufficiently high M is nevertheless able to return to its initial shape when heated above T_g or T_m, so that such deformation cannot strictly be considered irreversible.

As the extent of plastic deformation increases with the overall deformation, pronounced softening occurs, even when adiabatic heating effects are negligible (as will be assumed here). If fracture does not intervene, the force, f, on a specimen tested at constant speed in tension typically reaches a local maximum at a strain of a few percent, which is taken to correspond to a plastic yield stress, σ_Y (typically between 20 and 100 MPa in unoriented polymers). The subsequent yield drop is at least partly geometric in origin, although many glassy polymers also show an intrinsic yield drop [5]. The true stress on the polymer is given by $\sigma = A_o^{-1}\lambda f$, where A_o is the initial cross-sectional area of the specimen, so that Eq. (57) holds.

$$\frac{df}{d\lambda} = A_o\left(\frac{1}{\lambda}\frac{d\sigma}{d\lambda} - \frac{\sigma}{\lambda^2}\right) \tag{57}$$

Given that $d\sigma/d\lambda$ is a decreasing function of λ in the initial stages of deformation, as shown schematically in Figure 14.12, $A_o^{-1}f = \sigma/\lambda > d\sigma/d\lambda$ beyond a certain value of λ, and $df/d\lambda$ becomes negative. The resulting instability leads to localized necking in a tensile test, but if deformation continues beyond the yield point, the neck is stabilized by work hardening ($df/d\lambda$ becomes positive again) and propagates to the rest of the specimen at roughly constant f. Global work hardening then takes place, followed by rupture.

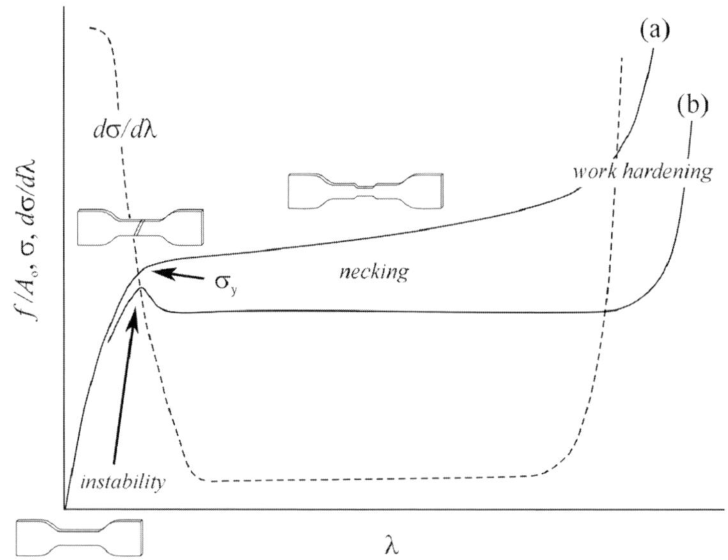

Fig. 14.12. Schematic of (a) the intrinsic stress-deformation curve of a polymer and (b) the force per unit area of an undeformed specimen observed as a function of total deformation in a tensile test.

In an amorphous glassy polymer, work hardening is considered to correspond to stretching of the entanglement network invoked to account for the rubbery plateau above T_g (see Section 14.3.3). This explains the recoverability of the deformation above T_g (in the absence of an applied force, the network retracts to its equilibrium conformation) and it is borne out by the observed correlation between the value of λ in the neck and the maximum extensibility of the network, $\lambda_{max} \approx N_e^{1/2}$. In semicrystalline polymers, the evolution of the crystalline texture may also contribute to work hardening, because the resolved shear stress on activated slip systems tends to decrease as deformation proceeds at constant stress, as will be discussed further (see Section 14.4.3).

The large strain response in the glassy or semicrystalline state is that of a nonlinear viscoelastic solid. However, both engineering and theoretical approaches to plasticity in polymers have largely developed as an independent discipline, in which σ_y plays a central role, in spite of its somewhat arbitrary definition (indeed it is not always possible to associate σ_y with a maximum in the force–deformation curve [5]). This is because in practice the yield point, rather than the ultimate strength, is usually considered to be the failure criterion for ductile materials.

For yield to occur at all, the global stress state must contain a deviatory component, so that pure hydrostatic stress states do not result in plasticity in a uniform specimen. Indeed, in many types of material, including metals, yield is usually well described by the Von Mises criterion, in which it is considered to be independent of the hydrostatic pressure, $p = (\sigma_{xx} + \sigma_{yy} + \sigma_{zz})/3$. With an appropriate frame of reference, any multiaxial stress state may be expressed as Eq. (58), where σ_1, σ_2, and σ_3 are the principal stresses (compare Section 14.2.1).

$$\boldsymbol{\sigma} = \begin{bmatrix} \sigma_1 & 0 & 0 \\ 0 & \sigma_2 & 0 \\ 0 & 0 & \sigma_3 \end{bmatrix} \tag{58}$$

The von Mises criterion is then Eq. (59).

$$(\sigma_1 - \sigma_2)^2 + (\sigma_2 - \sigma_3)^2 + (\sigma_3 - \sigma_1)^2 \geq 2\sigma_y^2 \tag{59}$$

It may be verified that σ_y is equal to the tensile yield stress by setting any two of the principle stresses to zero. For comparison, in pure shear, $\sigma_1 = -\sigma_2$ and $\sigma_3 = 0$, and the yield stress is $\tau_y = \sigma_y/\sqrt{3}$. Moreover, it is easily seen that Eq. (59) is independent of p. In fact, most polymers do show a pressure-dependent yield stress, an observation that is attributed to their relative compressibility, so that a negative value of p significantly reduces the space available for the molecules and hence reduces their mobility. However, this can usually be accounted for by replacing Eq. (59) with Eq. (60), where S and μ are materials constants [5].

$$(\sigma_1 - \sigma_2)^2 + (\sigma_2 - \sigma_3)^2 + (\sigma_3 - \sigma_1)^2 = S - \mu p \tag{60}$$

14.4.2
Models for Yield

The yield stress in disordered solids is strongly dependent on T and the deformation rate, and efforts have consequently been made to describe it in terms of a Ree–Eyring activated rate process, in which a rheological element "jumps" from place to place by overcoming local energy barriers [5, 22]. For an activation energy Q, and an attempt frequency ν_0, the jump rate $\nu = \nu_0 \exp(-Q/kT)$. In the absence of an applied stress, the jumps are in arbitrary directions, and there is no net deformation. However, an applied stress, σ, alters the effective energy barrier height for forward $(+)$ and backward $(-)$ jumps with respect to the stress axis, so that Eqs. (61) apply, where x is the displacement associated with each jump and A is the effective area of the rheological element.

$$\begin{aligned} \nu^+ &= \frac{\nu_o}{2} \exp\left(-\frac{Q - \sigma A x}{kT}\right) = \frac{\nu}{2} \exp\left(\frac{\sigma A x}{kT}\right) \\ \nu^- &= \frac{\nu}{2} \exp\left(-\frac{\sigma A x}{kT}\right) \end{aligned} \tag{61}$$

The strain rate in the stress direction is proportional to the net jump rate $\nu^+ - \nu^-$, leading to Eq. (62), where $V = xA$ typically corresponds to a few repeat units and is interpreted as an activation volume.

$$\dot{\varepsilon} \propto \frac{\upsilon}{2} \sinh\left(\frac{\sigma V}{kT}\right) \approx \dot{\varepsilon}_O \exp\left(-\frac{Q}{kT}\right) \exp\left(\frac{\sigma V}{kT}\right) \tag{62}$$

σ is then taken to equal σ_y when Eq. (62) is satisfied for a given strain rate $\dot{\varepsilon}$, which implies $\sigma_y \propto \log \dot{\varepsilon}$. This simple approach, which may easily be modified to include a pressure-dependent contribution to the activation energy, is often remarkably successful in predicting the strain rate dependence of σ_y, as shown in Figure 14.13(a) for PC. However, as shown in Figure 14.13(b), the agreement is more limited in polymers such as poly(vinyl chloride) (PVC). Because the Ree–Eyring model includes no assumptions regarding the nature of the rheological element, changes in the slope of σ_y versus $\log \dot{\varepsilon}$, such as those in Figure 14.13(b), have been attributed to secondary relaxations in the corresponding range of T and $\dot{\varepsilon}$. As pointed out in Section 14.3.1, the main secondary relaxation in PC occurs at $-80\,^\circ$C, but in PVC it is at about $50\,^\circ$C. This suggests that a complete description of yield may need to involve multiple activated rate processes [23].

Among the better-known efforts to provide a more microscopic description of yield within the general Ree–Eyring framework is the approach of Argon, based on metallurgical models, in which the energy barrier is associated with the elastic displacements necessary to accommodate the elementary shear processes [24]. Another well known, but somewhat different, approach is that of Robertson, who

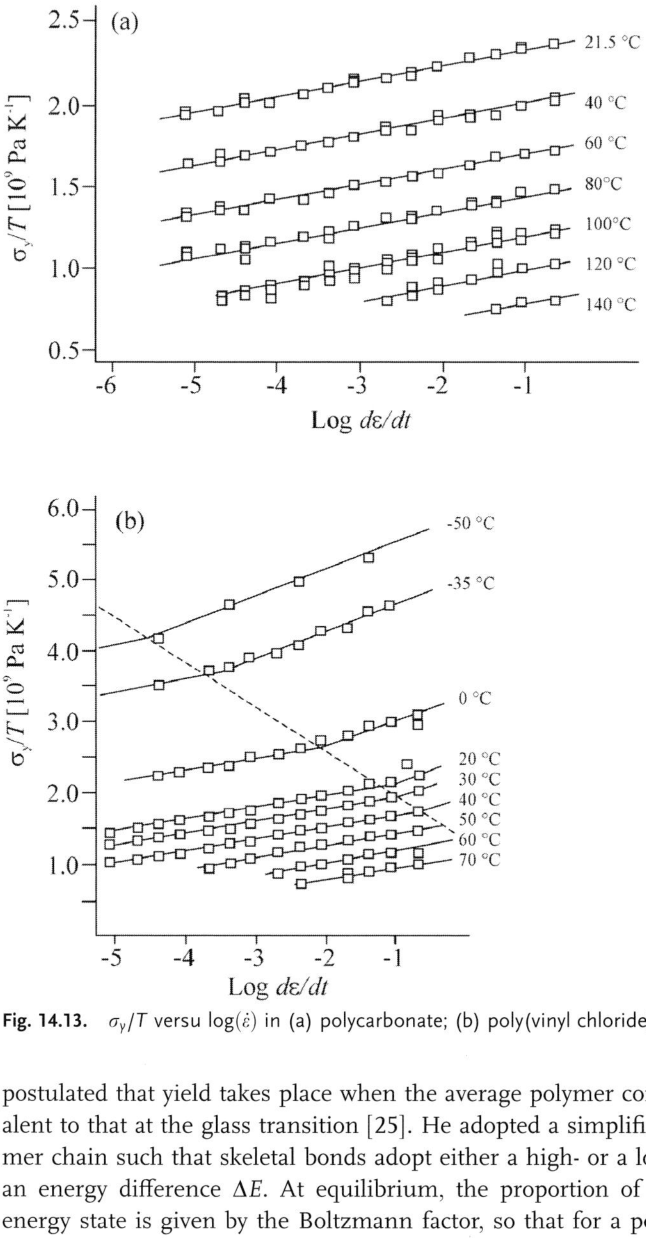

Fig. 14.13. σ_y/T versu $\log(\dot{\varepsilon})$ in (a) polycarbonate; (b) poly(vinyl chloride) (after Ref. 23).

postulated that yield takes place when the average polymer conformation is equivalent to that at the glass transition [25]. He adopted a simplified model for a polymer chain such that skeletal bonds adopt either a high- or a low-energy state with an energy difference ΔE. At equilibrium, the proportion of bonds in the high-energy state is given by the Boltzmann factor, so that for a polymer below its T_g. Eq. (63) holds, assuming the conformations at T_g to be substantially frozen in on vitrification.

$$\chi = \frac{\exp(-\Delta E/kT_g)}{1 + \exp(-\Delta E/kT_g)} \tag{63}$$

The value of χ in the presence of an applied stress, χ_{def}, is estimated by assuming the stress to modify the energy barrier to $\Delta E - \sigma v \cos \alpha$, where α is the local orientation of the bond with respect to the stress axis and v is the volume swept out during the transition from the low to high energy state. From the expression for χ_{def}, obtained by averaging the effective energy barrier over all possible orientations, a stress-dependent structure temperature, Θ, may be defined through Eq. (64).

$$\chi_{def} = \frac{\exp(-\Delta E/k\Theta)}{1 + \exp(-\Delta E/k\Theta)} \tag{64}$$

The WLF equation [Eq. (44)] may be used to calculate the viscosity as in Eq. (65), where η_g is viscosity at T_g.

$$\eta(\theta, T) = \eta_g \exp\left(2.303\left[\frac{C_1^g C_2^g}{\Theta - T_g + C_2^g}\frac{\Theta}{T} - C_1^g\right]\right) \tag{65}$$

The strain rate is then given by Eq. (66).

$$\dot{\varepsilon} = \frac{\sigma}{\eta(\theta, T)} \tag{66}$$

Robinson's model has met with some success in regimes of T close to T_g where the role of intramolecular barriers is relatively important. However, for T well below T_g, intermolecular barriers are expected to dominate, so that phenomenological or semi-phenomenological approaches of the Ree–Eyring type are arguably more appropriate to the observed T dependence, as borne out by Figure 14.13.

14.4.3
Semicrystalline Polymers

In amorphous polymers, plasticity is associated with deformation at $T < T_g$. In semicrystalline polymers, however, there is a further regime $T_g < T < T_m$, in which the amorphous regions are in the rubbery state and hence no longer contribute directly to yield. Plasticity is therefore dominated by the intermolecular barriers that oppose chain slip in the crystalline regions. This is of considerable practical importance, given that commodity semicrystalline polymers such as PE and isotactic polypropylene (iPP) are typically employed at $T > T_g$. Considerable effort has therefore been made to describe yield in semicrystalline polymers in terms of crystal plasticity theory. In isotropic polycrystalline materials, five independent crystal slip systems are generally necessary for global plasticity, but this requirement is relaxed in polymers owing to the presence of the amorphous phase. Indeed, yield in polymers is thought to involve a relatively limited number of slip systems with slip planes parallel to the chain axes (slip in planes perpendicular to the chain axes would require rupture of covalent bonds).

The classical approach of Young to yield in lamellar semicrystalline polymers is based on the nucleation of dislocations by thermal fluctuations [26]. The activation energy is calculated from the shear strain energy for a screw dislocation of width u [Eq. (67), where b is the magnitude of the Burgers vector (usually the repeat distance along the chain), r_o is the size of the dislocation core (about $4b$ according to computer simulations), G is the shear modulus, l_c is the length of the dislocation in the direction of the Burgers vector (taken to equal the lamellar thickness) and σ is the shear stress].

$$\Delta U = \frac{Gb^2 l_c}{2\pi}\left(\ln\left(\frac{u}{r_o}\right)\right) - \sigma bul \tag{67}$$

The activation energy for the formation of the dislocation, ΔU^*, corresponds to the maximum of $\Delta U(u)$, which leads to Eqs. (68) and (69).

$$u^* = \frac{Gb}{2\pi\sigma} \tag{68}$$

$$\Delta U^* = \frac{Gb^2 l_c}{2\pi}\left(\ln\left(\frac{u^*}{r_o}\right) - 1\right) \tag{69}$$

Given ΔU^*, one can estimate the critical stress for the formation of a dislocation σ_c. Usually, it is assumed that $\Delta U^* \approx 50kT$, which gives a σ_c comparable with experimentally determined values of the shear yield stress, τ_y. The main objection to the dislocation model is that the predicted temperature dependence of the yield stress is much weaker than that observed experimentally, particularly at higher T, and more recent efforts to describe yield have been based on thermally activated helical motions of polymer chains within their crystals, believed to be associated with the α relaxation, but which may occur below T_m [27].

In an isotropic polycrystalline polymer whose microstructure consists of stacked lamellae arranged in the form of spherolites, the slip systems activated depend on the local orientation of the lamellae with respect to the applied stress and, as deformation proceeds, these orientations are modified. To calculate the evolution of the crystalline texture, one can consider the polymer to behave as a crystalline aggregate. Although the entropic contribution of chain orientation in the amorphous regions may also need to be considered, the major contribution to work hardening in tension is rotation of the slip planes toward the tensile axis, so that the resolved shear stress in the slip direction diminishes. This results in a fiber texture in the limit of large deformations, such that the crystallites are oriented with their c axis (the chain axis) parallel to the stretch direction. Despite the relative success of such models, they do not explicitly address the micro-mechanisms involved in the transformation of the spherulitic texture into a fiber texture. One possibility is that the repeated shearing of lamellar fragments renders them thermodynamically unstable because of their small size. Models for the development of crystalline textures during drawing, based on this idea, thus explain the establishment of a new lamellar long period in highly deformed regions, which depends only on the deformation T and not on the initial lamellar thickness [27].

14.4.4 Crazing and Fracture

Rather than undergo homogeneous (non-cavitational) ductile necking, glassy amorphous polymers often show brittle behavior in tension, in that they fail abruptly prior to yielding and hence at a relatively low strain. In uncrosslinked or lightly crosslinked polymers, this phenomenon is usually associated with crazing [28, 29]. As shown in Figure 14.14, crazes are crack-like defects spanned by highly drawn fibrils with lateral spacings of the order of 10 nm and a constant draw ratio, close to λ_{max}, which implies that fibril extension takes place by ductile drawing at the craze–bulk interface. The presence of the fibrils means that crazes are capable of bearing loads normal to their faces. Indeed, the stress normal to the faces of an isolated growing craze, σ_c (the fibril drawing stress), is approximately equal to the global applied stress along most of the craze length. Hence, isotropic bulk samples deformed in uniaxial tension may contain a high density of crazes whose planes are parallel and perpendicular to the tensile axis. Even so, plastic deformation associated with crazing remains highly localized and contributes relatively little to the overall deformation. Given that crazes may act as preferential sites for crack initia-

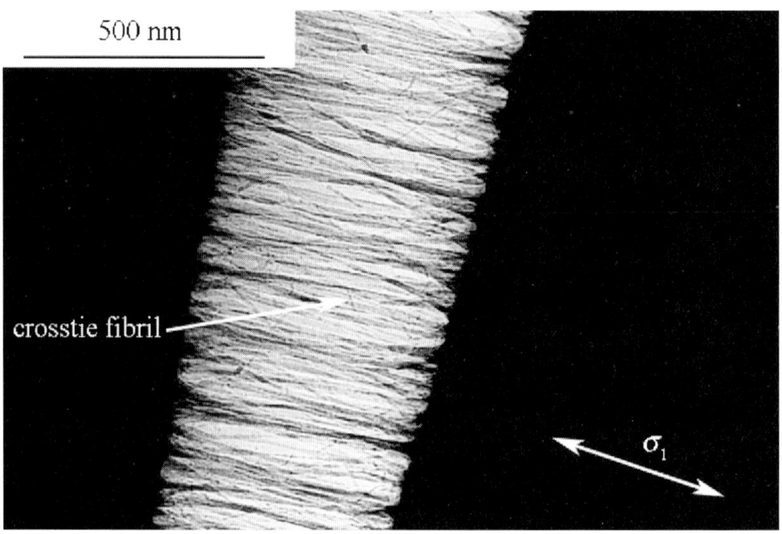

Fig. 14.14. Craze microstructure in a thin film of amorphous polycarbonate deformed in tension at −100°C.

tion, this accounts for their association with brittle macroscopic behavior as defined above. Crazing and ductile necking may nevertheless co-exist, depending on the relative stability of the crazes with respect to crack initiation.

Empirical criteria for the formation of crazes in multiaxial stress states, analogous to the von Mises criterion for yield [Eq. (59)], are based on the observation that crazing is absent in both compression and simple shear, which is reasonable given that one would expect cavitation to be favored by large values of p. A critical strain to craze, $\varepsilon_c = A + B/p$, is generally found to provide a reasonable description of craze nucleation, so that in terms of the principal stresses the criterion is given by Eq. (70), where A, B, C, and D are constants.

$$\sigma_1 - \nu\sigma_2 - \nu\sigma_3 = C + \frac{D}{\sigma_1 + \sigma_2 + \sigma_3} \tag{70}$$

Given that highly triaxial stress states are unfavorable to yielding [compare Eq. (59)], crazing tends to dominate in highly constrained geometries, such as notch tips, or in the vicinity of local stress concentrations in bulk specimens. This is not inconsistent with fibrillation by ductile drawing, because the void formation associated with crazing releases constraints on plastic deformation at the microscopic level.

The very small size of craze fibrils means that their surface energy is expected to play an important role in craze formation. Hence, although the large hydrostatic stress gradients associated with closely separated voids favor fibril drawing, they are offset by stress gradients arising from the surface tension at the void tips, Γ. In kinetic models for fibril formation, the fibrils are therefore argued to adopt the characteristic spacing D_o that maximizes the overall stress gradient and hence the rate of fibril extension. It has been shown on this basis that the craze stress $\sigma_c \propto \Gamma^{1/2}$, for a fixed deformation rate [28].

There is evidence that polymers with high entanglement densities, n_e, craze less readily than polymers with low n_e. One possible explanation for this is in terms of the creation of surface associated with fibrillation. For a fixed entanglement network, the creation of free surface must involve chain scission, as shown in Figure 14.15. The higher n_e, the more entanglements are lost during fibrillation, the higher the energy cost in creating the fibrils and hence the higher σ_c. This may account for the ductility of amorphous, high n_e "engineering thermoplastics" such as PC, in which crazing tends to be suppressed in favor of homogeneous deformation (σ_y is not directly dependent on n_e). It follows that highly crosslinked polymers, in which the influence of n_e and n_x may be considered additive, do not show crazing, although their ductility, and hence their toughness, is limited by the limit extensibility of the network [28]. The competition between homogeneous deformation and crazing is also likely to be strongly influenced by secondary relaxations, so that the relatively low β relaxation temperature of PC (see Section 14.4.2) may also contribute to its ductility. Indeed, shifting the secondary relaxation to higher T by chemical modification at constant n_e has been found to lead to embrittlement in PC [30]. Similarly, polymers that craze easily, such as aPS and PMMA, typically have relatively high secondary relaxation temperatures.

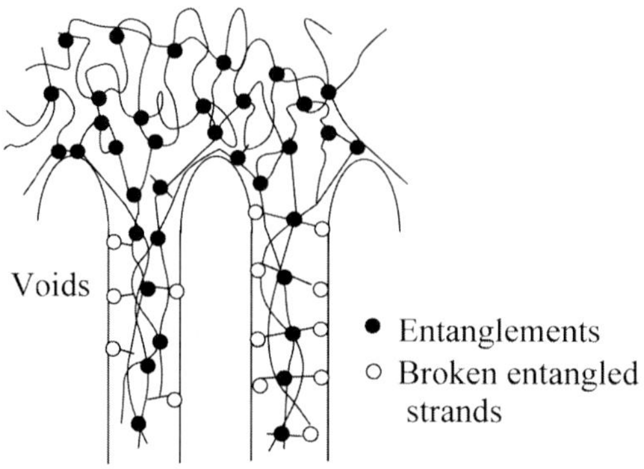

Fig. 14.15. Schematic of craze widening at the craze–bulk interface, illustrating the geometrically necessary entanglement loss.

The macroscopic property of immediate practical concern is often not so much craze formation as fracture resistance. In a cracked specimen, stress is concentrated at the crack tip, and the local stress state depends on the crack length. For small global deformations, it may be expressed in terms of a stress intensity factor, K, given by Eq. (71), where a is the crack length and F is a dimensionless function of the specimen geometry [31, 32].

$$K = F\sqrt{\pi a}\sigma \tag{71}$$

In many fracture mechanics-based approaches, crack advance is taken to occur when K reaches some critical value K_c (equivalent energy-based criteria are also widely used). K_c may then be measured using a pre-cracked specimen, in which a and hence K are well defined. In some cases (PMMA, for example) crack advance is observed to proceed via breakdown of a single craze at the crack tip. By modeling a craze as an orthotropic linear elastic body it has been shown that K_c is given by Eq. (72), where ν is Poisson's ratio, α is related to the craze anisotropy, σ_c is the draw stress normal to the craze-bulk interface, v_f is the fibril volume fraction in the craze, and σ_f is the stress to break a craze fibril [33].

$$K_{\mathrm{c}} \approx \alpha^{-1/4}\left(\frac{1-v_{\mathrm{f}}}{1-\nu^2}\frac{E}{\sigma_{\mathrm{c}}}\right)^{1/2}\sqrt{\pi D_{\mathrm{o}}}\sigma_{\mathrm{f}} \tag{72}$$

This approach incorporates the stress-concentrating effect of cross-tie fibrils, widely observed in crazes in glassy polymers (compare Figure 14.14). In the absence of any stress-concentrating effect, that is, for $\alpha \to 0$, a time-independent fibril failure criterion σ_f implies crack advance can never occur, because the stress in a given fibril can never exceed σ_c. This result has been confirmed by more detailed micromechanical modeling, and is important in that it provides a direct link between the

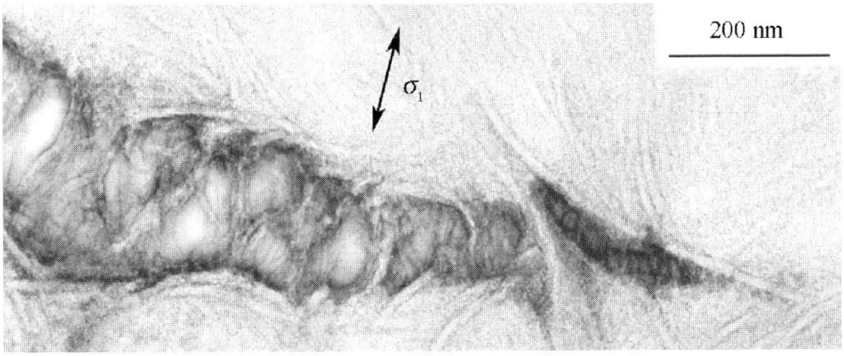

Fig. 14.16. Fibrillar deformation zone at a crack tip in a bulk specimen of semicrystalline high-density polyethylene deformed at room temperature (thin section stained with RuO_4).

macroscopic fracture behavior and microscopic quantities, which can be estimated by direct observation of craze microstructures. Although Eq. (72) is not strictly applicable to the more usual case of multiple crazing, or mixed crazing and shear at the crack tip, the guiding principle remains valid and serves to underline the importance of entanglement, not only for craze formation but also for ultimate failure.

Bulk semicrystalline polymers deformed in tension also show considerable localized micro-necking at T between T_g and T_m, in the form of either crazes or craze-like deformation zones, such as that shown in Figure 14.16. As with amorphous polymers, these are thought to have a strong influence on the ultimate failure properties. Although the details of the lamellar structure may vary widely, qualitative models for micro-necking in melt-crystallized semicrystalline polymers above T_g show many features in common. Deformation is usually assumed to initiate in interlamellar amorphous regions, which stretch and cavitate, so that cavity sizes are commensurate with the lamellar separation. The intervening material is then drawn down to form the mature fibrillar structure. Again, a high degree of entanglement and strong anchoring of the chains by the lamellae (that is, high M) are expected to promote toughness in semicrystalline polymers [5, 34, 35].

To improve the fracture resistance of polymers that craze readily, but which may have other advantages, such as cheapness in the case of aPS, a common strategy is to increase the density and extent of localized damage (crazing, for example) by creating local stress concentrations [36]. The greater the extent of local damage, the greater the energy dissipation during crack advance, and hence the greater the crack resistance. The presence of rubber inclusions in a glassy or semicrystalline matrix results in considerable local mismatch in stiffness. This increases their stress-concentrating effect and provides local concentrations in the von Mises stress or the hydrostatic stress, which will favor shear and craze yielding respectively. Cavitation of the modifier particles is also thought to be important in toughening and may in some cases change the local stress state sufficiently to promote a change in mechanism (for example, shear to craze transitions in glassy polymers).

14.5 Conclusion

The aim of this contribution has been to link the basic macroscopic phenomena associated with polymers with the unique features of their structure, the most obvious being the presence of long, flexible molecular chains. The important role of the conformational entropy of flexible chains, not only for rubber elasticity but for polymer dynamics in general, has been demonstrated. Moreover, the concept of an entanglement network, which underpins much of the theory of polymer dynamics in the melt, has also been shown to have important repercussions for the high strain behavior of solid polymers, namely plastic deformation, crazing, and fracture.

Notation

A	Helmholtz free energy [J]
A_o	cross-sectional area of a tensile specimen [m^{-2}]
a	Helmholtz free energy per unit volume [$J\ m^{-3}$]
a	statistical segment length of the primitive path (tube model) [m]
a_T	WLF shift factor
b	equivalent bond length (Kuhn length) [m]
b	magnitude of Burgers vector [m]
C_l	specific heat at fixed specimen length [$J\ kg^{-1}$]
C_1^g	WLF constant
C_2^g	WLF constant [K]
C_∞	characteristic ratio
D	tensile compliance [Pa^{-1}]
D_o	craze fibril spacing [m]
D_R	Rouse self-diffusion coefficient [$m^2\ s^{-1}$]
E	Young's (tensile) modulus [Pa]
E'	tensile storage modulus [Pa]
E''	tensile loss modulus [Pa]
E^*	complex tensile modulus [Pa]
$\mathbf{E}_{ijkl}$	stiffness tensor [Pa]
f	force [N]
f_g	fractional free volume at T_g (dimensionless)
f_v	fractional free volume (dimensionless)
G	shear modulus [Pa]
G'	shear storage modulus [Pa]

G''	shear loss modulus [Pa]
G^*	complex shear modulus [Pa]
G_e	plateau modulus [Pa]
J	shear compliance [Pa^{-1}]
K	bulk modulus [Pa]
K	stress intensity factor [Pa $m^{1/2}$]
K_c	critical stress intensity for crack advance [Pa $m^{1/2}$]
k	Boltzmann constant $= 1.38066 \times 10^{-23}$ J K^{-1}
L	tube contour length (tube model) [m]
l	length [m]
l_c	lamellar thickness [m]
l_o	mean length of skeletal bonds in a real macromolecular chain [m]
M	molar mass [g mol^{-1}]
M_b	molar mass per statistical segment [g mol^{-1}]
M_c	critical molar mass [g mol^{-1}]
M_e	entanglement molar mass [kg mol^{-1}]
m	number of beads in Rouse model (dimensionless)
N	number of statistical segments in a freely jointed chain (dimensionless)
N'	number of statistical segments between beads (Rouse model) (dimensionless)
N_l	number of skeletal bonds in a real macromolecular chain (dimensionless)
N_1	first normal stress difference (dimensionless)
N_A	Avogadro's number $= 6.022 \times 10^{23}$
N_e	number of statistical segments in a chain of molar mass M_e (dimensionless)
N_L	number of statistical segments in the primitive path (tube model) (dimensionless)
n	number of chains per unit volume [m^{-3}]
n_e	entanglement density [m^{-3}]
n_x	number of crosslinks per unit volume [m^{-3}]
$P(\mathbf{R}, N)$	probability distribution of **R** in a chain with N links
p	hydrostatic pressure [Pa]
Q	activation energy [J]
Q	heat transfer (into a thermodynamic system) [J]
$\mathbf{R}$	end-to-end vector of a linear chain [m]
R_{max}	maximum end-to-end distance of a real macromolecule
$\mathbf{r}_n$	vector corresponding to the nth bond in a chain [m]
r_o	dislocation core radius [m]
S	entropy [J K^{-1}]
T	temperature [K, °C]
T_g	glass transition temperature [K, °C]
T_m	melting point [K, °C]
U	internal energy [J]
u	width of screw dislocation [m]

u^*	critical screw dislocation width [m]
V	activation volume (Eyring model) [m^{-3}]
v_f	craze fibril volume fraction (dimensionless)
W	work (done on a thermodynamic system) [J]

Greek

α	craze anisotropy factor
α_f	coefficient of thermal expansion of fractional free volume [K^{-1}]
Γ	surface tension at craze void tips [N m^{-1}]
γ	shear strain
ΔE	energy difference (Robertson model) [J]
ΔU	shear strain energy of screw dislocation [J]
ΔU^*	activation energy for screw dislocation [J]
δ	phase angle
δ_c	critical crack opening displacement [m]
ε_c	strain to craze
$\boldsymbol{\varepsilon}_{ij}$	engineering strain
ε_{ij}	components of the engineering strain
ζ	friction coefficient [N s m^{-1}]
ζ_o	monomeric friction coefficient [N s m^{-1}]
η	viscosity [N s]
η_o	zero shear viscosity [N s]
η_g	viscosity at T_g [N s]
η_∞	limiting high shear rate viscosity [N s]
Θ	structure temperature (Robertson model) [K]
λ	deformation ratio (dimensionless)
μ	pressure coefficient (stress-dependent von Mises criterion) (dimensionless)
ν	jump rate (Eyring model) (dimensionless)
ν	Poisson's ratio (dimensionless)
ν	scaling exponent (Zimm model) (dimensionless)
ν_0	attempt frequency (Eyring model) (dimensionless)
ρ	density [kg m^{-3}]
σ_c	craze fibril draw stress [Pa]
σ_f	stress to break a craze fibril [Pa]
σ_{ij}	components of the engineering stress [Pa]
$\boldsymbol{\sigma}_{ij}$	engineering stress [Pa]
σ_y	tensile yield stress [Pa]
τ	relaxation time [s]
τ_c	critical stress for the formation of a screw dislocation [Pa]
τ_D	reptation time [s]
τ_R	Rouse relaxation time [s]
τ_y	shear yield stress [Pa]
$\Omega(\mathbf{R})$	number of conformations at fixed **R** (dimensionless)

ω	angular frequency (dimensionless)
χ	proportion of high energy states (Robertson model)

Acronyms

aPS	atactic polystyrene
iPP	isotactic polypropylene
K-BKZ	Kaye-Bernstein–Kearsley–Zapas
PC	bisphenol A polycarbonate
PE	polyethylene
PMMA	poly(methyl methacrylate)
PVC	poly(vinyl chloride)
WLF	Williams–Landel–Ferry

References

1 P. J. Flory, *Statistical Mechanics of Chain Molecules*, John Wiley, New York, **1969**.

2 W. L. Mattice, U. W. Suter, *Conformational Theory of Long Chain Molecules*, John Wiley, New York, **1994**.

3 L. J. Fetters, D. J. Lohse, W. W. Graessley, *J. Polym. Sci., Part B: Polym. Phys.*, **1999**, *37*, 1023.

4 M. Rubenstein, R. H. Colby, *Polymer Physics*, Oxford University Press, Oxford, **2003**.

5 I. M. Ward, D. W. Hadley, *An Introduction to the Mechanical Properties of Solid Polymers*, John Wiley, New York, **1993**.

6 U. W. Gedde, *Polymer Physics*, Chapman and Hall, London, **1995**.

7 L. R. G. Treolar, *The Physics of Rubber Elasticity*, Clarendon Press, Oxford, **1958**.

8 J. E. Mark, B. Erman, *Rubber-Like Elasticity, a Molecular Primer*, John Wiley, New York, **1988**.

9 B. Erman, J. E. Mark, *Structures and Properties of Rubberlike Networks*, Oxford University Press, Oxford, **1997**.

10 J. D. Ferry, *Viscoelastic Properties of Polymers*, 3rd ed., John Wiley, New York, **1980**.

11 N. W. Tschoegl, *The Phenomenological Theory of Linear Viscoelastic Behaviour*, Springer Verlag, Berlin, **1989**.

12 K. L. Ngai, D. J. Plazek, in *Physical Properties of Polymers Handbook* (J. E. Mark, editor) AIP Press, Woodbury, NY, **1996**.

13 M. Doi, *Introduction to Polymer Physics*, Oxford University Press, Oxford, **1996**.

14 M. Doi, S. F. Edwards, *The Theory of Polymer Dynamics*, Oxford University Press, Oxford, **1986**.

15 R. G. Larsen, *The Structure and Rheology of Complex Fluids*, Oxford University Press, New York, **1998**.

16 R. B. Bird, C. F. Curtiss, R. C. Armstrong, O. Hassager, *Dynamics of Polymeric Liquids, Vol. 2, Kinetic Theory*, 2nd ed., John Wiley, New York, **1987**.

17 J. des Cloizeaux, G. Jannink, *Polymers in Solution: Their Modelling and Structure*, Clarendon Press, Oxford, **1990**.

18 P. G. DeGennes, *Scaling in Polymer Physics*, Cornell University Press, Ithaca, **1979**.

19 L. J. Fetters, D. J. Lohse, R. H. Colby, in *Physical Properties of Polymers Handbook* (J. E. Mark, editor) AIP Press, Woodbury, NY, **1996**.

20 H. A. Barnes, J. F. Hutton, K. Walters, *An Introduction to Rheology*, Elsevier, Amsterdam, **1989**.

21 R. G. Larson, *Constitutive Equations*

for Polymer Melts and Solutions, Butterworths, Boston, MA, **1988**.
22 B. Crist, in *Materials Science and Technology, Vol. 12* (E. L. Thomas, editor), VCH, New York, **1993**.
23 J. C. Bauwens, C. Bauwens-Crowet, G. Homès, *J. Polym. Sci. A2*, **1969**, *7*, 735.
24 A. S. Argon, *Phil. Mag.*, **1973**, *28*, 839.
25 R. E. Robertson, *J. Chem. Phys.*, **1966**, *44*, 3950.
26 R. J. Young, *Introduction to Polymers*, Chapman and Hall, London, **1981**.
27 A. Galeski, *Prog. Polym. Sci.*, **2003**, *28*, 1643.
28 E. J. Kramer, *Adv. Polym. Sci.*, **1983**, *52*, 1.
29 E. J. Kramer, L. L. Berger, *Adv. Polym. Sci.*, **1990**, *91/92*, 1.
30 L. P. Chen, A. F. Yee, E. J. Moskala, *Macromolecules*, **1999**, *32*, 5944.
31 J. G. Williams, *Fracture Mechanics of Polymers*, Ellis Horwood, Chichister, **1984**.
32 H.-H. Kausch, *Polymer Fracture*, Springer Verlag, Berlin, **1985**.
33 H. R. Brown, *Macromolecules*, **1991**, *24*, 2752.
34 C. J. G. Plummer, H.-H. Kausch, *J. Macromol. Sci. – Phys.*, **1996**, *B35*, 637.
35 J. J. Benkoski, P. Flores, E. J. Kramer, *Macromolecules*, **2003**, *36*, 3289.
36 C. B. Bucknall, *Toughened Plastics*, Applied Science, London, **1977**.

15
Polymer Degradation and Stabilization[1]

Tuan Quoc Nguyen

15.1
Introduction

Most synthetic plastics and natural polymers are principally constituted of the elements C, H, N, and O, all of which are primary components of organic molecules. As with smaller organic molecules, these atoms are attached by relatively weak covalent bonds that are susceptible to chemical attack, particularly to oxidative degradation. During their potential lifetime, engineering plastics are exposed to diverse environmental factors which can act alone or in combination to adversely affect the initial material properties. These changes generally occur over several years, but can be accelerated or retarded in the presence of internal and external elements temperature, and temperature variations. Environmental factors that contribute to degradation include mechanical stresses, temperature variations, humidity, sunlight, oxygen, atmospheric pollutants, and microbial enzymes. Depending on the structural level at which material changes occur, it is usual to distinguish between physical, physicochemical, and chemical degradation.

- Physical degradation, commonly known as physical aging, results from changes in the thermodynamic state of the system (temperature, pressure, molar volume) during the daily use of the material. During processing, chains may be oriented and the sample inhomogeneously cooled, resulting in internal stress. With time, the chains have the opportunity to relax from their nonequilibrium conformations to a state of lower free energy. Partly immiscible blends, for instance, can remain in a homogeneous metastable state for an extended period of time. The presence of stress or heat can accelerate the phase separation process. The resultant changes in orientation, free volume, internal stress, crystalline content, and

1) The symbols used in this chapter are listed at the end of the text, under "Notation".

Handbook of Polymer Reaction Engineering. Edited by T. Meyer, J. Keurentjes

ISBN: 3-527-31014-2

phase transitions are known globally as physical aging. In principle, physically aged samples do not affect chemical structure and can be rejuvenated by proper thermal treatment.

- Physicochemical degradation involves diffusion of small to medium size molecules either from the polymer (plastifier or other additives) or into the polymer (water absorption, solvent, surface-active liquids), resulting in changes in polymer properties. Exudation of plastifier and surface migration of additives followed by washing are among the common problems encountered in natural weathering. In the presence of surface-active agents, environmental stress cracking can also develop. Although preserving the structural integrity of the polymer, physicochemical degradation is in general irreversible as a result of local changes in chemical composition, and can result in severe alterations of material properties.
- Chemical degradation concerns irreversible changes in the chemical structure of the polymer initiated by a number of activation agents frequently encountered in outdoor weathering:
  - oxygen (causing oxidation), reactive pollutants (ozone, nitric oxides), enzymes (causing biodegradation);
  - heat (causing thermal degradation);
  - photons (from UV or γ-radiation), electrons and high energy particles (causing photodegradation and radiative degradation);
  - mechanical stress (causing mechanochemical degradation).

In this review, we will focus exclusively on chemical degradation, with particular emphasis on environmental degradation. It should be kept in mind, however, that under practical weathering conditions, physical, physicochemical and chemical degradation may happen simultaneously and in a synergetic way, in some cases resulting in premature material failure.

Although the term "degradation" usually has a negative connotation, some polymer applications require degradability as an asset or even a prerequisite as, for instance, in biodegradable detergents in some biomaterials, and in the manufacture of microelectronic circuitry. In fact, the growing importance of degradation and stability assessment in plastic materials is largely motivated by two apparent contradictary interests:

- on one hand, there is an increased consumer demand for quality and durability to avoid the expensive labor costs of replacing deteriorated components;
- on the other hand, for some applications, the plastic should be able to degrade spontaneously and safely when disposed of, to reduce the solid waste burden.

Obviously, this "degradability-on-demand" can be realized only with an adequate understanding of the degradation and stabilization mechanisms.

15.2 General Features of Polymer Degradation

Over 95 wt.% of polymers are synthesized from eight vinyl monomers, and most of the remainder are from another couple of dozen monomers with diverse chemical structures. Despite this relative simple chemistry, the problem of degradation and stabilization of synthetic polymers is exceedingly complex. This complexity stems from the structural organization of plastic materials, which occurs at different levels:

- At the molecular level, most commercial plastics are mixtures of macromolecules with diverse substances, which can be traces of solvent or catalyst residues, pigments, fillers, and a variety of additives.
- The macromolecules themselves can be blends of homopolymers or copolymers.
- As to the differences in small organic molecules, even a simple homopolymer presents a diversity in chain length and configuration. The presence of chain branching, tacticity, monomer linking, or chain defects may all have important consequences for the material's stability.
- Molecular chains can arrange themselves into amorphous, oriented, and crystalline domains of different sizes, shapes, and distributions. This intricate morphology is globally known as supramolecular organization.

Studying degradation requires not only identification of the diverse chemical reactions which induce structural changes in the polymer, but also the interactions between the numerous chemical species initially present in the sample or formed in the course of degradation. Interdependence of the change in material morphology that occurs upon aging, and its influence on the rate of degradation, should be considered additionally.

15.2.1 Degradative Reactions

From the complexity of chemical processes, and the intricate relationship between chemical, morphological, and environmental factors which control degradation, it is safe to state that there are as many degradation mechanisms as commercial plastic formulations. Obviously, examining in detail all the effects mentioned for a given polymer system requires considerable effort. Fortunately, a large group of polymer degradation problems share a number of common features. Most degradative reactions of polyolefins, for instance, can be envisioned as the reverse process of vinyl polymerization, with four steps: radical initiation, depropagation (the reverse of propagation), chain branching under oxidative conditions (produced by intermolecular and intramolecular hydrogen transfer followed by β-scission), and radical termination. Except for the mode of initiation, the thermal, photolysis, and radiolytic degradation processes are similar in many respects and can be described

Initiation	$PH + X\bullet \rightarrow P\bullet + XH$	(1)
Radical conversion	$P\bullet + O_2 \rightarrow PO_2\bullet$	(2)
Propagation	$PO_2\bullet + PH \rightarrow POOH + P\bullet$	(3)
Chain branching	$POOH \rightarrow PO\bullet + \bullet OH$	(4)
	$PO\bullet + PH \rightarrow POH + P\bullet$	(5)
	$HO\bullet + PH \rightarrow H_2O + P\bullet$	(6)
	$PO\bullet \rightarrow$ various chain-scission reactions	(7)
Termination	$P\bullet + P\bullet \rightarrow$ inactive products	(8)
	$PO_2\bullet + PO_2\bullet \rightarrow$ inactive products	(9)

Scheme 15.1. General oxidative degradation mechanism for polyolefins (PH designates the polymer, P• a macroradical, and X• a radical of unspecified nature).

by the same reaction scheme, a simplified version of which, based on polyolefin degradation but applicable to other types of polymers with a carbon backbone, is given as Eqs. (1)–(9) in Scheme 15.1 [1].

15.2.1.1 **Initiation**

Covalent bond scission, with formation of free radicals or charged species, is the starting process in any type of chemical degradation. This primary event requires energy, which can be supplied by heat (thermal degradation), UV photons (photodegradation), high-energy photons or particles (radiolysis), mechanical forces (mechanochemical degradation), or reactive chemicals (solvolysis, enzymatic degradation). Heterolytic bond scission is restricted to degradation by reactive chemicals and by enzymes in biodegradation. For the other modes of degradation, bond scission is mainly homolytic, except in some particular situations described in Section 15.2.2. Apart from its free-radical nature, homolytic initiation is generally a complex and less well-known process (radical X• of reaction (1)). The exact mechanisms of initiation depend on the polymer structure, the presence of internal and external impurities, and the degradation conditions. Also, the mode of initiation may change during the course of reaction. Under oxidative conditions, for instance, formation of hydroperoxides followed by their decomposition rapidly becomes the prevalent mode of initiation.

15.2.1.2 **Propagation**

Once initiated, the alkyl macroradical can proceed through a number of possible reaction pathways. In the absence of oxygen, some macroradicals can depolymerize by unzipping, or they can abstract a hydrogen atom from a nearby molecule or from another group inside the same molecule. In oxidative environment, the free radical combines readily with molecular oxygen to produce peroxy radicals, which can subsequently abstract H atoms to re-form an alkyl radical. Radical conversion reaction is extremely rapid and is diffusion-controlled, owing to the diradical nature of molecular oxygen, with $k_2 \sim 10^8$–10^9 L mol^{-1} s^{-1}. Hydrogen abstraction reactions, on the other hand, require a high activation energy and are the rate-determining step with $k_2/k_3 > 10^6$. The activation energy for k_3 depends on the enthalpy of reaction $\Delta H = D_e(\text{P–H}) - D_e(\text{POO–H})$, where D_e is the bond dissociation energy. With $D_e(\text{POO–H}) \approx 360$ kJ mol^{-1}, peroxide abstraction reactions are endothermic with a rate constant decreasing rapidly with increasing C–H bond dissociation energy, according to the sequence: tertiary C–H ($D_e = 385$ kJ mol^{-1}) > secondary C–H (400 kJ mol^{-1}) > primary C–H (410 kJ mol^{-1}). This behavior has been verified in the case of polyolefins, for which it is found that HDPE is more resistant to oxidative degradation than LDPE, which is itself more resistant than isotactic polypropylene (iPP), owing to the increasing presence of tertiary C–H bonds in the polymer [2].

15.2.1.3 **Chain Branching**

Hydroperoxide decomposition is the key step in polymer oxidative degradation and has been extensively studied in a number of systems, using low MW analogs of the polymers. Corresponding rate constants and activation energies can be found in Ref. 2, p. 383. The fate of the hydroperoxides formed is diverse. The O–O bond energy is only approximately 170–195 kJ mol^{-1} and is easily cleaved by heat or UV light below 340 nm according to Eq. (10).

$$\text{PO–OH} \xrightarrow{\Delta, h\nu<340\ \text{nm}} \text{PO}\bullet + \bullet\text{OH} \tag{10}$$

Cations of transition metals (Cu, Fe, Mn, Co, Ti) which may exist in the polymer as contaminants or as catalyst residues can induce the decomposition of hydroperoxides and contribute to an increase in the oxidation rate at low temperature (Haber–Weiss reaction, Eqs. (11) and (12)).

$$\text{PO–OH} + \text{M}^{n+} \rightarrow \text{PO}\bullet + {}^{-}\text{OH} + \text{M}^{(n+1)+} \tag{11}$$

$$\text{PO–OH} + \text{M}^{n+1} \rightarrow \text{PO}_2\bullet + {}^{+}\text{H} + \text{M}^{n+} \tag{12}$$

At high hydroperoxide concentrations, a bimolecular dissociation mechanism [Eq. (13)] between hydrogen-bonded molecules (for instance, between adjacent peroxide groups formed by an intramolecular abstraction reaction in PP) can also occur.

$$2\text{PO–OH} \rightarrow \text{PO}_2\bullet + \text{PO}\bullet + \text{H}_2\text{O} \tag{13}$$

Apart from homolytic cleavage, the hydroperoxides can undergo a number of other reactions such as addition to double bonds that may have been initially present in the polymer or formed in the course of degradation [Eq. (14)] and radical-induced decomposition [Eqs. (15) and (16)].

$$\text{PO–OH} + \text{–HC=CH–} \rightarrow \text{PO}\bullet + \text{–}\underset{\text{OH}}{\text{HC}}\text{–}\dot{\text{C}}\text{H–} \tag{14}$$

$$\text{P}\bullet + \text{PO–OH} \rightarrow \text{PO}\bullet + \text{POH} \tag{15}$$

$$\text{HO}\bullet + \text{PO–OH} \rightarrow \text{PO}_2\bullet + \text{H}_2\text{O} \tag{16}$$

The decomposition of hydroperoxides creates more radicals than are initially formed by initiation, resulting in an autoacceleration of the degradation process (chain branching). Additionally, initiation by hydroperoxides is autoregenerated in the sense that, overall, the same quantity of peroxides is recovered at the end of the decomposition cycle.

Polyalkoxy and polyperoxy radicals formed from hydroperoxide decomposition are unstable intermediates, owing to the presence of an unpaired electron which weakens the dissociation energy for bonds in the β-position by some 100 kJ mol^{-1}. Monomolecular β-scission is the main mechanism for a decrease in MW and accounts for the variety of oxygenated products observed during oxidative degradation (see Section 15.4.4.6).

15.2.1.4 Termination

Under anaerobic conditions, the alkyl macroradicals can recombine by combination and dismutation [Eq. (17)].

$$\text{P}\bullet + \text{P}\bullet \rightarrow \text{P–P} \quad \text{or} \quad \text{P–H} + \text{P(–H)} \tag{17}$$

In the presence of oxygen, hydroperoxy radicals are the most abundant; this is a result of their relative low reactivity in solid polymers. Recombination of peroxy radicals can proceed through a tetroxide intermediate, giving peroxides, polymer alkoxy radicals, and oxygen [Eqs. (18a) and (18b)], or it can take place through a six-membered transition state which would lead to an alcohol, singlet oxygen, and triplet excited carbonyl, which, upon de-excitation, produces the chemiluminescence observed in many oxidative degradations (compare the Russell mechanism, Section 15.3.6).

$$2\text{PO}_2\bullet \rightarrow \{\text{P–O–O–O–O–P}\} \rightarrow \{\text{PO}\bullet + \text{O}_2 + \bullet\text{OP}\}_{\text{cage}} \begin{cases} \nearrow \text{PO–OP} + \text{O}_2 & (18a) \\ \searrow 2\text{PO}\bullet + \text{O}_2 & (18b) \end{cases}$$

As with smaller radicals, the reactivity of a macroradical much depends on the electron density of the atom bearing the unpaired electron. If reactivity is low, the chance of a radical being trapped before abstraction can occur increases with decreasing reactivity. For radicals formed by degradation, the lifetimes follow the order $P\bullet < PO\bullet < PO_2\bullet$. Free isolated $PO_2\bullet$ in polypropylene may have a half-life-time as long as 500 s at ambient temperature. To account for this exceptional stability, and in analogy to the post-gel effect in free-radical polymerization, it may be appropriate to include a monomolecular mechanism for radical recombination, in addition to the classical bimolecular scheme [Eq. (19)] [3].

$$(P\bullet, PO\bullet, PO_2\bullet) \rightarrow \text{trapped radicals} \tag{19}$$

The various steps described by the general auto-oxidative degradation of Scheme 15.1 will be illustrated by the oxidative degradation of polypropylene (Section 15.4.4).

15.2.2 Some Nonradical Degradation Mechanisms

Although the majority of polymers degrade thermally or photochemically by a free-radical process, there are a few examples where covalent bond scission occurs by electron transfer around a six-membered transition state. Thermal degradation of poly(ethylene terephthalate) [Eq. (20)] and side group cleavage in poly(methyl acrylate) and poly(*tert*-butyl methacrylate) are some examples where β-scission of the alkyl–oxygen bond takes place from a cyclic transition state without free-radical formation [4].

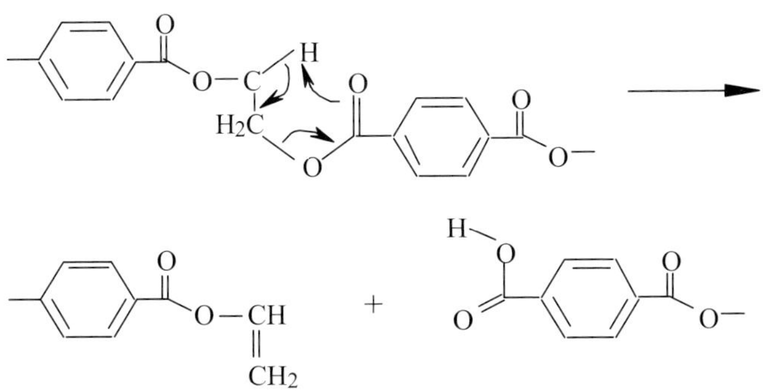

(20)

Cyclization of side chains may also occur without bond scission as in polyacrylonitrile (see Section 15.4.6.3).

15.2.3 Physical Factors

Polymer degradation is highly dependent on chemical and physical factors. The chemical factors will be examined in relation to thermal degradation (see Section 15.4.2). This section will be focused exclusively on the influence of physical factors on polymer degradation.

Polymer degradation can be observed in solution, in melt or in the solid state. Degradation in solution is principally used with "model" compounds for kinetics studies, whereas degradation in melt is frequently associated with processing con-

Tab. 15.1. Factors relevant to chemical reactions in solid polymers [5].

Principal factor	*Influences*
Mobility of reactants	translational and rotational diffusion of reacting groups as function of • temperature and macromolecular motions (T_g, T_β, T_γ, ...) • size and shape of reacting groups • free volume and conformation • interactions between reacting groups and matrix cage effect
Heterogeneity of the system	macroheterogeneity (gradients of absorbed light, oxygen concentration, ...) morphological heterogeneity (crystalline and amorphous domains, phase separation in block copolymers and blends) microheterogeneity (distribution of reactive sites, free volume)
Change in matrix structure during reaction	selective depletion of reactive sites crosslinking, chain scission, change in absorbance change in crystallinity
Factors other than mass transfer	excited energy transfer and migration electron transfer transfer of active sites due to chain reaction

ditions. Obviously, solid-state experiments are the most relevant to normal polymer weathering conditions. It is also in this state that degradation kinetics reveals its highest degree of complexity, as a result of interplay between physical, physico-chemical, and chemical variables. From the chemical viewpoint, elementary reactions should depend on temperature and not specifically on the state of aggregation. Chemical reactions in the solid do differ, however, from those in the gas or in the liquid phase by the rate of transport of reacting species, a factor which is further exacerbated by the heterogeneity of the reacting medium. Some prevalent factors which may affect chemical reactions in solid polymers are summarized in Table 15.1.

15.2.3.1 Glass Transition Temperature

In contrast to small organic molecules, long-chain polymers are characterized not only by the center of mass diffusion, but also by small-scale diffusion of a few monomer units. It is generally observed that even monomolecular reactions could happen only if these motions are unfrozen. In the wider sense, the dependence of reaction efficiency on polymer morphological structures can be described in terms of the free volume concept, and of diffusion constants [6]. These molecular characteristics are themselves dependent on the thermal transitions in polymers, the most important of which being the glass transition temperature.

The glass transition corresponds to the onset of large-scale motions of long segments of the macromolecules (10–20 monomer units). These liquidlike motions require more free volume, resulting in a larger volumic thermal expansion coefficient above the glass transition temperature (T_g). Below T_g, intra- and intermolecular rearrangements are strongly hindered. This restricted mobility also limits the diffusion of oxygen, which is a controlling factor in oxidative degradation. Reactions which require a notable change in activation volume should depend much on the T_g. This is the case for the Norrish II process, in which the transition state involves a six-membered ring conformation (Section 15.5.5.2). In the photolysis of vinyl ketone polymers and ethylene–carbon monoxide copolymers, it was observed that the quantum yield in solid film above the T_g was identical to that in solution, but decreased with temperature below T_g, to become negligible below the β-transition temperature. Reactions which involve free radicals generally require little change in activation volume and should proceed equally well below or above T_g, as is the case for the photo-Fries rearrangement (Section 15.5.5.3) [7].

Chain mobility has important implications for bimolecular reaction efficiency. Bimolecular reactions depend on (i) the frequency of encounter and (ii) the concentrations of the reactive species. Upon bond cleavage, macroradicals are formed in pairs and can initiate degradative reactions only if they are sufficiently flexible to move apart. If the macroradicals are held rigid, as in thermosets or below the T_g, they may recombine by the "cage" effect without any detectable molecular change.

Mass transfer limitations are a function of the diffusion coefficient for each species. Two situations can be encountered [6]:

- At least one reactant is a small molecule (or radical). Diffusion of small species, such as oxygen or methyl radicals, in solid glassy polymer can be much more rapid than predicted from the bulk viscosity of the medium. The internal viscosity of solid polymers can be very low and bimolecular reactions in the solid state, even below the T_g, can be almost as efficient as in solution.
- Both reactants are polymeric species. In this case, the effective encounter rate is highly limited by mass transfer and shows a marked increase above the T_g. Below the T_g, all the large-scale motions are suppressed although motion of short segments or of side groups of the main polymer may persist and have an influence on the rate of degradation. It is observed, for instance, that recombination of macroradicals in PMMA, PS, and PVC can occur well below the T_g. A valency migration mechanism [Eq. (21)], in which the unpaired electron is transferred between neighboring polymers until another macroradical is encountered, has been proposed [6].

$$-{\bullet}CH- + -CH_2- \rightarrow -CH_2- + -{\bullet}CH- \quad (21)$$

The rate constant for this radical "relay transfer" reaction has the same activation energy as that of the β-relaxation in the corresponding polymer. This result again suggests that the kinetics of macroradical reactions in solid polymers is sensitive to small-scale molecular dynamics characterized by the secondary $(\beta, \gamma, \ldots)$ transitions.

15.2.3.2 Polymer Morphology

Structural organization of a solid polymer can be characterized by three principal supramolecular parameters: the content of the amorphous phase, the degree of chain orientation in this phase, and the degree of crystallinity in semicrystalline polymers. These parameters depend to a large extent on its processing history, and on the glass transition (T_g) and melting temperature (T_m).

Crystallinity It is generally considered that only the amorphous regions in semicrystalline polymers can participate in the degradation reactions. The chemical inertness of the crystalline phase originates from the close packing of the polymer chains (minimum free volume), from the rigidity of the crystal lattice, and from the lack of oxygen to initiate oxidative degradation. Only small species such as hydrogen atoms can penetrate the crystal lattice. Any macroradical which may be created is virtually trapped in the crystallites as an inert species. From the effects mentioned, it is obvious that degradation stability parallels the degree of crystallinity. In some infrequent circumstances, it has been observed that increasing the degree of crystallinity in a polymer sample may increase the rate of photodegradation. Closer scrutiny of the results has indicated that multiple light scattering by the crystallites, which increases the effective optical length, and hence the number of absorbed photons for degradation, is the proper explanation for this unusual behavior. In addition to the degree of crystallinity, the size of the crystallites can also influence the polymer stability through the dual action of chain orientation and presence of tie molecules. A quenched LDPE, for instance, has a higher resistance to photooxidation than an annealed sample of the same degree of crystallinity, as a result of smaller crystallite sizes and a greater number of tie molecules [8]. In iPP, a reverse trend has been observed when the rate of oxygen diffusion becomes the control step (Section 15.4.4). It has been suggested that oxygen diffusion may depend on the type of amorphous phase, which is distinguished in PE as intralamellar, interlamellar, and interspherulitic (Ref. 6, p. 292).

Although it is well established that oxidative degradation of semicrystalline polymers initiates and spreads in the amorphous interphase without affecting the crystalline domains, it does change the density and crystalline content through a process known as chemicrystallization. The best documented account of this phenomenon is found in isotactic polypropylene. At the beginning of degradation, chain shortening from bond scission facilitates molecular rearrangement, leading to an increase in crystallinity. For long degradation times the crystallinity decreases, sometimes with a shift toward lower T_m, even if the degradation temperature is too low to allow for a change in molecular conformation [9].

Chain orientation Drawing orients the polymer chains, changes the sample morphology and can influence the rate of degradation in two ways:

- Orienting chains increases the intermolecular interactions, while at the same time decreasing the free volume and segmental mobility. Most of the orientation experiments have been performed with semicrystalline polymers (iPP or HDPE).

In many respects, the behavior of oriented amorphous phases shows qualitative similarity to that of the crystalline phase. The reduction of molecular mobility results in a decrease in the rate constant for chain propagation reactions. A decrease in free volume results in a reduction in the rate of oxygen diffusion.

- Even at the same concentration of absorbed oxygen, an oriented sample retains its mechanical strength better than an isotropic polymer and can survive a longer period of degradation.

Hydrogen bonding Polar polymers such as the polyamides, aliphatic or aromatic, are strongly hydrogen-bonded in the solid state. Polymers devoid of polar groups can nevertheless develop hydrogen bonding during the course of degradation as a result of formation of oxidation products (hydroperoxy, hydroxy, carbonyl, and carboxyl groups) which can then interact through the acidic hydrogen and the basic oxygen groups. The effect is similar to crosslinking, with a reduction in segment mobility and a decrease in gas permeability. In addition, hydrogen bonding changes the activation energy for some chemical reactions, such as hydroperoxide decomposition (Section 15.4.4).

Externally applied stress Most polymers in practical use are subjected to some type of deformation process. Depending on the amplitude, the duration, and the geometry of these deformations, the polymer morphology may change in a number of complex (although generally predictable) ways, such as chain orientation, shear banding, and crazing. How these stress-induced morphologies interact with environmental degradation factors is, however, less well documented.

In photo-oxidation, it is found that the rate of decomposition increased with applied stress. The main factors which affect the rate of decomposition following application of tensile stress are an increase in oxygen diffusion and an acceleration of chain scission. It is believed that the latter effect may result from a decrease in radical recombination rate, following the pulling apart of the formed radicals by the applied stress before they can recombine again. Several investigations in the 1970s tended to favor a molecular interpretation for stress-activated chemical degradation (Section 15.7.1). However, recent investigations of polyolefins tend to indicate that chemical evolution does not depend on mechanical stresses, which modify only the consequences of the chemical evolution and not the kinetics of the chemical reactions [10].

15.3 Degradation Detection Methods

A few effects of outdoor weathering, such as discoloration and embrittlement, are readily detectable by visual or manual inspection. Quantitative investigations nevertheless require modern analytical techniques capable of probing material changes at different levels, from determining macroscopic bulk properties to un-

Tab. 15.2. Hierarchy of degradation investigative methodologies.

Scale	*Aspect investigated*	*Methodology*
Macroscopic	decrease in mechanical properties with time	tensile test, fracture energy measurements
Microscopic	development of surface microcracks	optical and electronic microscopy
Supramolecular	change in morphology	DSC, TEM
Macromolecular	change in molecular weight distribution	GPC
Chemical	stable degradation products	oxygen uptake, FTIR, Raman, NMR, impedance spectroscopy
Reactive intermediates	free-radical formation	ESR, chemiluminescence

derstanding molecular mechanisms. The hierarchy of methodologies which has been applied to degradation investigation is summarized in Table 15.2.

Many analytical techniques used to inspect the cited properties are common to the field of polymer characterization: vibrational spectroscopy (FTIR, Raman), magnetic resonance spectroscopy (NMR, ESR) and liquid chromatography (GPC, HPLC). A few methods, such as oxygen consumption and chemiluminescence, are more specific to oxidative degradation. Mechanical tests are frequently used in combination with other analytical tools to asset the effects of degradation on mechanical properties.

15.3.1 Mechanical Tests

Many plastic materials are used for their load-bearing properties, and mechanical testing occupies a dominant position among the different criteria for degradation. Typical molecular consequences of degradation processes are chain scission and crosslinking. These effects have a profound influence on the mechanical properties of the material. In general, a predominance of crosslinking over chain scission increases the tensile strength while decreasing the elongation at break and the viscoelastic flow of the material. The reverse effects are observed for chain scission. The loss of toughness in semicrystalline polymers arises from oxidatve chain scission in the amorphous region, which involves:

- scission of tie molecules
- recrystallization of lower MW polymer fraction following scission
- densification from increasing polarity.

In thermal aging of polypropylene, for instance, a rapid transition from ductile to brittle behavior was observed before the appearance of carbonyl groups in the FTIR spectra. Recent data nevertheless indicate that some of the delay in >C=O buildup

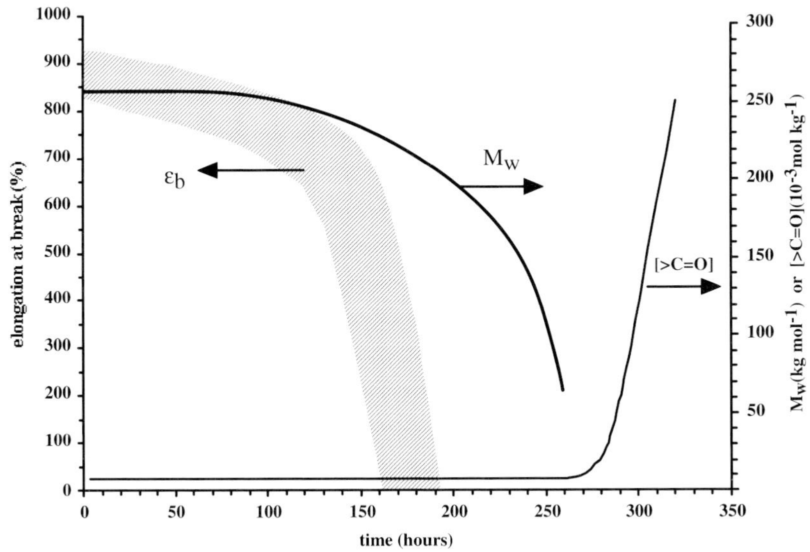

Fig. 15.1. Change in elongation at break, weight-average MW (M_w), and carbonyl content as a function of degradation time, in the thermo-oxidative degradation of unstabilized iPP.

may result from the loss, during the initial degradation stage, of volatile ketones and carboxylic acids which hence elude detection by FTIR. Gel permeation chromatography reveals the presence during the induction period of chain scission, most likely β-scission of alkoxy radicals, which accompanies loss in mechanical strength when the weight-average molecular weight M_w drops below some value of the order of 10^5 g mol^{-1} (Figure 15.1) [11].

Degradation, particularly photodegradation, is a surface phenomenon which affects essentially only the outer few hundred microns from the surface. Scanning electron microscopy invariably reveals the appearance of multiple cracks and microcracks at the surface of degraded polymers, as in Figure 15.2.

The morphology of these cracks depends on the type of polymer and the degradation conditions. Crack formation normally occurs on unstrained material according to the following sequence:

$$\text{Oxidation of the superficial layer} \rightarrow \text{change in material density}$$
$$\rightarrow \text{volume contraction} \rightarrow \text{differential deformation} \rightarrow \text{crack formation}$$

Oxidized polymers show an increase in density owing to the presence of polar functions. Densification of the degraded material creates internal stresses between the outer layer and the intact interior. Ultimately, cracks will appear when the mechanical strength of the degraded layer falls below the differentiial tensile stresses. Water absorption from ambient humidity or from rain, followed by surface evapo-

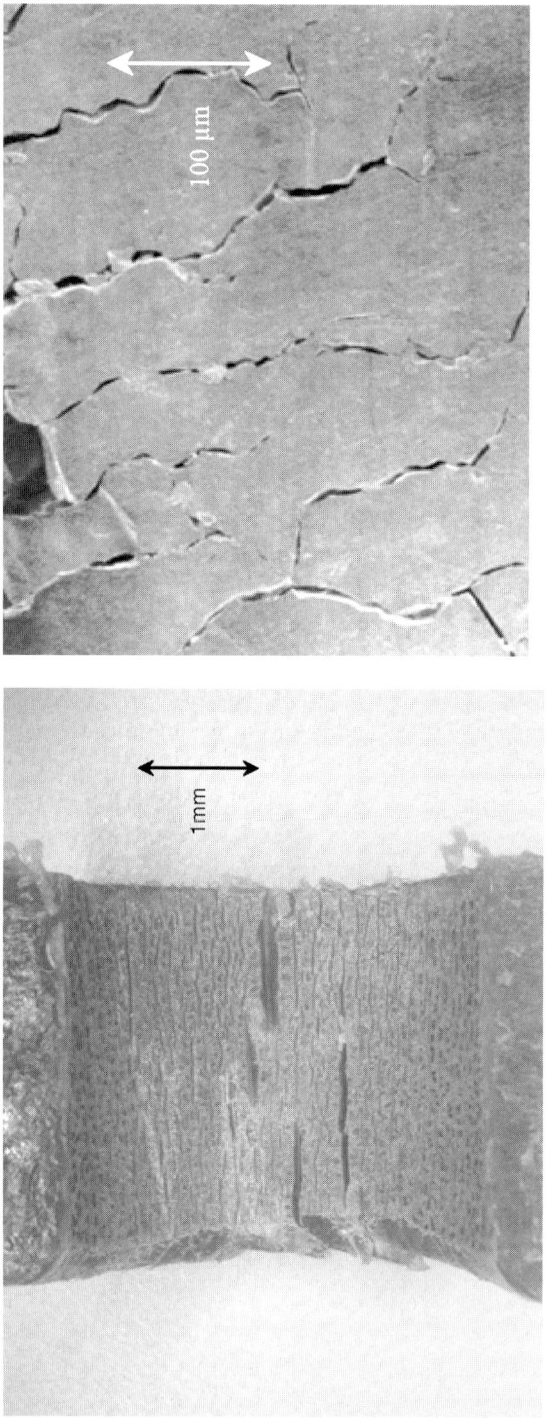

Fig. 15.2. Thermally degraded polypropylene samples. Left: macroscopic aspect, showing the interior of a water pipe degraded in hot water (six months, 90°C). Right: microscopic (SEM) view of an HAS-stabilized sample thermally aged in an oven (54 days at 135°C [11]).

ration, is another source of cyclic mechanical stresses which can favor crack formation in certain polymers. The spatial irregularities in oxidation are another source of differential stresses between degraded and undegraded domains.

Fracture usually occurs from surface cracks. Starting from the degraded layer, these surface microcracks can easily propagate into the internal intact layer by stress concentration at the crack tip. Some mechanical tests are more sensitive to the outer layers than to the bulk properties, and this peculiarity should not be overlooked when interpreting experimental results. Tensile strength of degraded elastomers, for instance, shows a different behavior when compared with tensile elongation, because the former depends on the entire cross-section of the material, whereas ultimate elongation is correlated with surface modulus because cracks initiated at the hardened surface immediately propagate through the sample (Ref. 10, p. 557).

15.3.2
Gel Permeation Chromatography

Polymer degradation involves cleavage of bonds, resulting in a decrease in MW, branching, cyclization, and crosslinking. These changes in the MW and MWD, and their evolution with the extent of degradation, can give valuable insights into the degradation mechanism. Before the advent of GPC, different MW averages such as M_n, M_v or M_w, were used routinely in evaluation of polymer degradation processes. Determination of the number-average MW (M_n) has the major advantage that it is directly related to the inverse of the number of molecular chains per unit polymer mass [Eq. (22), where n_i is the polymer molar fraction of molecular mass M_i].

$$M_n = \left(\sum n_i \cdot M_i\right) \tag{22}$$

In polymer degradation experiments, it is convenient to define the scission index s as the number of broken bonds per initial macromolecule. Because each main-chain scission produces an additional molecule, the indice s is directly given by the change in number-average MW regardless of the initial MWD and degradation mechanism [Eq. (23), with n representing the number of polymer chains per gram of polymer, and the superscript "°" referring to the initial conditions].

$$s = [n - n^\circ]/n^\circ = [M_n^\circ/M_n]^{-1} \tag{23}$$

Pure chain scission seldom occurs without other competitive radical reactions, except perhaps during the early stage of oxidative degradation. When chain scission and crosslinking happen simultaneously, the polymer molecular weight may increase or decrease depending on the relative importance of each process. For random scission and crosslinking, Flory and Charlesby have shown that a three-dimensional network begins to occur at the "gel point", when the crosslink density

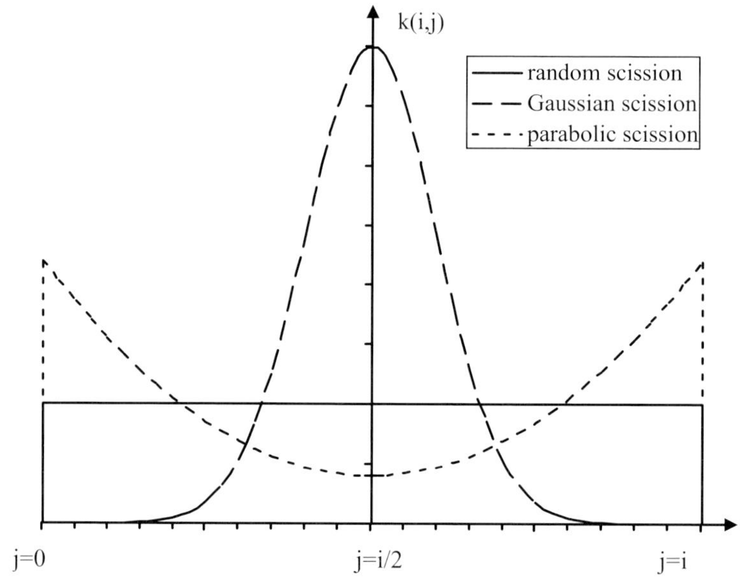

Fig. 15.3. Some commonly encountered scission probability distribution functions.

reaches one unit per weight average molecule. The soluble fraction (s) after the gel point can be used to determine the relative yield of chain scission to crosslinking during degradation (see Section 15.6.2).

Gel permeation chromatography (GPC), particularly in its multidetection version, is now a mature and well-accepted technique for MW characterization. The capability of GPC to determine changes throughout the MWD, in addition to the different MW averages, opens new possibilities for polymer degradation studies. The scission probability as a function of position along the molecular chain, for instance, could be inferred from the MWD of degraded polymer [12].

In a given polymer where most of the covalent bonds have on the average the same bond energy, it is expected that the rate of bond scission may not depend on its position along the chain (random scission). Although this is generally true for the thermal and photodegradation of several polymers, other bond scission statistics can also be observed, as shown in Figure 15.3 [12].

The Gaussian scission probability distribution function, with a preference for mid-chain scission, is frequently encountered in shear-induced mechanochemical degradation. The parabolic distribution, on the contrary, indicates a preference for chain-end degradation. This phenomenon has been reported in the hydrolysis of dextran, as a result of chain branching.

A fourth situation, not depicted in Figure 15.3, is encountered in some polymers which have weak links at a chain end or a low ceiling temperature. In the former situation, initiation starts at the chain end bearing the weak bond with volatilization of a monomer out of the reaction medium. In the second case, exemplified by polyacetals, initiation also occurs at a terminal position, but the process continues and the monomers keep being evolved until complete volatilization of the polymer (unzipping).

15.3.3
Fourier Transform Infrared Spectroscopy

Identification of the degradation mechanism, and its evolution with time, are mainly based on analysis of the intermediate and final chemical products. FTIR proves to be particularly suited to that purpose. The infrared absorption intensity A_S is related to the square of the change in dipole moment (μ) during molecular vibration according to Eq. (24), where ν is the frequency of the band center, and Q, the normal coordinate of the vibration.

$$A_S \propto \nu \cdot (\partial\mu/\partial Q)^2 \tag{24}$$

The dynamic dipole moments of oxidized compounds are high, owing to the polar nature of the oxidation products, and FTIR provides a sensitive and quantitative technique for the investigation of oxidative degradation.

Oxidized polymers show changes in the IR absorption spectrum over the whole range of measuring wavelengths. Most noticeable are the appearance of complex overlapping bands in the carbonyl (1600–1800 cm^{-1}) and hydroxyl (3200–3600

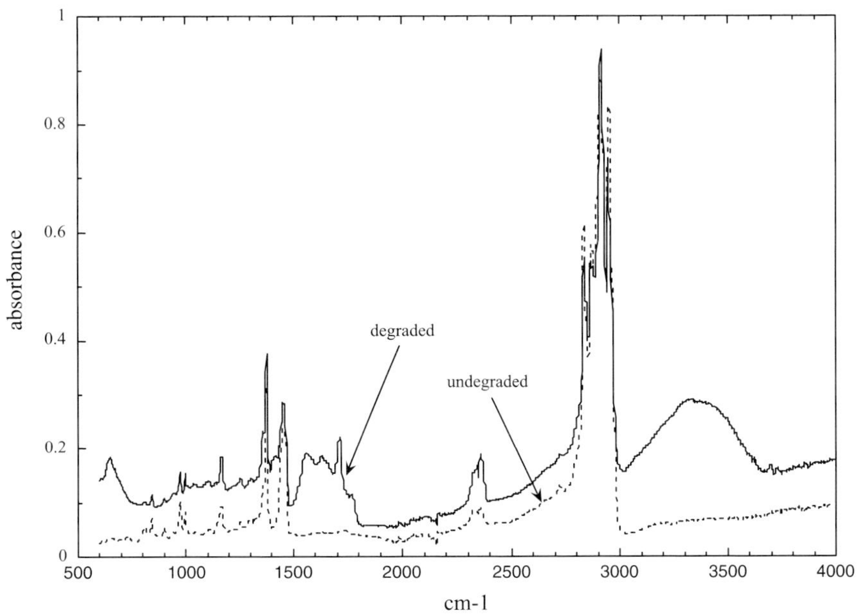

Fig. 15.4. Evolution of the ATR-FTIR spectra of an unstabilized commercial iPP, before (—) and after (----) 3 h of thermal treatment at 120°C.

cm^{-1}) regions (Figure 15.4). The content of carbonyl groups is an indirect indicator of the number of scission events. The average ratio of the absorption coefficient of hydroxyl (–O–H) to carbonyl (>C=O) functions is approximately 3.4:1, and this value can be used to estimate the relative proportion of hydroperoxides to other oxidation products of degradation.

IR absorption bands in amorphous solids are rather broad, with extensive overlap. Direct identification and quantification of the numerous oxidation products, most often of similar chemical structures, in a degraded polymer are difficult. More precise conclusions about the nature of absorbing species can be obtained by selective chemical derivatization. Selective modification of functional groups with reactive gases, such as SF_4, NH_3, SO_2, or NO, results in a shift in absorption band positions, which can then be compared with model compounds to allow for a better chemical assignment of the absorbing species (Table 15.3) [13].

Chemical derivatization can also be used for structure elucidation in complex mixtures. For instance, dimethyl sulfide has been used to differentiate peracids from other peroxy compounds. The rate of conversion of dimethyl sulfide into dimethyl sulfoxide is almost instantaneous with peracids [Eq. (25)], whereas it is slow with *sec*- and *tert*-hydroperoxides [Eq. (26)].

$$\mathrm{R(R')C(=O)\text{–}O\text{–}OH} + \mathrm{CH_3\text{–}S\text{–}CH_3} \rightarrow \mathrm{R(R')C(=O)\text{–}OH} + \mathrm{CH_3\text{–}S(=O)\text{–}CH_3} \quad (25)$$

$$\mathrm{RR'CH\text{–}O\text{–}OH} + \mathrm{CH_3\text{–}S\text{–}CH_3} \rightarrow \mathrm{RR'CH\text{–}OH} + \mathrm{CH_3\text{–}S(=O)\text{–}CH_3} \quad (26)$$

Tab. 15.3. Some common derivatization techniques for identification by FTIR.

Functional group	*Derivatization reaction*	*IR frequency [cm^{-1}]*
Carboxylic acid	$\mathrm{R\text{–}C(=O)\text{–}OH + NH_3 \rightarrow R\text{–}COO^-}$	1564
Carboxylic acid	$\mathrm{R\text{–}C(=O)\text{–}OH + SF_4 \rightarrow R\text{–}C(=O)\text{–}F}$	1848 1841 (α-branching)
Hydroperoxides	$\mathrm{R\text{–}O\text{–}OH + NO \rightarrow R\text{–}ONO_2}$ the sharp and intense absorption bands of nitrates and nitrites can be used to differentiate primary from secondary and tertiary compounds	1276 + 1290
Alcohols	$\mathrm{R\text{–}OH + NO \rightarrow R\text{–}NO_2}$	778
Aldehydes	$\mathrm{R\text{–}CH{=}O + 2Ag(NH_3)_2OH \rightarrow RCOO^-NH_4^+ + 2Ag\downarrow + 3NH_3 + H_2O}$ (Tollens' reagent does not react with ketones)	1550–1555

The carboxyl group, in the presence of hydroxyls, may be differentiated by reaction with silver trifluoracetate [Eq. (27); Ref. 7, p. 76] or with trifluoroanhydride [Eqs. (28) and (29)].

$$-COOH + CF_3COOAg \rightarrow -COOAg + CF_3COOH \quad (27)$$

$$-OH + (CF_3CO)_2O \rightarrow -O-CO-CF_3 \quad (28)$$

$$-COOH + (CF_3CO)_2O \rightarrow -COO-CO-CF_3 \quad (29)$$

15.3.4 Magnetic Resonance Spectroscopy

15.3.4.1 Nuclear Magnetic Resonance (NMR)

NMR is an absolute and unique method of resolving the microstructure of polymer species. This technique has been instrumental in the elucidation of degradation mechanisms in several polymer systems, including poly(vinyl chloride) (PVC).

At around its glass transition temperature (81 °C), unstabilized commercial PVC starts to discolor with release of hydrochloric acid, according to the global reaction of Eq. (30).

$$-(H_2C-CHCl)_n- \rightarrow -(HC{=}CH)_n- + n\,HCl \quad (30)$$

Because the thermal stability of PVC is substantially lower than one may expect from its nominal structure, numerous studies have been initiated to identify the defect sites responsible for this anomaly. After considerable controversy concerning the role, nature, and importance of possible irregular structures, the most active sites for initiating PVC degradation were finally identified by ^{1}H- and ^{13}C-NMR as internal allylic (Figure 15.5, **I**, **II**) and tertiary chlorine structures (**III**, **IV**) [14].

It was believed for a long time that head-to-head radical addition to monomers is a major route for formation of labile structures. Kinetic studies, in association with NMR measurements, reveal that formation of internal allylic and tertiary chlorine structures actually proceeds through an intramolecular or intermolecular chain-transfer reaction to polymer [Eqs. (31), (32); VC = vinyl chloride].

$-HC{=}CH-CH_2\underline{Cl}$ (**I**) $-HC{=}CH-CH\underline{Cl}-$ (**II**)

$CH_2-CHCl-CH_2-CH_2Cl$ (**III**)
|
$-CH_2-C\underline{Cl}-CH_2-$

$CH_2-CHCl-CH_2-CHCl-$ (**IV**)
|
$-CH_2-C\underline{Cl}-CH_2-$

Fig. 15.5. Major labile structural defects in emulsion PVC.

$$\mathrm{P\bullet \rightarrow -CH_2-\dot{C}Cl-CH_2-CHCl-CH_2-CHCl \rightarrow -CH_2-\underset{\displaystyle |}{C}Cl-CH_2-} \quad (\text{branch: } \mathrm{CH_2-CHCl-CH_2-CH_2Cl}) \qquad (31)$$

$$\mathrm{P\bullet + -CH_2-CHCl-CH_2- \rightarrow -CH_2-\dot{C}Cl-CH_2- \xrightarrow{+VC}\xrightarrow{+VC} -CH_2-CCl-CH_2-} \quad (\text{branch: } \mathrm{CH_2-CHCl-}) \qquad (32)$$

15.3.4.2 Electron Spin Resonance (ESR)

Most mechanisms for the degradation of polymers during processing, storage, and long-term use include free radicals. The technique of ESR is therefore a logical choice for the study of polymer degradation. Two kinds of free radicals are of key importance in determining degradation kinetics: (i) the carbon-centered macroalkyl C–C•–C; and (ii) the oxygen-centered alkylperoxyl ROO•, alkoxyl RO•, and acylperoxyl R(CO)OO• radicals. Most of these species have unique ESR spectra which allow their unambiguous identification under favorable conditions. In addition to structure identification, the decay signal of the radicals with time provides information about termination rates and reaction order.

At room temperature and in solutions, free radicals are too reactive, and their concentrations too low, to be conveniently detected by ESR spectroscopy. To observe and quantify transient radicals, it is customary to transform the unstable species into stable nitroxide radicals with the spin-trapping technique. Spin-trapping of a macroalkyl radical with pentamethylnitrosobenzene, for instance, can be represented schematically by Eq. (33). The structure of the parent radical could be deduced from the ESR spectrum of the spin adducts, using standard rules of spin coupling and experimental coupling constants.

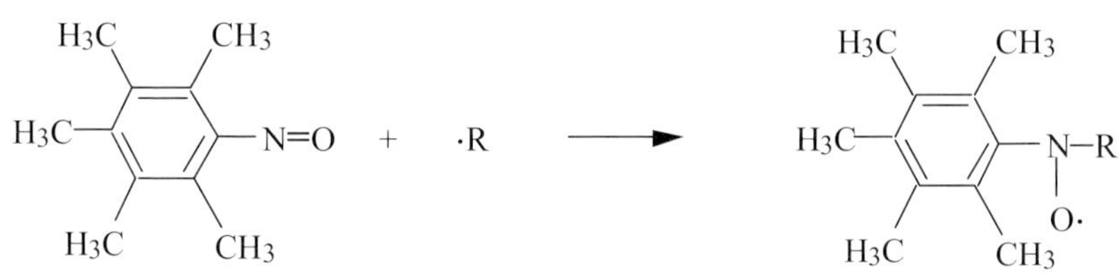

(33)

15.3.5 Oxygen Uptake

Oxygen uptake provides a simple, yet highly sensitive means to monitor oxidative degradation. A weighed amount of polymer is hermetically enclosed with a fixed amount of oxygen under controlled pressure. Oxygen consumption can be monitored by measuring volumetric change, or the decrease in oxygen content by gas chromatography. The first method cannot discriminate oxygen consumption from the release of volatile gases (CO, CO_2, H_2O, and so on) and tends to underestimate the amount of oxygen consumed. The second technique benefits from the high

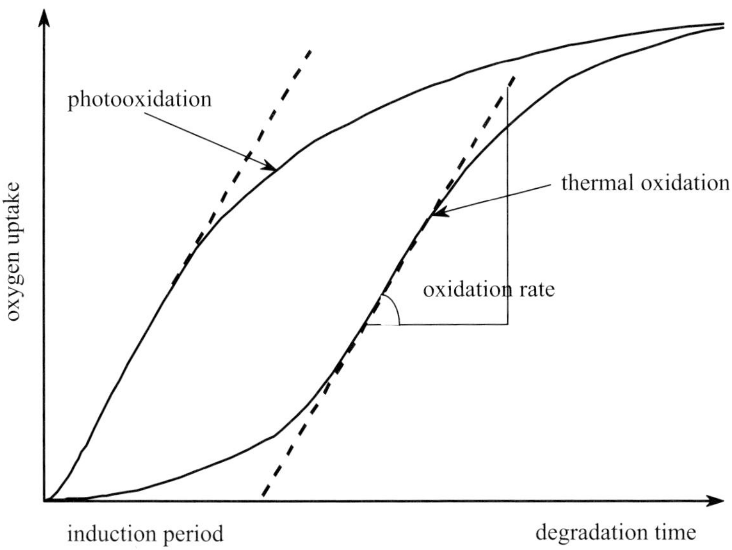

Fig. 15.6. Oxygen uptake curves during photo-oxidation and thermal oxidation of unstabilized polymers.

sensitivity and separation capability of gas chromatography which, with proper care, allows determination of oxygen consumption rates down to 10^{-13} mol g^{-1} s^{-1}. The gas chromatography approach has been used to determine the instantaneous oxidation rate of elastomers even at room temperature, and reliable lifetime can be predicted under ambient conditions without having recourse to extrapolation from accelerated testing [15].

Measurement of oxygen consumption is the starting point for several analyses of polymer oxidation kinetics. The oxygen uptake curve for most polyolefins shows an induction period with a very slow oxidation rate, before reaching a "steady state" with constant slope. The value of this slope has been often interpreted as the limiting oxidation rate of the polymer. The induction period is generally much shorter in photodegradation than in thermal degradation (Figure 15.6). The main difference stems from the different modes of initiation. The activation energy of primary photochemical processes, 8–40 kJ mol^{-1}, is generally much less than the energy required for thermal activation (120–200 kJ mol^{-1}). Because photodegradation generally occurs at low temperature, photodecomposition of peroxides is slow and this class of products tends to accumulate rather than being decomposed to initiate chain branching reactions as in thermal degradation.

During oxygen uptake, part of the absorbed oxygen is transformed into hydroperoxides, while the remaining fractions are found in hydroxyl- and carbonyl-containing compounds formed by competitive reactions of P• and POO• radicals. The relative importance of each process depends on the oxygen partial pressure, and various kinetic equations have been derived to relate the rate of oxygen absorption ($d[O_2]/dt$) to the formation of oxidation products, under low and high oxygen pressure conditions. Kinetic information, and rate coefficients for propagation and termination, have frequently been obtained from the fit of the oxygen consumption curve at relatively low conversions. This procedure, however, is complicated by the facts that intermediate oxidation products may be orders of magnitude more susceptible to oxidation than the starting material, and that degradation in solid polymers is fundamentally heterogeneous. Therefore, any obtained value can yield only average information about the degradation process.

15.3.6 Chemiluminescence

As with oxygen micro-uptake, chemiluminescence is an ultrasensitive technique for determining the rate of oxidation at the earliest stage of degradation. Oxidative degradation of polymers is generally accompanied by weak photon emission, with a very low quantum yield (approximately 10^{-9}). The light-emitting reaction which accompanies free-radical oxidation is generally attributed to an exoenergetic termination reaction of peroxy radicals according to the Russell mechanism [Eq. (34)].

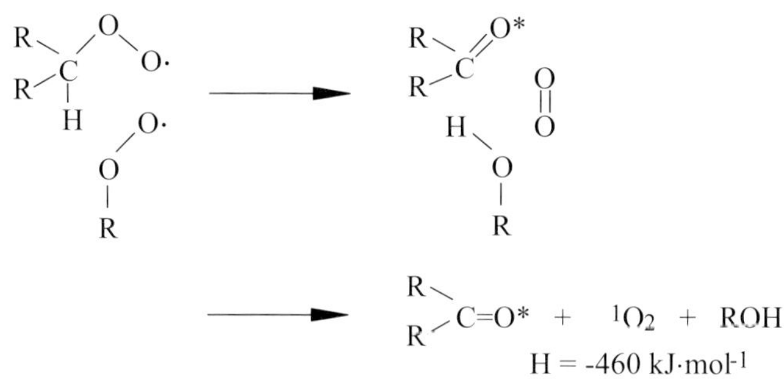

(34)

The Russell mechanism requires one of the terminating radicals to be either primary or secondary so that a six-membered transition state can be formed. Such a mechanism may be prevalent in PE and PA, but not in PP, where the chain-carrying radical is tertiary. For this polymer, more complex alternative routes for chemiluminescence-producing reactions have been proposed [Ref. 10, p. 175].

15.4 Thermal Degradation

Industrial plastics are frequently exposed to elevated temperatures during melt processing and in many engineering applications, some of which will be described in Section 15.4.6. The glass transition temperature (T_g) and the melting point (T_m) of semicrystalline polymers are the most important thermal characteristics of polymers. These two parameters alone are generally not sufficient to predict the material's service temperature, if thermal stabilization is not properly controlled. Several polymer systems produce toxic low molecular weight compounds when subjected to heat, which can create hazardous conditions if they come into contact with a human body, either by ingestion or by breathing. Flammability of polymers used in construction is another concern when the polymers are exposed to high temperatures. All these factors combined make thermal stability appraisal a necessity in most application developments.

15.4.1 Thermal Stability

The notion of thermal stability is a vague concept, since it depends primarily on the time scale of observation. A PMMA sample, for instance, can be stable at 300°C for a few seconds, but can withstand only a temperature of 150°C if the heating time spans several hours.

To be able to compare thermal stability between polymers of different structures, it is necessary to rely on some standardized system, such as the temperature of half-decomposition ($T_{1/2}$). The temperature of half-decomposition is defined as the temperature at which the polymer loses half of its weight when heated in vacuo for 30 min. Experimentally, $T_{1/2}$ can be conveniently determined by thermal gravimetry (TG). From the TG curves obtained at different scan speeds, an Arrhenius plot at constant weight-loss ratio is derived. The pre-exponential factor and activation energy determined are then used to calculate $T_{1/2}$.

15.4.2 Polymer Structure and Thermal Stability

Intrinsic chemical factors which influence heat resistance include primary bond strength, secondary or van der Waals interactions, hydrogen bonding, resonance stabilization, the mechanism of bond cleavage, structure regularity, intrachain rigidity, crosslinking, and branching. Owing to the presence of multiple secondary reactions, the effect of chemical structure on degradation kinetics is not easily rationalized from chemical first principles. The actual degradation kinetics is highly variable and depends not only on the polymer structure, but also on the reaction conditions: sample size, internal or external unstable structures, and additives. More specifically, it is experimentally observed that the degradation temperature and the product distribution can be controlled by changing the heating rate of the polymer, as a result of competitive pathways for degradation. Apart from the primary effects cited, differences in thermo-oxidative resistance can be discerned between different stereoisomers of the same compound (PMMA, PVC, and PP).

Notwithstanding this complexity, some semiempirical rules could be identified by analogy with small organic "model" molecules, and by considering that thermal degradation is a homolytic bond scission process initiated by thermally activated molecular vibrations:

- The rate of thermal degradation is related to the number of pendent groups present on the polymer chain. Thus polyisobutylene (PIB) degrades faster than isotactic PP, which itself decomposes faster than HDPE.
- A rigid polymer backbone has less possibility of rearrangement and fragmentation and can withstand higher thermal energy.
- The heat resistance increases with the number of covalent bonds per repeat unit: a crosslinked or ladder polymer can be broken only after scission of two or more covalent bonds.
- The bond of lowest energy is the first bond to be cleaved. Conversely, polymers with multiple bonds and aromatic structures are less prone to thermal degradation.

Poly(*para*-phenylene), with its rodlike structure composed of highly delocalized π-electron orbitals, satisfies most of the requirements for high thermal stability. Possessing $T_{1/2} > 400°C$ and an estimated melting point of 1400°C, PPP constitutes a reference for heat-resistant polymer. Poly(*para*-phenylene) can be synthesized by a number of routes, with the best heat-resistant material obtained by polymerization of 1,3-cyclohexadiene, followed by dehydrogenation of the polymer formed according to Eq. (35) [16].

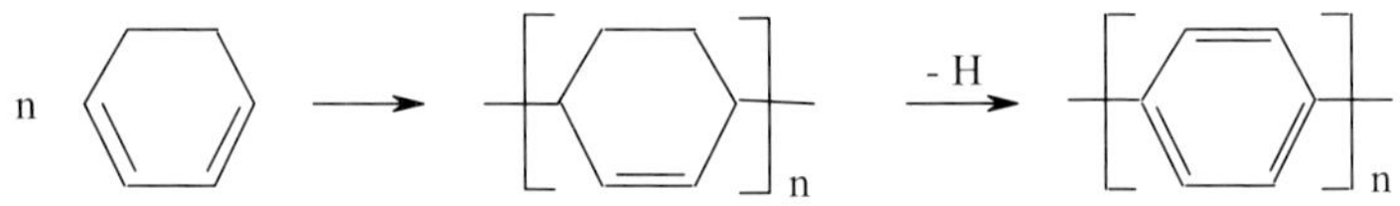

(35)

Moldable polymers can be obtained by increasing chain flexibility with substitution of lateral groups, at the expense of decreasing thermal stability. Polyimides and polyetherimides (**V**), with a semi-rigid structure and electronically stabilized aromatic rings, form an important class of melt-processible polymers with outstanding heat resistance (Table 15.4).

Apart from the decomposition temperature, the propensity for char formation is another polymer thermal characteristic of practical importance. The char-forming tendency generally shows a negative correlation with flammability: a sample with the greatest char residue is also the least flammable polymer and the presence of flame retardant additives also increases char formation. In the case of fire, the amorphous crust of char material that forms on the fluid surface reduces the heat flux to the burning fluid, which in turn reduces the combustion rate.

15.4.3 Computer Simulation

Thermal degradation involves scission of covalent bonds which can be formally written as Eq. (36).

$$A \xrightarrow{k} B + C \tag{36}$$

Tab. 15.4. Chemical structure and thermal stability by thermogravimetry of some representative polyetherimides

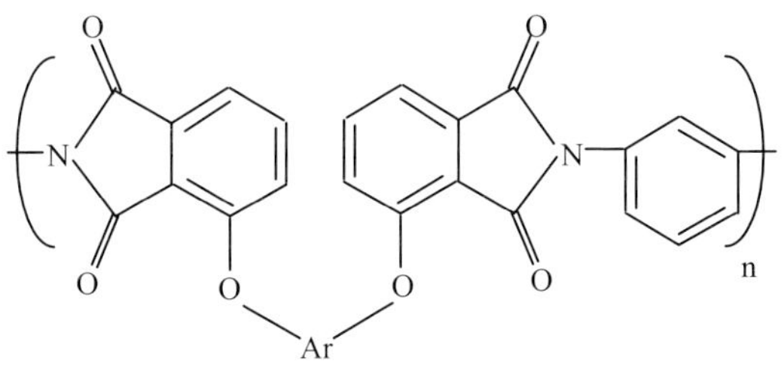

(V)

–Ar–	*Thermal stability [°C]*				
	T_g [°C]	*1 wt.% loss (air)*	*1 wt.% loss (N_2)*	*5 wt.% loss (air)*	*5 wt.% loss (N_2)*
4,4′-biphenyl	247	500	515	530	543
4,4′-benzophenone	239	490	485	530	523
4,4′-diphenyl ether	227	480	480	513	523
4,4′-diphenyl sulfide	209	470	470	504	505

According to Eyring's transition-state theory, the rate constant k is given by Eq. (37), with k_B = Boltzmann's constant, h = Planck's constant, ΔE_0 = molecular energy difference at absolute zero between the activated complex and the reactant, and $q_\ddagger$ and q_A the molecular partition functions of the transition state and reactant, respectively. The transition state is distinguished by a single negative vibration, a characteristic which has been used to identify true transition states in computer modeling.

$$k(T) = (k_B T/h)(q_\ddagger/q_A)\exp(-\Delta E_0/k_B T) \qquad (37)$$

With the increasing capabilities of computers and development of new numerical methods, it is now possible to predict polymer properties computationally. In addition to saving time, computer-aided chemistry can sometimes provide new insights into some decomposition mechanisms which are difficult to obtain by experimental techniques. Computer modeling has been used in an increasing number of ways to simulate thermal degradation. A few representative examples are described below.

- Ab-initio calculations were performed on a series of model structures to predict the effect of lateral alkyl substituents on the thermal stability of degradable polycarbonate. The optimized transition structures revealed that the C_α–O bond dissociates first, followed by abstraction of the β-hydrogen atom, developing a carbocation character in the transition state on the C_α atom. Substituents which stabilize the transition state will also accelerate the degradation rate [17].

- Molecular dynamics simulation has been conducted to understand the significant reduction in flammability of polymer nanocomposites. Using graphite as a model compound, it was found that polymer in an intercalated structure loses fewer fragments generated by the degradation and retains its shape longer as a result of repulsive nonbonding interactions between the polymer and the graphite layers. The fragments generated by thermal activation collide with the graphite and bounce back into the central unit cell, where they undergo recombination reactions with other free radicals. The decrease in polymer fragments which can escape as combustible fuel accounts for the observed improvement in fire resistance [18].
- Force field calculations with the MOPAC PM3 package [19] allow the prediction of thermal degradation mechanisms. First, the smallest representative section of the polymer to be investigated was selected. The program then calculated the heat of formation and Gibbs free energies at different temperatures of the polymer and for different radical fragments. The kinetic barrier for bond dissociation was estimated by stretching each individual bond to its breaking point, and plotting the heat of formation (enthalpy) as a function of bond extension. Calculations on the PA-6,6 structure identify the $>CO{-}CH_2{-}$ bond as the one having the lowest transition state energy, and the carbonyl carbon as the most susceptible site for radical attack. Upon dissociation, the methyl radical created folds back and attacks the $>C{=}O$ bond, forming cyclopentanone, in accord with experimental findings [20].

15.4.4 Thermal Oxidative Degradation of Polypropylene

Isotactic polypropylene (iPP) is a major commodity plastic material which cannot be utilized without thermal stabilizers. With a moderately complex structure, iPP is frequently used as a "model system" to test the different theoretical and experimental approaches to macromolecular degradation.

15.4.4.1 Initiation

The homogeneous oxidation of PP follows the free-radical auto-oxidation mechanism depicted in Scheme 15.1. Under isothermal conditions, the oxygen uptake curves display a pseudo-induction period during which oxidation is autoaccelerated. The duration for the induction period t_i depends on the sample purity and preoxidation history. For clean samples, t_i is reasonably reproducible and is of the same order of magnitude as the POOH lifetime, determined by iodometric titration. The temperature dependence of the induction time obeys an Arrhenius law with an apparent activation energy of 105 ± 15 kJ mol^{-1}, which is the same as for the decomposition of hydroperoxides. The corresponding rate constants are much lower than for the other degradation processes and account for the induction period during which hydroperoxides accumulate before reaching a maximum. It has long been recognized that residual Ziegler–Natta polymerization catalysts, gener-

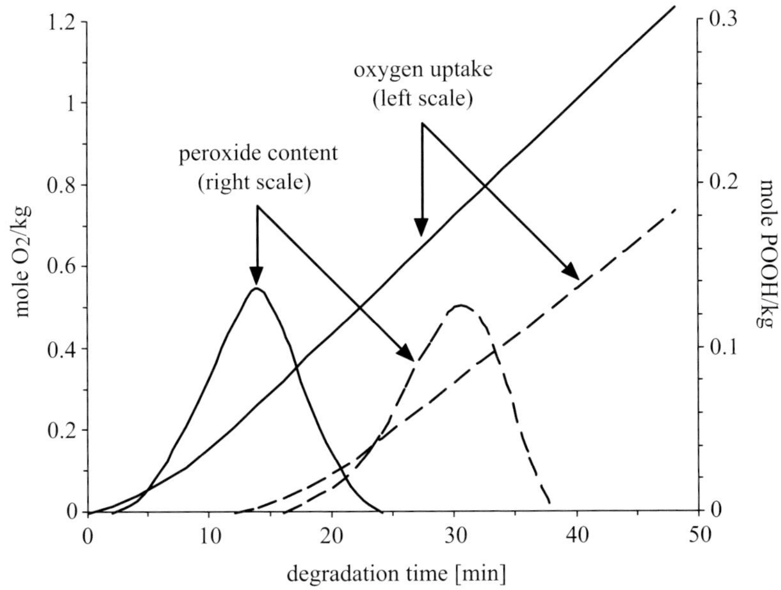

Fig. 15.7. Oxygen uptake (left-hand scale) and hydroperoxide formation (right-hand scale) in the thermo-oxidative degradation of iPP films at 130°C, with small (—) and large (----) spherulites.

ally at the 1–20 ppm level, accelerate the solid-state degradation of PP. The negative influence of polymerization catalyst residues depends not only on the type of catalyst, but also on its concentration. The induction period stage is complicated by the morphology of the sample. In iPP, for instance, a sample with small spherulite sizes (< 100 μm), obtained by rapid quenching, has a short induction time (Figure 15.7). A large spherulite sample (350–500 μm), obtained by prolonged annealing, results in a significant increase in the induction period. The difference was explained by a difference in the oxygen diffusion rate in the oriented amorphous regions, which are more strained in large spherulite structures [21].

It was shown from decomposition kinetics and by treatment with dimethyl sulfide that peroxides consist of two types: a fast-decomposing one composed of peracids, and a slowly decomposing one consisting of hydroperoxides and hydroperesters. During the induction period, the slowly decomposing hydroperoxides accumulate and the oxidation rate is controlled by the rate of decomposition, which may be finally catalyzed by metal ion residues. The autoacceleration stage is controlled by the fast-decomposing peracids [22].

Regardless of the exact mechanism of bond scission, the mechanistic step for initiation in PP can be described schematically by Eq. (38).

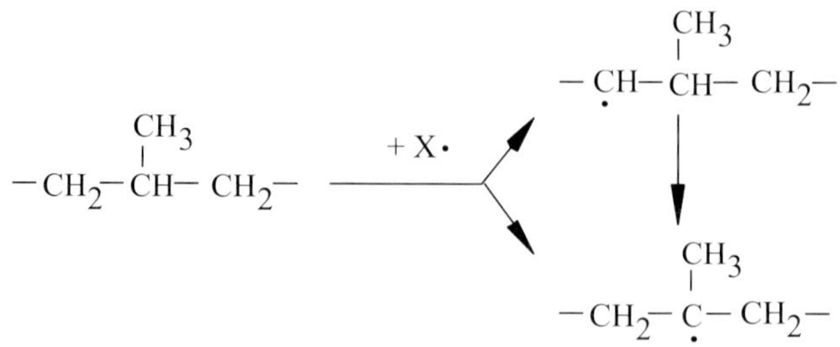

(38)

15.4.4.2 **Propagation**

For PP, the radical conversion step with oxygen is given by Eq. (39).

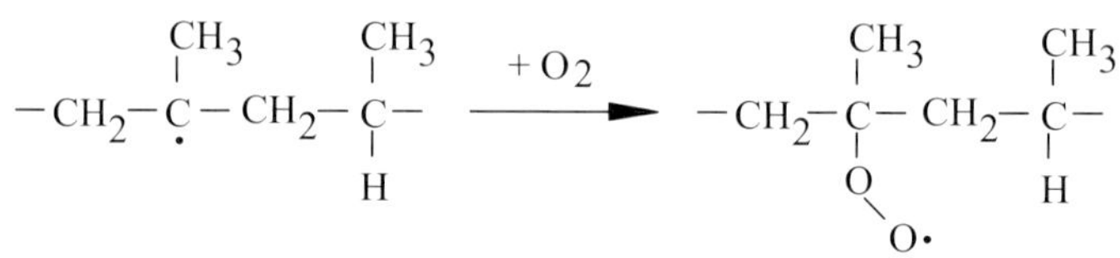

(39)

Hydrogen abstraction from tertiary carbon has an activation energy lower by approximately 15 kJ mol^{-1} than an abstraction reaction on secondary carbon (Ref. 2, p. 386), and is the predominant mode of formation of hydroperoxides in PP. In addition to energy considerations, the hydrogen abstraction rate constant is dependent on steric factors and polymer conformation. It is found, for instance, that k_3 (Scheme 15.1) in solution is lower in a theta solvent than in a good solvent, owing to increased steric repulsion in a contracted molecular coil [7].

Abstraction reaction in polymers can occur intramolecularly or intermolecularly. The former possibility is important in polymers which possess lateral groups (PP and PS), whereas it is nonexistent in linear polymers (HDPE). The possibility of intramolecular H abstraction has been advanced as one of the reasons for the high sensitivity of PP (in comparison to HDPE) toward oxidative degradation. Infrared studies have revealed that more than 90% of hydroperoxides in PP are hydrogen-bonded in sequences of two or more. This result supports an intramolecular hydrogen abstraction reaction, facilitated by a six-membered ring stereochemical arrangement [Eq. (40)].

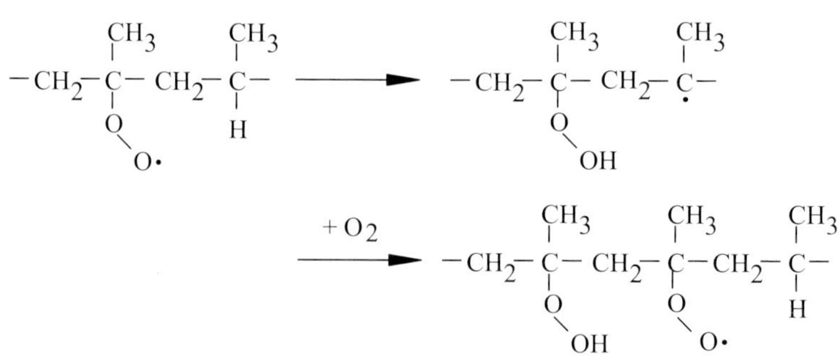

(40)

15.4.4.3 Chain Branching

Chain branching by homolytic decomposition of polymer hydroperoxides results in formation of highly reactive polymer alkoxy (PO•) and hydroxyl (•OH) radicals. The small and mobile •OH can readily abstract hydrogen from a nearby polymer chain, creating a secondary or tertiary macroalkyl radical. The tertiary alkoxy radical can abstract hydrogen intramolecularly, forming a primary alkyl radical as in Eq. (41).

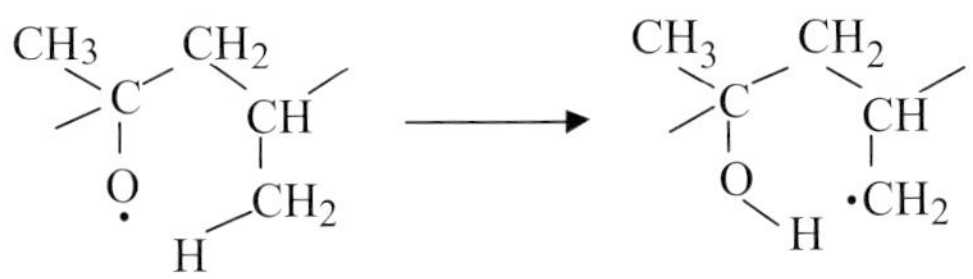
(41)

The polyalkoxy radicals can decompose further by β-scission, yielding ketones, aldehydes, and alkyl radicals, depending on the initial position of the radical [Eqs. (42a), (42b), and (43)].

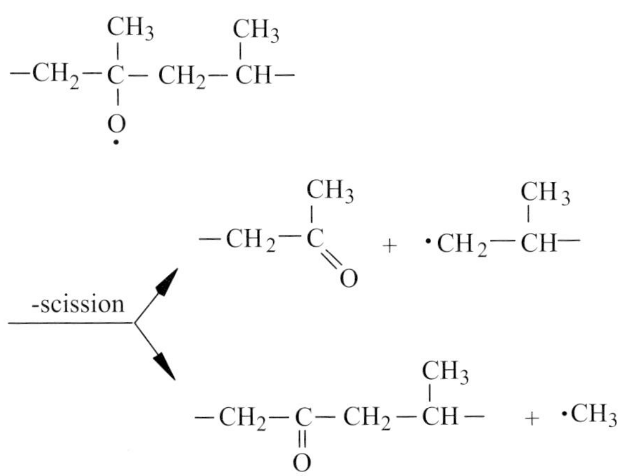
(42a)

(42b)

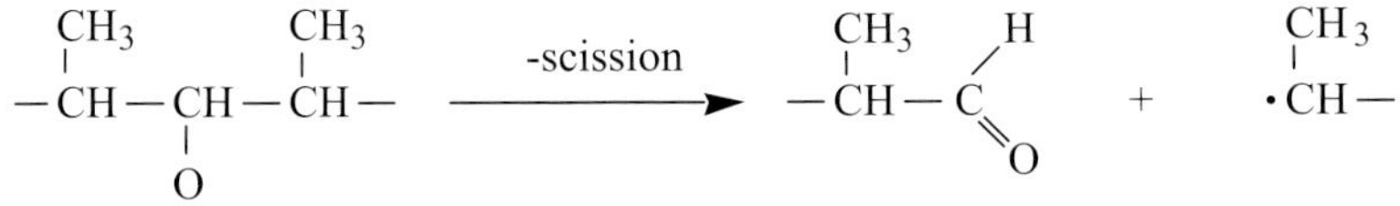
(43)

Similarly, the cleavage of peroxy radicals results in the formation of double bonds along with aldehydes and ketones [Eqs. (44) and (45)].

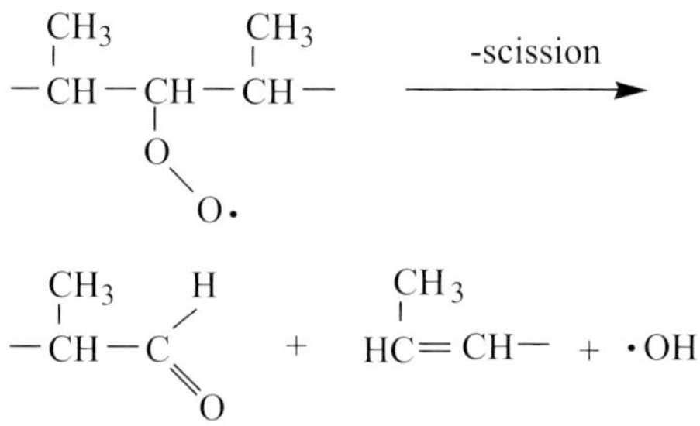

(44)

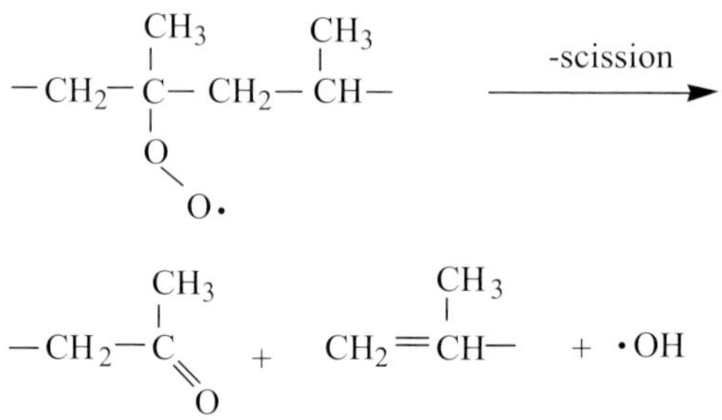

(45)

15.4.4.4 **Termination**

In an oxygen-deficient atmosphere, vinylidene and vinyl compounds may be formed from the disproportionation of polypropylene radicals [Eqs. (46) and (47)].

$$2P{-}\underset{\displaystyle CH_3}{\underset{|}{CH}}{-}CH_2\bullet \rightarrow P{-}\underset{\displaystyle CH_3}{\underset{|}{C}}{=}CH_2 + P{-}\underset{\displaystyle CH_3}{\underset{|}{CH}}{-}CH_3 \qquad (46)$$

$$2P{-}CH_2{-}CH\bullet \rightarrow P{-}CH_2{-}CH{=}CH_2 + P{-}CH_2{-}CH_2{-}CH_3 \qquad (47)$$

In the presence of oxygen, the majority of recombinations occur between tertiary peroxy radicals, to give dialkyl peroxides and oxygen [Eq. (48)].

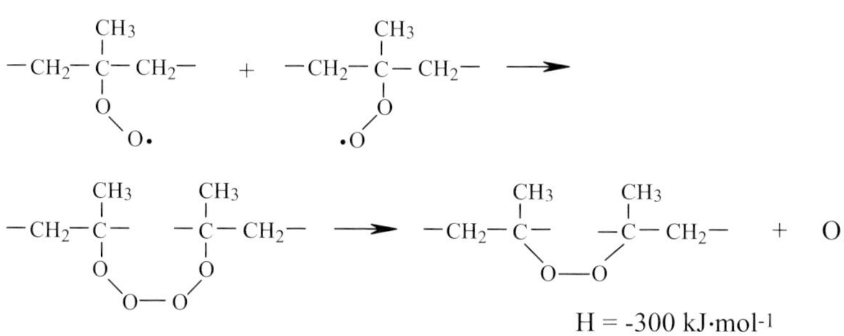

(48)

15.4.4.5 **Secondary Reactions**

In oxidative degradation, it is convenient to make a distinction between primary chemicals formed by reactions of the polymer peroxy radicals, and secondary products created in subsequent reactions of these primary compounds. Ketones and aldehydes formed by β-scissions may react further with hydroperoxides to form a variety of oxidation products. The formation of peracids, for example, proceeds via a multistep oxidation of aldehydes [Eq. (49)].

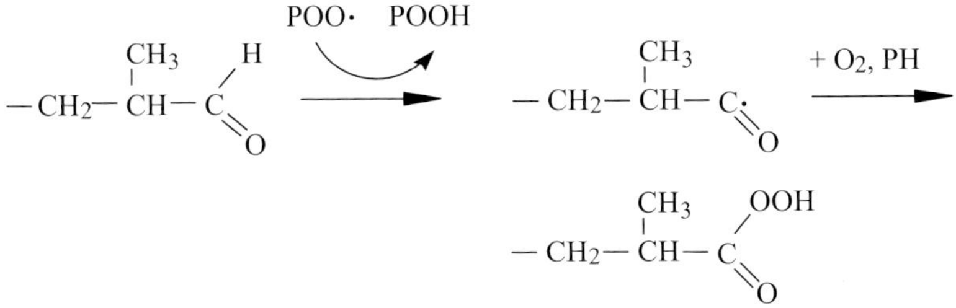

(49)

It should be emphasized that the reaction scheme of Eq. (49) is unlikely if hydroperoxides are distributed homogeneously over the whole sample. Owing to restricted mobility below T_m, the oxidation products cannot diffuse away. Locally, hydroperoxides can reach sufficiently high concentrations for secondary reactions to occur at a significant rate. In a similar fashion, the formation of carboxylic acids, esters and γ-lactones proceeds through a complex series of oxidation reactions of alcohols and ketones.

The aldehydes and ketones formed by β-scission can undergo Norrish type I [Eq. (50a)] and Norrish type II reactions [Eq. (50b)] during photodegradation (see Section 15.5.3).

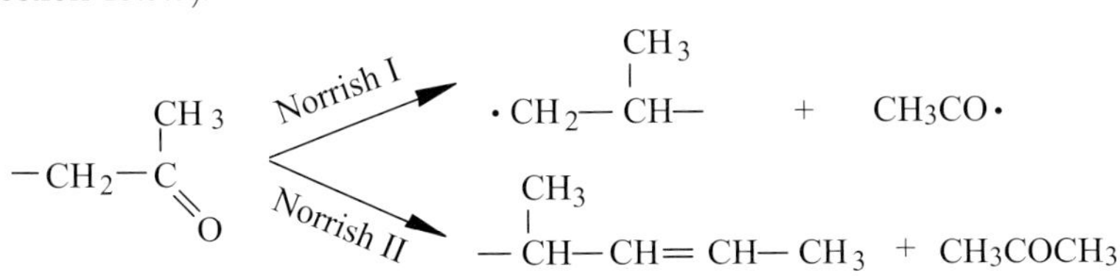

(50a)

(50b)

For a long time, the similarity between thermal and photolytic products was a mystery because Norrish photoprocesses are absent in the former situation. By using low MW model compounds, chemical derivatization, and ^{13}C-NMR techniques, it has been assessed that a large fraction of the carbonyl-containing compounds formed during oxidative degradation of PP are α-methylated acids. Such a carboxylic structure can only originate from the oxidation of macroalkyl radicals [Eq. (51)], which can be formed either by a Norrish I photoprocess or by β-scission of alkoxy radicals (Ref. 10, p. 583).

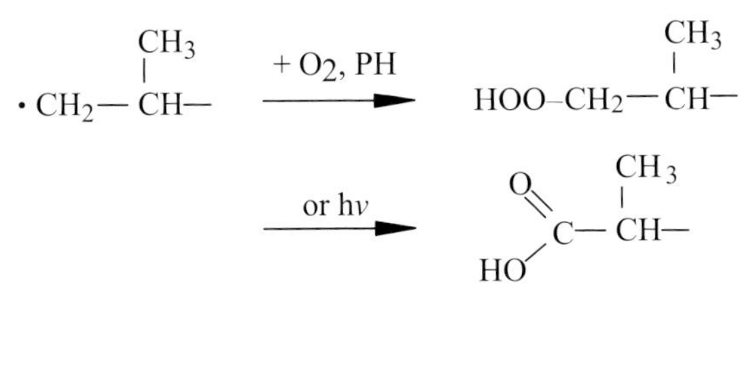

(51)

Once the structure of the carboxylic acids had been elucidated, the IR absorption band which appears as a shoulder at 1740 cm^{-1} in thermal or photo-oxidation of PP, and was initially attributed to ester functions, was reassigned to an acidic group hydrogen-bonded to a vicinal hydroperoxide (**VI**).

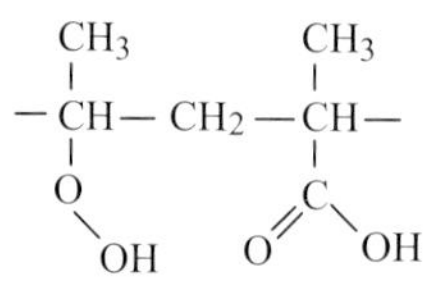

(VI)

15.4.4.6 Formation of Volatile Compounds

At least 40 different volatile compounds have been identified during the oxidative degradation of iPP, some of which are important for spreading the degradation according to the "infection" mechanism (Section 15.4.5). The mode of formation of the principal volatile products is indicated in Eqs. (52)–(55) [23]. It is proposed that CO and CO_2 are formed by the decomposition of carbonyl and carboxyl radicals following hydrogen abstraction from the parent compounds [Eqs. (52) and (53)]. Peracids formed by oxidation of formaldehyde can decompose into formic acid, water, carbon dioxide, and triplet oxygen according to Eqs. (54) and (55).

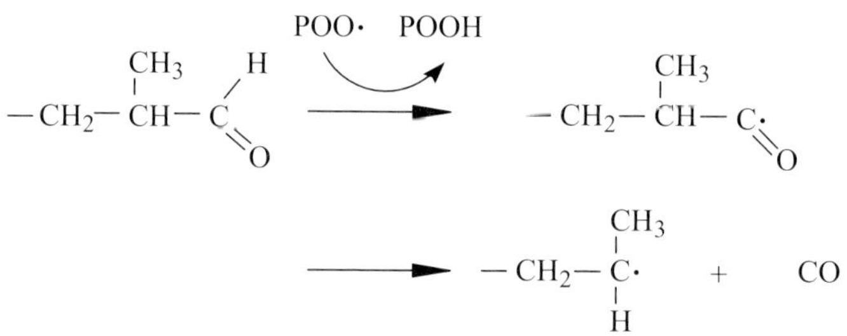

(52)

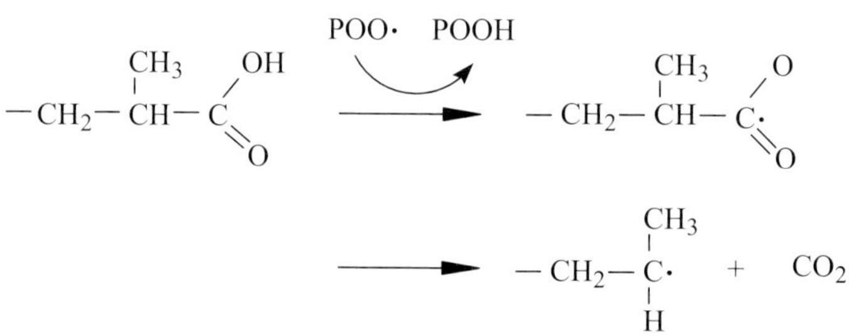

(53)

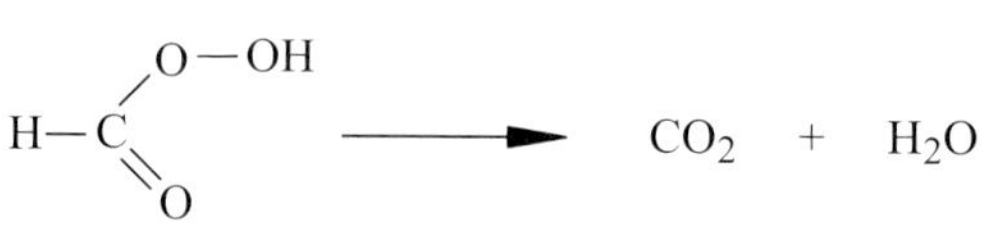

(54)

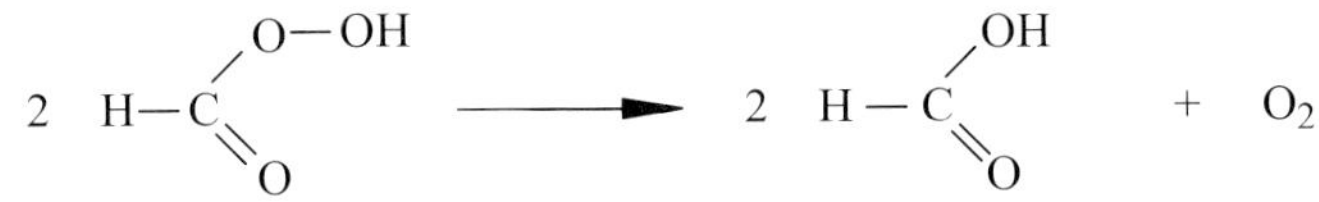
(55)

15.4.5
Homogeneous versus Heterogeneous Kinetics

Degradation reactions in the solid state are spatially heterogeneous, not only across the sample but also within the sample. At least two factors can account for the heterogeneity of the degradation:

- Nonuniform distribution of the reactants (oxygen, additives, impurities): uptake of oxygen and formation of reactive chemicals are the highest near the surface. Additives distribution can also be inhomogeneous as a result of inadequate blending or migration.
- Heterogeneous structure of the sample: oxygen (or any other chemical reactant) can diffuse more readily into amorphous than into crystalline domains, thus favoring oxidative degradation of the former regions.

Superimposed on this macroheterogeneity, which occurs on a micron scale, is microheterogeneity resulting from the low diffusivity of macromolecular species. In solid polymer, it is expected that most macromolecular constituents formed from radical reactions are localized in microdomains of size not exceeding 10–30 nm.

The preceding considerations indicate that degradation kinetics of solid polymers cannot be studied solely within the framework of homogeneous free-radical kinetics. In spite of these complications, it has been found in many instances that the homogeneous kinetics treatment of thermal oxidation gives a surprisingly good fit with experimental data [24]. It has been argued that the homogeneous kinetic model may be applied to an inhomogeneous system by considering the parameters as average values. It should be kept in mind, however, that the real concentrations of degraded species may be much higher locally than the mean values averaged over the bulk sample, and that extrapolating microscopic kinetics from macroscopic average data is a nontrivial problem.

The evidence for heterogeneous oxidation of solid polypropylene was revealed by chemiluminescence measurements. Chemiluminescence from single polymer powder particles revealed a large spread in the induction periods. The oxidation behavior changed with particle separation, and the result was interpreted as a result of physical "infection" from a particle with the shortest induction period to its nearest neighbors, causing rapid oxidation of the complete sample. The origin of this infection was traced back to the formation of volatile compounds, more specifically acetic acid and formaldehyde, which have the propensity to accelerate oxidative degradation [25]. Highly mobile radicals, such as •OH, can also diffuse far and spread the degradation.

Devising a chemical kinetics model which encompasses all sources of heterogeneity present in a solid polymer is a challenge. The starting point in most heterogeneous kinetics models is to consider that oxidation is initiated nonuniformly at a few sites, presumably around catalyst residues or other impurity centers in the amorphous region of the polymer. One approach to heterogeneous modeling was by Monte Carlo simulation, based on the random-walk diffusion of low MW reactive radicals. Another concept, called the "epidemic model", utilizes the analogy between the spreading of oxidation through the amorphous region of solid polymer, and the infectious contamination of a disease through a population. In this model, three distinct polymer populations are assigned: the "infectious" oxidizing fraction p_i, the remaining unoxidized fraction p_r, and the "dead" or oxidized fraction p_d. These populations may be described by a series of coupled differential equations, Eqs. (56)–(58), which are developed by noting that oxidation can spread only if there is unoxidized material within contact distance of the infected one.

$$-\mathrm{d}p_r/\mathrm{d}t = \beta p_r p_i \tag{56}$$

$$\mathrm{d}p_d/\mathrm{d}t = \alpha p_i \tag{57}$$

$$\mathrm{d}p_i/\mathrm{d}t = -\mathrm{d}p_r/\mathrm{d}t - \mathrm{d}p_d/\mathrm{d}t \tag{58}$$

Integration of the above system of differential equations using experimental β and α parameters give an excellent fit with the chemiluminescence curve when dealing with single particles, but the quality of the fit deteriorated with groups of particles, as a result of multiple-spreading processes [26].

By considering only the oxidized and the unoxidized fractions, a simplified spreading model was developed to explain the time dependence of hydroperoxide concentration during oxidative degradation of LDPE [27], expressed as Eq. (59), where n is the number of oxidized domains, N the total number of amorphous domains, and r the rate coefficient for spreading.

$$\mathrm{d}n/\mathrm{d}t = r \cdot n - r \cdot (n^2/N) \tag{59}$$

15.4.6 Applications of Thermal Degradation

Although the word "degradation" usually has a pejorative connotation, a number of important industrial applications depend on this class of reactions to develop new products or processes.

15.4.6.1 Analytical Pyrolysis

Pyrolysis of a polymer means thermal degradation in the complete absence of any external reactant. Analytical pyrolysis, defined as pyrolysis conducted in combination with some physicochemical separation technique (pyrolysis–GC–MS, for instance), has found a wide range of applications in biopolymers and synthetic polymers. Polymers pyrolyze by different combinations of three general mechanisms: random scission (PE), depolymerization (POM, PMMA), or elimination of small molecules other than monomer (for instance, HCl in PVC). Each polymer under strictly controlled pyrolytic conditions provides a typical pattern of degradation products, which can be used for fingerprint identification, composition analysis, structure determination, or mechanistic studies [28].

15.4.6.2 Introduction of New Chemical Functionalities

Controlled thermal degradation of polyolefins in the presence of peroxides has been carried out in extruders or in continuous mixers, to reduce polymer MW and sample polydispersity. The resulting material has more processing advantages than the undegraded one. Based on the same principle, random chemical functionalization of polyolefins with polar groups has been achieved by extruding the polymer in the presence of peroxy ketals or peroxy esters [29].

15.4.6.3 Chemical Modification of Polymer Structure

Carbon fiber is used in a variety of structural and electrical applications, and is probably the most well-known example of high-performance material produced by thermal degradation. Although carbon fiber can be produced in several ways, including alignment of molecules of pitch in its mesophase state, the most economical way consists of "hot-stretching" high MW polyacrylonitrile (PAN) fibers and while heating under controlled conditions. The filaments are first oxidized under tension at around 260°C in air, to convert the PAN precursor to a thermally stable structure by an exothermic reaction ($E_a = 3.5$ kJ g^{-1}, $\Delta H = -2.5$ kJ g^{-1}), according to Eq. (60) [30].

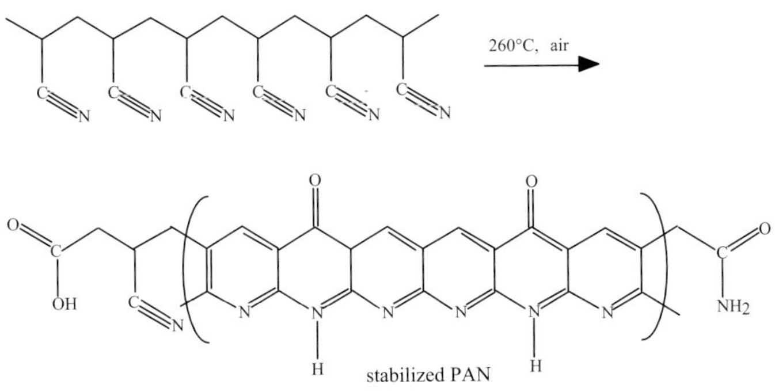

(60)

The carbonization step is carried out in an inert atmosphere (high-purity N_2) above 1000°C to yield carbon fibers containing 93–95% carbon [Eq. (61)] and a tensile modulus (E) of 140–200 GPa. This grade is mostly used in sporting goods and in composites.

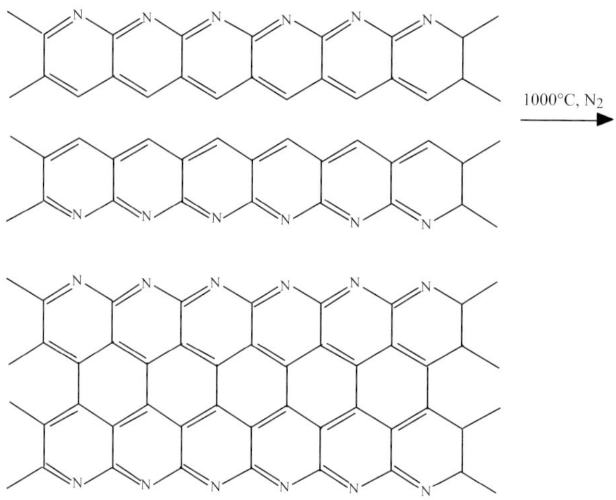

(61)

To achieve fibers in which the carbon crystals are further stretched and aligned, graphitization takes place around 2000–3000°C. These graphite fibers have a carbon content greater than 99% and a tensile modulus of 400–1000 GPa. For comparison, the tensile modulus of carbon nanotubes is 1000 GPa, and of diamond 1200 GPa.

15.4.6.4 Metal Injection Molding (MIM)

Polyacetals have a low ceiling temperature, and are readily depolymerized by unzipping at low temperature (0.4–0.8% min^{-1} at 222°C for POM). Owing to this low thermal stability, polyacetals can be used only if end-capped with stable groups (acetate or ether). This inherent thermal instability is exploited in an industrial method known as "metal injection molding", which allows fine metal powder mixed with a polymer binder to be processed by injection molding, in much the same way as thermoplastic materials [31]. In a procedure based on POM, the binder is removed by thermal devolatilization according to Eq. (62)

$$R{-}(CH_2{-}O{-})_nH \xrightarrow{HNO_3/150^{\circ}C} R{-}CH_2OH + (n-1)H_2C{=}O \tag{62}$$

15.4.6.5 Recycling

Polymer can be recycled by reuse of existing material (primary recycling), by regranulating the waste by mechanical means so that it can be melted and formed again (secondary recycling), or by transforming the waste into new chemical compounds (tertiary recycling) through chemical reactions. The last of these methods presents several economic advantages in comparison to primary or secondary recycling, since revalorizing steps can be avoided. Recently, thermolysis has been viewed as a viable alternative to recovery for polymer recycling and numerous studies have been directed toward this objective [32].

The term "feedstock recycling" has been used to describe this new class of plastics recycling technology, which breaks down solid polymers into a spectrum of basic chemical compounds that can be reused as raw materials for the chemical industry. Vinyl polymers, when pyrolyzed at temperatures from 200 to 500°C in the total absence of air, usually degrade to yield monomers (poly(methyl methacrylate), poly(α-methylstyrene), polystyrene, polyisobutylene) or a wide distribution of molecular fragments (polyethylene and polypropylene). In order to reduce the process temperature and to limit the range of products, particularly in the case of polyethylene, several catalyst cracking systems based on zeolites or clays have been developed. A newly developed technique in the recycling field is thermolysis coupled with reactive distillation.

15.5 Photodegradation

15.5.1 Absorption of UV Radiation by Polymers

Photo-oxidation is the most common mode of weathering for industrial polymers, and differs from thermal degradation principally in the mode of chemical activation. In place of vibrational energy, the energy of the photon provides the driving mechanism for free-radical generation. Photochemical processes are based on two fundamental principles: the Grotthus–Draper law, which states that only radiant energy that is effectively absorbed can activate molecules, and the Stark–Einstein law, which asserts that one absorbed photon can induce photochemical reaction of only a single molecule [33]. The first law implies the presence of an appropriate chromophore in the polymer, whereas the second law indicates that the formation of photoproducts is linearly dependent on the light intensity, and that a photoreaction can occur only if the energy of the photon is sufficient to overcome the corresponding activation energy.

For most organic molecules, absorption in the near-UV (190–400 nm) involves electronic transition from the highest occupied molecular orbital (HOMO) to the lowest unoccupied molecular orbital (LUMO). Commercial polymers can be broadly divided into two categories, depending on whether or not they contain delocalized π-electrons, sometimes combined with heteroatoms with nonbonding valence shell electron pairs (O, N, S) in their chemical structure. Polymers which contain carbonyls (>C=O), conjugated polyenes (>C=C–C=C<), ketenes (>C=C–C=O), and sulfones (–Φ–SO_2–Φ–) belong to the first group of chromophores and are capable of absorbing light in the near UV (190 nm $< \lambda_{max} <$ 400 nm); see Table 15.5. Those from the second group possess only single covalent bonds, such as C–C, C–H, C–O, C–Cl, and C–F, and can in principle absorb light only in the far UV ($\lambda_{max} <$ 190 nm).

Owing to vibrational and electronic couplings, UV absorption bands of polyatomic molecules in the condensed state are generally quite broad, with width at half-maximum commonly exceeding 60–80 nm. Even with this broadness, polymers such as polyolefins, polyacetals, poly(vinyl chloride), or polyacrylonitrile

Tab. 15.5. Principal chromophore groups in synthetic polymers[a].

Polymer	*Chromophore*	λ_{max} *[nm]*	ε_{max} *[L mol^{-1} cm^{-1}]*
Polyesters	>C=O	188 (279)	900 (15)
Polyaromatics	–Φ–	200 (256)	4 400 (226)
Poly(aryl ketone)s	–Φ–CO–Φ–	250 (350)	18 000 (120)
Polydienes	>C=C<	185 (230)	8 000 (2)
Conjugated polyenes	>C=C–C=C<	217	20 900
	$-(C{=}C)_3-$	263	$\sim 5 \times 10^4$
	$-(C{=}C)_{10}-$	432	$\sim 2 \times 10^5$
Ketenes	>C=C–C=O	220 (350)	2×10^4 (30)
Sulfones	$-\Phi-SO_2-\Phi-$		

[a] Values in brackets refer to the secondary absorption band.

should neither absorb nor degrade when exposed to light with wavelength above ~230 nm. In practice, however, the UV absorption spectra of commercial samples of all the cited polymers show a broad absorbance band which extends well above the expected limits (Figure 15.8). Generally, these extraneous absorbances are weak and can be determined only after proper signal processing. This situation is illustrated in Figure 15.9 for commercial additive-free films of HDPE and iPP.

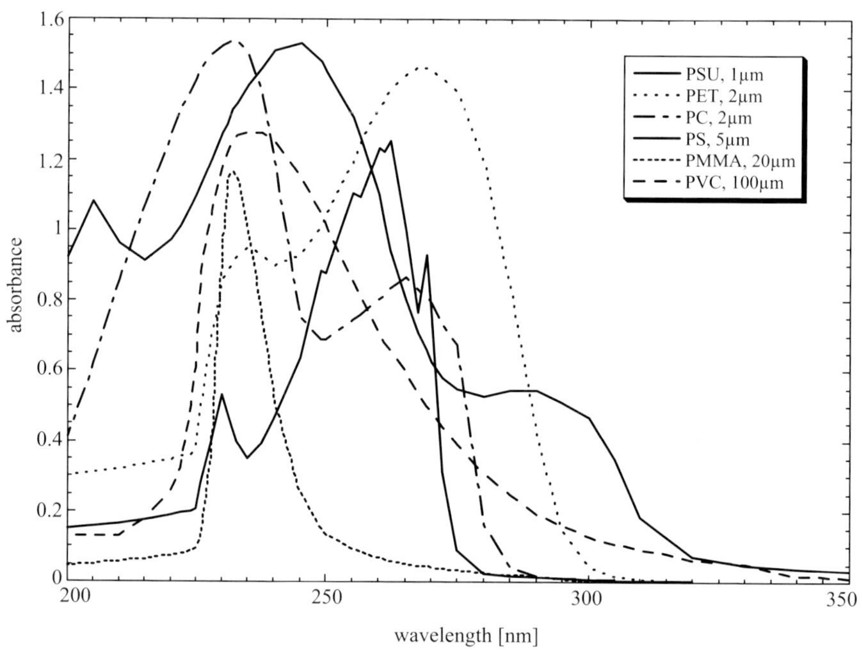

Fig. 15.8. UV absorption spectra of some common industrial polymer films, at the thickness indicated.

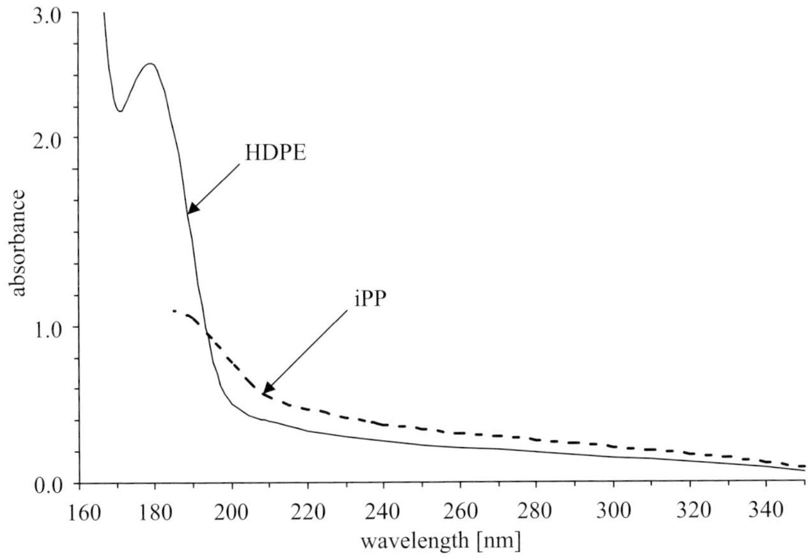

Fig. 15.9. UV absorption spectra of 50 μm-thick films of additive-free commercial HDPE (—) and iPP (---).

Far-UV spectroscopy indicates that the "absorption edge" of long-chain alkanes starts at about 155 nm. Absorption at much longer wavelengths observed in commercial polyolefins could be explained only by the presence of chromophore impurities and chemical defects formed during the synthesis, storage, and processing of the polymer. The absorption peak at 180 nm recorded in HDPE, for instance, has been attributed to vinylidene end groups, as revealed by independent FTIR measurements (Figure 15.9).

Impurities in polymers can be classified as internal or external [7].

Internal impurities These are part of the molecular structure, situated either along the chain or at chain end(s), and may consist of:

- Anomalous structural units, which result from the kinetics of polymerization.
- In-chain peroxides, formed during polymer synthesis. Free-radical polymerization is generally accomplished without strict exclusion of air and small amounts of oxygen dissolved in the monomer will be scavenged by the macroradicals and included in the polymer chain as peroxide linkages. Photolysis of these peroxide groups gives alkoxy radicals, leading ultimately to hydroperoxides. Alternatively, if polymerization has been carried at elevated temperatures, the peroxides incorporated will undergo thermal fragmentation and disproportionation to form phenyl alkyl ketone end-group chromophores.
- Carbonyl containing groups formed during material transformation. Processing, with elevated temperature and high shearing stresses, provides ample opportunities for thermal oxidation. Even at ambient temperature, some polymers such

as isotactic polypropylene or *cis*-polybutadiene, can be readily oxidized if unstabilized.

External impurities Such impurities are contained in the sample but not incorporated in the polymer structure.

During synthesis, processing and storage, the polymer can be contaminated or blended with a variety of external chemical species which may contain chromophore and photoreactive groups. Some typical external impurities found in commercial plastics are:

- residual catalysts and initiators,
- traces of solvents (aromatic solvents, in minute amounts, can act as photosensitizers (Ref. 26, p. 125),
- pigments, dyes, and additives,
- traces of metal, metal oxides, or metal salts from the reactor, processing equipment, and containers.

15.5.2
The Solar Spectrum

The majority of polymer bond dissociation energies are within the 290–420 kJ mol^{-1} bracket (allylic C–C being the lowest, and $-CH_2-H$ the highest). The activation energy of photochemical reactions in the gas phase usually lies just a few percent above the corresponding bond dissociation energy, while it can reach 40 kJ mol^{-1} for diffusion-controlled reactions in the condensed state. Therefore, depending on the type of bond to be broken, any absorbed photons with wavelength shorter than approximately 420 nm for the weakest bonds, to 290 nm for the strongest ones, could promote chain scission.

The importance of photodegradation in outdoor weathering depends much on the sunlight spectrum. The solar spectral irradiance at the top of the atmosphere is well-approximated by black-body radiation with a temperature of 5770 K. The Earth's atmosphere, on the other hand, is practically opaque to any wavelength shorter than about 290 nm as a result of ozone absorption. The UV range of solar irradiation is commonly divided into three spectral regions of decreasing wavelengths denoted by UV-A, UV-B and UV-C. Although UV-B light is the most efficient at initiating photodegradation, its intensity at the Earth's surface is fortunately very limited, owing to the screening effect of ozone, and accounts for $<0.5\%$ of the total radiant energy of ~ 1300 W m^{-2} at noon in the southern regions (Figure 15.10).

15.5.3
Photo-oxidation Profile

It has long been recognized that photochemical damage is more severe at the surface than in the sample interior. The reasons for this inhomogeneity are twofold:

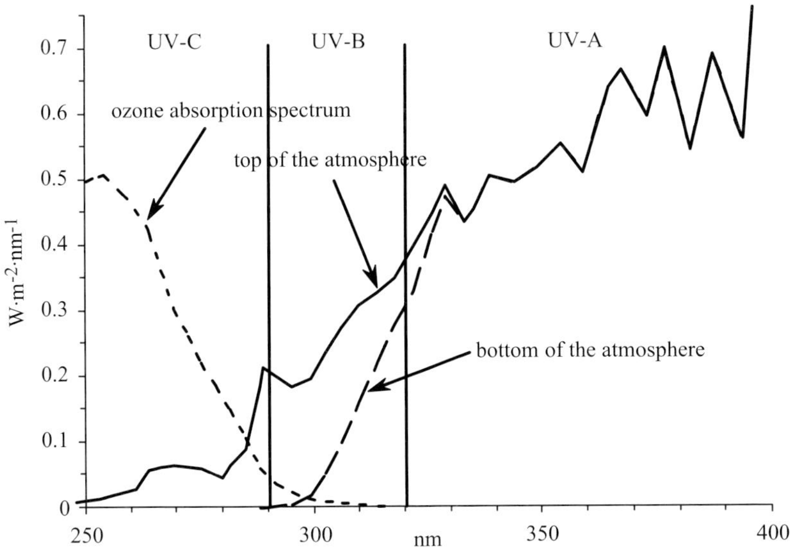

Fig. 15.10. Spectral irradiance of sunlight at the top (—) and bottom (––) of the atmosphere, in the summer at latitude 41°N. The difference results from ozone absorption.

Penetration distance of light For transparent systems, the light intensity $I(x)$ decreases with penetration distance x according to the Lambert–Beer law according to Eq. (63), where I_0 is the incident light intensity, and α the absorption coefficient, which is proportional to the molar absorption coefficient ε.

$$I(x) = I_0 \exp(-\alpha x) \tag{63}$$

Because most radiation is polychromatic, the range of available α is broad. For many materials, the propensity for degradation increases in parallel with the absorption coefficient when the light wavelength decreases. As a consequence, the depth of penetration decreases for the wavelengths which are most effective in promoting photodegradation. As the extent of degradation increases, more UV-absorbing substances are formed, which act as surface screeners and further increase the fall-off of the depth of the degraded layer.

For materials which scatter light, such as semicrystalline polymers, and for polymers which contain charges or pigments, the Lambert–Beer law should be replaced by the Kubelka–Munk equation [Eq. (64), with K = absorption coefficient, S = scattering coefficient and R = reflectance of the material].

$$(K/S) = (1 - R)^2/2R \tag{64}$$

Penetration distance of oxygen Oxidative degradation is a reaction between oxygen and the polymer. The solubility and diffusion coefficient of oxygen in solid polymers are typically one order and two orders of magnitude less, respectively, than in organic liquids. Because of this slow diffusion of oxygen, the oxidative rate decreases rapidly with depth from the surfaces. The effect of oxygen diffusion can be modeled by Fick's second law, modified by the rate of oxygen consumption (k). It is frequently assumed that the rate of oxygen consumption exactly matches the oxygen supply by diffusion, that is, $d[O_2]/dt = 0$. In this "steady-state" situation, the oxygen concentration profile does not change with time, resulting in a differential equation [Eq. (65)].

$$D(\partial^2[O_2]/\partial x) - k[O_2] = 0 \tag{65}$$

The oxygen diffusion constant, D, depends on the state of aging process, the reactant concentrations, and the sample morphology (orientation, crystallinity). It follows approximately an Arrhenius-type temperature dependence as in Eq. (66), where D_0 is a proportionality constant, and R the molar gas constant.

$$D = D_0 \exp(-E_a/RT) \tag{66}$$

In the case of iPP film, for instance, the values given by Eqs. (67)–(69) have been determined (Ref. 6, p. 169).

$$\text{Small-spherulite iPP:} \quad D = 26.0 \exp[-43\,900/RT) \text{ cm}^2 \text{ s}^{-1} \tag{67}$$

$$\text{Large-spherulite iPP:} \quad D = 3.5 \exp[-38\,500/RT) \text{ cm}^2 \text{ s}^{-1} \tag{68}$$

$$\text{Oriented iPP:} \quad D = 2.0 \exp[-38\,500/RT) \text{ cm}^2 \text{ s}^{-1} \tag{69}$$

Free radicals react with oxygen according to reactions (2) and (3) of the simplified degradation (Scheme 15.1). A simple approximation for $k[O_2]$ is given by Eq. (70), where C_1 and C_2 are constants [34].

$$k[O_2] = C_1[O_2]/\{C_2[O_2] + 1\} \tag{70}$$

By combining Eqs. (65) and (70), the oxidation distribution can be calculated by numerical methods and compared with experiments using some depth-profiling technique (microFTIR on microtome slices, ATR-FTIR, photoacoustic FTIR). The frequently encountered U-shaped curve indicates a "diffusional" regime, in which the rate of chemical reactions is limited by oxygen diffusion (Figure 15.11). The temperature coefficient of the oxidation rate, 105–125 kJ mol^{-1}, is much higher than the activation energy for oxygen diffusion (~40 kJ mol^{-1}). Thus, by lowering the temperature, one can reach the "kinetic oxidation" regime, characterized by an oxidation rate independent of sample size.

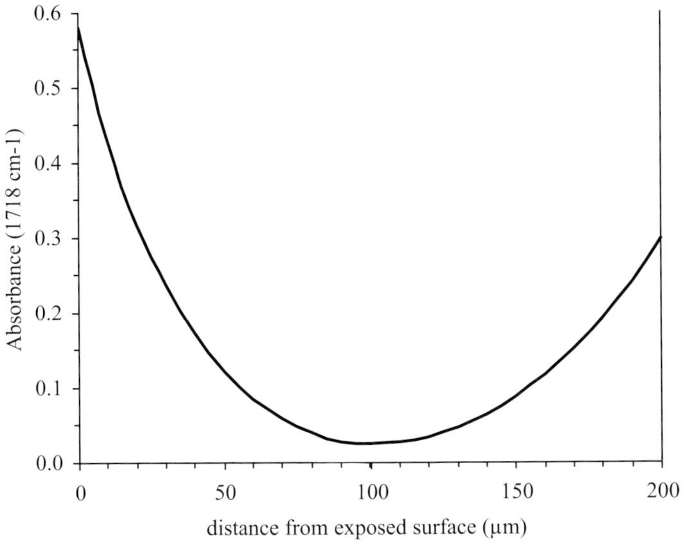

Fig. 15.11. Typical photodegradation profile for a weakly UV-absorbing, 200 μm-thick polymer film.

15.5.4 Influence of Wavelength: the Activation and Action Spectrum

The actinic effects of UV and visible radiation on material are wavelength-dependent, owing to the spectral selectivity in the absorption of the incident radiation, and the quantum efficiency of degradation. The absorption of a photon by a chromophore promotes an electronic transition between two molecular orbitals of different energies. Starting from the ground state, a transition to a different electronic state can be effected by changing the energy of the photon. It is therefore logical to expect that the irradiation wavelength may have a profound influence not only on the rate, but also on the mechanism of photodegradation. The concept of quantum yield, defined as the number of degraded molecules per absorbed photon, is widely applied in the photochemistry of small organic molecules. It is difficult to apply to industrial plastics, which are complex mixtures of species with variable photon absorption and different photophysical and photochemical properties. Early investigations relied on the notion of an activation spectrum to characterize the effect of the photon wavelength on the extent of degradation. The activation spectrum of a material is defined by (i) the spectral absorption properties of the material; (ii) the emission properties of the light source; (iii) the quantum efficiency of the photoinitiated processes; and (iv) the type of degradation being measured.

Although simple to measure, the activation spectrum does not consider the amount of incident radiant energy present at each wavelength, and is hence depen-

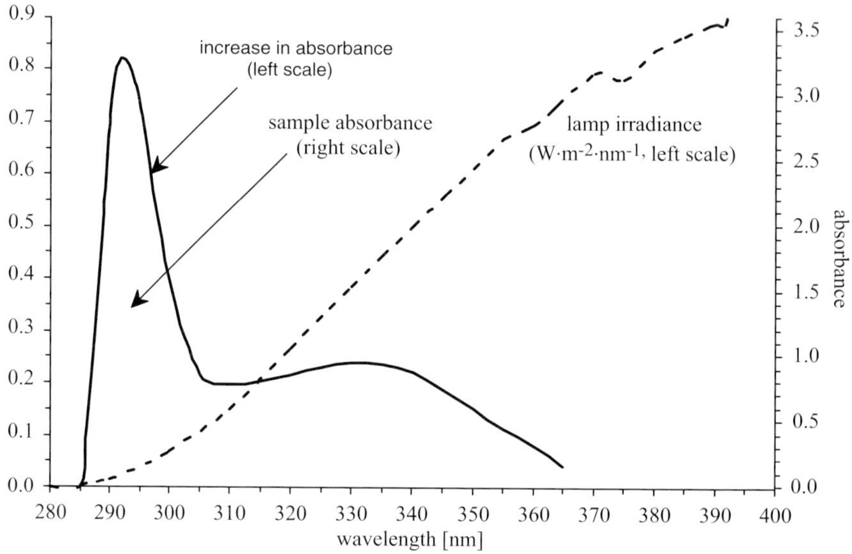

Fig. 15.12. Activation spectrum of a 700 μm film of unstabilized bisphenol-A polycarbonate, based on the increase in absorbance at 310 nm after accelerated aging with a borosilicate-glass filtered xenon lamp. Superimposed are the spectral irradiance of the irradiation source and the UV-absorption spectrum of the polymer (redrawn from data in Ref. [35]).

dent on the specific light source involved. The action spectrum is introduced to correct for this weakness by explicitly taking into account the efficiency of light-induced damage per *incident* photon, as a function of irradiation wavelength. Relatively few investigations have been performed on polymer action spectra until recently, as a result of the lack of suitable high-intensity, tunable, monochromatic source for irradiation. The use of sharp-cut filters in pairs with half-height bandwidths of approximately 20 nm does not usually provide sufficient wavelength resolution. The availability of large spectrographs now allows the recording of continuous activation and action spectra across the whole wavelength range of interest [34].

An action spectrum is a graph of the reciprocal of the radiant exposure required to produce a given effect (for instance, yellowing index or increase in absorbance at a standard wavelength) at each wavelength (Figure 15.12). All the data in such curves are normalized to the datum at the most efficient wavelength. This kind of plot gives crucial information on the spectral sensitivity of the material. Its knowledge is particularly important for proper comparison between outdoor weathering and accelerated artificial aging, for which the light source spectrum may differ significantly from solar irradiance.

In most instances, the action spectrum decreases empirically with increasing irradiation wavelength according to a logarithmic relationship (Figure 15.13B), as given in Eq. (71), with λ = irradiation wavelength, and $S(\lambda)$ = relative spectral damage [33].

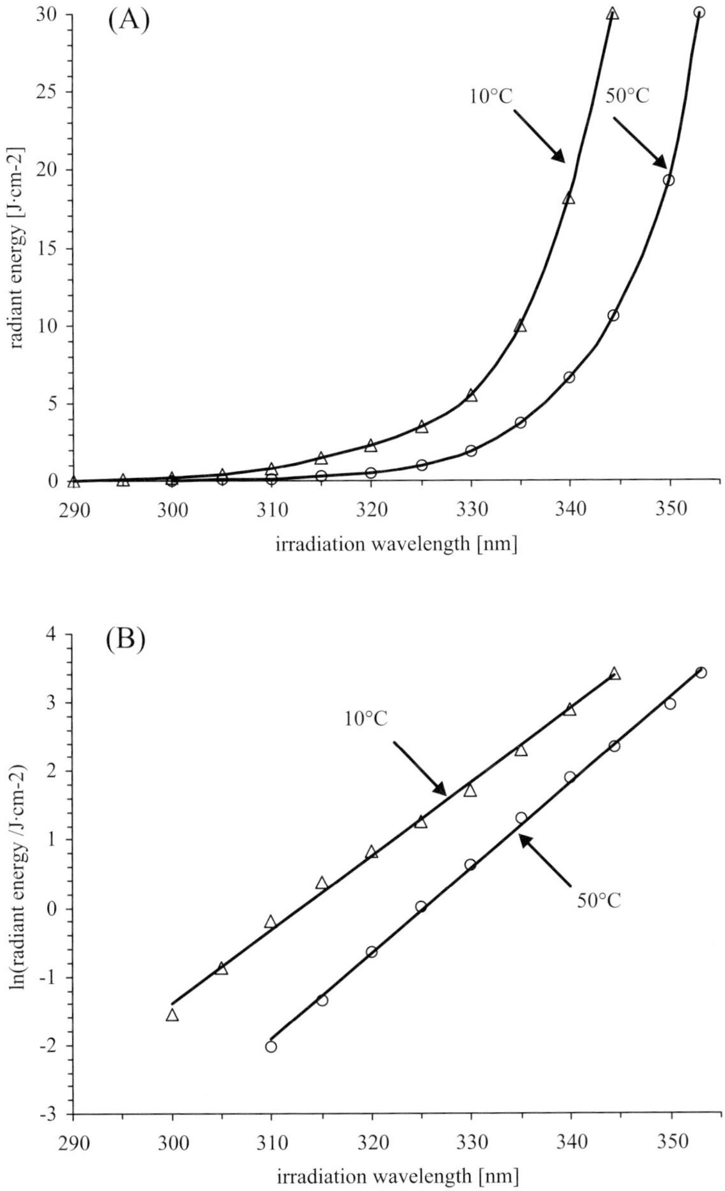

Fig. 15.13. Action spectrum of a 700 μm film of bisphenol-A polycarbonate, based on the radiant energy required for a 10% decrease in transmittance at 360 nm: (A) linear plot; (B) semilogarithmic plot.

$$\ln S(\lambda) = \ln(a) - b\lambda \tag{71}$$

Based on absorption and energetic considerations, photochemistry has long relied on the existence of a critical wavelength below which no reaction should occur (Section 15.5.1). The question then arises of the limit of applicability of the logarithmic relationship when the irradiation wavelength is much greater than the absorption maximum.

15.5.5 Photodegradation Mechanisms

Photodegradation differs from thermal degradation mainly in the mode of initiation. All the other mechanistic steps – bond scission, propagation, chain branching, and termination – are similar to thermal degradation. Indirect differences may exist, however, between thermal and photodegradation as a result of variations in experimental conditions, such as a higher temperature for the former mode of aging which may result in a lower concentration of dissolved oxygen, a lower rate of peroxide decomposition, or a change in material morphology.

15.5.5.1 Photoinitiation

It is practical to distinguish two types of initiation: primary, when the radiation is directly absorbed by the functional group of the polymer; and secondary, when the radiation is absorbed by certain of the oxidation products (peroxides, carbonyls). For saturated polymers, such as polyolefins, secondary initiation is the predominant mode of initiation, particularly in the latter stages of deterioration.

Formation of excited states following photon absorption constitutes the first step in any photochemical reaction. When light is absorbed by a chromophore, either "in-chain" or "external", the energy is used to promote an electron from the ground state to an excited state. The excited state can relax back to the ground state with release of extra energy in the form of a photon or heat. Photochemistry begins if the energy stored in the excited state is used to drive a chemical reaction. The rate at which this reaction occurs depends on the energy content (the exciting wavelength), the chemical nature of the absorbing chromophore, and its environment. The chemistry of the excited state in polymers has been studied with model compounds. In a real environment, however, the presence of additives, impurities, residual catalysts, and so on can exert a decisive influence not only on the degradation rates but also on the degradation mechanisms. The reactions are further complicated by the occurrence of additional chromophores which are formed during photodegradation. Following the diversity of impurities which can be encountered, and the complexity of ensuing reactions, there has been much controversy in the literature about the true photoinitiation mechanisms. This debate, however, is essentially academic, since the initiation mechanism does not influence the global rate of photo-oxidation.

Although acting as an external agent, oxygen deserves a special situation, owing to its presence in any practical weathering situation. Oxygen can be easily converted to the singlet state ($^1\Delta_g$ or $^1\Sigma_g^+$) by absorption of a photon, or it can be formed during decomposition of secondary compounds. An important mode of photoinitiation occurs via a charge-transfer complex between oxygen and the polymer (for instance, a polyolefin), with formation of hydroperoxides [Eq. (72)].

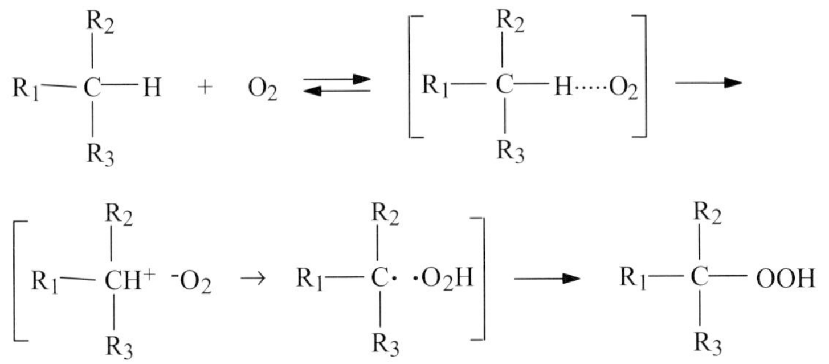

(72)

In spite of the diversity in photoinitiation mechanisms, two principal modes of reaction can be identified in most commercial polymers: the Norrish photoprocesses and the photo-Fries rearrangement.

15.5.5.2 The Norrish Photoprocesses

One common chromophore frequently encountered in a number of polymers is the carbonyl ketone which shows a weak π–π^* transition band centered at 280 nm but extending into the UV-B region. Polymers containing this chromophore undergo two types of photochemical reaction (Scheme 15.2):

- the Norrish type I reaction, resulting in an α-cleavage with formation of two end-polymeric radicals and carbon monoxide;
- the Norrish type II reaction, which involves an intramolecular H-abstraction via a cyclic six-membered transition state. The photoscission proceeds by means of a short-lived triplet state and a biradical intermediate.

15.5.5.3 Photo-Fries Rearrangement

A second large class of polymers contains phenyl moieties, either in the backbone (PPO, PC, PET, PSU) or as a side group (PS). Another common phenyl group photoreaction, when subjected to UV radiation, is the photo-Fries rearrangement (Scheme 15.3). Upon absorption of a photon, the n- or π-orbital of the chromophore is promoted to a singlet π^* excited state. Bond scission occurs primarily at aromatic ether C–O bonds, and causes rearrangement or degradation of the polymer backbone. Because the photo-Fries rearrangement proceeds in a "caged" environment, it is independent of free volume and is almost independent of T_g.

The photo-Fries mechanism can be promoted with light in the region above 300 nm (Figure 15.14), and accounts for the yellowing of the polymer observed at long wavelengths. Chain scission, on the other hand, is promoted by light at shorter wavelengths. Because photo-Fries products are easily photodegraded, only small

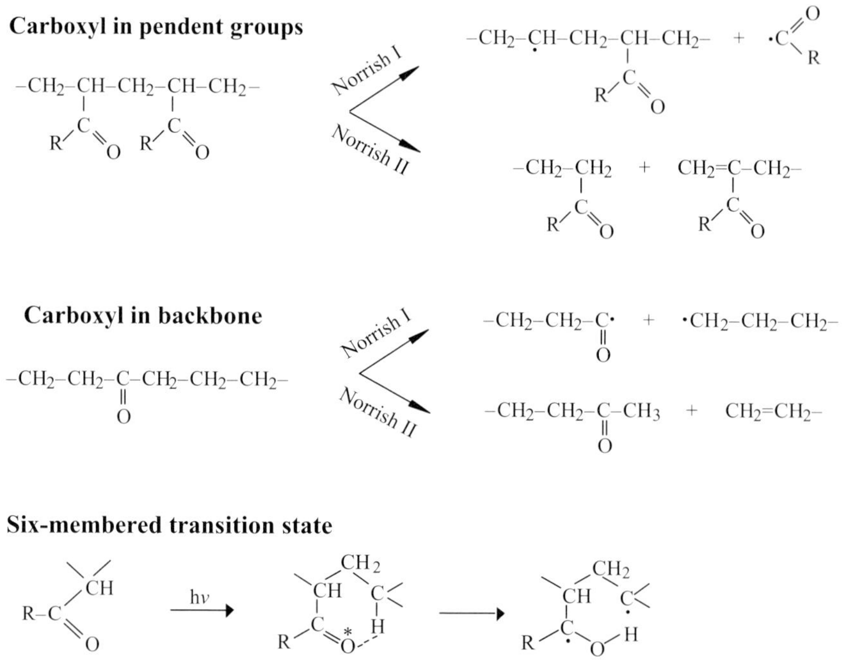

Scheme 15.2. Norrish photoprocesses in carbonyl-containing polymers.

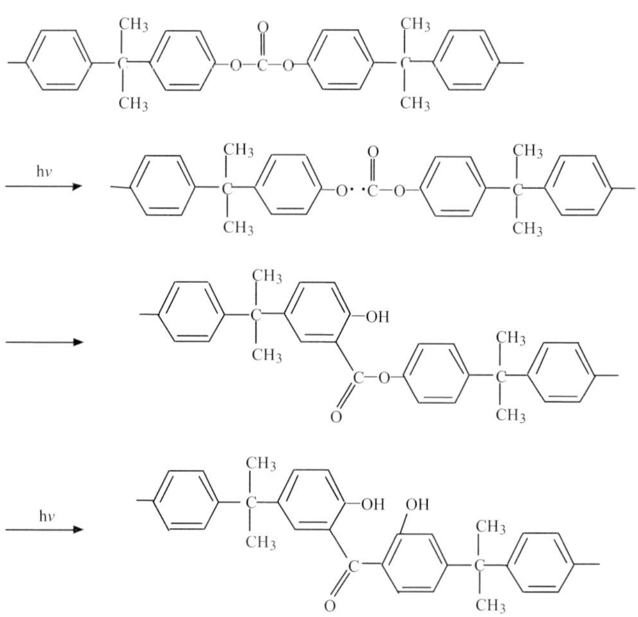

Scheme 15.3. Photo-Fries rearrangement in bisphenol-A polycarbonate.

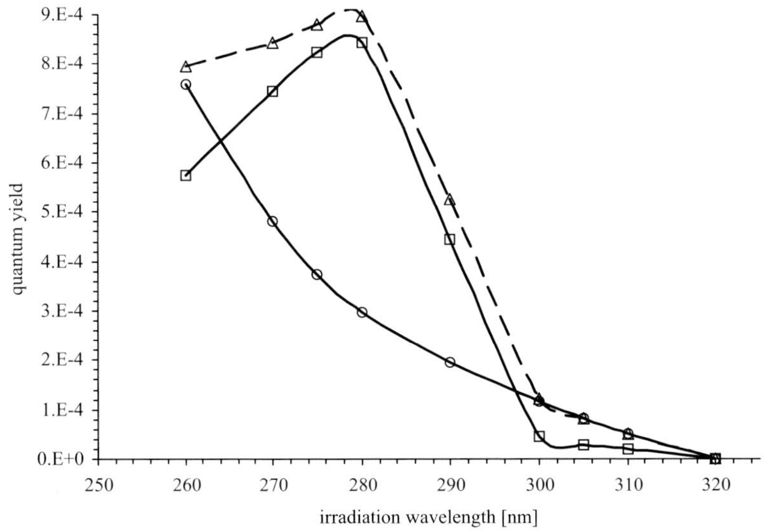

Fig. 15.14. Quantum yield for chain scission (–o–), efficiency of photo-Fries rearrangement (–□–), and absorption spectrum (--Δ--) of bisphenol-A polycarbonate [34].

amounts of them remain in aged material, which shows a predominance of side-chain oxidation products.

Once photoradicals are formed, they can either recombine in the cage or diffuse to the outside for further reactions. The radicals formed can undergo a variety of abstraction and decomposition reactions, resulting in a multitude of chemical products. The propagation and recombination steps are similar to those encountered in thermal degradation, and will not be discussed further here. In the presence of UV > 240 nm and oxygen, the phenyl groups in PS, PPO, and PC can undergo ring-opening oxidation with formation of mucondialdehyde (Scheme 15.4) [36].

15.6 Radiolytic Degradation

15.6.1 Interaction of High-energy Radiation with Matter

Understanding the effects of ionizing radiation on the properties of plastics materials is important in nuclear engineering, space research, radiation processing, and radiation sterilization.

Radiation composed of photons or particles with energy much higher than those encountered in electron bonding orbitals is referred as high-energy radiation (X- or γ-rays, high-energy electrons, protons, and charged particles). Owing to this excess

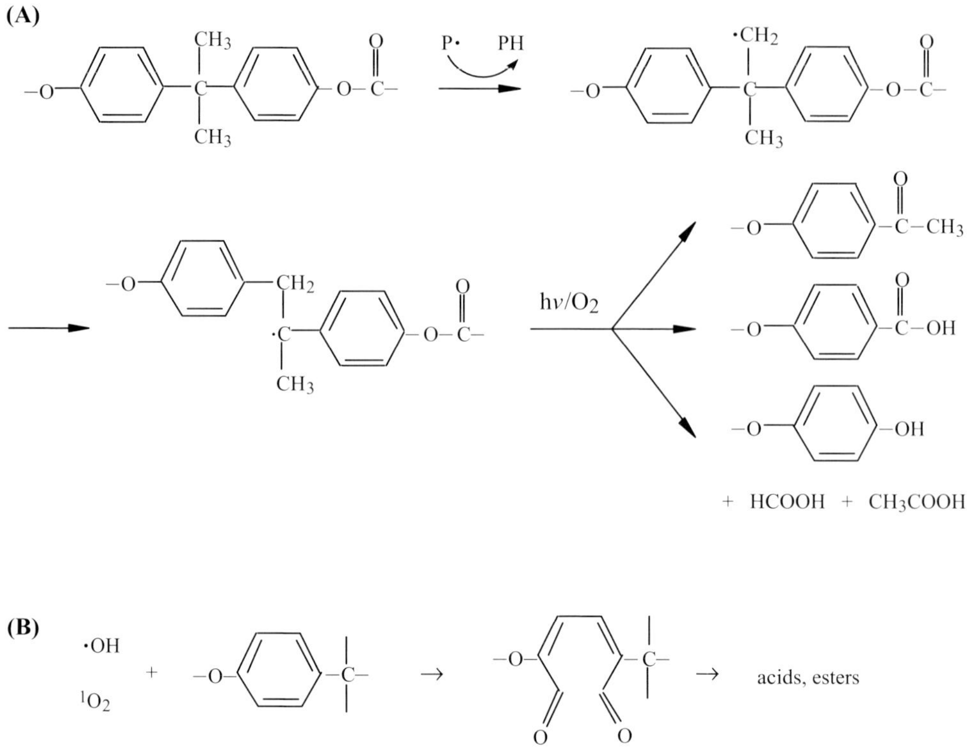

Scheme 15.4. Chain scission (A) and ring-opening (B) reactions in bisphenol-A polycarbonate.

energy, high-energy radiation can penetrate much deeper into material to create ions, superexcited states, and hot radicals. The degradative effects are also much more extensive than with UV. The principal sources of high-energy radiation are electron beam accelerators, which account for 90% of commercial radiation capacity; the remainder consist of ^{60}Co installations. ^{60}Co is unstable and decays to the stable ^{60}Ni according to Eq. (73).

$$^{60}\mathrm{Co}(t_{1/2} = 5.27\ \mathrm{y}) = {}^{60}\mathrm{Ni}^* + \beta^- \tag{73}$$

The nuclei of Ni atoms that result from this decay are in an excited state and immediately emit two γ-rays of energies 1.332 and 1.173 MeV. The low-energy β^- are absorbed by the ^{60}Co housing and all the radiolytic effects result from the γ-ray emission.

Electron accelerators can deliver higher dose rates, whereas ^{60}Co sources are characterized by a greater depth of penetration.

A fast electron loses most of its kinetic energy by inelastic collisions with electrons from the medium, producing energetic secondary electrons. Depending on the energy of the radiation, many secondary electrons of decreasing energy will be

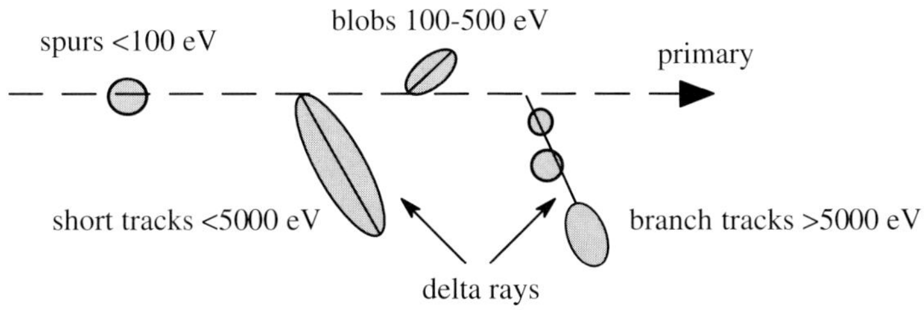

Fig. 15.15. Schematic representation of a fast electron track.

created (δ-rays). Owing to the augmentation of linear energy transfer (LET) with decreasing kinetic energy, the fast electrons lost most of their kinetic energy near the end of the track, resulting in local regions of highly reactive species called "short tracks", "blobs", and "spurs" (Figure 15.15).

In ^{60}Co irradiation, the primary interaction is Compton scattering between the high-energy photon and an atomic electron. Only part of the photon energy is transferred to the electron and the resultant photon of lower energy is scattered. The process will continue with creation of more scattered electrons and secondary electrons (Figure 15.16). Except for the primary event, the effects of γ-rays and accelerated electron irradiation are similar. From a single incoming γ-photon or high-energy electron, a shower of secondary electrons is generated which is responsible for most of the ensuing chemical reactions.

15.6.2
Radiation Chemistry

Radiation chemistry is now a well-established area of science that deals with chemical and physicochemical changes produced by the absorption of high-energy radiation (HER) by matter. The subject covers events that occur from the passage of the ionizing particle to the completion of chemical reactions. To encompass the enor-

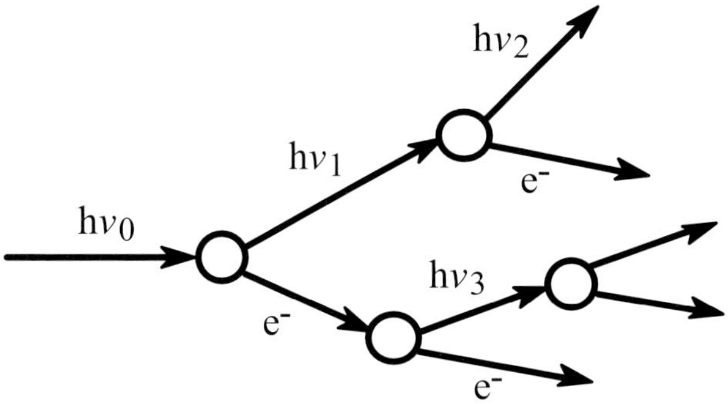

Fig. 15.16. Compton scattering and fate of the secondary electrons.

mous time span for the various processes, it is usual to distinguish between the "physical" stage (10^{-17}–10^{-11} s) when energy from the incident particle is deposited into localized regions of space ("spurr"), the "physicochemical" stage (10^{-12}–10^{-3} s) during which reactive species (ions, quasi-free electrons, excited molecules, hot radicals) are formed and react within the "spurs", and the "chemical" stage (10^{-9}–10^{-3} s in solution, several days in solid polymers) where stable species (radicals, trapped electrons, cations) diffuse and react outside the clusters. The standard unit of absorbed dose is the Gray (Gy), defined as the energy imparted by the high-energy radiation to a mass of matter equivalent to 1 J kg^{-1} (1 Gray = 100 rad). Molecular changes are characterized by a *G* factor, in units of μmol J^{-1}, defined as the event yield per 100 eV of absorbed energy.

Although much discussion has occurred in the past on the relative importance of radical and ionic reactions, it is now established that the major chemical changes in irradiated polymers are accounted for by free-radical reactions [37]. In the early stage of the reactions, the reactive species are concentrated in "spurs" and particle "tracks" in a manner similar to their parent ionized or excited molecules. The kinetics at this point must take into account the inhomogeneous distribution of the radicals, before they can diffuse away. Ionizing radiation is unique in the sense that reactions can be initiated randomly at any temperature. Cryogenic temperatures, at 4 K and below for instance, have been used extensively to prolong the macroradicals' lifetime for ESR measurements. Apart from the mode of initiation, all other material changes brought about by high-energy radiation are governed by radical reactions, in perfect analogy to those generated by other means, such as photochemical, thermal, or mechanochemical degradation. As a matter of fact, despite the enormous difference in HER energy (10^6 eV) and molecular binding energy (~5 eV), the chemical effects of HER can best be compared with those of UV light with energy in the 5–20 eV range.

The effects of ionizing radiation depend greatly on the structure of the polymer, the temperature, and the nature of environment. A 50% loss in ultimate elongation (a common measure of the effect of irradiation), for instance, can vary from doses as low as 3.5 kGy for PTFE, to more than 4000 kGy for PS, polyimide, or PEEK (Table 15.6). The unusual radiation sensitivity of PTFE is attributed to the unique stability of perfluoro macroradicals which favors chain scission over crosslinking. In the presence of air, these fluorine-containing radicals are converted into peroxy radicals which degrade readily into low MW fragments. PTFE can be crosslinked by HER when irradiated in an inert atmosphere above its melting point (330–340°C). Polymers containing phenyl groups owe much of their radiation resistance to excited-state energy transfer to the benzene rings, which act as excited-state quenchers (see Section 15.6.3). Although "energy transfer" is the widely accepted protection mechanism in the HER degradation of aromatic compounds, it does not explain why styrene does not show the same protective effect as polystyrene. One alternative suggestion is that H atoms resulting from the primary effect of radiation are added to the aromatic rings, and are no longer able to produce secondary macroradicals by abstraction.

Tab. 15.6. Relative radiation stability in air of major commercial polymers (based on a 50% decrease in ultimate elongation).

Polymer	*Dose [kGy]*
Polytetrafluoroethylene	10
Polytrifluorocholorethylene	30
Poly(methyl methacrylate)	300
Polycaprolactam	600
Isotactic polypropylene	1000
High-density polyethylene	1000
Poly(vinyl chloride)	1500
Poly(ethylene terephthalate)	2000
Poly(triethylene glycol dimethacrylate)	2000
Low-density polyethylene	3000
Polyurethanes	3000
Melaminoformaldehyde resin	4000
Polycarbonates	5000
Polystyrene	5000
Epoxy resin ED-10	15000
Epoxy resin ETZ-10	30000
Polyimides	100000

The two fundamental processes that result from radiochemical reactions are chain scission and crosslinking, characterized by G_s and G_x, respectively. If $G_s < 4G_x$, branched polymers can be formed and may eventually evolve into a three-dimensional network structure. Based on the assumptions that:

- the initial polymer MWD follows a random distribution,
- cross-linking occurs by H-linking, and
- crosslinking and scission occur with random spatial distribution (without clustering),

Charlesby and Pinner have shown that the sol fraction(s) should follow Eq. (74), where R [kGy] is the absorbed dose [37].

$$s + s^{0.5} = 0.96 \times 10^5/(R \cdot M_w \cdot G_x) + (G_s/2G_x) \tag{74}$$

The Charlesby–Pinner plot of $(s + s^{0.5})$ as a function of $1/R$ gives a straight line, with slope $\propto 1/G_x$, and an intercept on the ordinate equal to $G_s/2G_x$. Departure of the plot from a straight line may originate from deviation from a random distribution, or from chemical heterogeneities in copolymers. Regardless of the shape of the curve, the incipient dose for gelation is determined, by definition, from the sol curve at $s + s^{0.5} = 2$ (Figure 15.17).

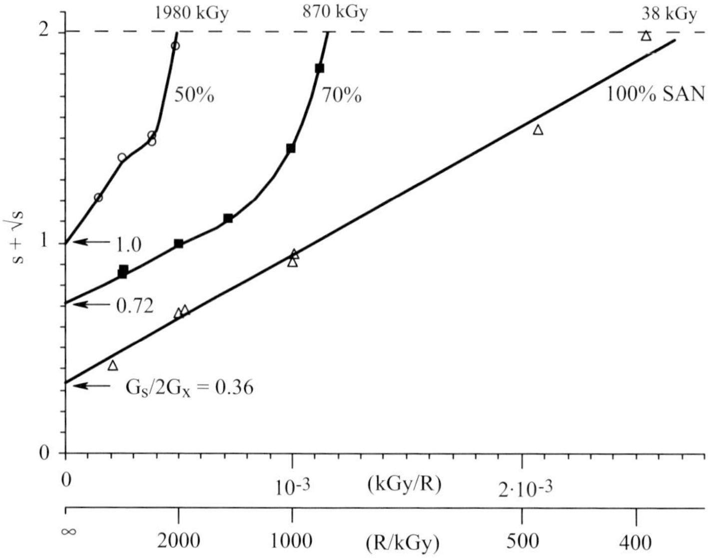

Fig. 15.17. Charlesby–Pinner plot of irradiated compatible PMMA–SAN blends. The sol and gel fractions refer solely to the SAN component in the mixture and are corrected for grafting to PMMA. Large curvatures of the plot for blends are interpreted as a result of chemical interactions between the two components (redrawn from Ref. [38]).

15.6.3
Radiolysis Stabilization

It has been known since the early days of radiation chemistry that some simple organic compounds, such as benzene, halogenated hydrocarbons, nitriles, amines, and alcohols, can protect the polymer from the deleterious effects of high-energy radiation. Many of these "antirad" substances interfere at some stage with the radiolytic degradation scheme, as depicted in Table 15.7, to reduce damage to the plastics. The most efficient present-day "antirad" agents are antioxidants which act essentially in the chemical stage by scavenging free radicals in a similar way to that in the other types of degradation. Aromatic compounds are highly efficient at quenching excess energy of excited states formed by geminal recombination. Because most commercial antioxidants have aromatic rings in their structures, they can also act as primary absorbers by diverting the radiation energy into harmless vibrational energy, as in Eqs. (75) and (76).

$$\mathrm{P} \xrightarrow{\gamma, e^-} \{\mathrm{P}^+ + e^-\}_{\mathrm{spur}} \longrightarrow \mathrm{P}^* \longrightarrow 2\mathrm{R}\bullet \tag{75}$$

$$\mathrm{P}^* + \mathrm{Q} \longrightarrow \mathrm{P} + \mathrm{Q}^* \tag{76}$$

Aromatic polymers owe much of their radiation resistance to this "internal protection" effect [40]. An efficient method of radiation protection would be to blend the

Tab. 15.7. Polymer protection scheme during radiolysis.

Radiolysis stage	*Physical or chemical event*	*Protection means*	*Protection mechanism*
Physical	energy absorption	sheets of lead or concrete	decrease in radiation intensity
Physicochemical	molecular ionization	positive ion scavengers	transfer of one electron to polymer cation without subsequent excitation
	geminate recombination	electron scavengers	acceptance of ejected electron and lower probability of dissociative recombination
	dissociative states	electronic excited-state quenchers	divergence of excitation energy into heat or longer-wavelength electromagnetic radiation
	cleavage of C–H bonds	H atom donors	transfer of H atom to macroradicals (reparation mechanism)
Chemical	diffusion and abstraction reactions of mobile H atoms	H atom acceptors	acceptance of H atoms, preventing formation of more radicals
	reactions of macroradicals	radical scavengers	reaction with macroradicals, formation of stable species
		surface coatings	decrease in oxygen diffusion
	reactions of macroradicals with oxygen	oxygen absorbers	React with molecular oxygen under irradiation
	benzophenone derivatives	Antioxidants	Disrupt oxidation chain reaction by converting peroxy radicals into stable products
	chain reaction of peroxy radicals (radio-oxidation)	peroxide decomposers	catalytic destruction of peroxy radicals

degradative polymer with a radiation-resistant one. Energy transfer (Förster type) is efficient only at short range and the effect is most noticeable with compatible blends, such as PMMA and SAN (Figure 15.17).

15.6.4 Applications

High-energy radiation can penetrate deeply into organic materials, and can initiate reactions at low temperature without added chemicals or catalyst. These unique features are exploited in an increasing number of industrial applications, particularly in the biomedical field, in which chemical contamination or thermal degradation should be avoided [41]. Radiation sterilization of medical commodities, one of the early achievements of radiation engineering, continues to increase its market share to the detriment of standard methods such as chemical sterilization with diethyl ether, or heat treatment. In the examples mentioned, the value of the finished products is generally high and the cost of radiation processing does not enter into consideration. In other applications, where the product is inexpensive radiation processing can still be economically viable if the quantity of radiation energy required is low. Radiation treatment of food and radiation-induced grafting or crosslinking of certain plastics belong to the latter category. Food is commonly irradiated at low doses in the 0.1–10 kGy range, in ordinary boxes or containers. Food preservation by irradiation is gaining acceptance for an ever-increasing number of agrochemical products such as spices, vegetables, and processed meat. This technique is currently viewed as the most effective of the available alternatives (cold storage, heat treatment, fumigants).

15.6.4.1 Radiation Sterilization

Disposable medical products are required by legislation to be sterilized prior to use. Among the four basic methods of sterilization – heat, ethylene oxide, gamma irradiation (^{60}Co), and electron beam techniques – the last has steadily increased its market share, largely at the expense of the diethyl ether method as a result of regulatory concerns about the quantity of chemical residuals. The Federal Drug Administration (FDA) in the US allows a device to be labeled as sterile if less than one device out of a batch of one million can show biological contamination. The standard dose to achieve this level is 25 kGy. Medical plastics that need to be sterilized must withstand at least this dose. The data given in Table 15.6 can be used as an initial estimate in material selection. Materials which are likely to degrade, such as PTFE or POM, should be avoided. Polymers containing aromatic groups have much greater radiation resistance than those with an aliphatic structure. Similar improvements can be obtained with crosslinks, and most thermosets and elastomers can withstand at least one radiation sterilization. The use of antioxidant additives can significantly offset the effects of radiation, and even a radiation-sensitive polymer such as polypropylene can withstand several radiation sterilization cycles when properly stabilized.

15.6.4.2 Controlled Degradation and Crosslinking

The reactions of chain scission, crosslinking, and grafting initiated by ionizing radiation have found many important applications in the plastics and rubber industries. Although high-energy radiation is destructive for most materials, this propensity can be usefully exploited in the controlled degradation of several polymers, such as PTFE, PEO, PP and cellulose. The most well-known example of degradation application is the manufacture of PTFE powders. Undegraded PTFE is too tough and slippery to grind. After submitting the polymer to a high dose range (500–1000 kGy), the degraded PTFE becomes brittle and can be converted to fine particles while conserving its low surface energy. The PTFE powder is used as a solid lubricant, or formulated as grease for high-temperature applications.

Radiation crosslinking is another way to tailor properties to specific applications. High-energy radiation provides a clean and cost-effective method to achieve this means. The technique has been applied in the crosslinking of high-purity medical products, such as gloves and condoms, orthopedic ultra-high MW PE hip joints, and biocompatible hydrogels. Other important industrial applications of radiation crosslinking include the production of heat-shrinkable products, wires, cables, and tubing.

15.7 Mechanochemical Degradation [42]

15.7.1 Initiation by Mechanical Stresses

15.7.1.1 Effect of Tensile Stress on Chemical Reactivity

The input of external energy, of either thermal or electromagnetic origin, is well known to the organic chemist as an efficient way to promote chemical reactions. Mechanical energy, although ubiquitous in everyday life, is rarely considered as a bona-fide source of chemical activation. The main reason probably stems from the inefficiency of this mode of excitation for small molecules where most of the applied mechanical energy is readily dissipated into heat and the work of bulk deformation, leaving little available as chemical potential. In polymer systems, on the contrary, mechanochemical reactions are ubiquitous as a result of the unique propensity of macromolecules to store free energy upon deformation and to sustain a high level of stress for a sufficiently long time for chemical reactions to occur. Mechanochemical degradation is ubiquitous in macromolecular systems and can be encountered in practically any field involving high MW polymers. The formation of free radicals by mechanical stress has been detected during polymer processing, analysis, weathering, and gel swelling. Chain scission has also been observed in applications as diverse as drag reduction, rubber mastication, and mechanochemical synthesis. It was known for centuries that the application of stress under certain conditions, such as the combing of wool, the mastication of rubber, or the kneading of dough, can improve the physical properties of a few materials which were later recognized as being polymeric.

Following the application of stress, a molecular system will attempt to minimize its total free energy by a series of configurational rearrangements, such as kink annihilation, chain disentanglement, and orientation. According to molecular mechanics, the excess energy of stress (E_s) in a chemical system comes from several sources, namely nonvalence and coulombic interactions, torsional deformation, and angular and valence-bond strain energies. Among the different modes of molecular deformation, bond-angle distortion and valence-bond stretching have the most profound influence on the electronic density distribution, and hence on the chemical reactivity. In the region of stresses of around 80 GPa, the valence angles for C–C bonds will become deformed from their equilibrium positions whereas bond stretching and rupture occur at some still-higher values in the vicinity of 740 GPa. Although stress-induced chemical reactivity can be explained only by quantum mechanical calculations, as was done for the acceleration effect of water during crack propagation of silica glasses, a qualitative picture can be gained based on mechanical arguments. In carbocyclic compounds, E_s originates principally from valence-angle deformation and it is well known that ring strain could affect the rate of some chemical reactions in a drastic way. For example, if the transition state necessitates distortion of the valence-bond angle from θ_0 to θ^*, the valence bond-angle contribution to the activation energy is given by Eq. (77), where k_θ is the elastic constant for bond-angle deformation.

$$E_{ang} = \tfrac{1}{2}k_\theta(\theta_0 - \theta^*)^2 \tag{77}$$

If, however, the same bond angle is already distorted from θ_0 to θ under mechanical stress, the activation energy will be given by Eq. (78), resulting in a difference of ΔE_{ang} [Eq. (79)].

$$E_{ang} = \tfrac{1}{2}k_\theta(\theta - \theta^*)^2 \tag{78}$$

$$\Delta E_{ang} = \tfrac{1}{2}k_\theta[(\theta_0 - \theta^*)^2 - (\theta - \theta^*)^2] \tag{79}$$

In solid polymers, rate enhancement under stress has frequently observed, for example in UV-photo-oxidation of natural silk, polycaprolactam, and poly(ethylene terephthalate). However, quantitative interpretation is difficult in these systems due to interdependence of several stress-dependent factors such as the rate of oxygen diffusion or changes in polymer morphology which supersede the elementary molecular steps. Similar experiments in the fluid state showed unequivocally that flow-induced stresses can accelerate several types of reaction, the best studied being the hydrolysis of DNA and polyacrylamide. In these examples, hydrolysis involves breaking of the ester O–PO and the amide N–CO bonds. The tensile stress stretches the chain, and therefore facilitates the formation of a transition state in which the bond length is increased for hydrolysis, thus enhancing the rate constant for the process. Other documented examples of stress-induced chemical reactions are the acylation of cellulose and the addition of 4-hydroxy-2,2,6,6-tetramethylpiperidine-1-oxy to rubber during mastication.

15.7.1.2 **Breaking Strength of a Covalent Bond**

Initiation in mechanochemical degradation involves homolytic scission, so one basic question would be what level of stress is necessary to separate two chemical moieties which have been attached by a covalent bond. The earliest account for this problem was reported by de Boer in 1936 [42]. The potential of a bond under equilibrium is normally approximated by the Morse potential, given by Eq. (80) where l is the length of the bond, l_0 is the equilibrium separation distance of the atoms, D is the bond dissociation energy, $a = (k_f/2D)^{0.5}$ is a parameter which defines the width of the potential minimum, and k_f is the bond force constant in the neighborhood of the equilibrium separation.

$$V(l) = D(1 - \exp[-a(l - l_0)]^2) \tag{80}$$

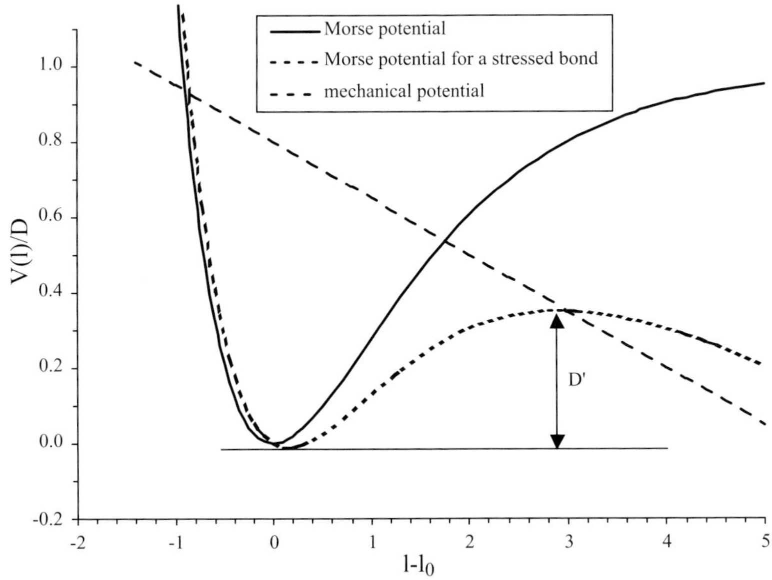

Fig. 15.18. The Morse potential of a bond under equilibrium (----) and in the presence of an applied force equal to $0.6 f_b$ (----).

If the bond is under tension with a constant force f_{ext} pulling on either end, the potential energy $V'(l)$ will be decreased by an amount equivalent to the work performed by the mechanical force over the stretching distance from the equilibrium position [Eq. (81); Figure 15.18].

$$V'(l) = V(l) - (l - l_e), \text{ where } C \text{ is a proportional factor} \tag{81}$$

The potential function $V'(l)$ has a minimum at $l_0' > l_0$, in accord with the intuitive expectation that the bond separation should increase in the presence of a tensile stress. The new activation energy (D') required to break the stressed bond could be calculated from the principle of virtual work performed on the bond in going from l_0' to l_b (Figure 15.19).

15.7.1.3 **Rate of Stress-activated Chain Scission**

From a chemical viewpoint, bond scission under stress is a particular case of a unimolecular dissociation reaction whose rate is enhanced by mechanical stress. As such, it could be treated with Eyring's transition-state theory [Eq. (37)], which permits one to bring the treatment of rate processes within the scope of thermodynamic arguments. By combining de Boer's thermodynamic formulation and the transition-state theory, Tobolsky and Eyring in 1943 developed the rate theory for thermally activated fracture of polymeric threads. When put into an Arrhenius-

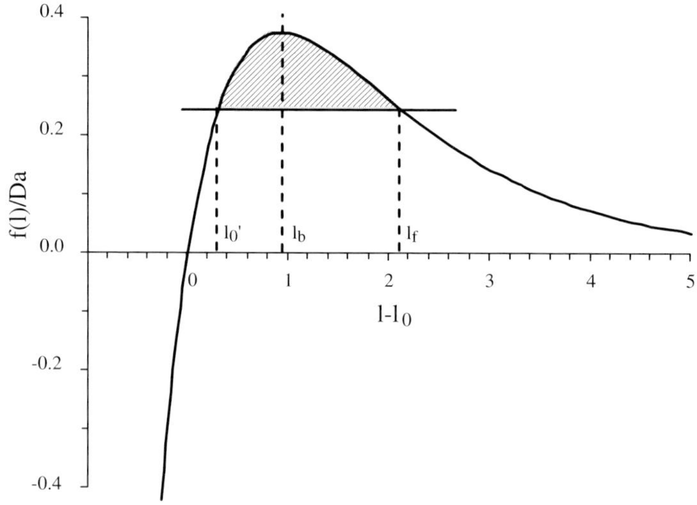

Fig. 15.19. Derivative of the Morse potential. The hatched area corresponds to the bond energy under stress.

like form, Eq. (82) was obtained, where k_c is the rate constant for bond scission, U_0, the thermal energy for bond rupture, and ψ the molecular stress.

$$k_c = A \exp[-\{U_0 - f(\psi)\}/k_B T] \tag{82}$$

Several attempts to relate the rate for bond scission (k_c) to the molecular stress have been reported over the years. The simplest and still the most popular is a linear relationship [Eq. (83)] between the decrease in activation energy and molecular stress [42].

$$f(\psi) = \beta\sigma \tag{83}$$

The factor β has the dimension of volume and is identified as the activation volume for the reaction.

15.7.2 Extrusion Degradation

To achieve useful properties, polymers are generally compounded with a variety of additives, transformed into granulates, before being processed into the final shape for commercial exploitation. During each step, the material is exposed to a high shear-rate and temperature. The two major sources of chemical degradation during processing are therefore mechanical stresses, which tend to decrease with increased temperature as with the melt viscosity, and thermal reactions. As a result of excessive pressure within the equipment in the course of processing, oxygen from the air generally has no access to the polymer melt. Except during the initial stage, when dissolved air can be used up at high rate, most of the degradation occurs under oxygen-deficient conditions. This feature has been verified during multiple extrusion of polypropylene [44]. In this experiment, as in other similar experiments, the concentrations of oxidation products detected by FTIR were very low. On the other hand, changes in the MW and MWD were always observed during extrusion. Because the MWD is of the utmost importance for the rheological properties, it is important to know the molecular parameters which may influence the rates of chain scission and crosslinking. Oxidative β-scission of alkoxy radicals is relatively unimportant in extrusion, as a result of lack of oxygen and low activation energy (59 kJ mol^{-1}). The majority of chain scissions are caused by the β-cleavage of alkyl radicals, with activation energies of 84–117 kJ mol^{-1}. Cleavage of a secondary alkyl radical or a tertiary radical produces a vinyl ($H_2C{=}CHR$) or vinylidene group ($H_2C{=}CRR'$), respectively. A kinetic model for PE crosslinking and scission involving alkyl radicals under processing conditions has been developed, based on and activation energy of 146 kJ mol^{-1} for chain scission, and 96 kJ mol^{-1} for crosslinking. As a result of the difference in activation energies, crosslinking tends to dominate at low temperatures and scission at high temperatures. Crosslinking reactions are attributed to the addition of alkyl radicals to vinyl groups, either initially present during synthesis or formed by thermal scission.

15.7.3
Applications

One of the earliest industrial applications of mechanochemical degradation was in the mastication of rubber, to reduce the MW of the latex to a level which could be processed. Rubbers vulcanized with a conventional high sulfur/accelerator system are known to be more resistant to fatigue than peroxide-cured rubbers and rubbers cured by sulfur-free or low-sulfur systems. This behavior was associated with the presence of polysulfidic ($-C-S_n-C-$) crosslinks in the conventional vulcanizate [45]. Because the S–S bonds are more labile than the C–C bonds (see Figure 15.18), it is postulated that they can be easily broken under fatigue, resulting in the release of local stresses during bond rearrangement.

Mechanochemical degradation creates free macroradicals in pairs, practically without any side reactions, and most potential applications of this technique are centered around the formation and subsequent reactions of these reactive species. Elongational flow-induced degradation breaks polymer chains exactly at their center [42, 46]. This remarkable propensity is being explored in the author's laboratory as a simple means of obtaining well-defined block copolymers. Polymerization and

–C–C–	–C–S–C–	–C–S–S–C–	–C–(S)n–C–
410 kJ·mol-1	350 kJ·mol-1	300 kJ·mol-1	260 kJ·mol-1

Fig. 15.20. Bond dissociation energies in vulcanized rubber.

synthesis of block or graft copolymers by mechanical forces have been the subjects of several reviews and books [47]. Although mechanochemical synthesis of inorganic materials has enjoyed wide industrial application, no similar development has yet been witnessed with plastic materials.

15.8
Control and Prevention of Aging of Plastic Materials

It has been shown in the preceding sections that raw polymers are highly susceptible to degradative oxidation. The success of plastic materials, which find applications in practically any aspect of life, relies heavily on the performance of polymer stabilizers, 70% of which are used for polyolefins. According to their principal protection activity, common polymer stabilizers are conventionally classified as antioxidants, photoantioxidants, photostabilizers, metal deactivators, antiozonants, and heat stabilizers for PVC.

15.8.1
Antioxidants

Antioxidants prolong the useful lifetime of the polymer by trapping free radicals ("primary" antioxidants) and by decomposing hydroperoxides ("secondary" antioxidants) formed during the course of degradation.

15.8.1.1 Radical Antioxidants

Since polymer degradation is initiated by free radicals, the most effective approach in both heat and light stabilization would be to reduce the number of reactive radical species by scavenging. Secondary aromatic amines and substituted phenol derivatives are among the most popular radical-scavenging stabilizers. The stabilization action of these compounds is based on the formation of aminoxy and phenoxy radicals, respectively, which can scavenge propagating radicals formed during degradation. Amine additives are converted into conjugated secondary products which absorb visible light. For this reason, hindered phenols are preferred in applications where discoloration is undesirable. Typical phenolic antioxidants are shielded with bulky alkyl substituents in the 2-, 4- and 6-positions to reduce the reactivity of phenoxy radicals toward hydrogen atom abstraction reactions on polymer and mutual coupling. The principal mode of action of hindered phenols is the transfer of a hydrogen atom to a propagating alkyl (R•) or alkylperoxy (ROO•) radical [Eq. (84)].

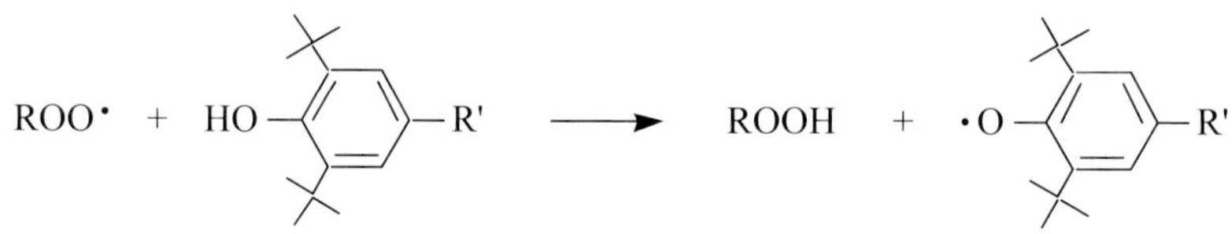

(84)

The phenoxy radicals are relatively long-lived and can undergo self-disproportionation, recombination with alkylperoxy radicals, or isomeric rearrangement followed by recombination. The resulting compounds may have some stabilization activity. Propionate-type hindered phenols constitute a special class in this category: during reaction, the phenol is transformed into phenolic cinnamates, which are known to be efficient chain-breaking antioxidants [Eq. (85)].

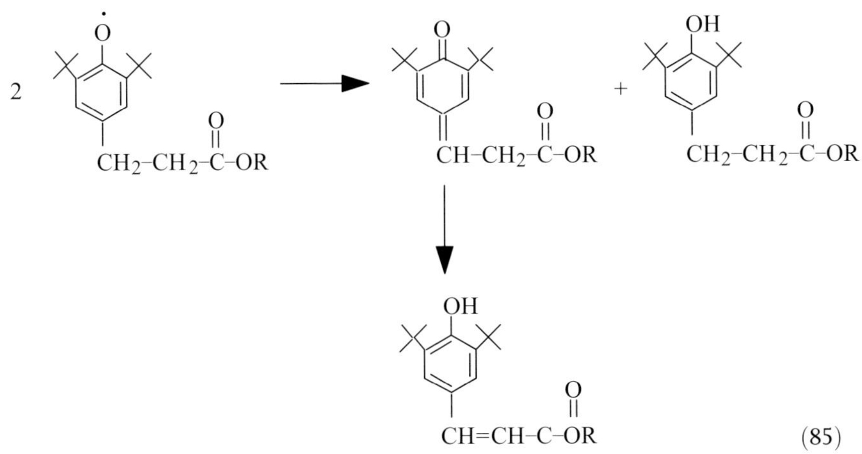

(85)

15.8.1.2 **Hindered Amine Stabilizers (HAS)**

With a few exceptions, such as hydroxybenzoate derivatives, most conventional phenolic antioxidants suffer from low photostabiliy and could not be used for photostabilization. This problem was solved with the introduction and development of hindered amine light stabilizers (HAS) [49]. Because this class of compounds was originally developed for photostabilization, they are frequently referred as "hindered amine light stabilizers" or HALS. HAS are commercialized under a variety of structures, most of which contain one or several 2,2,6,6-tetramethylpiperidine moieties linked together to form polynuclear species. The structure of a typical commercial HAS, Chimassorb® 944, is shown in Figure 15.21.

The protection mechanisms of HAS are based on a complex series of chemical transformations, with many unresolved details. The primary step includes oxidation in situ of secondary or tertiary HAS into nitroxides >NO• by the various oxidants present during oxidative degradation [Eq. (86)]. The substituent can be H, CH_3, OR, or $COCH_3$, and the oxidative species $^3O_2$, $^1O_2$, O_3, POO•, POOH, PC(O)OO•, or PC(O)OOH.

$$>N{-}R + \text{oxidative species} \rightarrow\rightarrow >NO^{\bullet} + \text{products} \tag{86}$$

Stable nitroxide radicals are traditionally considered as the principal species which account for the free-radical scavenging activity of HAS. Nitroxide radicals can trap alkyl, alkylperoxy and acylperoxy radicals. The probability of direct radical scaveng-

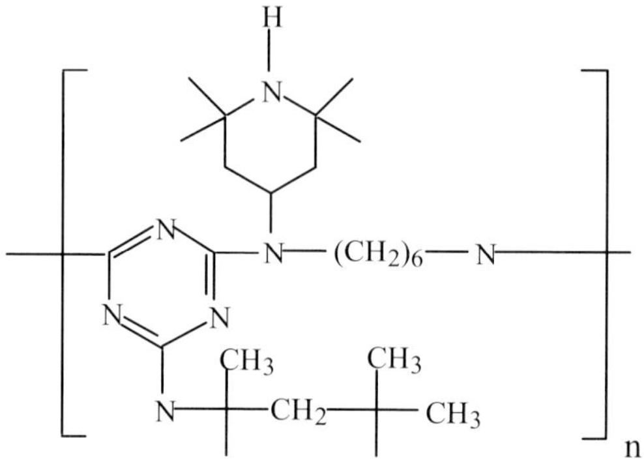

Fig. 15.21. Structure of a typical commercial hindered amine stabilizer (HAS).

ing, such as in Eq. (87), is nevertheless limited as a result of the higher rate for competitive reactions of R• with dissolved oxygen and by self-recombination.

$$>NO\bullet + \bullet R \rightarrow >NOR \tag{87}$$

As with the formation of peracids, it is suggested that scavenging occurs inhomogeneously and is enhanced in regions of accumulated hydrogen-bonded complexes, owing to the formation in situ of R• [Eq. (88)].

$$\{>NO\bullet \cdots HOOR\} \rightarrow \{>NO\bullet, HO\bullet, RO\bullet\} \rightarrow \{>NO\bullet, R'\bullet\} \tag{88}$$

The formation of hydrogen-bonded associates with hydroperoxides or phenolic additives is favored by the resonance-stabilized dipolar structure of nitroxides represented by Eq. (89).

$$>N{-}\dot{O} \longleftrightarrow >\overset{+\bullet}{N}{=}O^{-} \tag{89}$$

The transformed hydroxylamines and dialkylhydroxylamine ethers are assumed to play a key role in the mechanism of stabilization of HAS, through a regenerative inhibition cycle known as the Denisov cycle [Eqs. (90a)–(90c)] [50].

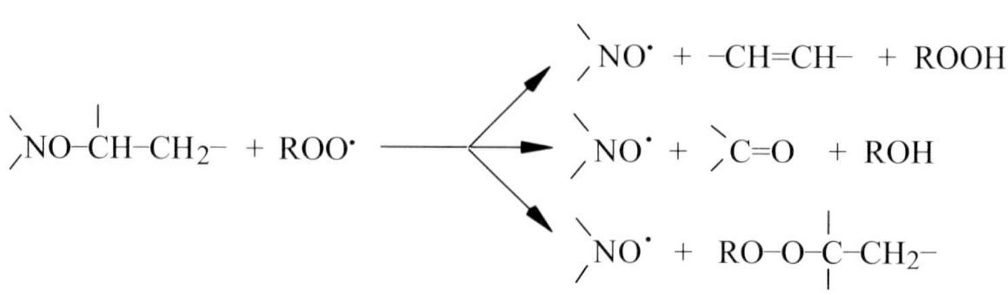

(90a) (90b) (90c)

Tertiary hydroxylamine ethers, the main constituent in degraded polypropylene, lead preferentially to the formation of alkyl peroxides [Eq. (90c)].

There is increasing evidence that other regeneration mechanisms are involved, such as scavenging of alkylperoxy or peracyl radicals by hydroxylamines and alkylhydroxylamines according to Eq. (91).

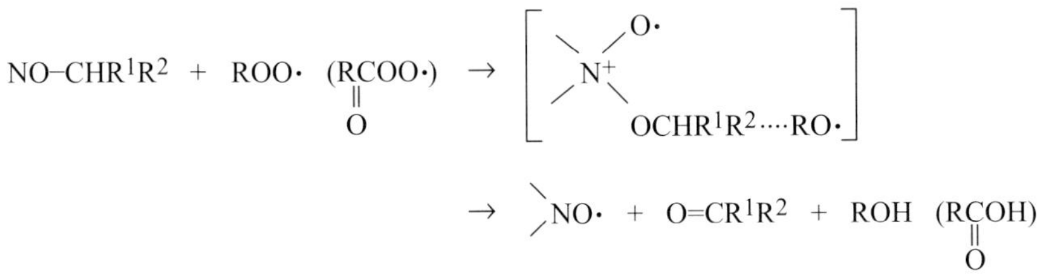

(91)

Another proposed mechanism of action involves long-range energy transfer from polymer–oxygen exciplexes to HAS or HAS–oxygen complexes [49]. This process is particularly efficient in inhibiting photoinitiation which takes place essentially at the crystalline/amorphous interface, where the HAS is concentrated.

Basicity of HAS Secondary and tertiary HAS are relatively basic and may catalyze hydrolysis of some polycondensation polymers, such as polycarbonate. HAS lose their activity if they come into contact with strong acids, and hence cannot be used to stabilize PVC or in formulations which contain halogenated flame retardants. HAS with an acylated amino group (>$NCOCH_3$) has a lower pK_a and can better resist an acidic environment.

Thermal stabilization with HAS As a result of its protection mechanism, HAS are particularly effective in photodegradation. Recently, it has been assessed that HAS can also be used for thermal degradation. In this case, however, HAS should be used in combination with hindered phenols because many degradation products of HAS, which are efficient antioxidants, can be formed only with light.

15.8.1.3 Peroxide Decomposers

Peroxides are responsible for the autoacceleration stage during oxidative degradation of polymers. Hydroperoxide decomposers, which cause the reduction of hydroperoxides to alcohols, comprise two main chemical classes: organic sulfur and trivalent phosphorus compounds, generally used in combination with phenolic antioxidants. The decomposition mechanism is complex and involves radical and nonradical processes. During the course of decomposition, a mixture of oxidation products, generally endowed with additional peroxidolytic properties, are formed. For instance, the decomposition of a thioether in the presence of alkyl hydroperoxides can be represented schematically by a series of reactions [Eq. (92)] which involve thiyl RS•, sulfinyl RS•(O), sulfonyl $RS•(O)_2$, and perthiyl RSS• radicals.

$$[\mathrm{R{-}O{-}C{-}(CH_2)_2{-}}]_2\mathrm{S} \rightarrow [\mathrm{R{-}O{-}C{-}(CH_2)_2{-}}]_2\mathrm{SO}$$
$$\rightarrow \mathrm{R{-}O{-}C{-}(CH_2)_2{-}SOOH} + \mathrm{H_2C{=}C{-}C{-}O{-}R} \rightarrow \cdots \rightarrow \mathrm{SO_2, H_2SO_4}, \ldots \quad (92)$$

Sulfur dioxide and sulfuric acid, both of which are also active hydroperoxide decomposers, are formed at the ultimate stage of transformation.

Organic phosphites are very effective hydroperoxide decomposers. Their peroxidolytic activity results from a sacrificial transformation of phosphite (trivalent) to phosphate (pentavalent), according to the general reaction scheme in Eq. (93).

$$\mathrm{R{-}OOH} + \mathrm{P(OR')_3} \rightarrow \mathrm{ROH} + \mathrm{O{=}P(OR')_3} \quad (93)$$

15.8.2
Photostabilizers

Outdoor weathering is the most common source of polymer degradation. The most direct and efficient way to avoid photodegradation would be to prevent the photons from reaching the polymer either by applying a paint or coating to the surface, or by blending the polymer with strong light-absorbing particles such as carbon black. This procedure, however, is inapplicable for situations where aesthetic appearance or transparency of the material is essential.

UV absorbers The use of organic compounds with a high absorption coefficient in the UV but transparent to the visible, known as UV absorbers, should act in a similar fashion to paints, by protecting the polymer while conserving the optical clarity of the material. Paints and coatings themselves are not stable to solar radiation and are often compounded with UV absorbers to prolong their protection efficiency. To be useful, UV absorbers must strongly absorb at wavelengths harmful to the polymer, must harmlessly dissipate the energy that they absorb, and must persist in the matrix for the expected lifetime of the article. Most commercial UV absorbers are derivatives of benzophenone, benzotriazole, triazine, oxanilide, or cyanoacrylate with aromatic substituents. All have a highly delocalized π-electron structure with a high extinction coefficient ($\varepsilon = 10\,000$–$40\,000$ L mol^{-1} cm^{-1}) in the near-UV region (290–350 nm). The photostability of benzophenone-based UV absorbers is conferred by intramolecular proton transfer in the excited state, and most likely by a charge-separation mechanism in the case of a cyanoacrylate structure.

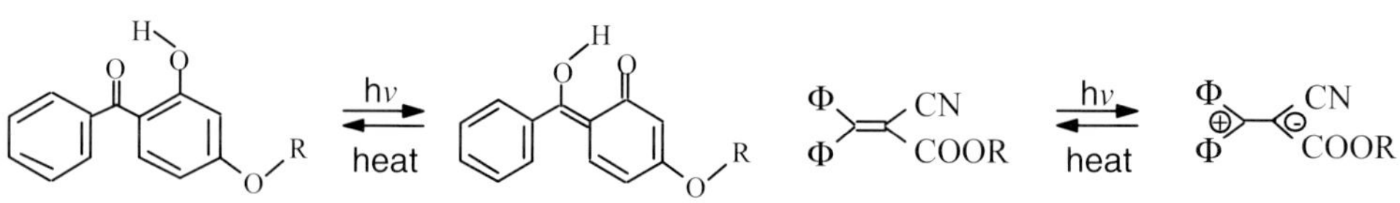

Scheme 15.5. Energy dissipation mechanism for a protic (left) and a cyanoacrylate (right) UV absorber.

Quenchers Quenchers are used to capture the excess energy of excited chromophores [Eqs. (94) and (95)], which is then dissipated as vibrational energy, before they can initiate harmful reactions.

$$P \rightarrow P^* \rightarrow \text{radicals} \tag{94}$$

$$P^* + Q \rightarrow P + Q^* \tag{95}$$

Effective quenchers are based on nickel complexes (phenolate, dithiocarbamate, dithiophosphate) and are used almost exclusively in polyolefins at a concentration of 0.2–0.5%. Nickel complexes impart a green coloration to plastic articles and their importance was significantly reduced with the introduction of hindered amine light stabilizers (HALS).

15.8.3 PVC Heat Stabilizers

In spite of its inherent thermal instability (see Section 15.3.4), PVC possesses several attractive properties, such as economy of production and processing, and the ease of variation of its properties, with appropriate blending with other polymers or additives, from hard and tough materials to elastomeric ones. Currently, PVC ranks second only to polyolefins in terms of worldwide production among industrial polymers. This technical success is the result of considerable research in the field of degradation and stabilization of this polymer since its introduction as a commodity plastic in the middle of the 20th century.

The low thermal stability of PVC originates from the presence of labile structures and of the autocatalytic deleterious effect of the hydrochloric acid evolved. Thermal stabilizers for PVC consist principally of metal carboxylates and organotin compounds (primary stabilizers), used in combination for preventive and curative functions. As with degradation, uncertainties continue to exist in the exact stabilizing mechanisms of these additives. There is evidence that organotin derivatives stabilize PVC by substituting the labile allylic chorine with a more thermally stable thioether group [Eq. (96)].

$$\begin{aligned}&(C_4H_9)_2Sn(SCH_2COO{-}C_8H_{17})_2 + {-}CH_2{-}CH{=}CH{-}CHCl{-}CH_2{-}\\&\quad\rightarrow {-}CH_2{-}CH{=}CH{-}CH(SCH_2COO{-}C_8H_{17}){-}CH_2{-}\\&\qquad + Cl{-}Sn(C_4H_9)_2(SCH_2COO{-}C_8H_{17})\end{aligned} \tag{96}$$

Hydrogen chloride is bound with metallic (Zn, Ca, Ba) organic acid salts with formation of the metal chloride and the corresponding free fatty acids. The formation of polyene sequences can be prevented by combination reactions with thiol or maleate derivatives. Although efficient in blocking the degradation, most of the inorganic stabilizers leave toxic residues and current research is focused on the development of new, less polluting, organic stabilizers.

15.8.4 Other Classes of Stabilizers

15.8.4.1 Metal Deactivators

Very often, thermoplastics contain minute amounts of metallic compounds which originate from polymerization catalysts, contaminated fillers, polymerization or processing equipment, or metal contact (wire and cable insulators) during the use of the polymer. The interactions between the polymer and metallic substances are complex, but generally result in the accelerated aging of the material. Most metal deactivators are bifunctional stabilizers with phenolic and nitrogen, or phenolic sulfide and phosphite, moieties in their structure, and act by a chelating action which reduces the harmful effects of the metal ions.

15.8.4.2 Antiozonants

Diene-type rubbers are particularly sensitive to ozone attack and should be protected by antiozonants. Common antiozonants are aromatic diamines capable of direct ozone scavenging by adduct formation.

15.9 Lifetime Prediction

Most of the additives used to protect the polymer from degradation are eventually consumed during the stabilization processes. In addition to chemical loss, the stabilizers can also be depleted by precipitation, migration to the surface, and extraction. For all these reasons, the concentration of additives diminishes with time up to a point where no protection effect could be observed. In general, the stabilization lifetime increases with the amount of added stabilizer, up to an optimum determined by the finite solubility of the additives in bulk polymers, or by secondary reactions which may have adverse effects on the stabilization mechanisms (Figure 15.22).

Excluding disposable devices, most plastic materials are expected to perform reliability for many decades. In general, outdoor weathering of stabilized samples is too slow to be useful for quality control or for formulation development. As a remedy for this flaw, a number of accelerated weathering tests have been devised. Depending on the type of degradation, accelerated tests can be obtained by increasing the temperature (thermal degradation), or increasing the light intensity (photodegradation) and/or the partial oxygen pressure (oxidative degradation). Owing to its economic relevance, photoaccelerated aging is enjoying the most dramatic development, with numerous commercial systems based on a UV source or mirrors designed to mimic the action of the Sun, with a several-fold increase in irradiation intensity. Most of the time, results of accelerated aging show poor correlation not

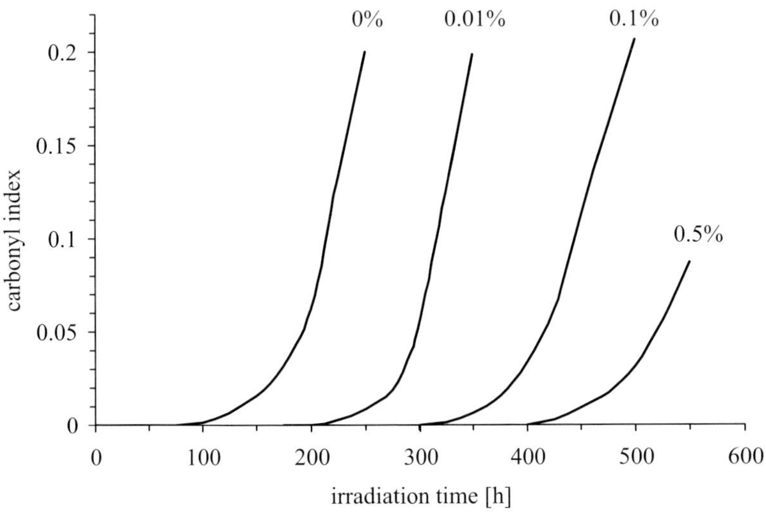

Fig. 15.22. Photo-oxidation of commercial polypropylene, in the absence and in the presence of the indicated concentrations of hindered amine stabilizer.

only with outdoor weathering, but also between different devices, or for identical devices between different laboratories. This lack of correlation is not entirely unexpected, in view of the wide variability inherent in weathering conditions: biological disturbances, pollution, climate (temperature, humidity, wet periods); and changes in the spectrum and intensity of the source with aging, and in the test duration. In addition to the external factors, more fundamental reasons for deviations could be traced back to the bimolecular reactions between intermediate species, whose concentrations in real conditions are too low to interact, or to diffusional processes of the reactants (oxygen, water, stabilizers) that become limiting factors in accelerated conditions. By using conditions which mimic exact weathering conditions in terms of spectral distribution, temperature, and oxygen uptake, but with a different photon intensity, it has been assessed that accelerated aging in polyolefins is not mechanistically different from natural weathering. Therefore, an extrapolation procedure between these two modes of degradation is possible, providing all the dynamics (diffusion) factors could be kept under control. Identifying and controlling all the dynamics factors in a real system is nevertheless a challenging task.

To account for the poor reproducibility in accelerated aging, statistical approaches have been proposed, such as the nonparametric statistics which relies on performance comparison with a control material. The control material should be as similar as possible to the tested polymer in terms of structures, size, and shape, and its weathering durability should be well established in the specified exposure test. One useful approach to reduce error analysis was to compare a test property, not as a function of time or absorbed radiation, but relative to another process going on in the aging material. The objective sought is to have the rates of the two processes occur in the same relationship relative to one another under a wide range of experimental conditions, from natural to accelerated test conditions. The process of interest should be easy to determine and sensitive to low levels of degradation, to minimize the extrapolation range. In thermal aging of nitrile rubber, for instance, it has been found that the ultimate tensile elongation shows good correlation with the rate of oxygen consumption over the experimentally accessible temperature range for mechanical measurements, which was 96–71.5°C. At lower temperatures, the time to failure was inconveniently long for measurement. Oxygen consumption, on the other hand, could be determined by gas chromatography all the way down to room temperature. Based on indirect evidence that the chemistry underlying oxygen consumption is responsible for changes in modulus and elongation values, a lifetime of 100 years was predicted for the elastomer in ambient conditions. It should be noted that an Arrhenius plot extrapolation, based solely on the mechanical properties at elevated temperatures, predicted a lifetime of 50 000 years, which is some 500 times longer than the correctly estimated value [15].

15.10 Conclusions

This chapter has given a general description of some recent advances in the field of polymer degradation and stabilization. Despite a considerable volume of work and equally impressive progress in the comprehension of the molecular mechanisms which underlie polymer degradation, many challenging problems are still open to further investigation. The ever-increasing complexity of commercial plastics, which includes the use of copolymers, polymer blends, and different combinations of additives in their formulations, adds another degree of difficulty to the already complex problem of polymer degradation. Each type of interaction between the different components in the formulation, both polymeric and nonpolymeric, may influence differently the degradation process. In particular, the antagonistic and synergistic effects between stabilizers in additive blends need better mechanistic comprehension.

Two main areas where major advances can be expected in the future include integral characterization of the degraded polymer, and realistic modeling of degradation kinetics.

Most degradation and stabilization mechanisms reported in the literature are inferred from the observation of a few end products (generally carbonyl and hydroxyl compounds, which can be easily monitored by FTIR). A comprehensive characterization of the starting material, and of the degradation products at different stages of aging, is still lacking in most investigations. Such a complete chemical identification, which appears to be feasible with the modern analytical techniques available, is a prerequisite to establish a correlation between effects of specific structures on the behavior of significant degradation intermediates.

The inhomogeneity of reaction sites, inherent in solid-state reactions, may lead to distributions of reactivity and to time-dependent rate constants. These effects should be recognized and taken into account when performing kinetics analysis. This is consequential to the problem of stabilization, since additives need to diffuse toward degradation centers for the stabilization process to be operative. Some heterogeneous reaction models aimed toward a better description of radical chain processes in solid polymers have appeared in the literature. One of these models, the infectious spreading reaction, has been examined in Section 15.4.5. Among other advances in solid polymer degradation, one can cite:

- the concept of "polychronal kinetics" which was introduced to explain the stepwise conversion of reactive species under isothermal conditions [6]; and
- the "spongy micelle" microreactor model in which the amorphous phase in a semicrystalline polymer is depicted as formed by "granule-core" zones, surrounded by "oriented domains" close to the crystallites [51]. By applying different chemical kinetics scheme to each zone, with allowance for "interzone" radical migration, the autooxidation curve of polyolefins can be faithfully reproduced. It is found that the induction period for oxygen consumption corresponds to the homogeneous oxidation stage, whereas the (exponential) autoaccelerated stage may be interpreted as a result of heterogeneous oxidation spreading of low MW peroxides. This model refutes the interpretation of chain branching as a source of oxidation autoacceleration.

All these theories need to be confirmed by experimental verification and tested for their general applicability to different polymer systems.

Notation

A_s	infrared absorption intensity [$cm \cdot mmol^{-1}$]
a	constant of the Morse potential [m^{-1}]
D	oxygen diffusion constant [$cm^2 \cdot s^{-1}$]
D_e	bond dissociation energy [$kJ \cdot mol^{-1}$]
E	tensile modulus [$GN \cdot m^{-2}$]
E_a	energy of activation [$kJ \cdot mol^{-1}$]
E_s	excess energy of stress [$kJ \cdot mol^{-1}$]
E_{ang}	valence-bond angle contribution to the activation energy [$kJ \cdot mol^{-1}$]
ΔE_0	molecular energy difference at absolute zero between the activated complex and the reactant [J]
ΔH	enthalpy of reaction [$kJ \cdot mol^{-1}$]
h	Planck's constant $6.6256 \cdot 10^{-34}$ [$J \cdot s$]
I_0	incident light intensity [$W \cdot m^{-2}$]
$I(x)$	light intensity [$W \cdot m^{-2}$]
K	Kubelk–Munk absorption coefficient [cm^{-1}]
k	rate of reaction [s^{-1}] (first-order), [$L \cdot mol^{-1} s^{-1}$] (second order)

k_f	bond force constant in the neighborhood of the equilibrium separation of two chemical moieties [$N{\cdot}m^{-1}$]
k_B	Boltzmann's constant $1.380{\cdot}10^{-23}$ [$J{\cdot}K^{-1}$]
k_c	rate constant for bond scission [s^{-1}]
k_θ	elastic constant for bond-angle deformation [$N{\cdot}rad^{-1}$]
l	length of a covalent bond [nm]
l_0	equilibrium separation distance of the atoms in a bond [nm]
M_i	molecular weight of ith polymer [$g{\cdot}mol^{-1}$]
M_n	number-average molecular weight [$g{\cdot}mol^{-1}$]
M_v	viscosity-average molecular weight [$g{\cdot}mol^{-1}$]
M_w	weight-average molecular weight [$g{\cdot}mol^{-1}$]
N	total number of amorphous domains
n	number of polymer chains per gram of polymer [g^{-1}]
n	number of oxidized domains
n_i	polymer molar fraction
p_d	"dead" or oxidized fraction of polymer
p_i	"infectious" oxidizing fraction of polymer
p_r	remaining unoxidized fraction of polymer
Q	normal coordinate of vibration [nm]
$q_\ddagger$	molecular partition function of the transition state
q_A	molecular partition function of the reactant
R	reflectance
R	molar gas constant 8.314 [$J{\cdot}mol^{-1}\ K^{-1}$]
R	absorbed dose of radiation [Gy or $J{\cdot}kg^{-1}$]
r	rate coefficient for spreading [s^{-1}]
S	Kubelka–Munk scattering coefficient [m^{-1}]
$S(\lambda)$	relative spectral damage
s	scission indice
T	temperature [°C]
$T_{1/2}$	temperature of half-decomposition [°C]
T_g	glass transition temperature [°C]
T_m	melting point [°C]
t_i	induction period [°C]
$t_{1/2}$	half-life [s]
U_0	thermal energy for bond rupture [$kJ{\cdot}mol^{-1}$]
$V(l)$	Morse potential [J]
x	penetration distance [m]

Greek

α	absorption coefficient [cm^{-1}]
β	activation volume for a reaction [$m^3{\cdot}mol^{-1}$]
λ	irradiation wavelength [nm]
λ_{max}	wavelength of light absorption peak [nm]
ε	extinction coefficient [$L{\cdot}mol^{-1}{\cdot}cm^{-1}$]

μ	dipole moment [C·m]
ν	frequency of band center [cm^{-1}]
θ	angle [rad]
ψ	molecular stress [$GN \cdot m^{-2}$]

Acronyms

ATR-FTIR	attenuated total reflection FTIR
ESR	electron spin resonance
FDA	Federal Drug Administration
FTIR	Fourier Transform infrared spectroscopy
iPP	isotactic polypropylene
GPC	gel permeation chromatogaphy
HAS	hindered amine stabilizer
HDPE	high-density polyethylene
HER	high-energy radiation
LDPE	low-density polyethylene
LET	linear energy transfer
HAS	hindered amine light stabilizer
HOMO	highest occupied molecular orbital
LUMO	lowest unoccupied molecular orbital
MIM	metal injection molding
MW	molecular weight
MWD	molecular weight distribution
NMR	nuclear magnetic resonance spectroscopy
PA	polyacrylamide
PAN	polyacrylonitrile
PC	polycarbonate
PEEK	poly(ether ether ketone)
PET, PETP	poly(ethylene terephthalate)
PIB	polyisobutylene
PMMA	poly(methyl methacrylate)
PMMA–SAN	PMMA–poly(styrene-*co*-acrylonitrile)
POM	polyoxymethylene
PP	polypropylene
PPO	poly(phenylene oxide)
PPP	poly(*para*-phenylene)
PS	polystyrene
PSU	poly(ethersulfone)
PTFE	polytetrafluoroethylene
PVC	poly(vinyl chloride)
SEM	scanning electron microscopy
TG	thermal gravimetry
VC	vinyl chloride

References

1 J. L. Bolland, G. Gee, *Trans. Faraday Soc.*, **1946**, *42*, 236.

2 *Handbook of Polymer Degradation*, S. Halim Hamid, ed., Marcel Dekker, New York, **2000**.

3 D. S. Achilias, C. Kiparissides, *Macromolecules*, **1992**, *25*, 3739.

4 A. Azapagic, A. Emsley, I. Hamerton, *Polymers, the Environment and Sustainable Development*, I. Hamerton, ed., John Wiley, London, **2003**, p. 34.

5 M. Hasegawa, K. Horie, *Prog. Polym. Sci.*, **2001**, *26*, 259.

6 N. M. Emanuel, A. L. Buchachenko, *Chemical Physics of Polymer Degradation and Stabilization*, VNU Science Press, Utrecht, The Netherlands, **1987**.

7 J. F. Rabek, *Photodegradation of Polymers, Physical Characteristics and Applications*, Springer Verlag, Berlin, **1996**.

8 A. Torikai, A. Takeuchi, S. Nagaya, K. Fueki, *Polym. Photochem.*, **1986**, *7*, 199.

9 M. Elvira, P. Tiemblo, J. M. Gomez-Elvira, *Polym. Degr. Stab.*, **2004**, *83*, 509.

10 J. Lemaire, J. L. Gardette, J. Lecoste, P. Delprat, D. Vaillant, in *Polymer Durability, ACS Adv. Chem. Ser.*, **1996**, *249*, 577.

11 B. Fayolle, L. Andoin, J. Verdu, *Polym. Degr. Stab.*, **2000**, *70*, 333; R. Gensler, C. J. G. Plummer, H. H. Kausch, E. Kramer, J. R. Pauquet, H. Zweifel, *Polym. Degr. Stab.*, **2000**, *67*, 195.

12 T. Q. Nguyen, *Polym. Degr. Stab.*, **1994**, *46*, 99.

13 L. Lacoste, D. J. Carlsson, *J. Polym. Sci., Polym. Chem. Ed.*, **1992**, *30*, 493.

14 W. H. Starnes Jr., *Progr. Polym. Sci.*, **2002**, *27*, 2133.

15 K. T. Gillen, M. Celina, R. L. Clough, J. Wise, *Trends Polym. Sci.*, **1997**, *5*, 250.

16 J. P. Critchley, G. J. Knight, W. W. Wright, *Heat-Resistant Polymers*, Plenum Press, New York, **1990**.

17 V. van Speybroeck, Y. Martelé, M. Waroquier, E. Schacht, *J. Am. Chem. Soc.*, **2001**, *123*, 10650.

18 M. R. Nyden, J. Gilman, *Comp. Theor. Polymer. Sci.*, **1997**, *7*, 191.

19 J. J. P. Stewart, *J. Comp. Chem.*, **1989**, *10*, 221.

20 D. A. Gallagher, *Scientific Computing & Automation*, June **1996**.

21 T. A. Bogaevskaya, T. V. Monakhova, Yu. A. Shlyapnikov, *Vysokomol. Soyed.*, **1972**, *14A*, 1552.

22 P. Gijsman, J. Hennekens. *Polym. Degr. Stab.*, **1993**, *42*, 95.

23 S. Commereuc, D. Vaillant, J. L. Philippart, L. Lacoste, J. Lemaire, D. J. Carlsson, *Polym. Degr. Stab.*, **1997**, *57*, 175.

24 S. Verdu, J. Verdu, *Macromolecules*, **1997**, *30*, 2262.

25 P. Eriksson, T. Reitberger, B. Stenberg, *Polym. Degr. Stab.*, **2002**, *78*, 183.

26 M. Celina, G. A. George, N. C. Billingham, in *Handbook of Polymer Degradation*, S. Halim Hamid, ed., Marcel Dekker, New York, **2000**, p. 159.

27 F. Gugumus, *Polym. Degr. Stab.*, **1996**, *52*, 159; *Polym. Degr. Stab.*, **1996**, *53*, 161.

28 T. P. Wampler, *J. Anal. Appl. Pyrol.*, **2004**, *71*, 1.

29 M. Saule, S. Navarre, O. Babot, W. Maslow, L. Vertommen, B. Maillard, *Macromolecules*, **2003**, *36*, 7469.

30 J. P. Riggs, in *High Performance Polymers and Composites*, J. A. Kroschwitz, ed., John Wiley, New York, **1991**, pp. 20–65.

31 B. A. Mathew, R. Mastromatteo, *Metal Powder Report*, **2002**, *57*, 20.

32 W. C. McCaffrey, D. G. Cooper, M. R. Kamal, *Polym. Degr. Stab.*, **1998**, *62*, 513.

33 R. L. Feller, *Accelerated Aging: Photochemical and Thermal Aspects*, Research in Conservation Series, The Getty Conservation Institute, **1994**.

34 H. W. de Bruijn, in *Handbook of Polymer Degradation*, S. Halim

HAMID, ed., Marcel Dekker, New York, **2000**, p. 611.
35 A. L. ANDRADY, N. D. SEARLE, L. F. E. CREWDSON, *Polym. Degr. Stab.*, **1992**, *35*, 235.
36 H. ZWEIFEL, *Chimia*, **1993**, *47*, 390.
37 T. Q. NGUYEN, Ph.D. thesis, No. 243, EPFL, Lausanne, **1976**; T. Q. NGUYEN, T. GÄUMANN, *Radiat. Phys. Chem.*, **1977**, *10*, 263.
38 A. CHARLESBY, *Atomic Radiation and Polymers*, Pergamon, Oxford, **1960**.
39 T. Q. NGUYEN, H. H. KAUSCH, *J. Appl. Polym. Sci.*, **1984**, *29*, 455.
40 V. S. IVANOV, *Radiation Chemistry of Polymers*, VSP, Utrecht, **1992**, p. 202.
41 I. KAETSU, in *Radiation Processing of Polymers*, A. SINGH, J. SILVERMAN, eds., Hanser, Munich, **1992**, pp. 149–186.
42 T. Q. NGUYEN, H. H. KAUSCH, *Adv. Polym. Sci.*, **1992**, *100*, 73–182.
43 S. N. ZHURKOV, V. E. KORSUKOV, *J. Polym. Sci.: Polym. Phys.*, **1974**, *22*, 881.
44 V. A. GONZÁLEZ, G. NEIRA-VELÁZQUEZ, J. L. ANGULO-SÁNCHEZ, *Polym. Degrad. Stab.*, **1988**, *60*, 33; H. ZWEIFEL, *Polymer Durability, ACS Adv. Chem. Ser.*, **1996**, *249*, 375.
45 M. S. S. COKER, G. SCOTT, H. A. A. SWEIS, *Polym. Degr. Stab.*, **1982**, *4*, 333.
46 T. Q. NGUYEN, R. POROUCHANI, H. H. KAUSCH, in *Flexible Chain Dynamics in Elongational Flow*, T. Q. NGUYEN, H. H. KAUSCH, eds., Springer Verlag, Berlin, **1999**, pp. 185–258.
47 A. CASALE, R. S. PORTER, *Polymer Stress Reactions*, Academic Press, New York, **1978**.
48 J. POSPISIL, S. NESPUREK, *Polym. Degr. Stab.*, **1995**, *49*, 99.
49 F. GUGUMUS, *Polym. Degr. Stab.*, **1993**, *40*, 167.
50 E. T. DENISOV, *Dev. Polym. Stab.*, **1980**, *3*, 1.
51 G. E. ZAIKOV, *Handbook of Polymer Degradation*, S. HALIM HAMID, ed., Marcel Dekker, New York, **2000**, p. 437.

16
Thermosets[1]

Rolf A. T. M. van Benthem, Lars J. Evers, Jo Mattheij, Ad Hofland, Leendert J. Molhoek, Ad J. de Koning, Johan F. G. A. Jansen, and Martin van Duin

16.1
Introduction

16.1.1
Thermoset Materials

The world of thermoset materials is characterized by a high degree of both diversity and complexity, often difficult for non-experts to oversee [1]. Let us first ask ourselves what thermoset materials have in common. Polymeric materials are divided into two classes, thermoplastic and thermosetting materials. The term "thermoplastic" refers to the ability of these polymers to become plastic, that is, to melt and flow upon heating. In terms of viscoelasticity, a thermoplastic material can be hard and elastic at room temperature, but it behaves like a liquid when heated to a certain higher temperature. The expression "thermoset" or "thermosetting" is sometimes erroneously understood to be exclusively a material which "sets" (becomes hard and elastic) upon heating, but there are many examples of thermoset materials which have "set" without a temperature trigger, for example through irradiation (see Section 16.8.5). For the discussions in this chapter, the term "thermoset" will merely refer to the inability of this class of polymeric materials to behave like a liquid when heated. A thermosetting material can be hard and elastic at room temperature and be heated above its glass transition temperature to become "softer", and deformable. The deformation, however, will be transient rather than permanent: the softened material is still fully elastic and behaves like a rubbery material.

1) The symbols used in this chapter are listed at the end of the text, under "Notation".

Handbook of Polymer Reaction Engineering. Edited by T. Meyer, J. Keurentjes

ISBN: 3-527-31014-2

16.1.2
Networks

The molecular origin of thermoset behavior lies in the formation of a polymeric network. Unlike thermoplastic materials, which consist of individual macromolecules that can move relative to each other over infinite distances in the melt, the polymeric chains in a thermosetting material are attached to each other and can therefore not move any more relative to each other over greater distances. They are bound together through so-called crosslinks and form an infinite, three-dimensional structure. A consequence of the presence of junctions between the polymeric chains, other than the inability to behave like a liquid and flow, is the inability of this material to dissolve in another medium such as a solvent. Solvent or plasticizer molecules can migrate into the network structure, to swell it, but this is limited to a maximum determined by the crosslink density.

The ideal network structure can be envisaged as a three-dimensional array of crosslink points, each crosslink point being connected to at least three other crosslink points via linear polymer segments, which are called elastically active network chains. In practice non-ideal network elements are also present, such as loops or dangling ends (Figure 16.1). Network density, or crosslink density, is expressed as the concentration of either the crosslink joints or the elastically active network chains (those chains that are part of the infinite structure and attached to crosslink junctions at both ends) per unity of volume of the unswollen material.

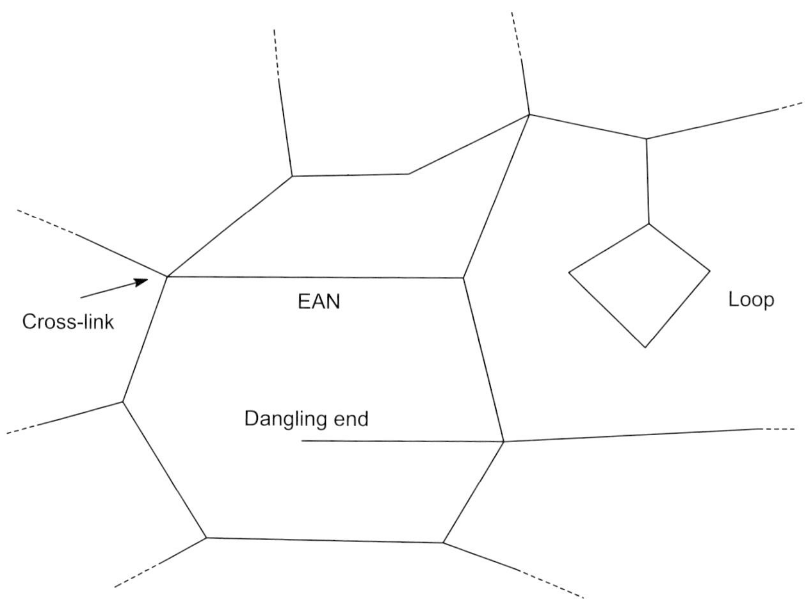

Fig. 16.1. Network nomenclature: crosslinks, elastically active network chains (EAN), dangling ends, and loops.

16.1.3
Advantages

There are two obvious advantages in the material property of a thermoset compared to a thermoplastic material: the network structure will ensure a higher resistance toward unwanted penetration of chemicals (solvents, dyes, water, acids, bases, and electrolytes) in use, and secondly, a thermoset has a high-temperature stability as it retains its shape and (some of) the properties at elevated temperatures. These are not the main advantages of a thermoset material, however. The wide processing window of thermosets has given them the present position in the performance materials market. A thermoset can be processed easily (shaped, cast, molded, poured) as a liquid; the liquid is transformed into the final thermoset material after the processing, through a series of chemical reactions. The final properties of the material are thus determined by the network formation reactions performed in situ. As a consequence, the chosen molecular weight of the starting material can be arbitrarily low. The viscosity, in relation to the molecular weight of the (short) polymeric chains, is decreased strongly as well. In contrast, a thermoplastic material derives its properties from the high molecular weight of its polymer chains which allows the chains to entangle with each other. This high molecular weight in most cases results in rather high melt viscosities for these polymers. A high processing viscosity is a disadvantage with respect to both processing speed (filling the mold, flowing out) and wetting phenomena: dispersion of pigments and fillers and even flow-out over the walls of the mold or the substrate.

16.1.4
Curing Resins

In thermoset nomenclature the low molecular weight polymers (oligomers) are referred to as resins. Apart from their relatively low molecular weight, typically between a few hundred and a few thousand grams per mole, resins are characterized by the presence of reactive groups in the chains, ideally at least two per molecule. These reactive groups form the chemical handles to connect the polymer chains together through covalent crosslink bonds, via a chemical reaction. The process of crosslinking is mostly referred to as "cure" or "hardening". When the reaction proceeds, the gel point is defined as the moment in time when the first molecule of infinite dimensions is formed in the polymer mixture. The viscosity of the mixture then approaches infinity and in this state the material is referred to as gel. The curing reaction can still proceed beyond this point to increase the crosslink density.

16.1.5
Functionality

The number of reactive groups present in one molecule is referred to as functionality. This term can reflect both on a monomer and on a polymer. For example, ethylene glycol is a two-functional monomer for a polyester (see Section 16.6), and a linear epoxy resin with two glycidyl groups on each end is a two-functional polymer (Section 16.4). An unsaturated polyester (Section 16.7) or EPDM rubber (Section 16.9) can have much higher functionalities, that is, unsaturations along the polymer chain. Functionality is an important concept to understand the role of a polymer chain in the final network. For example, a two-functional polymer of a certain molecular weight can only be built in as an elastically active network chain of that same length (at maximum, provided that it reacts at both groups). A three-functional polymer chain of similar molecular weight, containing a branch in the chain, can give a higher crosslink density because it has an internal network junction as well. The junctions in the network can originate from a branching point already present in the polymer, or from chemical reaction between polymer strands.

16.1.6
Formulation

The mixture of the different constituents of a formulation for a thermosetting material is always rather complex and referred to as a "system". Systems can be classified as either 1K (one component) or 2K (two components). In the first case, all the (co)reactive ingredients of the formulation are already mixed, but the network formation is only started when it is triggered chemically or physically, for example by a raise in temperature, by electromagnetic radiation, or by exposure to air (oxygen) or moisture. In the latter case the two components will immediately start their chemical reaction once they are mixed with each other, with or without an additional influence from outside (such as a rise in temperature to speed up the crosslinking reaction).

Independently of the system's reactivity (1K/2K), the combination of reactive groups required to form the network can range from very simple to highly complex. Reactions can be between co-reactive groups of different chains of one resin type, or between the reactive groups on one resin type with corresponding reactive groups of another compound, that compound either being a chemical (monomer) or another resin (polymer) that is added to the first resin. In some resins only one kind of reactive group A is present which is reactive with itself ($A + A \rightarrow AA$); in most resins, however, A is not reactive with itself and requires the presence of an additional compound (B) of complementary reactivity to give a chemical bonding reaction ($A + B \rightarrow AB$). Dual-cure reactions also exist ($A + B + C + D \rightarrow AB + CD$) between mixtures of resins of different reactivities. In the next sections the most important resin classes will be presented. Systems can be formulated from just one one of them but also from a wide, almost infinite, variety of combinations of these resins.

The systems are made more complex with the addition of additives to control the chemical reaction, either to enhance it (catalysts, accelerators, initiators) or to slow it down or inhibit it (stabilizers, inhibitors). Besides additives that control the network formation, other additives can also be used to influence the final properties: fillers, pigments, plasticizers). Formulation of thermosetting systems is a technology of its own and requires a deep insight into both the complex chemical reactions and the final network structure in order to steer toward the desired material properties.

16.1.7 Production

It is highly important to note, however, that the production of the resins themselves is in most cases far from as straightforward as would be expected for a simple low molecular weight polymer. Unlike the (linear) thermoplastic polymers, resins are designed to react to form networks; they contain several reactive groups and often are branched polymers. Either by uncontrolled reaction or uncontrolled branching, polymer molecules with a much higher molecular weight than intended, or even polymer networks, could already be formed during the production stage ("gelation"). This should be avoided at all times, as the production facility could be plugged and highly damaged in such a case. In each of the following section on the resin types a short description is given of the (mostly batchwise) production technology associated with them, and the measures for controlling unwanted, premature, reactions.

16.1.8 General Areas of Application

The areas of application of thermosetting materials are in coatings, rubbers, inks, adhesives, laminates, press moldings, composite (glass-filled) moldings, mineral wool, and electronic materials. A general overview of the main application areas of the resins discussed in the following sections is given in Table 16.1 [1]. They are found in a wide variety of markets (household, automotive, building and construction, electronics, paper and press) and have a collected industrial volume of several tens of millions of tons.

Tab. 16.1. General overview of thermoset applications.

Resin	*Coating*	*Structural parts*	*Wood and paper composites*	*Adhesives*	*Inks*	*Elastic rubbers*
Phenolic	+		++	+	+	+
Amino	+	+	++			
Epoxy	+	+		+		
Alkyd	++				+	
Saturated polyester	++					
Unsaturated polyester	+	++		+		
Acrylate	++			+	+	+
Rubber						++

16.2 Phenolic Resins

16.2.1 Introduction

Historically, phenol–formaldehyde resins (phenolic resins, phenoplasts) are the oldest synthetic thermoset materials: as early as in 1910 the first ones ("bakelite" to their inventor Baekeland) were introduced. Who does not remember the black telephones and light switches that were popular in the 1930s and 1940s? Phenolic resins are based on hydroxy-aromatic compounds (phenol and phenol derivatives) and aldehydes [2]. Furfural is occasionally used as aldehyde; but by far the most widely used aldehyde is formaldehyde. Formaldehyde can be administered as an aqueous solution (formalin), polymeric solid (paraformaldehyde) or solid adduct with ammonia (hexamethylenetetramine, HMTA). The resins are formed by a step-growth mechanism in aqueous solution.

Thermoset materials can be made from phenolic resins themselves, requiring an acid or base catalyst or additional formaldehyde source, or in combination with various other resins. An exhaustive list is not given here, but examples of combinations with all the resin types in the next sections are known (especially amino resins, Section 16.3; epoxy resins, Section 16.4; alkyd resins, Section 16.5; and saturated polyester resins, Section 16.6).

16.2.2 Chemistry

The reactive group involved on the hydroxy-aromatic compound is not the hydroxy group itself but the carbon atoms on the ortho (2,6) and para (4) positions relative to the hydroxy group (attached to carbon atom 1). Unsubstituted phenol can therefore react with formaldehyde from three different positions independently, so should be regarded as a three-functional monomer. If the phenol derivatives have substituents on the ortho and/or para positions, their functionality is decreased accordingly. For example (see Scheme 16.1), cresol (methylphenol) is two-functional when the methyl group is in the 2 or 4 position relative to OH (*o*-cresol and *p*-cresol, respectively) but three-functional when in the 3 position (*m*-cresol). Further, commonly used two-functional substituted phenol derivatives are *p*-*tert*-butylphenol and *p*-*tert*-nonylphenol. The aliphatic groups on these monomers decrease the solubility of the resins in water but increase the solubility (compatibility) of these resins with more hydrophobic solvents (hydrocarbons) and resins.

The aldehyde group of formaldehyde can react twice with an aromatic carbon atom of a phenolic monomer and should therefore be regarded as a two-functional monomer. After the first reaction, the carbonyl group has been changed into a hydroxy group. After the second reaction, the hydroxy group is either etherified or substituted for a carbon–carbon bond. The reaction pathways are displayed

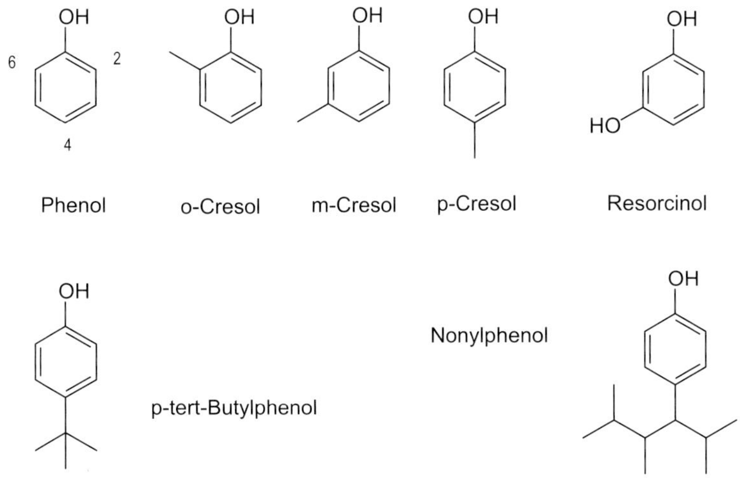

Scheme 16.1. Some phenolic monomers.

in Scheme 16.2 Starting from three-functional phenol and formaldehyde, a branched structure is obtained in which the phenol moieties are linked together by formaldehyde-derived bridges. These bridges can comprise either one methylene group ("methylene bridge") or two methylene groups connected by an ether oxygen atom ("ether bridge"). One formaldehyde molecule is needed to form a methylene bridge; two are needed to form an ether bridge. The relative amounts

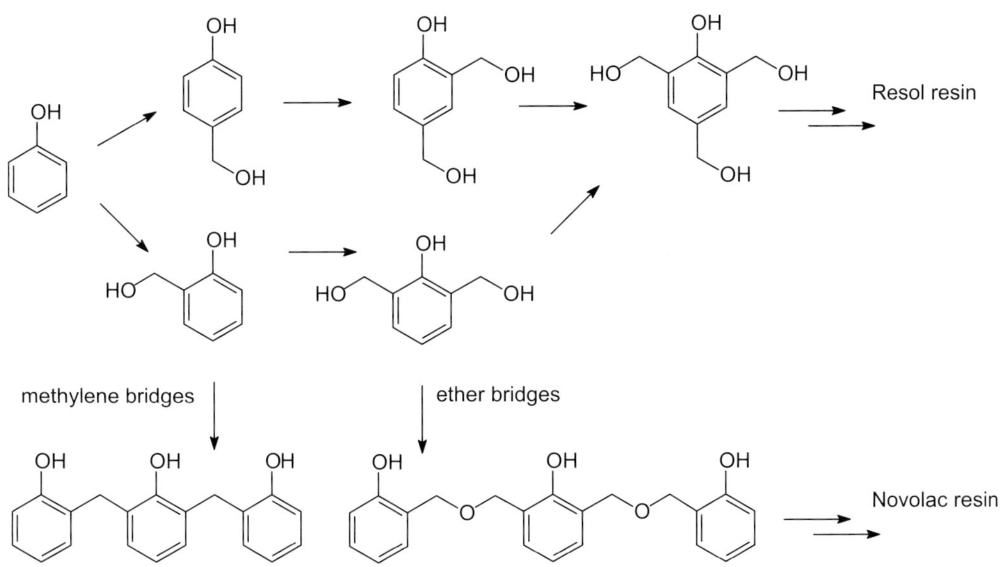

Scheme 16.2. Phenolic resin formation.

of these bridges are determined during the resin synthesis by the reaction parameters.

Two reaction parameters are of specific importance: pH and molar ratio. When phenol and formaldehyde are dissolved together in water, a pH of about 3–4 is reached (phenol is slightly acidic). After prolonged heating, hardly any reaction is noticeable. The reaction requires a catalyst, either an acid or a base, to reach a sufficient speed. Depending on the ratio between phenolic monomers and aldehyde, the phenolic resins are classified as either resol or novolac types.

16.2.2.1 **Resols**

In a resol synthesis reaction, formaldehyde is used in excess over the phenolic monomer. In the case of phenol (P), typical P/F (formaldehyde) ratios are between 1:1 and 1:3. Therefore the ready resin will still contain reactive groups derived from single-sided reacted formaldehyde: hydromethylene groups ("methylol groups"). Care must be taken not to let the reaction run too far, because the mixture can form a gel. Acid catalysis, for example, is therefore not applicable, because the reaction will be out of control (proceed spontaneously toward the gel point). With weak-base catalysis, however, it is possible to reach a more or less stable situation (at fairly low temperature, below 60°C) in which most formaldehyde molecules have reacted with phenols to form methylols, but hardly to form bridges. Catalysts often used are alkali and alkaline earth oxides and hydroxides (such as potassium hydroxide, calcium oxide). *ortho*-Methylols are slightly preferred over *para*-methylols. The higher the pH, the higher the para content and the more ether bridges are transformed into methylene bridges. If a higher molecular weight is desired, the reaction can be allowed to proceed further in a slow and controlled manner at 60°C, until the right viscosity is reached. A higher molecular weight can be obtained more easily without risk of gelation by using two-functional phenol monomers, partly (together with phenol) or exclusively. Resols can be cured to thermoset materials themselves by adding an acidic catalyst and/or increasing the temperature.

16.2.2.2 **Novolacs**

Novolac resins are obtained by reacting formaldehyde with an excess of the phenolic monomer (P/F $>$ 1:1). Either strong acids or strong bases can be used as catalysts. Acids are most widely used, but with bases a higher content of ortho–ortho linkages can be obtained if desired for mechanical properties. Unlike resols, which are mostly methylolated monomer species with some occasional polymeric species, novolacs immediately form true polymer chains of predictable molecular weight. In principle methylene bridges, being more thermodynamically stable than ether bridges, predominate in the polymer chains. So the P/F ratio (P being in excess) determines ideally the (number-averaged) molecular mass M_n. Novolacs cannot form extensive networks by themselves because of the formaldehyde deficiency. A gel could possibly be formed, however, because of branches in the polymer chains when a P/F ratio close to 1:1 is chosen with a three-functional phenolic monomer.

The higher the M_n of the resin and the higher the functionality of the phenolic monomers, the higher the risk of gelation during production. This can be calculated with the Flory–Stockmayer and related mathematical theories. One easy way out is not to use phenol (three-functional) but, for example, *p*-cresol (two-functional) as the monomer: only linear polymeric chains can then be formed.

Curing of novolacs requires addition of extra formaldehyde; mostly HMTA, typically 8–15%, is used as a solid formaldehyde source.

16.2.2.3 **Epoxy-novolacs**

Without extra formaldehyde, novolac resins are hardly reactive with other resins, as the aromatic hydroxyl groups are relatively inert. They can be used to introduce other reactive functionalities, however. The most important example of a functionalized novolac resin is in the reaction with epichlorohydrin (Scheme 16.3). The resulting resins contain glycidyl groups and can be used as epoxy resins (see Section 16.4). The advantage of these epoxy resins over classical bisphenol-A glycidyl resins is that they contain more epoxy groups per unit of weight and have a higher functionality as a resin (one glycidyl group per phenolic unit). Both phenol- and cresol based novolacs are used as epoxy resin precursors. Besides reaction with epichlorohydrin, phenolic resins are sometimes also reacted with alkyl chlorides or alcohols to etherify the aromatic hydroxy groups. In this way, their compatibility/solubility in combination with other resins (or solvents) is enhanced.

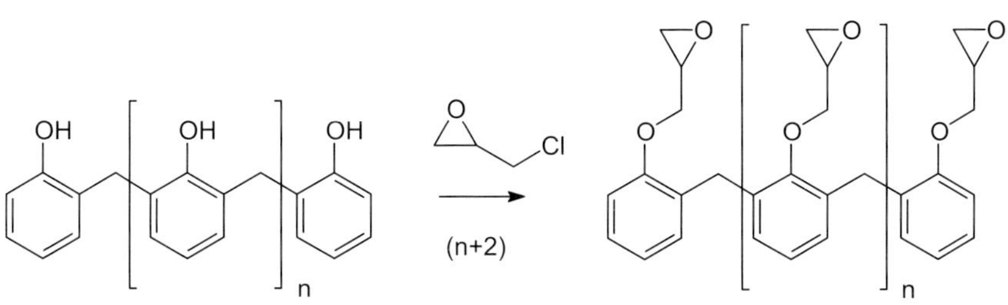

Scheme 16.3. Epoxy-novolac resins formed with epichlorohydrin (ECH).

16.2.2.4 **Discoloration**

The major disadvantage of phenolic resins, both resols and novolacs, is the strong yellow to black color of the resins, or the cured materials. The discoloration (the monomers and sometimes the resins are colorless in principle) takes place when the material comes into contact with oxygen, which is virtually impossible to prevent. By oxidation of methylene bridges, for example, *p*-quinone puffered structures are formed which have a high extinction coefficient for visible light (Scheme 16.4 top). Azomethine structures also can be formed when oxidation takes place in a resin which was cured in the presence of ammonia (for example, from "hexa"); see Scheme 16.4 bottom.

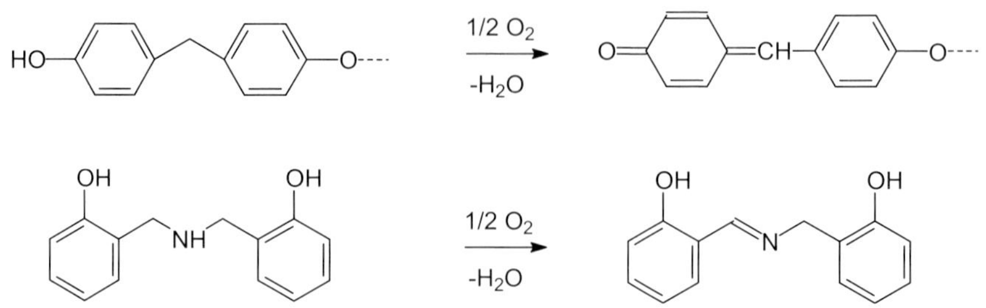

Scheme 16.4. Oxidation of phenolic resins leading to discoloration.

16.2.3
Production

Phenol resins, both resols and novolacs, are exclusively made in batch processes. In the production of novolacs, molten phenol is charged into the heated and stirred reaction tank together with the weak acidic catalyst (for example, oxalic acid). At 95°C, the aqueous formalin solution is slowly added, while the reaction heat is removed through distillation of water at slightly diminished pressure. When all the formaldehyde has reacted, the water is distilled off and the reaction with the excess phenol is continued at higher temperature (140–160°C). When the desired viscosity is reached, the excess of unreacted phenol is removed by distillation as well. The molten resin is let down onto a cooling belt, and broken into small glassy lumps.

Resol resins are made at temperatures from room temperature to a maximum of 60°C. As in novolacs, molten phenol is charged first into the reactor together with the basic catalyst (for example, calcium oxide), after which the formalin solution is slowly added with cooling of the mixture through water distillation. A higher vacuum (40–50 mbar) must be used than with resol resins because of the lower reaction temperature. When the desired viscosity has been reached, the reaction can be quenched by addition of a small amount of acid (to a pH of approximately 4–5). Then water is removed through distillation, either to dryness or to a concentrated aqueous solution. If the resol resin has sufficient molecular weight to be a stable solid at room temperature, the dry resin melt is processed to glassy lumps over a cooling belt, like a novolac resin. Very low molecular weight resol resins ($M_n < 200$ g mol^{-1}) are dispensed as aqueous solutions; resins having higher molecular weight (500–800 g mol^{-1}) are often diluted with alcohol solvents (for better solubility).

16.2.4
Properties and Applications

Phenol resins have found their ways to numerous applications in varying markets: wood glues, bonding agents for mineral and glass fibers, laminates, molded parts, coatings, inks, rubbers, matrix materials for grinding and brake composites. Their success is explained by their relatively low price and high degree of chemical and thermal inertness. Thermal decomposition sets in only between 220 and 250°C, depending on the composition. Phenol resins are quite stable to hydrolysis, even at high pH, and are therefore suitable for making particle board for contact with concrete castings. Once cured, phenolic materials are hard but brittle. To overcome their brittleness, they can be combined with other resins in thermoset formulations. Their compatibility with other resins can be tuned through the choice of phenolic monomer (resins based on nonylphenol can be dissolved even in hydrocarbons), their molecular weight, and the degree of etherification. Due to their discoloration they are mostly used in non-surface applications. For example, coatings based on phenol/epoxy systems are used as primers but not as topcoats. High-pressure laminates are composed of a pile of papers impregnated with resol resin, on top of which a décor paper impregnated with a colorless melamine–formaldehyde resin (next paragraph) is placed, the stack being cured together in a press.

Resols can also be used in combination with unsaturated rubbers such as EPDM (see Section 16.9) which are cured cationically.

16.3 Amino Resins

16.3.1 Introduction

Amino resins are based on reactions of compounds containing amino, imino, or amide groups with aldehydes [3]. As in phenolic resins, the most widely used aldehyde is formaldehyde. In an aqueous environment a condensation reaction between formalin and the amino, imino, or amide compound leads to highly branched, low molecular weight polymers. Introducing an acid catalyst is usually necessary to initiate further condensation reactions that lead to curing.

The commercial history of amino resins goes back to 1924 with the use of urea as the amino source. In 1935 melamine (2,4,6-triamino-1,3,5-triazine)–formaldehyde (MF) resins were introduced, as more expensive but higher quality resins than the urea–formaldehyde (UF) resins. These two resins are still the main amino resins used to date. Applications exist of pure UF or MF, but also of physical or chemical mixtures of melamine and urea resins, leading to melamine–urea–formaldehyde (MUF) resins.

16.3.2 Chemistry

In principle the condensation reaction between formalin and the amino group, referred to as methylolation and somewhat similar to the reaction of phenol and formaldehyde (see Section 16.2.2), can take place at four positions on a urea mole-

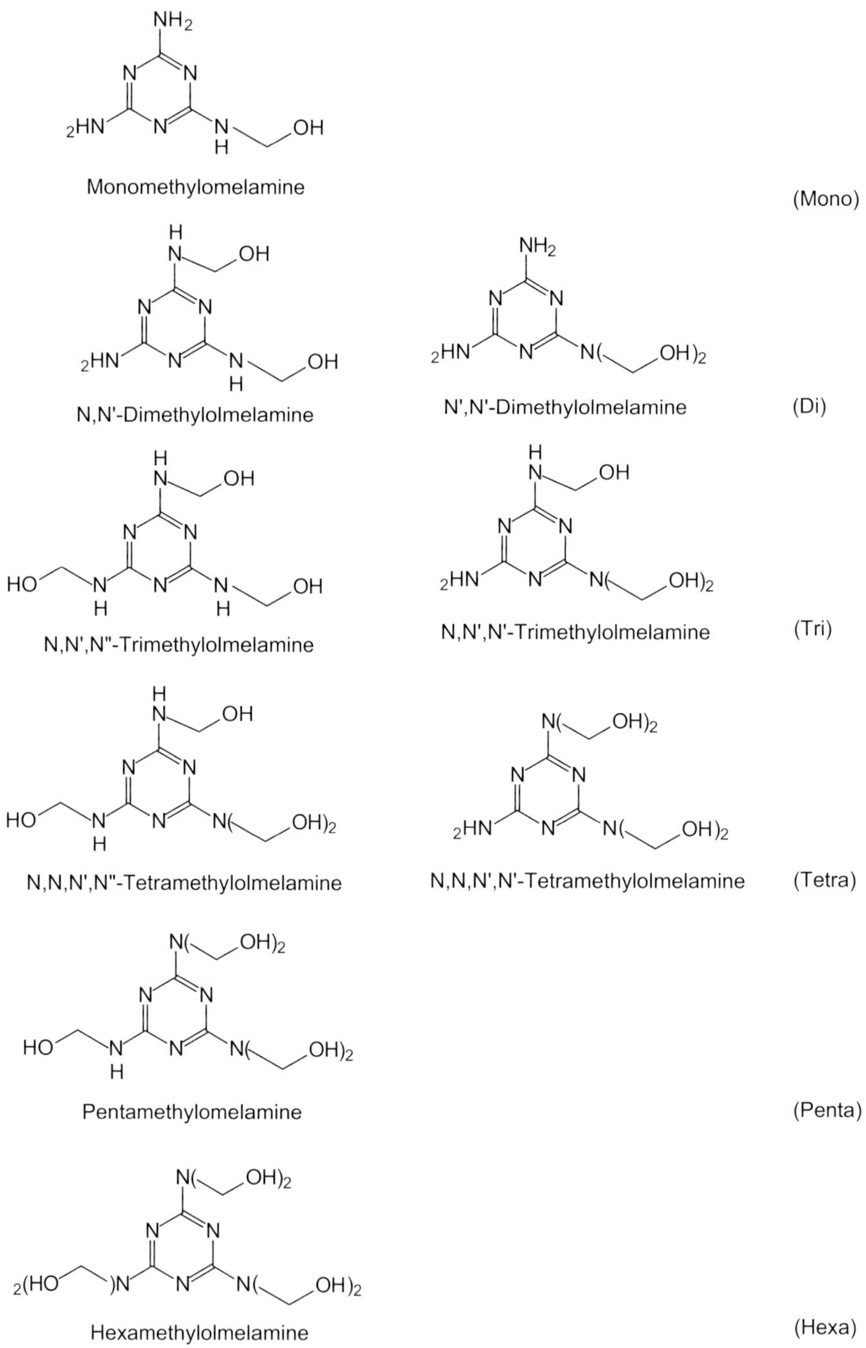

Scheme 16.5. The nine possible "methylol" adducts of melamine and formaldehyde.

cule and at six positions on a melamine molecule. The methylolation reactions of melamine and urea are fast and reversible at temperatures over 70°C. Melamine actually dissolves in the reaction mixture due to this methylolation reaction. For melamine nine different methylolated melamine molecules have been identified (Scheme 16.5), each with its own molecular equilibrium constant. Both primary ($HOCH_2NH$) and secondary methylols ($HOCH_2NCH_2OH$) are formed. In particular the tri- (primary) and hexa- (secondary) methylolated species are the most stable. In the polymerization reactions, depending on the molar ratio of melamine to formaldehyde, any substitution state of melamine can be found. Therefore, MF resins are usually highly branched. For urea four different methylolated ureas have been identified (Scheme 16.6). The existence of a fifth, tetramethylolurea, is uncertain. The primary dimethylolated species is the most stable one and in polymerization reactions urea tends to react primarily as a two-functional monomer. In accordance, UF resins with a low urea/formaldehyde ratio are only slightly branched.

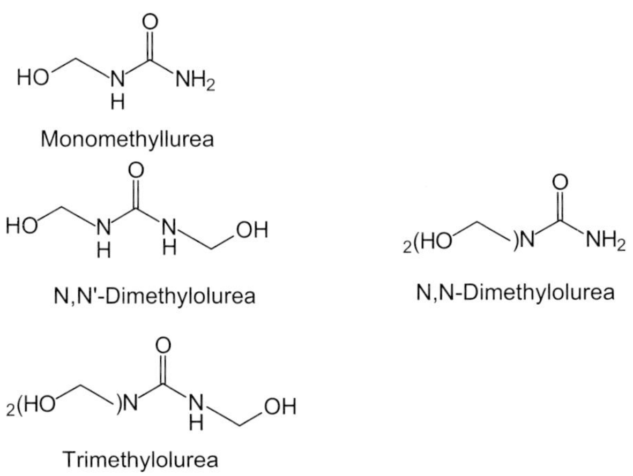

Scheme 16.6. The four possible "methylol" adducts of urea and formaldehyde.

16.3.2.1 **Polymerization Chemistry**

In MF resins, melamine methylol groups condense with melamine amino groups into methylene bridges linking two triazine moieties. Two melamine methylol groups condense to a methylene ether, in short ether bridges (see Scheme 16.7). The main reaction parameters are pH and the molar ratio of formaldehyde to melamine (F/M) and/or urea. Typical evolution of the chemical composition of a melamine formaldehyde resin during synthesis is shown in Figure 16.2.

In UF resins urea homocondensation results in methylene and ether bridges as well, as shown in Scheme 16.8. In MUF resins, next to the homocondensations of melamine and urea species, co-condensations can also take place, again via both ether and methylene bridges (see Scheme 16.9) [4]. The critical parameters influ-

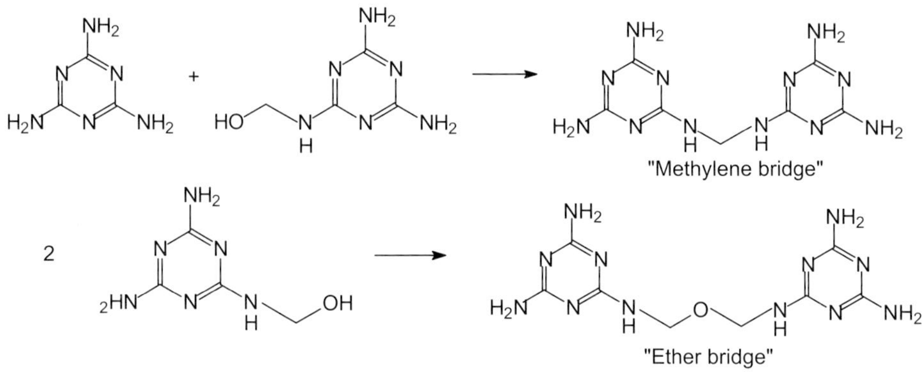

Scheme 16.7. Condensation reactions of amino resins: homocondensation of melamine.

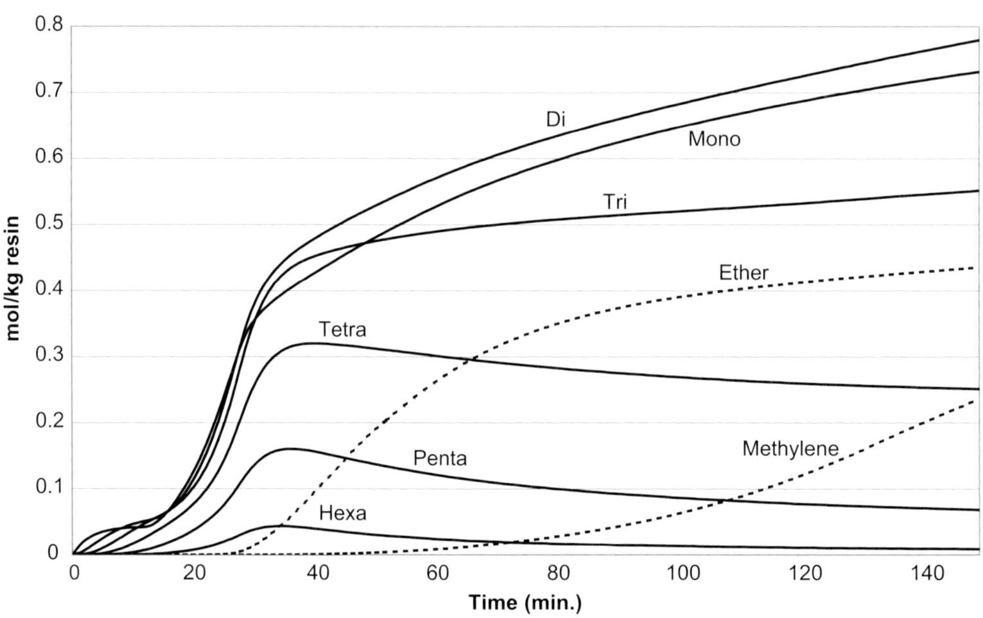

Fig. 16.2. The concentration of the methylol groups (solid lines) and the methylene and ether bridges (broken lines) as a function of synthesis time. (F/M = 1.65, 95°C, pH 9.5). For the labels on the methylol lines, see Scheme 16.5.

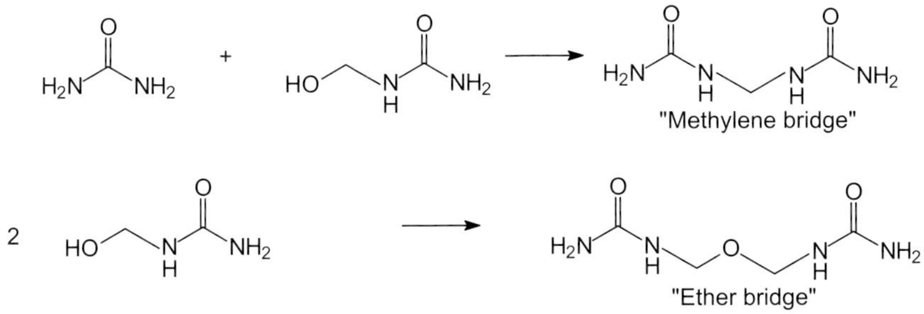

Scheme 16.8. Condensation reactions of amino resins: homocondensation of urea.

encing homo- and co-condensation, as well as homo- and co-condensation, are the molar amounts of M, U, and F during each phase of the resin synthesis, temperature, and pH. Changes in the pH have a direct and substantial influence on the condensation reaction kinetics and, thus, on the number of methylene and ether linkages formed. Changes in the pH of 0.3 can double the rate of an individual reaction. In moderately alkaline media (pH 7–9) methylene bridges are formed predominantly, while in stronger alkaline conditions (pH 9–10) primarily ether bridges are produced. Unlike methylene bridges, ether bridges are susceptible to hydrolysis. Equilibrium is reached after short reaction times. Therefore, a fairly constant concentration of ether bridges is often observed, caused by similar rates of formation and hydrolysis. Methylene bridges do not undergo this hydrolysis and thereby increase in concentration during synthesis; see also Figure 16.2.

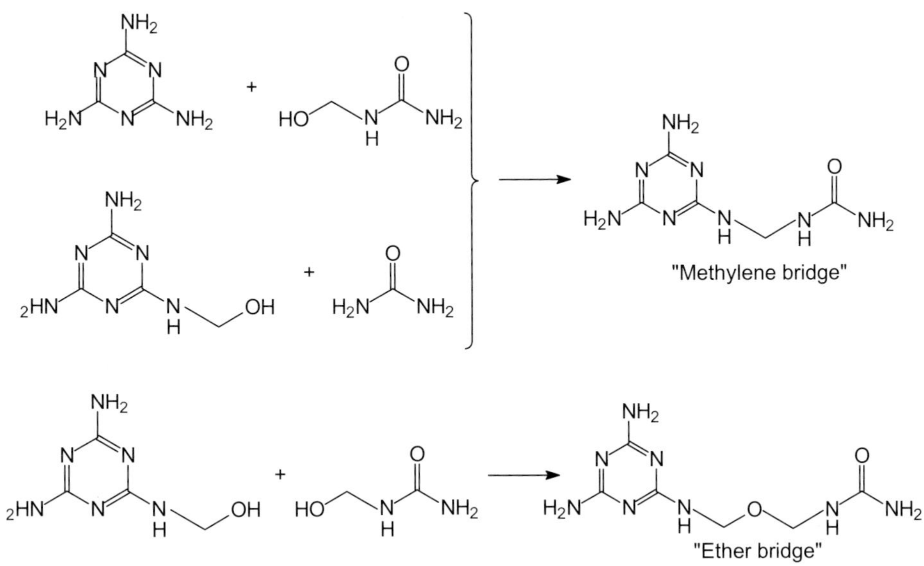

Scheme 16.9. Condensation reactions of amino resins: co-condensation of melamine and urea.

16.3.3 Production

Amino resins are produced in batch processes at temperatures of around 90°C in aqueous solution under alkaline conditions. The solid content (amount of dry substance: melamine, urea, and formaldehyde) in the reactor is around 50–70% by weight. For MF resins the pH during production is around 9–9.5 at the beginning of the process when all the raw materials are added to the reactor, and will stay around that value during resin preparation. For UF resins the first phase of the reaction is similar, alkaline methylolation, but after some hours the pH is brought down to around 5 to promote the bridge formation reactions.

When producing a MUF resin, different scenarios for adding the raw materials to the reactor are possible regarding the moment of dosing. One should choose different pH settings accordingly. Numerous recipes are described for synthesizing a MUF resin. It is possible to simply use a mixture of urea–formaldehyde and melamine–formaldehyde resins, but usually melamine and urea are introduced into the reactor at different times. Two major concepts have found general industrial acceptance. One starts with the synthesis of a UF resin, followed by a co-condensation step of melamine (UF+M); the other starts with a MF resin and co-condenses urea in the second stage (MF+U). Mostly, in each case a final urea dosage is used to lower the molar ratio. The UF+M concept is characterized by a backbone of UF chains and a lower branching efficiency of the melamine molecule. In the MF+U concept the branching degree of the melamine molecule is increased and the melamine is more efficiently incorporated into the polymeric structure, resulting in a reduction of the dominant character of the urea backbone. Both methods of production have their own merits in resin stability and in the performance in the different fields of application.

The desired degree of condensation of the resins is determined in the case of UF and MUF by the viscosity of the reaction mixture. Typically resins have viscosities in the range of 50–400 mPa s. UF resins are milky white due to small crystallites of linear UF segments; MUF resins can be milky white as well, but also clear. In MF resins the degree of condensation is determined via the so-called "water tolerance". This value is expressed as the relative amount of water that can be added to the resin while remaining clear, that is, not turning turbid because of phase separation of water-insoluble higher oligomers. Increasing condensation times give rise to increasing molecular weight and, hence, lower water tolerance.

Because of their reactivity and intrinsic gelating potential, amino resins are characterized by a limited storage stability. In general an aged resin (more than one to three months old) has a very high viscosity or is gelled and cannot be applied any more. Typically acceptable storage times are one month, depending on storage conditions but also on initial viscosity, and/or water tolerance, and/or the chemical structure of the resin. For some applications the resins are spray-dried and applied as dry solid or redispersed solutions. Solid amino resins do not show this fast ageing behavior because the polymerization reactions are halted by the strong decrease in molecular mobility (vitrification) of the polymeric system.

16.3.4
Properties and Applications

Curing of amino resins proceeds readily in an acidic environment; an acidic catalyst is usually added to the amino resin just prior to use. As catalysts, ammonium sulfate for UF and MUF and *p*-toluenesulfonic acid for MF are typical examples. A cured amino resin is hard and brittle. Some degree of flexibilization can be achieved by adding glycols or polymers containing hydroxyl groups to the resin. UF resins are much more sensitive to moisture than melamine resins. Applications which require a certain moisture resistance, either in swelling or hydrolysis, necessitate the use of melamine (MUF or MF).

The majority of amino resins are used in combination with wood or paper. The interaction with these materials is very good; amino resins are used as wood glues in wood composites like particle board, plywood, and engineered lumber products. Amino resins can also be the binder material in paper. Most money papers are reinforced by melamine resins. A larger application in combination with paper, especially for melamine–formaldehyde, is as impregnation resins. Paper is impregnated with resin and hot-pressed, giving decorative films for products like worktops and laminate flooring with a very durable surface (hard, scratch- and chemical-resistant).

Laminate flooring provides a very good illustrative example of the application of melamine in combination with mainly paper and wood fibers. This product has acquired its own place among other floor coverings such as carpets, tiles and parquet. Going from the bottom to the surface of a laminate floor, one finds: a backing paper impregnated with melamine resin; a substrate such as particle board where the wood particles are glued together with MUF resin; a wood décor printed paper impregnated with melamine resin, and a transparent overlay paper impregnated with melamine resins. This pile of materials is then hot-pressed (typically at 150–200°C and 40–60 bar) for between 30 s and 2 min. After sawing into pieces and sanding, the laminate decorative flooring is ready for application in homes and commercial remises, being very durable, scratch- and liquid-resistant, and easy to clean.

Other applications include binders for molding materials such as electrical plugs or dinnerware, binders for mineral and glass wool insulation and roofing, flame-retarding agents, and leather auxiliaries.

16.4
Epoxy Resins

16.4.1
Introduction

Epoxy resins also can look back at a long history of development, as the first ones became commercial in the early 1940s after their invention in 1938. The term

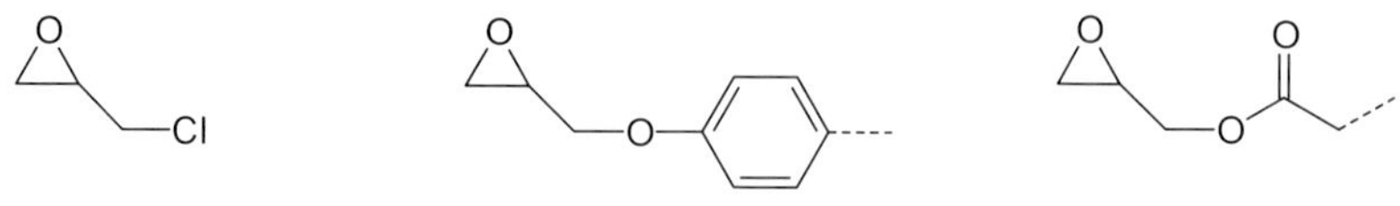

ECH Glycidyl ether Glycidyl ester

Scheme 16.10. Epichlorohydrin (ECH) and glycidyl compounds.

"epoxy" refers to the presence of an epoxide group in the compound (monomer or resin). This group is a three-membered cyclic ether (two carbon atoms and one oxygen atom in a ring) and is called an oxirane in official IUPAC nomenclature. Epoxy compounds can, for example, be made though oxidation of a carbon–carbon double bond. By far the largest industrial source of epoxy groups is epichlorohydrin (ECH). This compound contains a highly reactive, highly toxic, epoxy group that can easily be attached to other molecules, including polymers, by a substitution reaction at the carbon–chloride bond. Epoxy groups originating from reaction with ECH are called glycidyl groups; examples are given in Scheme 16.10. The epoxy novolac resins, polymeric and polyfunctional glycidyl ethers, have already been discussed (see Section 16.2.2). The most dominant class of expoxy resins comprises glycidyl ethers of bisphenol-A.

16.4.2 Chemistry

The synthesis of epoxy resins from bisphenol-A and epichlorohydrin (ECH) is, in simplified form, a net substitution reaction of the chlorine group of ECH by the aromatic hydroxy groups of bisphenol-A to give a diglycidyl ether with evolution of hydrochloric acid; 1 mol of base is required per hydroxy group to neutralize this acid. In more chemical detail, the aromatic hydroxy group is first deprotonated by the base and then reacts with the epoxy group of ECH. Then the chloride group is eliminated by an internal rearrangement by which a new glycidyl group is formed. The idealized reaction product is the diglycidyl ether of bisphenol-A (DGEBPA), a crystalline solid, as shown in Scheme 16.11. In industrial practice the compound is a liquid which contains a small amount of higher oligomers ($n = 0.13$), and is called "standard liquid". These higher oligomers are formed by reaction of aromatic hydroxy groups with the glycidyl group of DGEBPA. It is also possible to purposely synthesize higher oligomers ($n = 2$–20) by catalyzing this reaction with, for example, quaternary ammonium salts or triethanolamine as nucleophilic catalysts. The different ways of making higher molecular weight polymers is described below (see Section 16.4.3). The closer the molar ratio of ECH/bisphenol-A in the resin to 1:1, the higher the molecular weight. A higher molecular weight results in a higher glass transition temperature of the polymer and a lower number of epoxy groups per unit weight (see Table 16.2). In contrast to epoxy-novolac resins, an epoxy resin of this kind is always two-functional, with

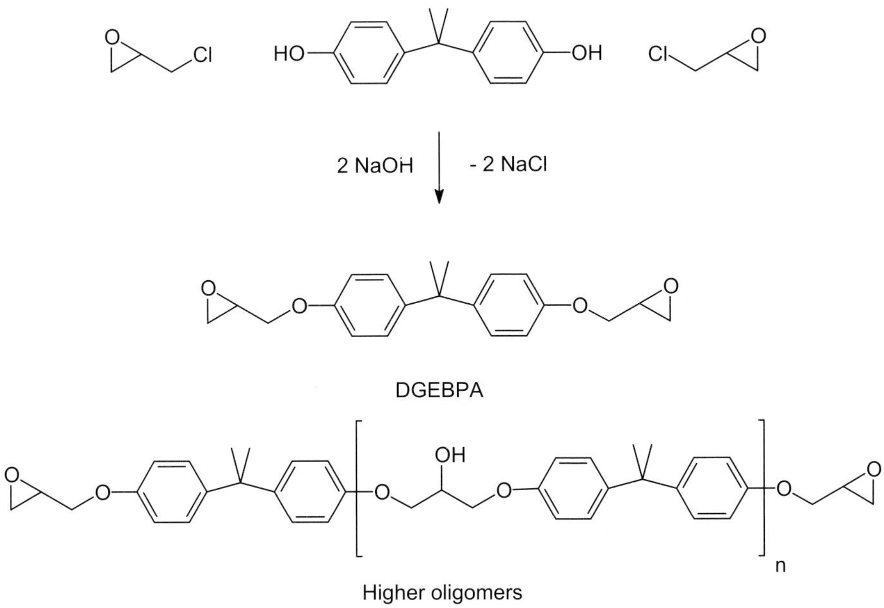

Scheme 16.11. Synthesis of the diglycidyl ether of bisphenol-A (DGEBPA), and higher oligomers.

glycidyl groups on both ends only. Some slightly higher functionalities can be the result of side reactions (branching as a result of the aliphatic hydroxyl groups reacting with glycidyl groups during the polymer synthesis). This branching can be minimized by using a selective catalyst such as triethanolamine.

16.4.2.1 Cure

For the chemistry of curing epoxy resins both the glycidyl groups and the aliphatic hydroxyl groups are of importance. The higher the molecular weight of the resin,

Tab. 16.2. Epoxy resin overview.

Resin	*n*	*ECH/BPA*[a]	*DGE-BPA/BPA*[b]	M_n	T_g [°C]
Standard liquid	0.16			360	<10 (liquid)
1001 type	2	1.33	2	900	65–75
1004 type	5.5	1.15	1.4	1900	95–105
1007 type	14	1.07	1.14	4300	125–135
1009 type	16	1.05	1.12	5000	145–155

[a] Molar ratio of building blocks in product. Applicable to Taffy process for the 1001 type.
[b] Molar ratio of monomers in Advancement process.

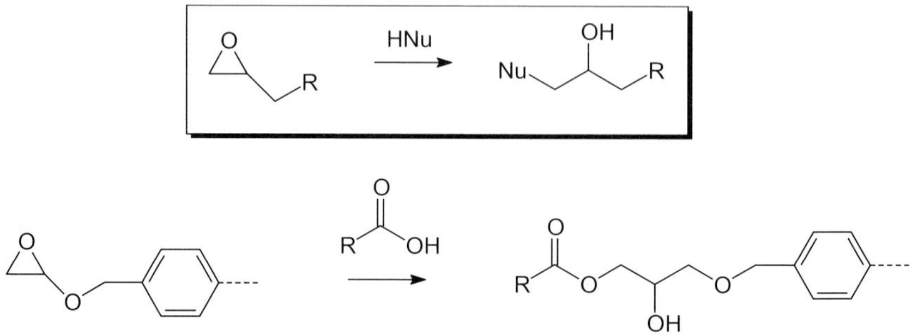

Scheme 16.12. Reactions of glycidyl groups with compounds containing active hydrogen groups, such as carboxylic acids.

the more hydroxyl groups are present in relation to the glycidyl groups. Epoxy resins can be cured themselves by addition of a (Lewis) acidic or basic catalyst. Tertiary amines and imidazolines are especially effective catalysts. They serve as a nucleophilic ring-opening initiator for a cascade of epoxy–epoxy reactions, resulting in ether linkages between the polymer chains. The glycidyl groups can also undergo ring-opening reactions with a wide variety of active hydrogen-containing compounds, in close analogy to the synthesis reaction where the aromatic hydroxyl groups acted as active hydrogen donors (see Scheme 16.12). Depending on the reactivity of the active hydrogen donor, a catalyst may be needed to effect a sufficiently fast cure reaction.

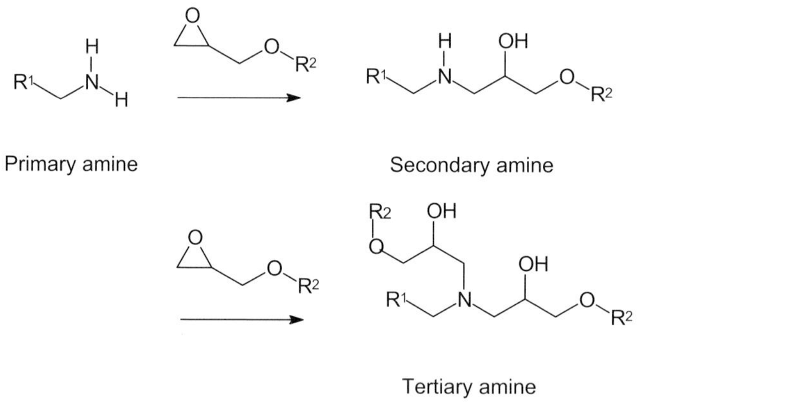

Scheme 16.13. Reaction of amine groups with glycidyl groups, formation of tertiary amines.

Amines generally do not need a catalyst, as the reaction with epoxy resins occurs spontaneously at room temperature. Therefore, amine curing agents can be used for epoxy resins only in two-component (2K) formulations. Low molecular weight polyamines such as diethylenetriamine and triethylenetetramine are frequently used as such. Because of their toxicity, they are also modified with amide groups to lower their volatility. To their advantage, the functionality of these amines is quite high. Not only do they contain several nitrogen groups in a small molecule, but each NH counts as an active hydrogen donor in the reaction with epoxy groups. A primary amine group, NH_2, is two-functional: after one reaction with a glycidyl group a secondary amine group, NH, remains which can undergo an additional reaction with a glycidyl group, as shown in Scheme 16.13.

Phenolic resins can react in two ways with epoxy resins. Using the right catalyst, the aromatic hydrogen groups can react with glycidyl groups, in analogy to the synthesis of higher epoxy resin oligomers. This applies to both resol and novolac type resins (see Section 16.2.2). Methylol groups, as present in resol resins, can also react with the aliphatic hydroxyl groups of bisphenol-A based epoxy resins, with formation of ether linkages.

Carboxylic acids and anhydrides also can be used for curing epoxy resins. Carboxylic anhydrides can react spontaneously at room temperature with the aliphatic hydroxyl groups of bisphenol-A based epoxy resins, with formation of carboxylic acid groups. The carboxylic acid groups can then react further with the glycidyl groups (see Scheme 16.12), at higher temperatures and using the right catalyst such as a quaternary ammonium or phosphonium halide. Polyester resins (see Section 16.6) can be used as a source of carboxylic acid functional polymers to cure the epoxy resins.

16.4.3 Production

Three variations on the synthesis of epoxy resins from bisphenol-A and ECH exist to date, differing in the ratio between these monomers and the viscosity of the product resins, as determined by the removal of the sodium chloride formed.

16.4.3.1 Standard Liquid

The diglycidyl ether of bisphenol-A (DGEBPA) is made with a large excess of ECH (6 equiv.) to minimize formation of higher oligomers. Bisphenol-A and ECH are charged together into the reactor. Powdered base (NaOH) is added slowly to this mixture at 60°C to make the reaction proceed, while salt (NaCl) is formed as a by-product. After the exothermic reaction is finished the temperature is raised to 100°C while water (also a by-product of the reaction) is distilled off under reduced pressure, and then the temperature is raised further to 120°C to drive the reaction to completion. The salt formed and residual base are removed from the product through washing with water. Finally all remaining volatile components, including the excess ECH, are removed under vacuum (30 mbar). This technical grade (average $n = 0.16$) contains approximately 90% DGEBPA ($n = 0$), 9% dimer ($n = 1$), and 1% trimer ($n = 2$).

16.4.3.2 **Taffy Process**

With a much lower excess of ECH, and in the presence of a catalyst, it is possible to make higher oligomers directly in a procedure similar to that for the standard liquid, called the Taffy process. Bisphenol-A and a stoichiometric amount of NaOH (10% aqueous solution) are charged into the reactor, and ECH is added with vigorous stirring at 45°C. The reaction mixture is heated to 95°C and is kept at this temperature for another 1–2 h. The product, a viscous liquid at this temperature, is washed with water to remove the salt and catalyst, dried, let down over a cooling belt and flaked. Because the viscosity of the product is a limiting factor in the water washing step, this process is applicable to lower oligomers only, maximum $n = 4$, for example the 1001 type.

16.4.3.3 **Advancement Process**

When even higher oligomers are desired, it is important to remove the salts from the reaction before the final molecular weight is obtained. This is achieved in the Advancement process (see Scheme 16.14). In this process the reactor is charged with standard liquid resin, free of salts. Fresh bisphenol-A, in a specific molar ratio to the DGEBPA for the desired M_n, and an appropriate catalyst are added. The polymerization is effected exclusively by consecutive reactions of the hydroxyl groups of the bisphenol-A with the glycidyl groups of the DGEBPA. It is carried out at higher temperatures than the Taffy process, well above the glass transition temperatures of the targeted polymer; see Table 16.2. After the desired viscosity has been reached, the product is let down over a cooling belt and flaked to glassy pellets. Unlike the Taffy process, in which oligomers with every value of n are formed ($n = 0, 1, 2, 3, 4, 5 \ldots$), the Advancement process yields polymers with predominantly ($> 90\%$) even numbers of repeating units ($n = 0, 2, 4, 6, 8, 10 \ldots$) as a result of the consecutive reactions between DGEBPA and bisphenol-A. The presence of about 10% $n = 1$ dimer in the standard liquid resin causes 10% of the Advancement product to have odd numbers of n.

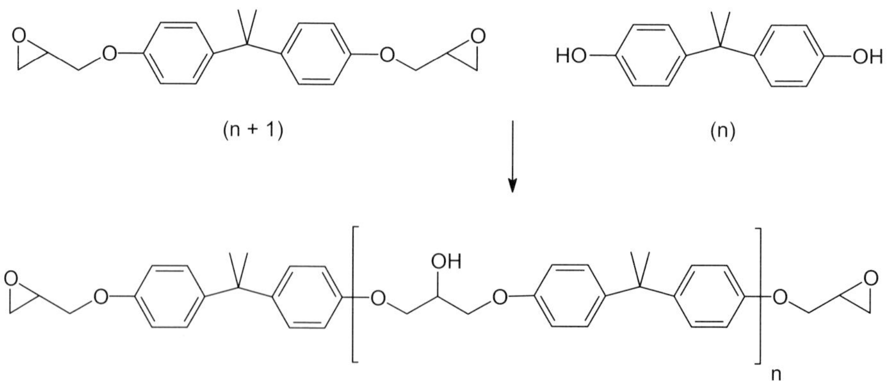

Scheme 16.14. Epoxy resin synthesis via the Advancement process.

16.4.4
Properties and Applications

Epoxy thermosets have excellent mechanical properties: they are characterized by a high hardness but relatively low brittleness (in comparison with phenolic and amino resins). This is caused by the β-transitions in the bisphenol-A moieties along the polymer chain. The higher the molecular weight, the higher the flexibility of the network (the longer the elastically active network (EAN) chain) and the lower the crosslink density when the cure reaction involves only the glycidyl groups. If a high flexibility is desired with a higher crosslink density, rubbery resins can be used to modify the lower molecular weight resins. During cure they separate as small rubber domains inside the thermoset matrix.

The lower molecular weight epoxy resins, when cured with, for example, amines or anhydrides, are used as 2K adhesives and in liquid coatings. When cured with phenolic resins, excellent hydrolysis resistance is obtained. Beverage and food packaging cans are coated on the inside with such systems for good sterilization performance. Epoxy resins are popular for their properties in many other coating applications also, but predominantly in primers and coatings for indoor use. The high T_g of the higher oligomers makes them very suitable for powder coatings (see Section 16.6.5), especially when cured with dicyanodiamide or acidic polyester resins. The presence of many aliphatic hydroxyl groups along the chain provides good chemical interaction with, for example, metal surfaces. This results in excellent adhesion to metal substrates, and a high resistance to corrosion. The disadvantage of epoxy resins based on bisphenol-A or novolac is that they are very prone to photooxidation at the aromatic ether bonds and are, thus, poorly resistant to UV light during outdoor weathering.

An example of epoxy resins that cure when a strong acid is liberated is given in Scheme 16.35 (see Section 16.8.5, on UV curing). Press moldings from epoxy resins, catalyzed by tertiary amines for cure, are used in automotive and electronic components. As they only react with their glycidyl end groups, they are relatively insensitive to over-cure. They are furthermore dimension-stable, and are very resistant to heat and chemical influences.

16.5
Alkyd Resins

16.5.1
Introduction

Alkyd resins can be described as polyesters containing unsaturated fatty acids. Many of the chemical and technological aspects coincide with those of saturated polyesters (see Secion 16.6). We discuss them here first because the alkyd technology, itself originating from early in the 20th century, stemmed from the use of vegetable oils in the painting and graphic arts industry. In comparison with simple vegetable drying oils, alkyds provide harder films and much faster drying. As vegetable oils, they are crosslinked through an autoxidative drying mechanism by ambient oxygen from the air. They are compatible with a wide range of solvents and other resins. They currently constitute a versatile and economic group of binders for the paint industry, with a world annual volume of approximately 1.5 Mt.

16.5.2
Chemistry

Alkyd resins are essentially built up of three groups of components: polyhydric alcohols, polyacids and (vegetable) monoacids. The polymerization process is by esterification (see Section 16.6.2). Examples of typical alkyd monomers are pentaerythritol (tetramethylolmethane), trimethylolpropane, and glycerol as polyhydric alcohols (Scheme 16.21, below), and mainly phthalic anhydride but occasionally isophthalic acid, maleic anhydride, and adipic acid as polyacids (Scheme 16.22, below). The most commonly used monoacids are polyunsaturated fatty acids (the main chemical species of which are oleic, linoleic, and linolenic acids; Scheme 16.15) that impart to the alkyd its drying properties, and benzoic acid which is sometimes used to raise the T_g. The main vegetable oils of industrial relevance are linseed oil, soybean oil, safflower oil, and sunflower oil. They all contain differ-

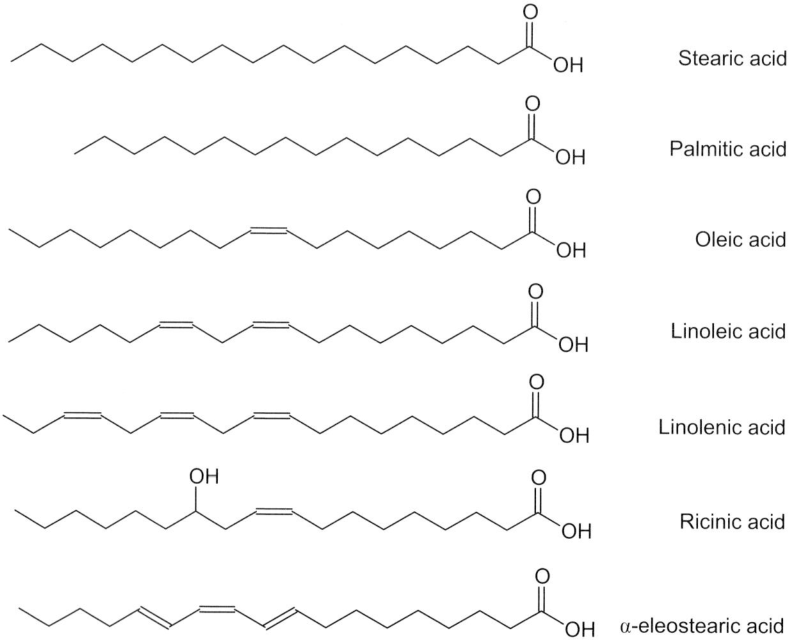

Scheme 16.15. Some vegetable fatty acids used in alkyd resins.

Tab. 16.3. Unsaturated fatty acid content of vegetable oils.

Fatty acid	*Linseed oil*	*Soy oil*	*Safflower oil*	*Sunflower oil*	*Tung oil*	*Castor oil*
Palmitic	6	11	8	11	4	
Stearic	4	4	3	6	2	
Oleic	22	25	13	29	4	9
Linoleic	16	51	75	52	8	83
Linolenic	52	9	1	2		
Ricinoleic						8
Eleostearic					82	
Iodine number	180	130	145	130	175	155

ent amounts of fatty acids (Table 16.3). The numbers given are merely indications, as seasonal influences affect the relative fatty acid contents. The concentration of unsaturated groups in the oil is expressed by the iodine number (g I_2 per 100 g oil), based on the known addition reaction of iodine to unsaturated carbon bonds.

The fatty acids can be incorporated, along with the other polyester monomers, into the resin as such. This is called the fatty acid process, but, if glycerol also is desired to be present in the resin, it can be more economical for the vegetable oil, that is, a triester (triglyceride) composed of glycerol and fatty acids, to react directly with the polyester ingredients. In that case the oil cannot be introduced in the polyesterification stage, as the oil itself is not compatible with the growing polyester. A kind of dispersion of oil-free polyester particles would then be formed in the oil. These undesired systems are called glyptals. To avoid the formation of glyptals, the fatty acids in the oil have to be randomly distributed over the glycerol of the oil and the polyhydric alcohols present in the formulation in a first transesterification step. The resulting hydroxyesters can intermix in a second step with the acids and then result in a homogeneous product. This "alcoholysis process" is depicted in Scheme 16.16.

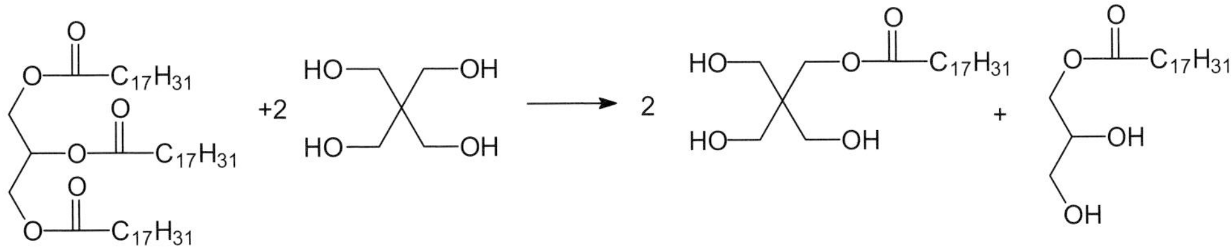

Scheme 16.16. The alcoholysis process for alkyd resins.

In a similar process, called acidolysis (Scheme 16.17), the glycerol from the oil is distributed over the fatty acids and isophthalic acid. This procedure is not possible

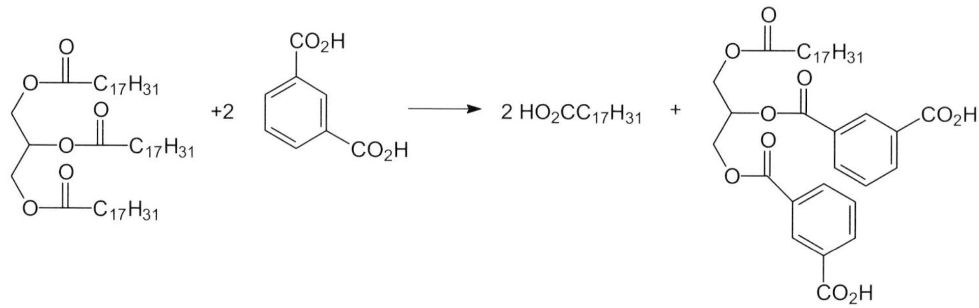

Scheme 16.17. The acidolysis process for alkyd resins.

with phthalic anhydride, since free carboxylic acid groups are required for this transesterification reaction. Orthophthalic acid is not stable at the acidolysis temperatures (260°C) and will form mainly the more stable cyclic anhydride. The process is therefore primarily used in the production of printing ink resins, where isophthalic acid is the diacid of choice.

If a higher T_g of the resin is desired, for example to obtain harder coatings, the chemistry of the alkyd resin can be modified after the polycondensation process by incorporation of phenolic (see Section 16.2) and/or colophonium resins.

16.5.2.1 **The Alkyd Constant**

In contrast to radical-type polymerization, in the making of branched polycondensates like alkyds the build-up of molecular weight can be controlled to a large degree. The starting point in these calculations is the understanding that an infinite molecular weight (gelation) can in principle only be obtained if the average functionality F of the monomers is equal to or higher than 2.

This average functionality is defined as: $F = (E_{alcohol} + E_{acid})/M_0$, in which $E_{alcohol}$ denotes the number of equivalents of alcohol groups, and E_{acid} the number of equivalents of acid groups, while M_0 denotes the total number of moles of monomers in the system. In practice, alkyd resins are cooked with a surplus of hydroxyl groups. This means that not all the hydroxyl groups will esterify and consequently the number of effective alcohol equivalents is equal to the number of acid equivalents; in other words, $E_{alcohol} + E_{acid} = 2E_{acid}$. Thus, $F = 2E_{acid}/M_0$. The alkyd constant K is defined as $2/F$, equalling M_0/E_{acid}, and should not be lower than 1. In practice, an alkyd constant of 1.02–1.06 is recommended to avoid gelation regimes at all times.

16.5.2.2 **Autoxidative Drying**

A basic reaction scheme of the autoxidative drying process is depicted in Scheme 16.18. A catalyst is needed to give this reaction sufficient speed; usually a cobalt(II) salt is used. The diene moieties of the linolenic acids, especially, are involved in this process. After formation of a conjugated hydroperoxide with molecular oxygen from the ambient air, the hydroperoxide decomposes into oxy-radicals that can fur-

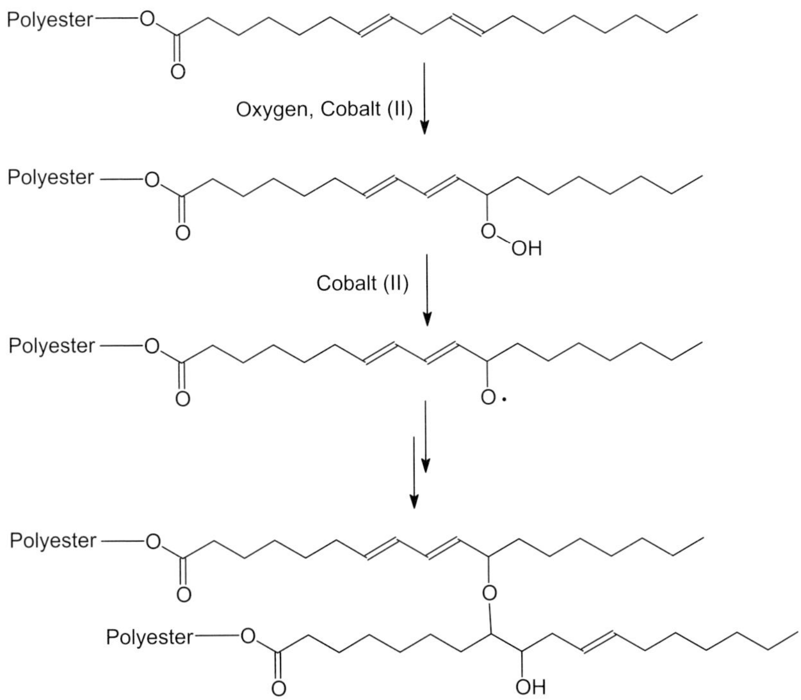

Scheme 16.18. Autoxidative drying of alkyds and oils.

ther react with unsaturations of other fatty acids. Mainly by addition and recombination reactions of these radicals the final network structure is formed. Side reactions during the autoxidation process can lead to yellowing.

16.5.3 Production

The alcoholysis process has lower raw material costs than the fatty acid process but higher production costs, since it is a two-step process. The decision whether to produce from oil or from fatty acids is thus largely based on the amount of oil/fatty acids in the formulation (see short oil/long oil resins, discussed in Section 16.5.4).

Regardless of the way fatty acids are incorporated, the main step of alkyd synthesis is the polyesterification. This process is quite similar to the process for saturated polyester resin production, and actually more or less the same equipment can be used (Figure 16.3). The polyesterification of the alkyd raw materials normally proceeds at 240°C with the aid of xylene as a mode of transportation of the water, by forming a binary azeotrope with it. The xylene and water vapors rise through the column on top of the reactor and are condensed on top of a separator. The xylene is pumped back into the reactor and the water is removed. Because of

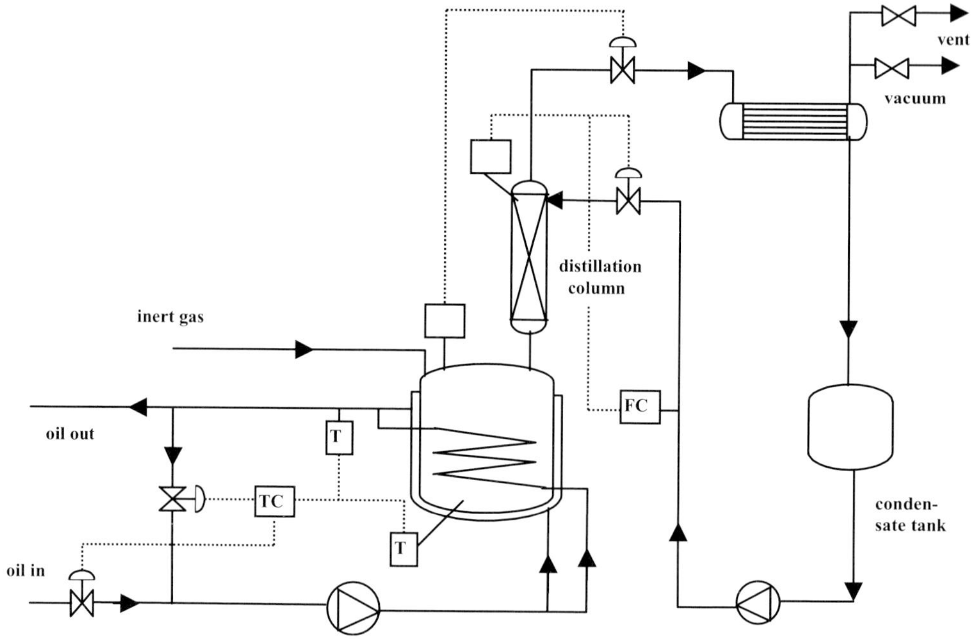

Fig. 16.3. Reactor design for production of polyester and alkyd resins. FC, flow control; T, temperature measurement; TC, temperature control.

undesirable volatile by-products that often form ternary azeotropes with water, the reaction water cannot be disposed of in the sewage system without incineration.

Variables measured at regular intervals to control and, if necessary, adjust the process are:

- Acid value: a measure for the amount of acid groups still present in the reaction mixture, expressed in mg potassium hydroxide needed to neutralize 1 g of solid resin. This value drops throughout the process as the esterification proceeds.
- Viscosity: an indication of the weight-averaged molecular weight, usually measured by the torque of the reactor stirring motor or that of a separate rotor/stator measuring device. The viscosity increases with time as a result of increasing molecular weight of the resin. If desired, a graph can be drawn plotting log viscosity against acid value. This is a more or less straight line and enables the operator to judge whether the resin will be within the specifications, which show up on the graph as a rectangle.

After reaching these specifications, potential modifications can be made, for example with silicones for better outdoor durability, or polyamides to impart thixotropy. Finally the resin is cooled, diluted in solvent and/or emulsified, and if necessary filtered.

16.5.4
Properties and Applications

16.5.4.1 Short Oil Alkyds

Short oil alkyds contain less than 35% oil by weight, calculated as triglyceride. These resins have a relatively high T_g and therefore will "feel" dry as soon as the solvent has evaporated from the paint. Chemical drying through the unsaturated fatty acids will be relatively modest and other crosslinking reactions will have to be considered. Examples are etherification of the OH groups with resol (see Section 16.2) or amino resins (Section 16.3), either thermally (stoving) or by acid catalysis. Applications are in non-yellowing wood varnishes and protective coatings, for example for steel constructions.

16.5.4.2 Long Oil Alkyds

Long oil alkyds can contain from 55% oil, even up to 80%. Long oil alkyds are typically used for trade (professional painting) and DIY (do-it-yourself) applications. Application is mostly by brush and roller and the paints are supplied in white spirit.

Their T_g is low and resins will feel sticky several hours after the solvent has evaporated. Chemical drying is intensive and as a result of this the molecular weight builds up gradually in time. For a typical long oil alkyd resin, full chemical drying is obtained in a period of weeks, and can be accompanied by some yellowing. Typical oils/fatty acids that are used are linseed and soybean. Since linseed contains triply unsaturated fatty acids, it has more crosslinking potential but will also smell and yellow more. For this reason linseed oil based alkyds are mainly used for primers. For topcoats the oil of choice is soybean but safflower and sunflower are also used, the latter being superior to soybean with regard to yellowing and smell. Other oils with very high unsaturation that are often used in primers for semi-industrial applications, applied on-site, are fish oil (marine applications) and tung oil (marine and industrial construction maintenance).

16.5.4.3 Medium Oil Alkyds

Medium oil alkyds contain, as the name already suggests, 35–55% oil by weight and are used in the higher-quality trade and DIY paints. In quick-drying properties and yellowing they outperform the long oil alkyds and they are mostly dissolved in white spirit.

16.5.5
Alkyd Emulsions

Alkyd paints dominated the architectural coating market for a long period until the appearance of polymer dispersions or the so-called latex paints. Specifically for wall application waterborne paints based on poly(vinyl acetate) homo- and copolymers, styrene–acrylics and pure acrylic latexes almost completely took over the market from the alkyd resins for both interior and exterior application. However, for wood and metal application alkyd paints still enjoy the majority of the market share, thanks to better overall paint performance. For the same application fields, alternatives to the conventional alkyd paints were desired for environmental reasons: namely, to avoid the use of (neurotoxic) solvents and solvent emissions to the air. From about 1980, aqueous alkyd emulsions have been developed and are now breaking through in the market as environmentally friendly systems. Alkyd emulsions can best be described as a dispersion of solventless alkyd resin droplets in water (o/w). When the paint dries, the water evaporates and the emulsion changes from oil-in-water (o/w) to water-in-oil (w/o). The paint properties of the dried and autoxidatively crosslinked film are largely comparable to those derived from solvent-borne alkyd paints [5].

16.5.5.1 The Inversion Process

In order to keep this emulsion stable in time during storage, the particle size of the droplets must be small enough (typically <2 μm) and surfactants must be used. Alternatively to the use of additional surfactants, incorporation of surface-active groups, such as polyethylene glycol derivatives, in the alkyd resin is also possible; it is sometimes even more desirable not to have free surfactant molecules in the resin from a paint quality point of view. The production of an alkyd emulsion with built-in surfactants starts with a conventional alkyd polyesterification process, in which acid- or hydroxy-functional surfactants are added along with the other monomers. Care must be taken to keep the viscosity as low as possible to facilitate the next step. After the polycondensation reaction is finished, the resin is cooled to <100°C and water is slowly added to the hot resin to form a w/o emulsion. With increasing water content, an inversion to an o/w emulsion will take place. The inversion is accompanied by a large increase in viscosity. After the inversion more water can be added to reach the right viscosity at room temperature.

16.6 Saturated Polyester Resins

16.6.1 Introduction

Thermosetting polyester resins are almost as old as alkyd resins. The first attempts to make polyesters derive from the second half of the 19th century. In the beginning of the 20th century more and better-defined polyesters were made by Vorlander and later by Smith. Most of the polyester resins were based on glycerin and aliphatic diacids as fumaric and adipic acids. From these times the first application of polyester resins in paints and coatings was developed. Besides the modification with fatty acids (alkyd resins), which are cured by air, thermosetting resins were developed which are cured by crosslinkers or other resin types. The choice of monomers, more specifically the ratio of polyacids to polyalcohols, determines the type of end groups, for a polyester these being either acid –COOH or hydroxy –OH. These functional end groups can react with epoxy resins (see Section 16.4), isocyanates, phenolic resins (Section 16.2), and melamine resins (Section 16.3).

16.6.2 Chemistry

Polyester resins consist of polyfunctional acids and polyalcohols [6]. These compounds esterify at higher temperatures with evolution of water (Scheme 16.19). The chemical equilibrium for this polycondensation reaction is unfavorable. It is important that water is removed from the reaction mixture continuously, shifting the equilibrium to the right (see Chapter 4). The temperature for this reaction lies normally between 200 and 260°C and the choice is dependent on the stability of the raw materials used. This direct esterification reaction is far the most used chemistry for saturated polyester resins. Polycondensation via transesterification, as in Scheme 16.20 and primarily used in the synthesis of thermoplastic polyesters, is only occasionally used for resins. By varying the type and the molecular ratio of the reactants, the molecular weight and structure of the polymer, and the type and number of functional end groups, can be varied. Excess of acid functional raw materials gives an acid functional polymer. Excess of polyfunctional alcohols gives a polymer with hydroxyl end groups. If branching is desired, polyfunctional raw materials are used. Mostly three- or four-functional polyhydroxyl compounds are used. In practice a large variety of raw materials are available for the formulator

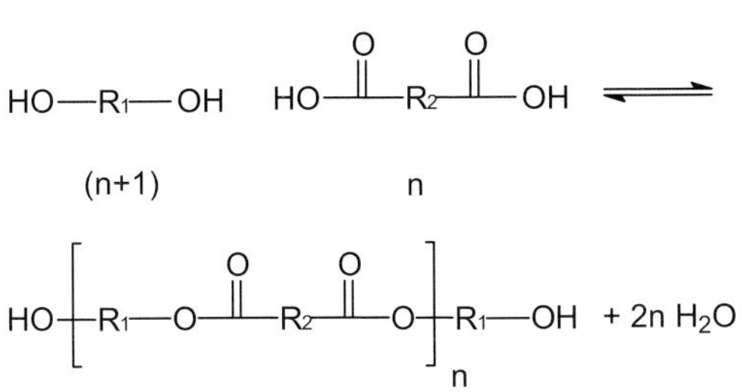

Scheme 16.19. General esterification scheme for OH-polyester resins.

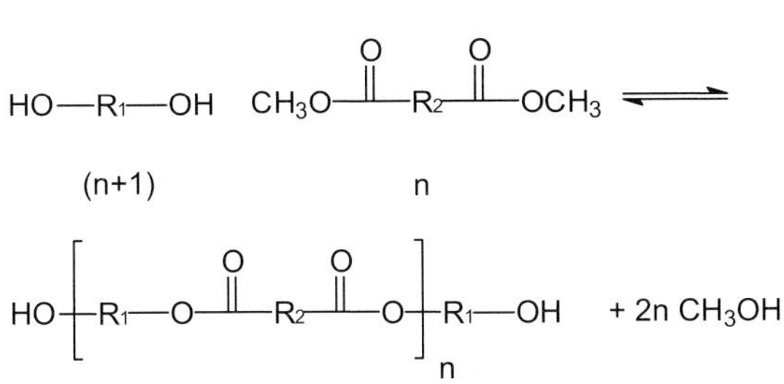

Scheme 16.20. General transesterification scheme for OH-polyester resins.

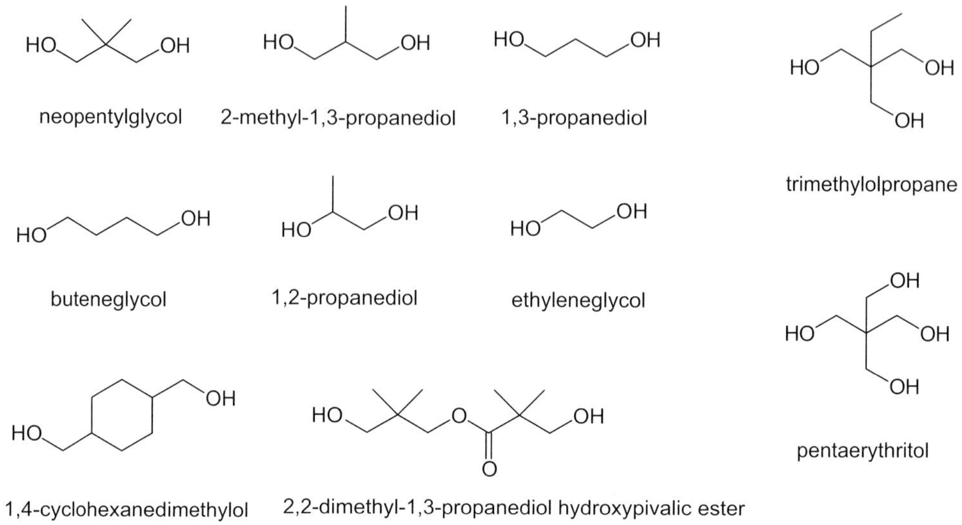

Scheme 16.21. Some polyol monomers used in polyester resins.

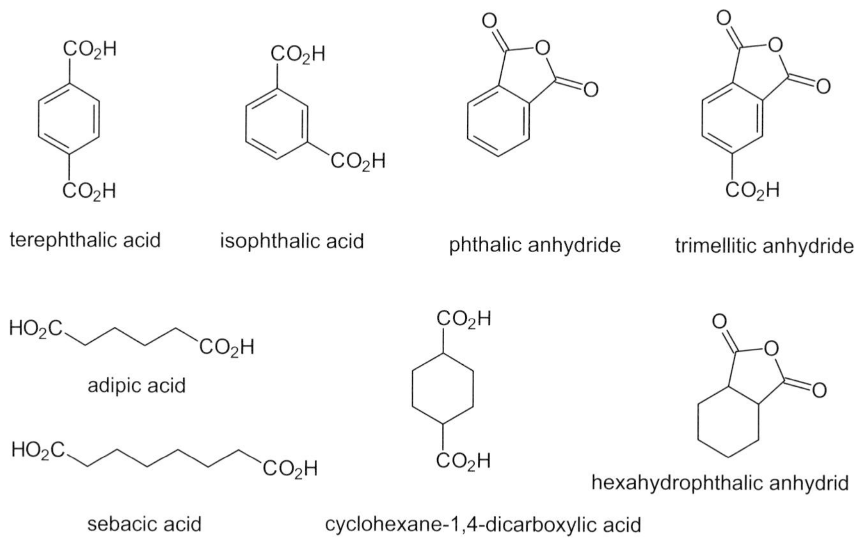

Scheme 16.22. Some polyacid monomers used in polyester resins.

to design the polymer molecule. Schemes 16.21 and 16.22 give an overview of the most frequently used polyol and polyacid monomers, respectively. In principle, the molecular weight distribution can be calculated, but for the more complicated formulations computer aid is needed.

The choice of diols depends on which properties the resin would get. Rigid and cyclic diols will increase the T_g, melting point, and hardness of the polyester resin. Long and flexible diols will decrease the T_g and melting point, and improve the flexibility. Diols which are substituted on the β-position toward the hydroxyl groups, such as neopentyl glycol, are normally more resistant against hydrolysis and photooxidation. Besides a wide variety of diacids, several cyclic anhydrides are available as well. Anhydrides react faster than the corresponding acids and require less water to be distilled off. Aliphatic diacids will lower the T_g and melting point, but give more flexibility. Aromatic diacids give a higher T_g and hardness to the coating film and are more resistant toward hydrolysis. Aliphatic diacids impart less UV absorbance to the resin and therefore enhance the UV resistance of coatings made from these resins. From aliphatic diacids, crystalline or semicrystalline polyester resins can also be made.

16.6.3 Production

Most saturated polyester resins are produced in the melt phase. At temperatures of about 200–260°C and in the presence of a catalyst the diacids and the diols are esterified by removal of water. The high temperature during the synthesis is normally sufficient to remove the water efficiently from the reactor by distillation, under reduced or normal pressure. To remove the last percentages of water and to complete the reaction, a stronger vacuum is applied at the end of the reaction.

In cases where the resins are to be supplied in a solvent anyway, azeotropic distillation can be used to remove the water more efficiently and at lower temperatures. Figure 16.3 describes the reactor design for polyester resin production, which can also be applied to alkyd resin production (see Section 16.5). The product resin can be either cooled and discharged via a cooling belt in case a solid delivery form is required, or thinned with solvents.

16.6.3.1 Monitoring the Reaction

The conversion and progress of the esterification reaction are controlled by analyzing samples that are taken from the mixture. For the application of thermosetting resins the amount of end groups is very important and has to be measured during production. The concentration of end groups (–COOH and –OH) is measured by titration; several titration methods have been developed. The concentration is expressed as the "acid value" or "hydroxyl value" in mg KOH per gram resin (by back-titration in the case of –OH). The removal of water by distillation is not 100% efficient and some diols, diacids, small oligomers, and side products of the reaction at high temperature can be removed from the reactor with the water. In order to reach the right specifications, small amounts of either diols or diacids can be added ("corrections"), as even in the early stages of the reaction a misbalance of the OH/COOH groups can be detected. Another important specification is the viscosity of the melt or of the solution, as an indication for the molecular weight. Both parameters are plotted for a typical polyester resin in Figure 16.4.

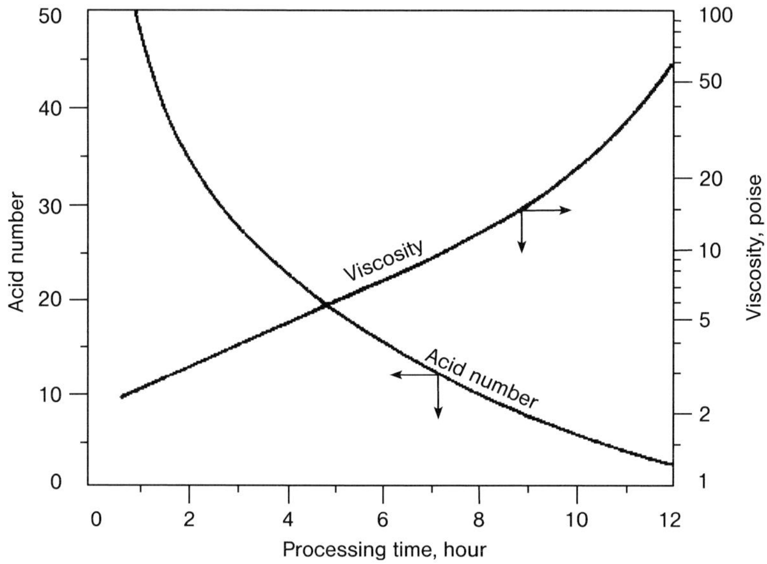

Fig. 16.4. Evolution of acid number and viscosity with time during polyester resin synthesis.

16.6.4
Properties and Applications

The ratio of "hard" to "soft" monomers determines the T_g of the polymer. The reactivity during the curing reaction and network formation is determined by the overall amount of end groups and the overall functionality.

The main application of these types of polyester resins is in thermosetting coatings. Thermosetting paints are formulated by mixing polyester resins (in solid form, melt, or solution) with a crosslinker, pigments, and additives. These paints will be cured to a coating film, generally with stoving. Surface coatings based on saturated polyester resins distinguish themselves from other resins by good adhesion on metals, good flexibility (ability to be deformed together with the substrate after painting), and hardness. Their resistance to hydrolysis is rather poor; their outdoor durability is modest in comparison to acrylic binders but superior to epoxy resins. Saturated polyester resins can be applied in all kinds of paint types. The major application fields today include coatings for beverage and food cans, and coated steel coils. Saturated polyester resins further play a dominant role in the binders for thermosetting powder coatings, described in the next paragraph.

16.6.5
Powder Coatings

Powder coatings are, as the name suggests, paints in solid, powdery form [7]. All the ingredients are solid materials, which are mixed in a hot-melt compounding

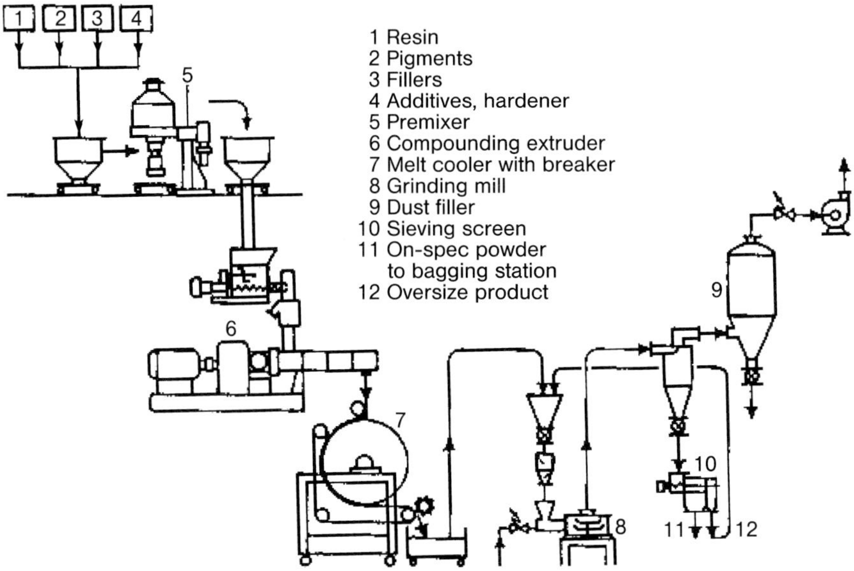

Fig. 16.5. Powder coating production.

process (extrusion). Figure 16.5 depicts the powder coating preparation process. Powder paints consist of a binder (polymer, crosslinker and curing catalyst), pigments, and additives. All these solid ingredients are mixed thoroughly in a premixer and fed into an extruder. The mixture is molten and homogenized in a short period of time, typically in half a minute, to prevent premature crosslinking reactions. After the extrudate is cooled down, it is crushed and milled to the required particle size, which is dependent of the application but is normally in the range of about 50–100 μm.

16.6.5.1 **Application**

Powder paints are commonly applied by electrostatic spraying on metal substrates. The powder particles are charged inside the spray gun (Corona or Tribo) and adhere to the earthed metal object. Powder coating is ideal for coating complicated three-dimensional objects because the powder particles can easily reach recessed surfaces or even the backside of surfaces, following the electrical field lines from the gun to the substrate. Powder paint which does not reach the object or disassociates from the surface ("overspray") can easily be gathered, recycled, and re-used without loss of quality (Figure 16.6). The substrate, covered with a layer of powder paint, is subsequently heated in a hot-air oven. The powder particles melt, coalesce, and flow out into a film.

In some special cases the powder paints can be applied by dipping. The object is preheated and dipped into a fluid bed of powder paint particles. The powder particles adhere to the preheated surface of the object and melt into a film. Further

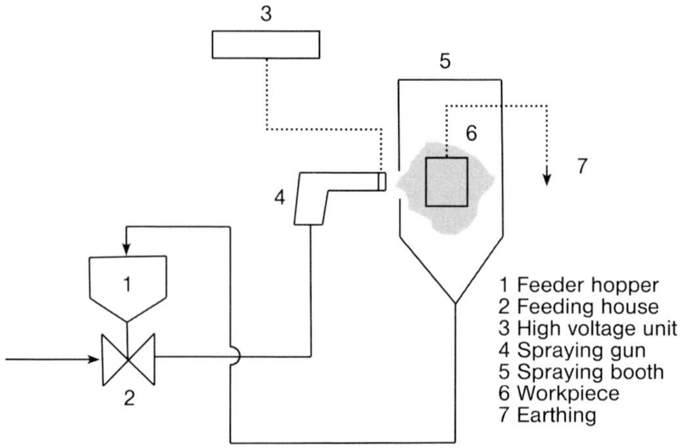

Fig. 16.6. Powder coating spraying and recycling system.

heating causes the crosslinking reactions in the film to give the required thermoset properties. The processes of melting, flowing, and curing take place in the same cycle and this makes it difficult to obtain a completely flat film before the cure prohibits further flow. Powder coatings are therefore characterized by a certain visual "orange peel" effect.

16.6.5.2 Crosslinking

Although many examples exist of thermosetting powder coatings (using mainly polyamide, polyethylene, and polypropylene), most powder paints are thermosetting systems, which are cured at much higher temperatures (typically 180–200°C) than the extrusion temperature (90–110°C). Because of these high curing temperatures, powder coatings find their application mainly on metal surfaces.

The first thermosetting powder coating systems were based on bisphenol-A epoxy resins, which were cured by either amide/amine or carboxylic anhydride crosslinkers (for the chemistry, see Section 16.4.2). The epoxy-based powder coatings show excellent application properties, such as flow, mechanical strength, and chemical resistance. However, they have limited exterior durability and they show a severe tendency toward yellowing. These disadvantages have greatly stimulated the development of powder coatings based on saturated polyesters with carboxylic end groups.

When cured with bisphenol-A resins, these polyester resins result in coatings with much improved exterior durability with less yellowing. When cured with triglycidyl isocyanurate (TGIC) excellent outdoor durability is obtained and polyester resin coatings are used, for example, in the building industry (façades). The mutagenic properties of TGIC have stimulated the search for safer crosslinkers, such as β-hydroxyalkylamides, which are now taking over the market.

Hydroxyl-functional polyester resins are also used; these polyesters are cured with aliphatic (for outdoor use) or aromatic (indoor use) polyisocyanates.

16.6.5.3 **Advantages**

The advantages of powder coating compared to the more classic type of coatings in solution are clear: they are 100% solvent-free and thick layers up to 500 μm can be applied in one step. Moreover, they are very cost-efficient owing to a very high application efficiency, close to 100%, as over-sprayed powder can be easily recycled. Last but not least, their mechanical properties are usually equal or superior to those of their solvent-borne counterparts.

Powder coating can also be favorably combined with UV curing (see Section 16.8.5.2). In that case, the powder coating can also be applied to heat-sensitive substrates such as wood, paper, and plastics because the powder can melt and flow from about 100°C. In addition, the melt/flow-out phase is easily separated (hot air or infrared radiation) from the cure phase (irradiation with UV light) resulting in a good flow. Binders for UV-curable powder coatings are either based on acrylated saturated polyester resins (see Section 16.8.3.2) or on unsaturated polyesters (Section 16.7) in combination with vinyl ether crosslinkers (Secheme 16.34).

16.7 Unsaturated Polyester Resins and Composites

16.7.1 Introduction

Unsaturated polyester resins (UP resins) cover a broad range of products serving a wide variety of market segments. They can be considered the most versatile of the thermoset resins. By varying the resin composition, one can produce an amazing range of properties in polyesters. Additionally, polyesters are capable of being processed by a number of different methods, varying from simple hand-lay-up contact molding to highly automated compression molding. Glass fiber reinforced UP resins are extensively used in building and construction; and in the transportation, electrical, and electronic industries; and for sanitary and domestic applications. The biggest use of these resins is in the construction industry, followed by transportation. The worldwide consumption of UP resins in 2002 was approximately 2.5 million metric tons.

16.7.2 Chemistry

The unsaturated polyester is a macromolecule with an M_n value between roughly 1000 and 3000 g mol^{-1}, obtained from a polycondensation reaction (that is, a polymerization with the formation and elimination of water) between unsaturated- and saturated dicarboxylic acids (or their corresponding anhydrides) and diols [8]. This reaction and the most commonly used raw materials are represented schematically in Figure 16.7. Mono- or trifunctional acids or alcohols or other monomers may be incorporated optionally into the polyester backbone to tailor properties.

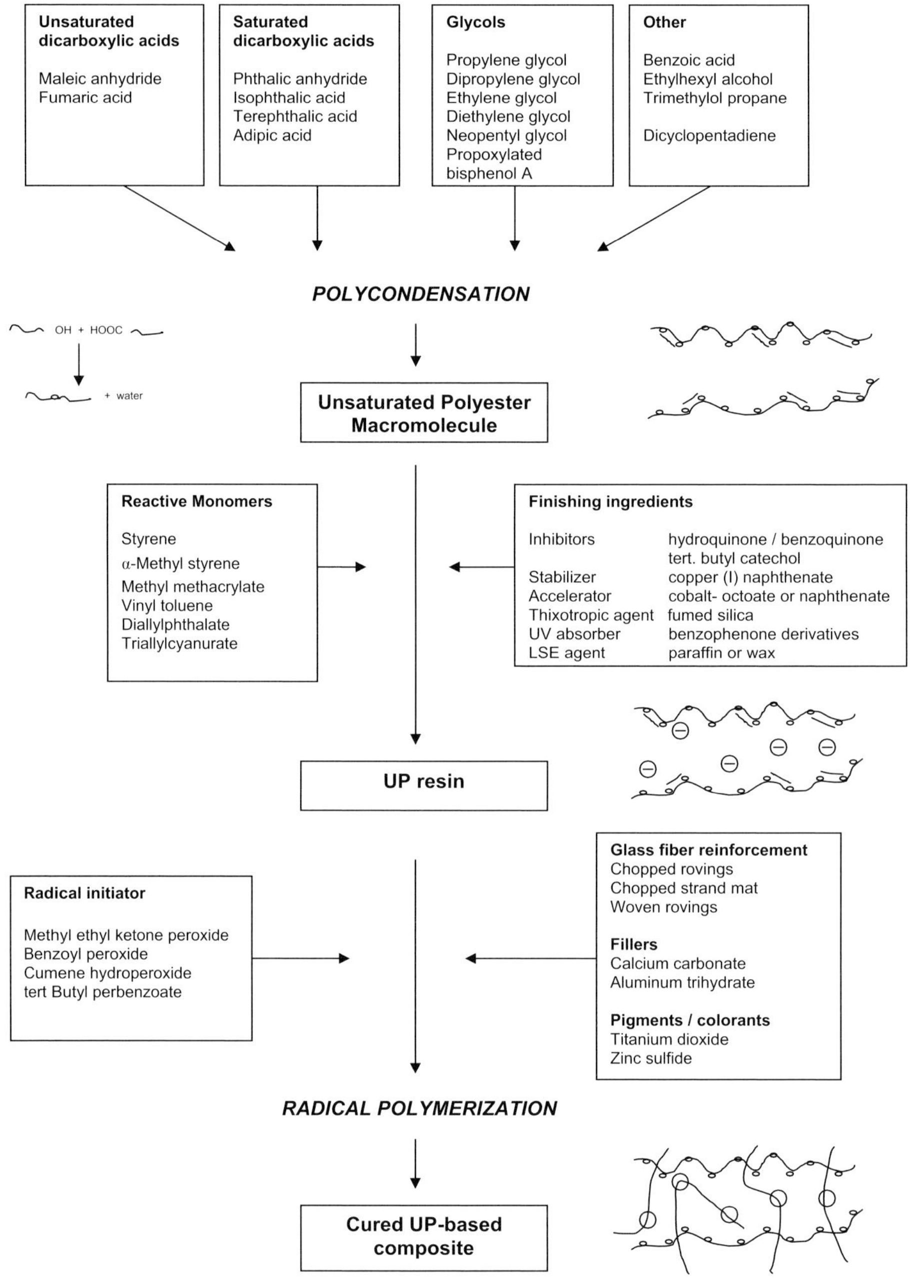

Fig. 16.7. Unsaturated polyester resins: from raw materials to cured composites.

The preferred unsaturation in the polyester backbone is the fumaric double bond, which results from using either fumaric acid or maleic anhydride as the unsaturated dicarboxylic acid. Under the reaction conditions normally applied to synthesize the polyester, most of the maleic (cis) double bonds isomerize to fumaric (trans) double bonds. The unsaturated polyester macromolecule itself is either a very viscous liquid or a solid. Usually this polyester is dissolved in a reactive monomer containing a double bond capable of copolymerizing through a radical polymerization with the (fumaric) unsaturations in the polyester backbone. The workhorse of the reactive solvents is styrene. The low-viscosity polyester solution (UP resin) thus obtained is formulated with a number of ingredients that provide critical properties, such as storage stability, reactivity, thixotropy, UV resistance, and color, before being sold by the polyester producer.

16.7.2.1 Crosslinking

The processor converts the liquid polyester by means of a radical polymerization, in the presence of glass fiber reinforcement and/or fillers, into a three-dimensional composite network as displayed in Figure 16.7. This so-called crosslinking or curing reaction takes place between the fumaric unsaturations in the polyester backbone and the unsaturation in the styrene monomer. The reaction is typically initiated by free radicals generated by the decomposition of peroxides. Depending upon the type of peroxide (and optionally the accelerator used) the crosslinking reaction may take place at temperatures as low as 10°C or as high as 160°C.

The rate at which these radicals are generated is one of the factors determining how fast a resin gels and cures. The temperature development of a mass of resin against time from the moment of (accelerator and) peroxide addition is an important characteristic of a polyester resin. This is illustrated for a typical curing process at ambient temperature in Figure 16.8. The portion AB is the induction period during which free radicals produced by decomposition of the peroxide are being absorbed by the inhibitor present in the resin. This is the so-called pot life of the resin, during which processing of the part can take place. This time may vary from a few minutes up to 20 h. At point B the inhibitor concentration has become low enough for an appreciable proportion of the free radicals produced to initiate resin polymerization. Due to heat of polymerization, the temperature rises. The exothermic peak D does not necessarily correspond to completion of the curing process. As the resin solidifies during the latter part of the curing process, the rate of reaction is greatly reduced (diffusion-controlled). A post-cure at elevated temperatures may be needed to reach full cure.

16.7.2.2 Styrene Emission

Actually more than 50% of UP resin consumption still involves open-mold techniques, hand lay-up and spray-up being the most important. During processing and curing, styrene emission is considerable. Styrene (boiling point 145°C) is classified as a dangerous substance according to EC Directive 67/548/EEC and is labeled with the following R (Risk) and S (Safety) phrases: R10 (flammable), R20 (harmful by inhalation), R36/38 (irritating to eyes, skin, and respiratory tract), S2 (keep out

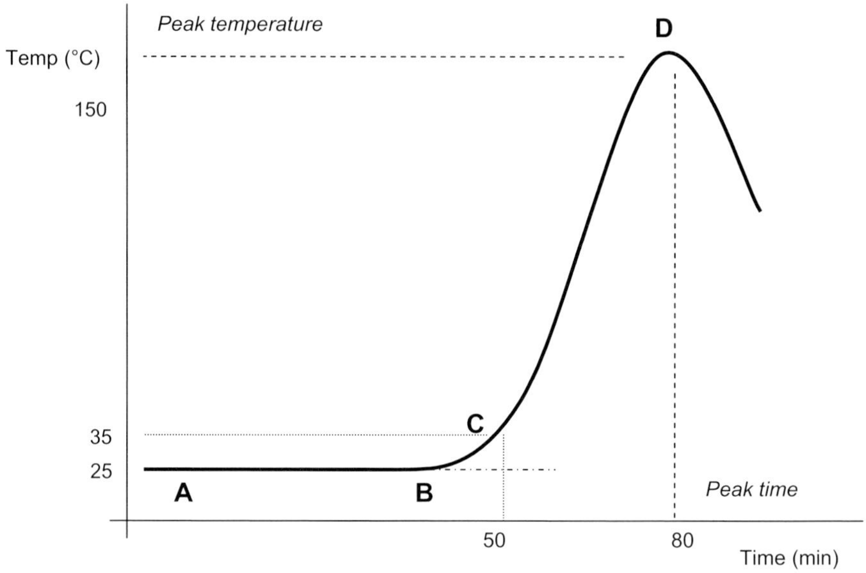

Fig. 16.8. A typical curing profile of unsaturated polyester resins at ambient temperature.

of reach of children), S23 (do not breathe gas/fumes/vapor/spray). In 1992 the European Center for Ecotoxicology and Toxicity of Chemicals concluded in a study, *Styrene Toxicology*, that the carcinogenic potential of styrene is so low that occupational or environmental exposure to styrene is unlikely to present any genotoxic or carcinogenic hazard in humans. Styrene has, however, a proven neurotoxic effect. Occupational exposure to high styrene concentrations may lead to symptoms like drowsiness, headache, and tiredness. As a consequence the threshold limit value (TLV) has been lowered significantly since the 1980s in virtually all countries.

The UP industry reacted by developing so-called LSE (low styrene emission)- and LSC (low styrene content) resins. An LSE resin contains paraffin or wax in such a concentration that upon only a little styrene evaporation it forms a thin barrier film on top of the laminate, curtailing styrene emission. However, one aspect that warrants attention is the interlaminar adhesion of parts built up from several laminate layers. Note that these LSE resins are only active in the static state when the surface of the laminate is at rest. During processing, when the surface is regularly disturbed, the effect of the film-forming additive is far less.

LSC resins have a styrene concentration between 25–35% versus the 40–50% of classical UP resins. Because of the lower styrene content, the evaporation during processing and cure is considerably less. As the processing technique is determining the viscosity range of the resin that can be used, this property cannot be changed. So, keeping the same resin viscosity at lower styrene content implies that the viscosity of the unsaturated polyester must be reduced. This can be achieved by reducing the molecular weight of the unsaturated polyester but, unfortunately, this leads to a strong deterioration of mechanical and chemical proper-

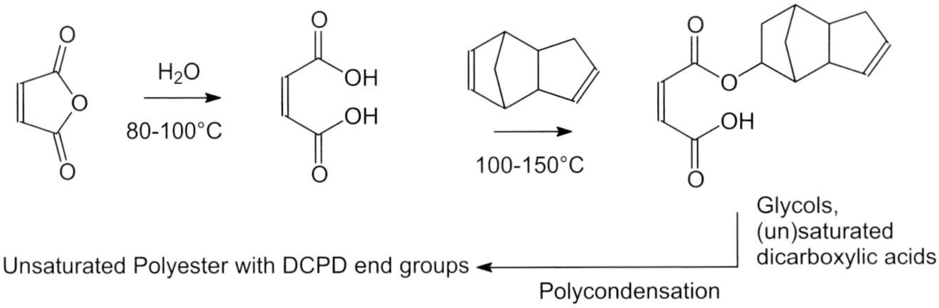

Scheme 16.23. Synthesis of DCPD end-capped unsaturated polyester resins.

ties. Another way is to replace the polar end groups of the polyester by apolar end groups, which show good solubility in styrene. Dicyclopentadiene (DCPD) end-capped polyesters (see Scheme 16.23) are therefore the resins of choice to reach the required properties. Of course, addition of paraffin to these resins will further reduce the styrene emission.

The most obvious way to reduce styrene emission is to use another monomer. However, there appears to be no real alternative that fulfills all the technical and economical requirements as well as being less toxic, and consequently styrene is at most partially replaced in some cases.

Other elements which alone, or in combination, will contribute to less styrene emission/exposure in the workplace are:

- use of specially designed spray guns (for example, flow coaters), which reduce the surface area of the resin that is sprayed, and consequently reduce emission;
- a change from open mold techniques to closed or semi-closed mold techniques; in particular, the vacuum-assisted resin infusion technique, whereby a flexible film is used as a counter-mold, is gaining popularity;
- proper and adequate ventilation;
- use of robots.

16.7.2.3 Vinyl Ester Resins

A special class of unsaturated polyesters, developed for composite applications in corrosive environments, is the so-called vinyl esters and vinyl ester urethane resins.

The best-known vinyl ester is made by reacting the diglycidyl ether of bisphenol-A (DGEBA; see Section 16.4.2) with a monocarboxylic unsaturated acid such as methacrylic or acrylic acid; see Scheme 16.24. This addition reaction occurs at a temperature of 120–140°C. The vinyl ester urethanes, on the other hand, are synthesized from a low molecular weight (unsaturated) polyester diol, mostly based on bisphenol-A, a diisocyanate, and a hydroxyalkyl (meth)acrylate (see also Schemes 16.30 and 16.31, Section 16.8.3.3). As for standard UP resins, the most commonly used reactive solvent is styrene. Both types of products have only a few ester linkages (which are the points of greatest vulnerability to chemical attack) in the mole-

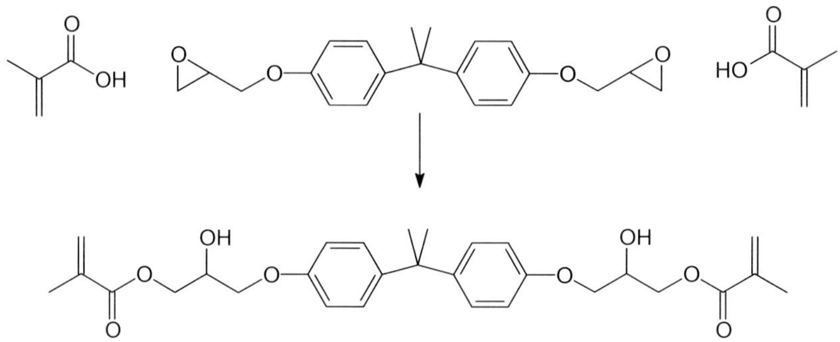

Scheme 16.24. Synthesis of vinyl esters from methacrylic acid and DGEBPA.

cule and hence show improved chemical resistance, whilst the network built up from the terminal (meth)acrylate double bonds results in a tougher, more resilient resin structure.

In Table 16.4, the UP, vinyl ester, and vinyl ester urethane resins are classified in order of chemical resistance.

16.7.3 Production

Unsaturated polyesters are produced batchwise in agitated stainless steel reactors of up to 50-ton capacities (see Figure 16.3). The reactors are equipped with heating jackets and internal heating coils to bring the raw materials to the required reaction temperature (190–230°C).

The production starts by charging the raw materials into the reactor by the use of pumps and meters for liquids such as most of the glycols, molten maleic anhydride, and molten phthalic anhydride. Solid raw materials, such as isophthalic acid, are added manually by emptying big bags or automatically by pneumatic transport from silos. All the reactants may be charged in one go except for a few dicarboxylic acids which are somewhat more difficult to esterify. In the latter case a two-stage method is applied in which the first step is the reaction of such a dicarboxylic acid (for example, isophthalic or terephthalic acid) with a large excess of glycol until a clear solution with a low acid number is obtained, followed by addition and reaction of the other raw materials. When the reactor is brought up to its reaction temperature the contents are periodically sampled to determine the acid number (a measure of the acid groups that have not yet reacted) and viscosity; this may be done either off-line or in-line. The synthesis is continued until the specified acid number and viscosity values are achieved. The distillation column performs the separation of water from unreacted (low-boiling) glycols. During synthesis, inert gas (usually nitrogen) flows continuously through the reactor to prevent gelation due to oxidation and to aid removal of water. An alternative method used to remove the water formed from the reaction mixture is to apply vacuum.

Reaction times vary from 8 to 30 h, depending on the raw materials used, the degree of condensation or molecular weight desired, and the reaction temperature employed.

After cooling, the unsaturated polyester is discharged from the reactor into the so-called thinning tank, which has a preweighed styrene charge in it. The resin is then finished with additives, tested for quality control, and pumped to storage.

16.7.4 Reinforcement

The majority of the UP resins are used in glass fiber reinforced (GRP) applications. Because E-glass offers sufficient strength at low cost it accounts for more than 90% of all glass fiber reinforcements. It also provides good electrical resistance (hence its name). Sizing is applied to the filaments immediately after their formation to ease processing, to protect the fibers from breakage during processing, and to create better bonding between the fiber and the resin matrix. Although it comprises only 0.25–5% of the total fiber weight, sizing is a dynamic force behind the performance of any fiberglass product. Sizing chemistry distinguishes each manufacturer's product and determines the grade of fiberglass for different processing techniques. Sizing can improve the wettability of the fiber during resin impregnation, thereby reducing part-manufacturing time. Coupling chemistry to enhance bonding between the glass fiber and the matrix resin can also improve the overall strength of the final composite.

Glass filaments are supplied in bundles called strands, rovings, or yarns. A strand is a collection of more than one continuous glass filament. A roving generally refers to a bundle of untwisted strands wound in parallel to form a cylindrical, flat-ended package, similar to thread on a spool. Single-end roving contains only one continuous strand of multiple glass filaments. Multiple-end roving contains numerous wound strands. Yarns are collections of filaments or strands that are twisted together.

Rovings, the most common form of glass fiber, can be chopped or woven to create secondary fiber forms used in composite manufacturing.

Mats are non-woven fabrics that provide equal strength in all directions. Chopped mats contain randomly distributed fibers cut to lengths typically ranging from 4 to 6 cm and held together with a chemical binder. These mats provide low-cost polymer reinforcement primarily in hand lay-up, continuous laminating, and some closed-mold applications. Continuous-strand mat is formed by swirling fiber onto a moving belt finished with a chemical binder that holds the fiber in place. Continuous-strand mat is primarily used in compression molding, resin transfer molding, and pultrusion applications.

Woven rovings are made by weaving the untwisted rovings into cloth. Bidirectional woven fabrics provide good strength in the 0°/90° directions and allow fast composite fabrication. To prevent laminates failing by interlaminar shear between layers of woven rovings, these are generally interspersed with layers of chopped-strand mat.

Tab. 16.4. Classification of UP, vinyl ester, and vinyl ester urethane resins according to their chemical resistance (European standard EN 1321).

Resin group	*Resin type*	*Types of glycols*	*Types of acids*	*Styrene content [wt.%] max.*	*T_g [°C] min.*	*HDT [°C] min.*	*Tensile strength [MPa] min.*	*Elongation at break (tension) [%] min.*	*Flexural strength [MPa] min.*
1A	UP	standard glycols[a,b]	orthophthalic acid, ethylenedicarboxylic acids	45	85	60	60	2.0	90
1B	UP	standard glycols[a,b]	orthophthalic acid, ethylenedicarboxylic acids	45	120	90	50	1.5	75
2A	UP	standard glycols[a,b]	isophthalic acid, HET acid, ethylenedicarboxylic acids	50	85	60	60	2.0	90
2B	UP	standard glycols[a]	isophthalic acid, HET acid, ethylenedicarboxylic acids	50	120	90	50	1.5	75
3	UP	standard glycols[a]	terephthalic acid, ethylenedicarboxylic acids	50	140	110	75	3.0	120
4	UP	neopentyl and halogenated neopentyl glycol (min 80 mol%)[c] and a diol with at least one secondary OH group (max 20 mol%)[c]	isophthalic acid, orthophthalic acid, ethylenedicarboxylic acids	55	120	90	65	3.0	110
5	UP	hydrogenated bisphenol A[d] and cyclohexane dimethanol	orthophthalic acid, ethylenedicarboxylic acids	45	120	90	50	1.5	100
6	UP	dipropoxy-bisphenol A and/or dipropoxy halogenated bisphenol-A	ethylenedicarboxylic acids	55	130	110	60	2.0	110
7A	VE	epoxidized bisphenol-A and/or epoxidized halogenated bisphenol-A	methacrylic aid and/or acrylic acid	55	110	90	75	4.0	130

7B	VEU	dialkoxy-bisphenol-A and/or dialkoxy halogenated bisphenol-A (min. 90%), alkoxy (meth)acrylate	ethylenedicarboxylic acids	50	120	105	75	3.5	130
8	VE	epoxidized novolac	methacrylic aid and/or acrylic acid	50	150	120	75	2.5	130

[a] Standard glycols: ethylene, 1,2-propene, diethylene, dipropene glycols; neopentyl glycol, 1,3-butanediol, 1,4-butanediol and corresponding halogenated glycols.
[b] May also contain cyclic unsaturated hydrocarbons.
[c] Related to the sum of the diol components.
[d] This diol deviates from the one mentioned in EN 13121; the diol mentioned in this table is more commonly used.
Note: The chemical resistance of the resins increases on going from group 1 → group 8.

Tab. 16.5. A comparison of reinforcing fibers.

Reinforcing fiber	*Tensile strength [MPa]*	*Tensile modulus [GPa]*	*Specific gravity [g cm^{-3}]*	*Properties*
Aramide	2800–3000	85–130	1.44	low density, good specific properties, moderate cost
Boron	3500	400	2.55	high modulus, high cost
Carbon	2500–3200	210–700	1.75–1.96	high modulus, electrical conductivity, high cost
Glass: E-glass	2400	72	2.54	good strength and processability, low cost
S-glass	4500	85	2.49	high strength and good processability
Polyester	1050	10	1.38	good impact resistance and chemical properties
Polyamide	1000	5	1.16	good impact resistance, alkali resistant
Ultra high molecular weight polyethylene	3400	110	0.975	low density, good impact

Surfacing mat, or veil, is used in conjunction with reinforcing mats and fabrics to provide good surface finish. It is effective in blocking out the fiber pattern of the underlying mat or fabric. Surfacing mats are also used as the inside layers of corrosion resistant composites, providing a smooth, resin-rich surface.

Fibers can also be produced from carbon, boron, and aramid materials. Generally these fibers exhibit higher tensile strength and stiffness than do their glass counterparts, as can be seen in Table 16.5. However, these specialty fibers are more expensive. Therefore they are typically reserved for applications demanding exceptional fiber properties, for which the customer is willing to pay a premium.

16.7.5 Fillers

Fillers not only reduce the costs of composites, but also frequently impart performance improvements which might not otherwise be achieved by the reinforcement and resin ingredients alone. Fillers can improve fire and smoke resistance by reducing organic content in composite laminates. Other important properties that can be improved through the proper use of fillers include water resistance, weathering, surface smoothness, stiffness, dimensional stability, and temperature resistance. Calcium carbonate is the most widely used inorganic filler. It is available at low cost in a variety of particle sizes and surface treatments.

Aluminum trihydrate is used as a filler when improved fire/smoke performance is required. When exposed to high temperatures this filler gives off water in a highly endothermic reaction, thereby reducing the flame spread and development of smoke.

16.7.6 Processing

Fabricators of unsaturated polyester resins use either open-mold or closed-mold processes. The main open-mold processes are hand lay-up, spray-up, continuous lamination, filament winding, and centrifugal casting. The most used closed-mold techniques are hot-press molding and resin-transfer molding. Pultrusion and cold-press molding may be considered as semi-open or semi-closed processes. An overview of the processing techniques for UP resins, their possibilities and limitations, is given in Table 16.6.

Irrespective of the process used to produce glass-reinforced composites there are three basic principles in selecting materials for a particular end use [9]:

- mechanical strength;
- chemical, electrical, and thermal properties;
- cost/performance.

Mechanical strength The type of fiber, the amount in the composite, and the way in which the individual strands are positioned determine the direction and level of strength that is achieved. There are three generic types of fiber orientation:

- unidirectional oriented fibers, providing the greatest strength in the direction of the fibers;
- bidirectional oriented fibers, with some fibers positioned at an angle to the rest, as with woven fabric: this provides different strength levels in each direction of fiber orientation;
- multidirectional or randomly oriented fibers: this arrangement provides essentially equal strength in all directions of the finished part.

Chemical, electrical, and thermal properties The chemical building blocks of the polyester, the molar ratio of maleic anhydride to saturated dicarboxylic acid, and the choice and content of the monomer dictate to a large extent the mechanical properties (rigid or flexible), the glass transition temperature (T_g) or HDT (heat deflection temperature), and the thermal and chemical resistance of the cured products. Additives are available for use with many resin systems to provide properties such as color, high surface finish, UV resistance, and low shrinkage.

Cost/performance If a large number of parts are to be made from one set of molds, the lowest unit production cost is likely to be obtained by using compression or injection molding. In these processes, presses, tooling, and automated handling equipment can be applied to speed production cycles and reduce direct labor

Tab. 16.6. Comparison of common composite process considerations.

	Hand lay-up and spray up	*Continuous lamination*	*Filament winding*	*Centrifugal casting*	*Pultrusion*	*Cold-press molding*	*Resin infusion (vacuum)*	*Light resin transfer molding ($p > 1$ bar)*	*Resin transfer molding ($p > 1$ bar)*	*Hot-press molding (BMC/SMC)*
Type of reinforcement	chopped rovings, mat, woven rovings	mat, chopped rovings	rovings, mat, woven rovings	mat, chopped rovings, woven roving	rovings, mat	mat, woven rovings	mat, woven rovings, cont. fibers, preform	mat, woven rovings, cont. fibers, preform	mat, woven rovings, cont. fibers, preform	chopped rovings, mat
Glass [wt%]	25–35	25–35	55–80	15–40	50–75	25–50	40–65	25–55	25–50	25–70
Laminate thickness [mm]	2–25, usually 2–10	2–5	2–25	2–25	2–25	1–10	2–50	2–6	2–6	1–10
Cure temp. [°C]	ambient to 40	60–110	ambient to 100	ambient to 100	100–150	25–50	ambient to 60	ambient to 60	ambient to 60	100–170
Styrene emission	high	low	high	high	low	low	low	low	low	low
Type of mold	single GRP, wood, etc	metal rollers or sheet	steel or around plastic liner	steel	hardened steel die	double GRP or metal	bottom, GRP or light metal; top, foil	bottom, stiff GRP, top, flex. GRP	double GRP or light metal	matched metal

Size limitation	in principle, none	width of the machine	generally 6 m long, 6 m diameter	generally 6 m long, 6 m diameter	continuous, die dimensions (600 mm × 250 mm)	press size	in principle, none	mold dimensions	mold dimensions	press size
Molded in											
ribs	yes	no	yes, externally	no	no. of cross-sections	yes	yes	generally no	generally no	yes
inserts	yes	no	no	no	no	yes	yes	generally no	generally no	yes
foam panels	yes	no	yes	no	no	generally no	yes	yes	yes	no
No. of moldings to justify the investment	1–1000	>500 km y^{-1}	1–2000	>1000	>300 km y^{-1}	500–5000	1–1000	100–5000	500–5000	>10 000
Production rate	low	up to 12 m min^{-1}	moderate	moderate	up to 4 m min^{-1}	moderate	low	moderate to high	moderate to high	high
Labor content	high	low	moderate	low	low	low	moderate	moderate	moderate	low
Quality of molding	operator-dependent – one smooth surface	good	good: inside smooth	good: two smooth surfaces	good	good; two smooth surfaces	good; one smooth, one semi-smooth surface	good; two smooth surfaces	good; two smooth surfaces	excellent; two smooth surfaces
Typical products	boats, garden ponds, containers, shelters	roofing lights, corrugated sheets	tanks, pipes, tubes	pipes and tubes	rods, tubes, profiles	automotive, industrial, electrical parts, ventilator housings	boat hulls, wind mill wings	truck parts, hatches, covers	radomes, panels	automotive, industrial, electrical parts

costs. Conversely, if only a limited number of parts are required, the simpler processes such as hand lay-up or spray-up may be the cost-effective methods, since they usually involve low-cost tooling and minimal fabrication equipment. Occasionally, the shape of the part dictates that a particular process is used. For instance, cylindrical objects can often best be made by centrifugal casting or filament winding, and components having constant cross-sections that require high strength by pultrusion.

The main conversion techniques for UP resins are explained in Sections 16.7.6.1–16.7.6.10.

16.7.6.1 Hand Lay-up and Spray-up

Both hand lay-up and spray-up processes essentially involve placing reinforcement and liquid resin onto the surface of an open mold. As the two names suggest, hand lay-up involves applying the resin and reinforcement (for example, glass fiber chopped-strand mats) by hand, while spray-up uses spray equipment to deposit resin and reinforcement (for example, chopped glass fibers) onto the mold. These relatively cheap techniques are often used to produce large, complicated, strong composite parts such as boat hulls.

16.7.6.2 Continuous Lamination

Continuous lamination is another open-mold process. In this method resin is doctored onto a film of cellophane or poly(vinyl alcohol) and glass mat or chopped rovings placed on top of the resin. A second film is then placed on top of this, and the sandwich obtained is passed through rollers, which compact and remove air. They can also shape, as with corrugated type laminates. The sandwich is then passed through a heat source for curing.

16.7.6.3 Filament Winding

Filament winding is a technique used for the manufacture of pipes, tubes, cylinders, and spheres and is frequently used for the construction of large tanks and pipe work for the chemical industry. In the process glass fibers are drawn through a resin bath to impregnate them with resin. The impregnated rovings are then wound under tension round a rotating mandrel. Generally the feed head supplying the rovings to the mandrel traverses backward and forward along the mandrel.

By suitable design, the structures obtained, having a glass content of up to 80%, can withstand very high pressures.

The filament winding process may be combined with spray-up. This hybrid process is called "chop hoop". The reinforcement of the resulting laminate than consists of a blend of continuous and randomly chopped fibers.

16.7.6.4 Centrifugal Casting

In centrifugal casting, resin, fillers, and reinforcement are deposited against the inside surface of a rotating mold. Centrifugal force holds the material in place until the part is cured. The interior surface of the centrifugally cast parts can be given an additional layer of pure resin to improve their surface appearance and to provide additional chemical resistance. Large-diameter composite pipes are produced by this technique.

16.7.6.5 Pultrusion

Pultrusion is a technique used for producing continuous-fiber reinforced sections in which the orientation of the fibers is kept constant during cure. The composites are made by pulling reinforcement through a resin-impregnating bath and then through a heated die, the interior of which has the desired shape. Pultruded composites, due to their generally high reinforcement levels (60–75%), have exceptionally high mechanical properties parallel to the direction of pultrusion. Examples of pultruded products are ladder rails, electrical equipment, insulator rods, light poles, window profiles, and gratings.

16.7.6.6 Cold-press Molding

Reinforcing material, which may be glass mat or woven fabric, is cut to shape and placed in the GRP or light metal mold. Then the resin is poured in and the press closed with a strong deceleration toward the end of the stroke. The resin displaces air and impregnates the reinforcement. The heat of polymerization generated accelerates the cure of the resin, giving a relatively short cycle time.

16.7.6.7 Resin Infusion

In this process one GRP mold is used and a plastic bag or film mounted along the mold flange. Resin is pulled through the reinforcement that is placed between the rigid and flexible molds, using vacuum. Applications include large parts, which are normally produced using contact molding. Resin infusion is a closed process and thus limits the evaporation of styrene into the atmosphere.

16.7.6.8 Resin-transfer Molding

This method employs a male and a female GRP or metal mold, which are designed to fit tightly together with a rubber gasket seal. The reinforcement is placed between the molds. After the molds have been clamped together, catalyzed resin is forced into the mold by pressure (or vacuum) until resin leaves the mold through a vent hole. An overspill area is built into the mold to ensure that after trimming a good-quality molded edge is left. Low-viscosity resins are necessary to keep the pressure requirements moderate and to facilitate wetting-out of the reinforcement.

16.7.6.9 Hot-press Molding

With hot-press molding the most convenient method is to use a preformed molding compound to which all the necessary ingredients have been added. Two commonly used molding compounds are SMC (sheet molding compound) and BMC (bulk molding compound).

SMC is composed basically of five principal ingredients: unsaturated polyester resin, shrink-reducing thermoplastic, reinforcement, fillers, and additives (see

Tab. 16.7. Typical compositions of SMC and BMC.

Ingredient	***Function***	***Formulation [parts by weight]***	
		SMC	***BMC***
Unsat. polyester + styrene	resin	100	100
t-Butyl peroxybenzoate	radical initiator	2–3	1–2
Zinc stearate	mold release	3–4	3–4
Magnesium oxide	thickening	2–3	0–3
Pigment dispersion	color	0–8	0–15
Calcium carbonate	filler	150–200	200–250
Thermoplastic additive	shrink reduction	5–20	5–20
Chopped rovings	reinforcement	50–100	45–75

Table 16.7). SMC technology comprises two distinct manufacturing steps: compounding and molding. In the compounding operation all the ingredients, except the fibers, are mixed together to form a highly viscous paste. The paste is then applied to two carrier films (usually polyethylene) to form a sandwich layer with the glass fibers in the middle (see Figure 16.9).

The compounded sheets are then stored to mature in a controlled environment; in a period of three to seven days the viscosity of the compound increases from about 25 Pa s to 25 000 Pa s or higher. This thousand-fold increase in viscosity is obtained through a so-called thickening reaction. In chemical terms the added magnesium oxide reacts with the carboxylic end groups of the unsaturated polyester to obtain longer polyester chains through complex salt formation. Moreover, the obtained magnesium carboxylate salts cluster in an ionic lattice, giving rise to a fur-

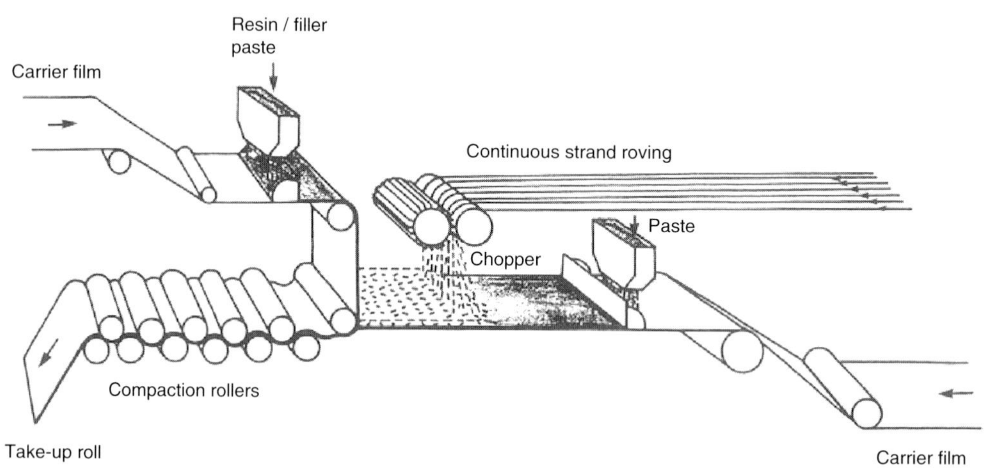

Fig. 16.9. Schematic representation of the SMC machine.

ther viscosity increase. The first part of the reaction, the salt formation, is irreversible, while the ionic forces break down within the lattice again upon heating in the molding process and allow the compound to flow.

When the compound is ready for molding it is cut into pieces of predetermined dimensions. The pieces are then stacked in a specific arrangement (charge pattern) in the mold so that the flow of the material is optimal. The flow is achieved by the compression action of the mold, which is normally a matched set of steel dies heated to a temperature of about 150°C. During the molding cycle (at a pressure of about 7.5 GPa), which is about 1–3 min long, the resin cures completely. When the mold opens, the part is removed and trimmed before it is sent to secondary operations, which may include cutting holes and openings using a water jet, router, punch, or drill. Very often, different SMC parts are glued together to improve structural integrity as well as to provide space for hardware assembly.

In the past, molding compound growth into many high volume applications was excluded because of high polymerization shrinkage resulting in surface defects, warpage, internal cracks, and notable depression ("sink") on the surface opposite reinforcing ribs and bosses. The ultimate solution to the problem has been the addition of certain thermoplastics to the formulation of the composite. These thermoplastics, also called low-shrink- or low-profile additives, are usually present at only 2–5 wt% of the final molded part or 7–20 wt% of the organic portion of the formulation.

In all effective formulations two phases are formed in the cured composite: a crosslinked thermosetting phase and a separated thermoplastic phase well dispersed throughout the resin matrix. Upon heating of the composite during the molding process the thermoplastic is far above its glass transition temperature and the thermal expansion counteracts the volume shrinkage of the crosslinking unsaturated polyester. Upon cooling, micro-voids are created in the composite due to a higher thermal shrinkage of the thermoplastic phase than the crosslinked thermoset. The best shrinkage control systems (low-profile and class A systems), those with the most phase separation and micro-voiding, do not provide uniform pigmentation, particularly in darker colors. Shrinkage control systems that are less effective (low-shrinkage systems) provide reasonable pigment acceptance but at the cost of reduced dimensional stability and surface smoothness.

A parallel technology to SMC is bulk molding compound (BMC). BMC consists basically of the same ingredients as SMC (see Table 16.7). However, its production method, appearance, and properties differ.

BMC is commonly manufactured by using high-speed plowshare mixers. The different components of the paste are fed into the mixer and mixed until the compound is fully homogeneous. The chopped-glass fibers are introduced at the end of the process to prevent too much fiber degradation.

BMC may be used in compression as well as in injection molding. In the latter process the compound is conveyed from the hopper to the nozzle of the barrel by screw rotation or plunger displacement. The metered charge is fed into the heated mold by fast translation of the screw or plunger. Similarly to SMC-based composites, BMC products are characterized by stiffness, temperature resistance, and dimensional stability. Due to the shorter glass fibers, however, strength and impact are lower.

16.7.6.10 Casting, Encapsulation, and Coating (Non-reinforced Applications)

Although the majority of UP resins find their way to reinforced composites, there are also some non-reinforced applications, such as the polyester resins widely used for button manufacture. Here the resin needs to be colorless and color-stable after cure. Furthermore, when cured the resin should exhibit good hot water resistance, hardness, and impact resistance. Such resins are also used for encapsulation purposes. In most non-reinforced applications, however, use is made of resins with high filler loadings of up to 85 wt%. Cultured marble, polyester concrete, and auto-repair putties are examples of such applications. Yet another non-reinforced application is that of coatings: gel coat and topcoat are typical coatings used for polyester-based laminates, while special UV-curable unsaturated polyester resins are used in high-gloss coatings for furniture.

16.7.7 Design Considerations: Mechanical Properties of Composites

Composite designers choose from a variety of fiber reinforcement and resin systems (both thermoset and thermoplastic) to develop a part. Basically, the reinforcement provides mechanical properties such as stiffness and tensile and impact strength. The matrix material transfers the loads to the fibers, and also protects the fibers from abrasion and chemical attack.

Thermoset matrices have *E*-moduli between 3000 and 4000 MPa while those of E-glass reinforcements are about 70 000 MPa. The tensile strength of glass fibers is around 2400 MPa and that of an UP resin about 50–60 MPa. The glass fiber content of the composite is usually between 25 and 75%.

The fibers may be oriented randomly within the resin, but it is also possible to arrange for them to be oriented preferentially in the direction expected to have the highest stress. Such a material is said to be anisotropic (having different properties in different directions), and control of the anisotropy is an important means of optimizing the material for specific applications.

Consider a typical region of composite of unit dimensions constructed from parallel and continuous arrangements. This region can be idealized as shown in Figure 16.10 on the left-hand side, by gathering all the fibers together, leaving the matrix to occupy the remaining volume.

The total load applied in the direction of the fibers to the composite ($F_{//}$) is carried partly by the reinforcing fibers and partly by the matrix according to the generic Eq. (1).

$$F_{//} = F_{\mathrm{f}} + F_{\mathrm{m}} \tag{1}$$

The fiber and matrix phases act in parallel to support the load. In terms of stress

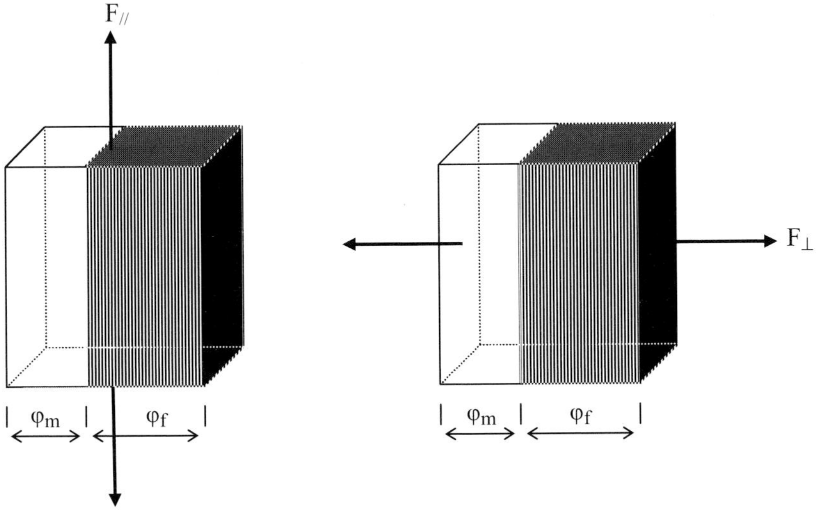

Fig. 16.10. Loading of the composite parallel (left) and perpendicular (right) to the direction of the fibers.

(σ) equation (1) can be rewritten as Eq. (2), where $\sigma_{//}, \sigma_f$, and σ_m are respectively the average stress in the tensile bar, the stress in the fiber, and the stress in the matrix. A is the cross-section of the test bar, A_f the total cross-section of the fibers, and A_m the remaining cross-section of the matrix.

$$\sigma_{//} \cdot A = \sigma_f \cdot A_f + \sigma_m \cdot A_m \tag{2}$$

To what extent the stress is carried by the fibers depends on the "grip" the matrix has on the reinforcement. In the worst case, due to a total lack of adhesion between the fiber and the matrix, there is no "grip" at all. The matrix solely then governs the properties, and consequently the composite shows a poor mechanical performance. The perfect "grip", on the other hand, results from infinitely long fibers with an excellent adhesion to the matrix. In this ideal case the strain (ε) in the fibers is equal to that of the matrix (iso-strain conditions). Knowing that for a material with a linear elastic behavior $\sigma = E\varepsilon$, wherein E is the stiffness or Young's modulus, we can derive Eq. (3) from Eq. (2).

$$E_{//} \cdot \varepsilon \cdot A = E_f \cdot \varepsilon \cdot A_f + E_m \cdot \varepsilon \cdot A_m \tag{3}$$

For infinitely long fibers the volume fraction is directly represented by the respective cross-sections. Bearing in mind that the strain is the same in all cases, we rewrite Eq. (3) to get Eq. (4), where $E_{//}$, E_f, and E_m are the moduli of the composite, fiber and matrix, and φ_f and φ_m are the volume fractions of the fiber and matrix ($\varphi_f + \varphi_m = 1$).

$$E = E_f \cdot \varphi_f + E_m \cdot \varphi_m \tag{4}$$

This relationship is known as "the rule of mixtures", predicting the overall modulus in terms of the moduli of the constituent phases and their volume fractions.

Using essentially the same approach, but for iso-stress conditions, it is possible to calculate the corresponding relationship for the modulus perpendicular to the fiber direction. In that case the "total grip" of the matrix is not sufficient to deform the reinforcement, which usually has a much higher modulus than the matrix. Here, as a first approximation, the transverse modulus, $E_\perp$, is determined by considering the matrix and reinforcement acting in series (see Figure 16.12, right-hand side). In this situation it is not the strains of the reinforcement and the matrix that are equal, but the stresses (an idealization), as Eq. (5) states.

$$\sigma_\perp = \sigma_f = \sigma_m \tag{5}$$

The deflection of the fibers and the matrix add to give the overall transverse deflection [Eq. (6)].

$$\varepsilon_\perp = \varepsilon_f + \varepsilon_m \tag{6}$$

Since $\sigma = E \cdot \varepsilon$ for small deformations, the Eq. (7) for the transverse modulus can be derived.

$$1/E_\perp = \varphi_f/E_f + (1 - \varphi_f)/E_m \tag{7}$$

Inspection of these simple equations reveals that at the usual levels of fiber addition the longitudinal modulus is dominated by the fiber modulus but the transverse modulus is more influenced by the matrix modulus. The prediction of transverse modulus is considered less reliable, in spite of its occasional agreement with experiment.

In more complicated composites, for instance those with fibers in more than one direction or those having particulate or other non-fibrous reinforcements, Eq. (4) provides an upper bound to the composite modulus, while Eq. (7) is a lower bound.

The formulas above are related to continuous fibers but in practice the fibers have a finite length. There is a critical length, equal to the shortest length which will allow the stress in the fiber to reach the tensile fracture stress. This length depends upon the ratio of the moduli of the two phases, the strength of the interfacial bond, and the shear strength of the polymer. Typical values for glass and carbon fibers are about 1–2 mm; in fact it is not the fiber length by itself but, more precisely, the aspect ratio, that is, the ratio of length to diameter of the fiber, that governs the properties.

More detailed calculations of the mechanical properties of composites can be found in Ref. 10.

16.8 Acrylate Resins and UV Curing

16.8.1 Introduction

The term "acrylate resins" refers to acrylate-functional polymers. This means that the reactive group, which can be either an acrylate or a methacrylate functionality (Scheme 16.25), is still present in the polymer. (This is also the distinction from acrylics, which are polymers prepared from acrylate monomers and in which the acrylate functionality is no longer present.) These polymers are generally dissolved in a reactive diluent or in a diluent mixture. Depending upon the application these resins are cured with a peroxide initiator or with the combination of light and a photoinitiator, the so-called "UV curing" [11]; see Section 16.8.5. When curing occurs with a peroxide, the peroxide is mixed with the resin and the mixture is then cast into on a mold.

Acrylate Methacrylate

Scheme 16.25. Chemical structures of acrylate and methacrylate functionality.

The reactive diluent is required for tuning the resin formulation to the application viscosity. The nature of the diluent is partly dependent on the structure of the resin. In the case of methacrylate-functional resins, methacrylate-functional diluents are mostly used, whereas in the case of acrylate-functional resins, acrylate-functional diluents are commonly applied.

Between acrylates and methacrylates there is a great difference in the homopolymerization rates. As a rule of thumb, it can be said that acrylates are ten times more reactive than their methacrylate analogs. Mixtures of acrylates and methacrylates are not often used, as acrylates are not able to increase the rate of a methacrylate resin. In that specific case, considering the reactivity ratios of this copolymerization, the methacrylate will dominate both the rate and the composition of the polymer formed.

The composite resin industry is the main exception with respect to the selective choice of reactive diluents. Within this industry, styrene is used as well as reactive diluent (known in combination with unsaturated polyesters). Methacrylates are then often referred to as – chemically erroneously – vinyl esters.

16.8.2
Chemistry

The basic chemistry upon curing is the homopolymerization of a (meth)acrylate functionality as is depicted in Scheme 16.26. This polymerization is a radical chain polymerization. The propagating radical is carbon-centered, and therefore this polymerization is sensitive to oxygen inhibition as oxygen can quench carbon-centered radicals very effectively. As a result of this oxygen inhibition the top layer is generally not as thoroughly cured as the bulk of the material.

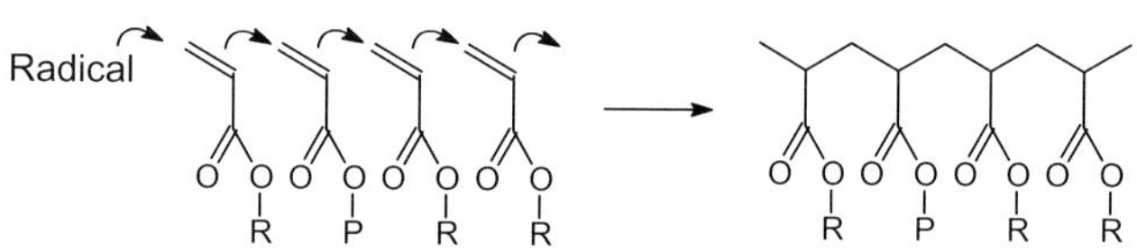

Scheme 16.26. Chain polymerization of acrylate groups.

The three-dimensional crosslinked network which is formed upon curing (Scheme 16.27) consists of two main segments: the polyacrylate chain and the polymer, to which some acrylate groups are attached. This schematic picture clearly illustrates the methods by which the properties of the network can be adjusted: the polyacrylate chain, the side groups, the number of crosslinks, and the acrylate-functional polymer.

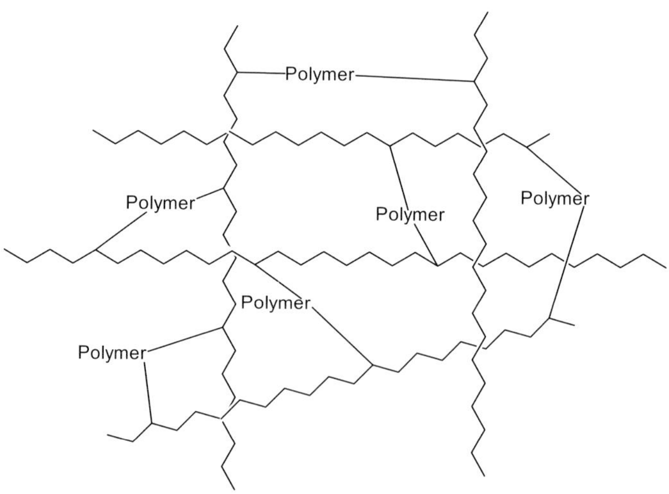

Scheme 16.27. A network formed by polymerization on an acrylate resin.

16.8.3
Production

Within the acrylate-functional resins many distinctions can be made, for instance for functionality, polymerization speed, polarity, and so on. However, these distinctions generally do not result in different methods of production. The production can be divided into three main classes, from each of which a characteristic example will be given.

16.8.3.1 Epoxy Acrylates

The easiest acrylates to produce industrially are the epoxy acrylates; their preparation (see Scheme 16.28) starts with an epoxide-functional resin (see Section 16.4.2). In principle any epoxide-functional material can be chosen. In this reaction (meth)acrylic resin is added to the epoxide at elevated temperatures. (around 90–130°C). The (meth)acrylic acid adds to the epoxide in a ring-opening reaction resulting in an ester alcohol group. Basically this reaction is similar to the reactions used in the preparation of epoxy resins (see Section 16.4). The reactions can be either acid- or base-catalyzed; base catalysis is the more frequently used, since it limits the number of possible side reactions (for instance, transesterifications). Although these reactions can be carried out in solvent, industrially they are most frequently performed in bulk. Generally these preparations are performed in a batch-type process.

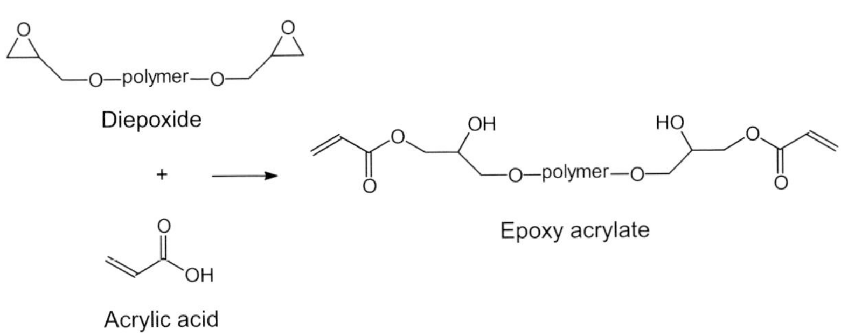

Scheme 16.28. Preparation of epoxy acrylates.

Acrylated oils are a special type of epoxy acrylates. They are formed from epoxidized oils such as linseed oil, and (meth)acrylic acid.

16.8.3.2 Polyester Acrylates

The synthesis of polyester acrylates (Scheme 16.29) is a batch process as well. In contrast to the preparation of epoxy acrylates this reaction is performed in a solvent, and generally acid-catalyzed. The preparations differ from the synthesis of unsaturated polyesters (see Section 16.7), mainly because (meth)acrylates have very reactive double bonds, which can easily homopolymerize. Consequently high reaction temperatures cannot be used. Generally the maximum temperature that

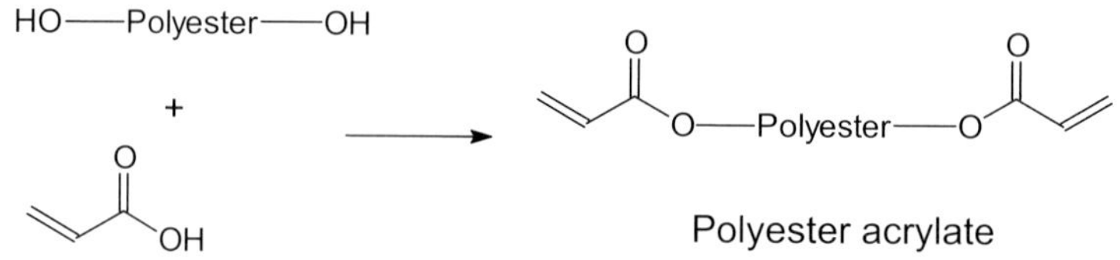

Scheme 16.29. Synthesis of polyester acrylates.

can be applied in these reactions is 130°C. In order to remove the water which is formed in this condensation reaction in an efficient manner at these low temperatures, a solvent is selected which forms an azeotropic mixture with water. In most cases toluene is the preferred solvent.

After the preparation of the polyester acrylates, the mixture is washed with water and/or dilute base in order to remove the catalyst and excess acrylic acid. Next the toluene is evaporated and the resin is obtained.

Silicone acrylates also are usually prepared via an esterification reaction between a silicone diol and (meth)acrylic acid.

16.8.3.3 **Urethane Acrylates**

Urethane acrylates are produced in a batch-type fashion as well. Two procedures can be followed:

- Inside-out (Scheme 16.30), or

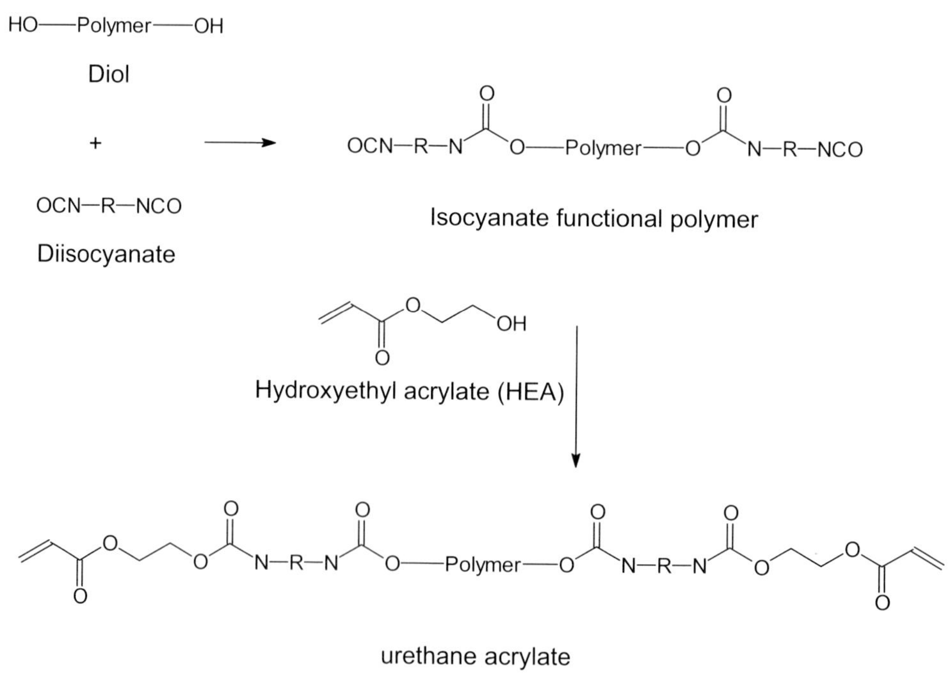

Scheme 16.30. Inside-out synthesis of a urethane acrylate.

• Outside-in (Scheme 16.31).

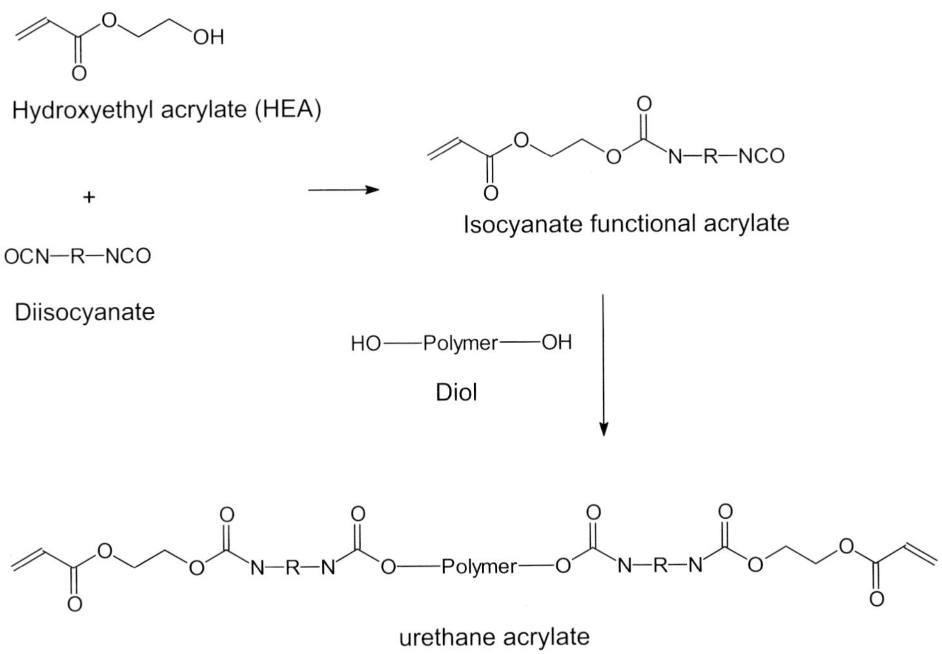

Scheme 16.31. Outside-in synthesis of a urethane acrylate.

Each procedure has its own advantages and disadvantages, which will be briefly discussed after a description of both methodologies.

For both procedures the reaction temperatures normally never exceed 90°C. Many diisocyanates can be used in these reactions. In those cases in which more than just statistical control is needed, isophorone diisocyanate (IPDI) is generally employed which has two different isocyanate groups (primary/secondary). The alcohol–isocyanate reactions are generally catalyzed with Lewis acids such as dibutyltin dilaurate (DBTDL). Tertiary amines can be used as well.

With respect to IPDI, it should be noted that with DBTDL as catalyst the secondary isocyanate is approximately 10–20 times more reactive than the primary isocyanate. With a tertiary amine as catalyst, however, the primary isocyanate is about five times more reactive than the secondary isocyanate.

16.8.3.4 Inside-out

The preparation of a urethane acrylate according to the inside-out technology starts with the polymeric diol. The reactor is charged with the diisocyanate, catalyst, and stabilizer. Now the diol is added slowly and a strong exothermic reaction starts. The rate of isocyanate addition is generally used to control the reaction exotherm in such a way that the temperature does not exceed 45°C. The control of this exotherm is especially important when IPDI is used as diisocyanate. For a higher selectivity, then sometimes even a maximum temperature of 35°C is chosen. After complete addition of the isocyanate, the intermediate isocyanate-functional polymer can be isolated, although normally the subsequent reaction step is performed in the same reactor. Next 2-hydroxyethyl acrylate (HEA) is added at such a rate that the temperature rises to 60–70°C. After addition of the HEA the reaction is normally stirred at 80°C until all the isocyanate has been converted.

16.8.3.5 **Outside-in**

The basic reactions are similar to the Inside-out methodology. However, the reaction sequence starts with reaction of HEA (the outside of the polymer) with the diisocyanate instead of the polyol (the inside). The reaction vessel is charged with diisocyanate, catalyst, and stabilizer to which the HEA is added, resulting in an acrylate-functional isocyanate. Next the polyol is added, and a urethane acrylate can be obtained.

16.8.3.6 **Comparing Inside-out with Outside-in**

The main advantages and disadvantages are summarized below.

The Inside-out advantages are: higher reactor filling; and the reactive group is introduced in the last reaction step, thereby reducing the risk of gelation. The disadvantages are: high viscosities in the reactor from the start so that temperature control becomes difficult; and oligomerization of diols via reaction of two diols with a diisocyanate can occur, resulting in higher viscosities and a broader M_w distribution.

The Outside-in advantages are: reduced viscosity, and therefore improved cooling and stirring; and consequently higher addition rates (shorter batch times) and a higher selectivity. The main disadvantage is the higher risk of gelation as the reactive acrylate is in the reactor from the start.

For many urethane acrylates both methods can be used without any problems, and in practice they are both used. Some disadvantages of the Inside out technology related to the viscosity can be overcome by performing the reaction in a reactive diluent. However, in that case, there is a higher risk of gelation.

In special cases only one method is available, namely in those cases in which the higher selectivity of the Outside-in method is required. For instance, in the case of a triol the Inside-out method will result in a gel, whereas such a three-functional urethane acrylate can be made by the Outside-in method. An extreme example is that the Outside-in technology makes it possible to use hydroxy-functional acrylics (polymers with at least ten OH groups per chain) as the polymers in the preparation of urethane acrylates.

16.8.3.7 **Stabilization**

All the acrylate resins suffer a severe risk of homopolymerization. One can even safely state that without correct stabilization these resins cannot be made. The most often applied stabilization package consists of a phenolic stabilizer such as 2,6-di-*t*-butyl-4-methylphenol in combination with air; it is especially the oxygen in air that is required. Oxygen is a very efficient acrylate polymerization inhibitor and this is one aspect in which the preparation of acrylate resins differs from most other resin synthesis.

Many resin syntheses are performed under nitrogen (no discoloration, and so on) while here, without oxygen, the synthesis would almost be impossible. A direct consequence is that the evaporation of the toluene in the production of polyester acrylates is the critical step due to the reduced amount of oxygen and not the synthesis. For those cases an anaerobic inhibitor like phenothiazine can be used.

16.8.3.8 Dilution

The viscosity of the acrylate resins is generally not so high that dilution is required for the synthesis or for discharging the reactor at ambient temperatures. Consequently most acrylate resins can be obtained without reactive diluent, leading to a larger formulating freedom for the end formulator.

The selection of the reactive diluent is generally based on factors such as dilution power, reactivity, tensile properties, surface tension (wettability), shrinkage, volatility, odor, color, and stability.

16.8.3.9 Safety

Almost all acrylates and many methacrylates are sensitizers. This means that an allergic reaction can occur due to improper handling; over the years many examples have been reported where workers have become allergic due to repeated skin contact. In order to minimize skin contact, gloves should be worn at all times and all (meth)acrylates should be treated carefully, unless the producer has proven by testing that a specific (meth)acrylate is not a sensitizer.

16.8.4 Properties

The properties of the cured resin can be varied to a large extent. However, as these resins are regarded as expensive compared to the unsaturated polyester resins, they are generally used in areas in which a higher demand or control on the properties is required. An example of such an application is the chemical anchoring in which methacrylate resins are used to attach steel for balconies, bridges and suchlike to a concrete wall, floor or even ceiling.

In tuning the properties the reactive diluent plays an important role as well. For instance, when the side chain of the acrylate diluent is changed from linear aliphatic to cycloaliphatic, the hardness of the network will increase without affecting the flexibility of the network. The hardness increases as well when the overall functionality of the diluents is increased; so going from a monofunctional to a difunctional diluent results in a large increase in hardness. However, in this case the network becomes harder and more brittle.

When cure speed (or polymerization time) is irrelevant, the choice between an acrylate and a methacrylate gives another parameter for adjusting the hardness. In general the hardness of a polymethacrylate is 100°C higher than that of the corresponding polyacrylate.

The polarity of the network also can be varied by manipulation of the side chains of the reactive diluent. Both apolar rubbers and polar hydrogels can be obtained through the correct choice of side chain substituent.

In the above-mentioned examples, it appears that most of the mechanical properties can be achieved by correct choice of side chain. This is, however, not correct. Many of the basic mechanical properties, such as elasticity, elongation at break, and modulus, originate from the polymer in the acrylate resin and not from the side chain functionality.

Generally, epoxy acrylates yield hard, tough materials, while urethane acrylates give elastic networks and polyester acrylates possess properties which lie between urethane and epoxy acrylates.

16.8.5 Introduction to UV Curing

16.8.5.1 General Introduction to UV-initiated Radical Polymerization

A special area of application in which the (meth)acrylate resins described above are used predominantly is UV curing. A UV curing resin formulation comprises a photoinitiator, that is, an organic compound that generates radicals upon UV irradiation [12]. The basic reaction scheme is depicted in Scheme 16.32.

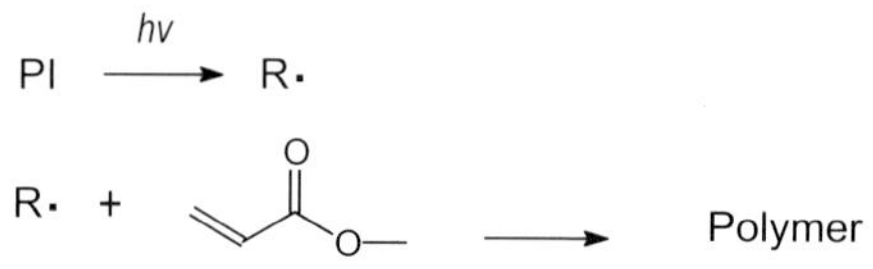

Scheme 16.32. A UV-initiated radical polymerization.

This cure on demand, combined with flexibility with respect to the properties of the network, makes this system the most efficient process for the rapid production of polymeric crosslinked materials with well-defined properties. This very efficiency is the reason why photoinitiated radical polymerization is widely employed in high-performance applications where emphasis is put on the mechanical as well as the optical properties. For instance, modern dental restorative fillers are UV-curable materials. The aspherical lenses in CD applications are prepared via a UV curing process. Contact lenses are nowadays prepared in the same way. The Internet could not exist without UV curing, as the optical fibres used for data transport are protected by two UV-curable coatings. In stereolithography, three-dimensional objects are prepared via UV curing, which makes this type of curing a valuable aid in modern medicine. UV curing is not limited to apparently high-tech applications, however. Most common glossy magazines have a UV-cured coating.

It should be noted that, due to the time control of curing, these resin systems are ideally suited for fundamental kinetic investigations, especially combined with analytical techniques like real-time infrared spectroscopy. These investigations have led to the conclusion that the radical polymerization is dominated by a reactive diffusion-controlled process and that bimolecular termination is the main termination mode.

A standard UV-curable acrylate resin formulation consists of at least three ingredients: a photoinitiator, an acrylate resin, and a reactive diluent.

The UV curing process is strongly dependent on the photoinitiator. In fact, by the choice of photoinitiator the chemistry can be altered completely. There are fundamentally different photoinitiators that can be used for three different chemistries:

- radical curing, employing a radical photoinitiator;
- cationic curing, employing a cationic photoinitiator; and
- base-mediated curing, employing a photolabile base.

Around 90–95% of all UV curing processes are radical initiated polymerizations. Consequently, radical UV curing will be discussed in more detail below, while cationic and base-mediated curing will be discussed briefly in separate sections.

16.8.5.2 **Photoinitiators for Radical Polymerization**

The photoinitiator makes the resin formulation UV-curable. Only the development of thermally stable but photolabile compounds enabled the development of UV curing. For an efficient UV curing, the absorption of the photoinitiator should match, at least partly, the emission spectrum of the light source used. Two main types of photoinitiators are used: photoinitiators based on an α-cleavage process (Norrish type I), and photoinitiators based on an electron transfer followed by a hydrogen abstraction process (Norrish type II). α-Cleavage photoinitiators are monomolecular initiators, of which α-hydroxy or α-alkoxy ketones and benzoyl phosphine oxides are well known examples shown in Scheme 16.33.

Norrish type II photoinitiators are bimolecular initiators. Generally an aromatic ketone is used in combination with a tertiary amine. Both aliphatic and aromatic tertiary amines can be used. A well-known example of such an initiating system is benzophenone with dimethylaminoethanol.

Irrespective of the initiating system, type I or type II, it is clear that at least two different radicals are formed. Initiation proceeds in the case of type I via both radicals although at different rates. However, in the case of a type II initiating system only the amine radical initiates and the ketyl radical can be considered inert with respect to initiation. Consequently type I initiations are preferred when speed is an issue. The advantage of a type II system is that it is less sensitive to oxygen inhibition than a type I, as the ketyl radical can react with oxygen thereby reducing the active amount of oxygen.

16.8.5.3 **Resin**

All the acrylate resins described above in Section 16.8.3 can be employed in UV curing. Low molecular weight materials are mostly used since they possess a lower viscosity. In many UV applications thin layers are employed, which makes the resin viscosity an important processing parameter.

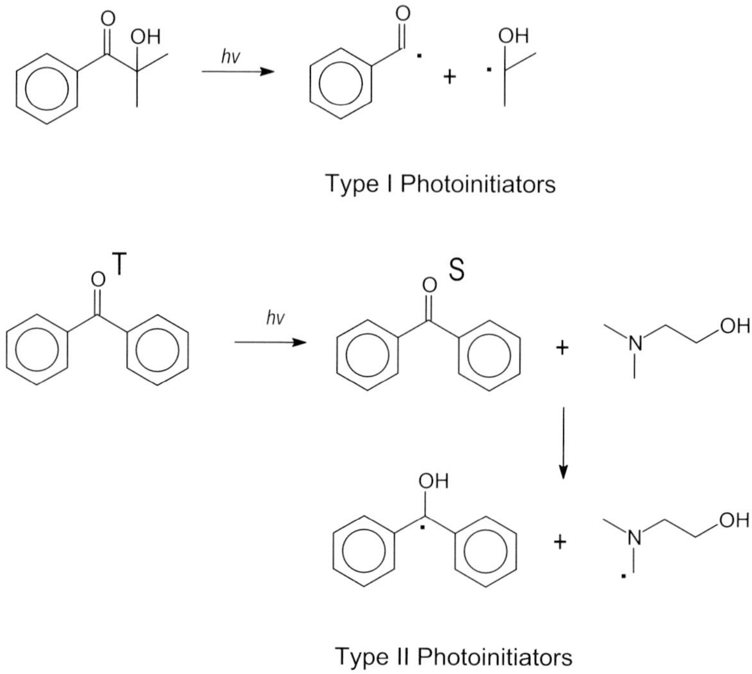

Scheme 16.33. Different types of radical photoinitiators. S, singlet state; T, triplet state.

However, the urethane acrylates, which generally possess higher viscosities, also have higher polymerization rates.

Unsaturated polyesters (see Section 16.7) can be used in UV curing as well, but only in combination with a special reactive diluent system, which is based on vinyl ethers. Kinetic analysis has revealed that in that specific case the fumarate forms a charge transfer complex with the vinyl ether and a homopolymerization of this charge-transfer complex takes place as depicted in Scheme 16.34. An advantage of a charge-transfer complex radical polymerization is that it is less susceptible toward oxygen inhibition than methacrylate systems. This is illustrated by the fact that the first commercial UV powder system, in which the powder is sprayed (much oxygen), melted, and then UV-cured, was based on a fumarate–vinyl ether system.

16.8.5.4 Reactive Diluent

In liquid UV curing systems there is even more emphasis on the judicious choice of the reactive diluents, due to the application's viscosity requirements. The reactive diluent also has a large impact on the cure speed. An example concerns polarity: apolar reactive diluents polymerize more slowly than more polar analogs. Hydrogen bonds play a role as well, which is illustrated by the fact that monomers which possess hydrogen bonds react faster than their analogs without hydrogen bonding. However, monomers with hydrogen bonds generally possess a higher

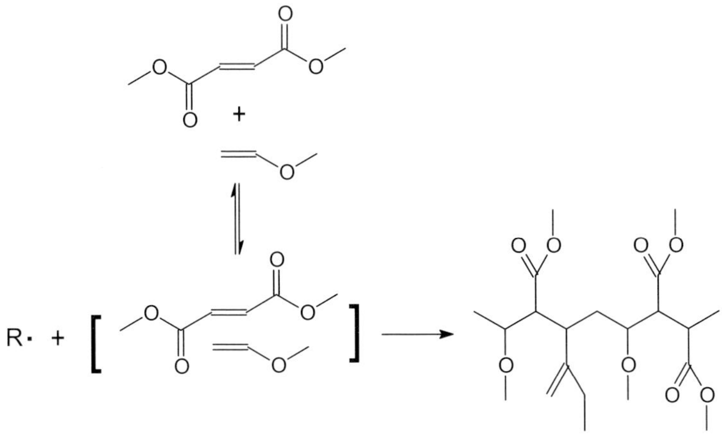

Scheme 16.34. Copolymerization of fumarate with vinyl ether via a charge-transfer complex.

viscosity. The functionality is another important factor, as higher-functional materials polymerize faster than lower-functional materials. The hardness or T_g of the resulting polymer is also important, as reactive diluents which give rise to high-T_g polymers polymerize faster than monomers which result in low-T_g materials (assuming the same functionality). Obviously the basic functionality is important; methacrylates react approximately ten times more slowly than acrylates.

N-Vinyl amides such as *N*-vinyl caprolactam are sometimes used as reactive diluents as well, especially in those cases in which a high demand is put on the polymerization speed. An example of this is the fiber optic industry, in which the cure process is performed at draw speeds around 30 m s^{-1} (irradiation times are in the order of milliseconds).

When an unsaturated polyester is used as resin, only reactive diluents containing vinyl ether can be applied. Moreover, the molar ratio of fumarate bonds to vinyl ether bonds should be around 1:1 because the crosslinking reaction is a strictly alternating copolymerization.

16.8.5.5 **Curing Process**

The cure process of a sheet covered with a UV-curable resin is depicted schematically in Figure 16.11. The basic process is very simple: a sheet is put on the conveyor belt and the resin is cured under the UV lamp. After irradiation the resin is

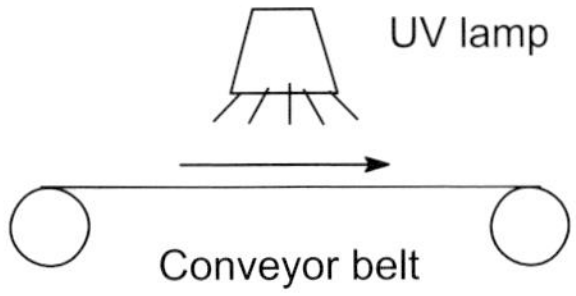

Fig. 16.11. Representation of a basic UV curing process.

fully cured. When very thin layers or high speeds are employed, curing in air becomes more difficult due to the inhibiting effect of oxygen. In these cases nitrogen inerting, that is, flushing the area under the UV lamp with nitrogen, can be used to overcome this problem. Other gases such as carbon dioxide can be used as well.

A main disadvantage of acrylate polymerization is shrinkage. However, this shrinkage process is well defined since it is directly related to the number of double bonds converted. In the lens applications the quartz molds are adjusted for this shrinkage.

16.8.5.6 Cationic Curing

Besides photoinitiators which generate radicals, photoinitiators also exist which generate a strong acid. These initiators enable the on-demand curing of epoxide and hydroxy/epoxide mixtures. The initiators are sulfonium or iodonium based. A cationic curing of an epoxy/alcohol system is depicted in Scheme 16.35.

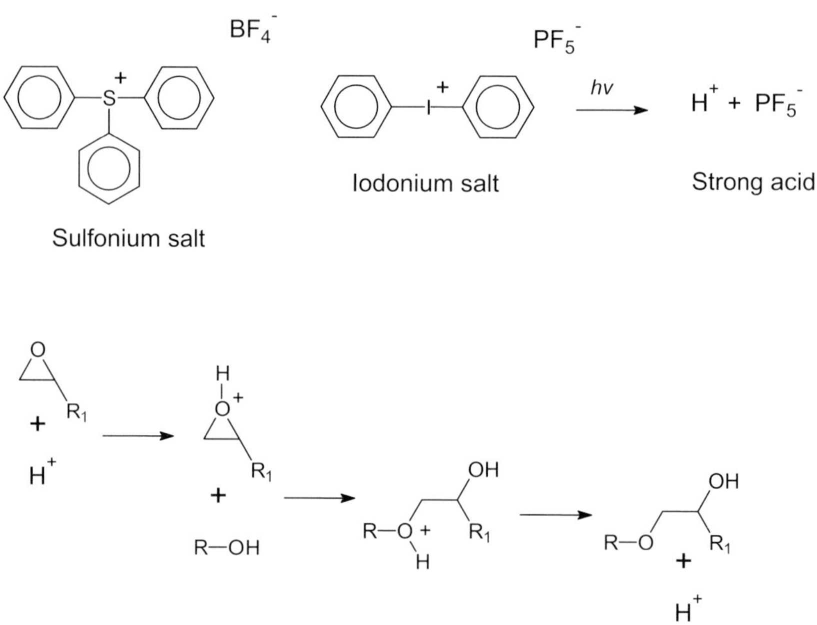

Scheme 16.35. Cationic photoinitiators and cationically photoinitiated polymerization of epoxide resins.

The acid strength of the resulting acid is strongly dependent on the counter ion. The acid strength of the system can be directly related to the polymerization rate (stronger acid results in faster polymerization). Consequently water can reduce or even inhibit these polymerizations. However, minimal amounts of alcohols or even water can enhance the cure speed. The highest rates are generally obtained in those cases in which the molar ratio of alcohol to epoxide is 1:5–1:10.

Besides an epoxide system, vinyl ether systems also exist. The cationic vinyl ether polymerization is among the fastest polymerizations known in the UV curing industry, especially since termination is almost absent.

16.8.5.7 Base-mediated Curing

Base-mediated UV curing is a recent development, since UV-labile bases have only recently become commercially available. Depending on the base liberated and its base strength, various kinds of organic chemistry can be used for the network formation.

Weak UV-labile bases are commonly used in lithographic processes, in which their main function is as an acid scavenger. In these processes the photochemically liberated base scavenges a thermally formed acid and no polymerization will occur on the irradiated areas.

16.9 Rubber

16.9.1 Introduction

In this final section a class of thermoset materials will be introduced, elastomers or rubbers, which has some distinct differences from the thermoset resins presented in the preceding sections [13]. First, most elastomers are characterized by very flexible polymer chains and, as a result, the glass transition temperature of elastomers, indicating the transition between a hard, rigid, glassy phase and a soft, mobile, rubbery phase, is well below room temperature. Crosslinked elastomers are relatively soft materials in their actual application (typically with a modulus of 1–10 MPa; that is, in the Sh A (Shore A) hardness range), whereas most crosslinked thermoset resins are hard materials (1 000–10 000 MPa). Secondly, elastomers are high molecular weight polymers (typically with a weight-averaged molecular weight M_w ranging from 50 000 to 500 000 g mol^{-1}), resulting in a high melt viscosity. Therefore, the processing temperature is higher than for thermoset resins and the processing equipment is much "heavier". Thirdly, most elastomers have a large number of sites per chain available for crosslinking (typically a functionality of 10–10 000 sites per chain). Consequently, the gel point is already reached at a low crosslinking conversion and any crosslinking must be avoided during mixing, processing, and shaping. Many rubber properties are directly related to the crosslink density; for example, the modulus, the elastic recovery, and the swelling resistance all increase upon increasing the crosslink density. However, these properties are not determined only by the number of chemical crosslinks, but are also affected significantly by the number of physical entanglements which are trapped upon crosslinking.

Despite these differences in chain flexibility, molecular weight, and functionality, elastomers have the formation of a three-dimensional network in the final processing step in common with the thermoset resins such as the phenolic, epoxy, and acrylate resins. Crosslinking of a rubber renders it insoluble and, in addition, yields a material which has strongly enhanced physical properties, such as tensile properties (modulus, tensile strength, and elongation at break) and elastic recovery (under both compression and tension).

16.9.1.1 Types of Rubber

The most important rubber is natural rubber (NR), with an average volume of roughly 50% of the total world rubber production (approximately 15 500 kton y^{-1}) over the decade 1990–2000. NR is a relatively low-cost rubber collected as a latex from *Hevea brasiliensis*, and yields excellent physical and dynamic properties, explaining applications ranging from tires (low heat buildup) to thin-walled, soft products like gloves and balloons. Of the synthetic rubbers polybutadiene (BR) and styrene–butadiene copolymers (SBR) have also found major applications in the tire industry, while the latter is also used in foams and footwear. Acrylonitrile–butadiene copolymers, that is, nitrile rubber (NBR), have a high swelling resistance because of their high polarity, making them suitable for applications with oil and solvent contact. These so-called polydiene rubbers have unsaturation in the polymer backbone, making them relatively sensitive to oxidation, ozonolysis, and thermal degradation. Ethylene–propene–diene terpolymers (EPDM) are much more stable in this respect. Some other rubbers worth mentioning are butyl rubber, which has a low air permeability used in inner-tire tubes, and silicone rubbers and fluor rubbers, which are high-performance rubbers (high heat and solvent resistance).

In summary, the range of commercial rubbers is quite large, with each rubber type being available in a large number of different grades. Because it is beyond the scope of this overview to discuss all these rubbers in detail, we will zoom in on one rubber only. EPDM (Scheme 16.36) was chosen for that purpose for two reasons. First, EPDM has a very wide structural variation in

- molecular weight (distribution),
- long-chain branching (LCB) (distribution),
- chemical composition (including ethylene and propene levels, type and level of diene, but also monomer sequence distribution),

yielding an almost countless number of combinations, many of which have specific application advantages, and making EPDM a good representative for rubbers in general. Secondly, EPDM has a very good balance between processability, properties, and price, and has evolved since the early 1990s from a high-performance rubber into a mature commodity-plus rubber. With its annual world usage of approximately 900 kton y^{-1}, EPDM is today the fourth rubber volume-wise after NR, SBR, and BR, and its usage is actually the largest for a non-tire rubber.

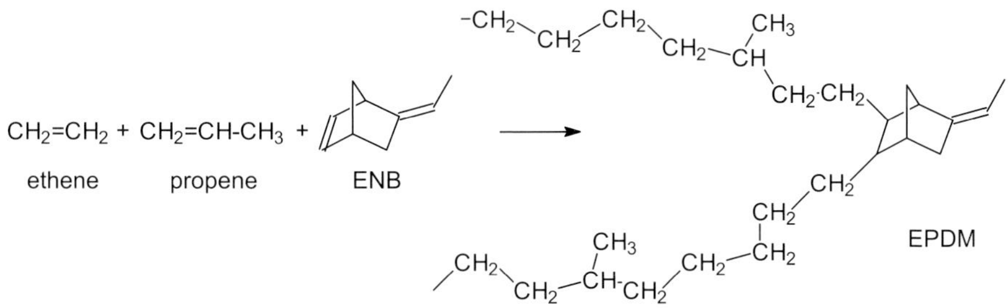

Scheme 16.36. Polymerization and structure of ethylene/propene/diene terpolymer (EPDM), with 5-ethylidene-2-norbornene (ENB) given as an example for the diene.

16.9.2 Polymerization

EP(D)M is produced via insertion polymerization of ethylene, propene, and a diene [14] (Scheme 16.36). Traditionally, Ziegler–Natta-type vanadium-based catalysts are used in combination with metal alkyl cocatalysts and sometimes promoters in order to obtain an efficient, well-controlled polymerization. The EPM copolymer cannot be crosslinked with sulfur due to the absence of unsaturation, and the efficiency for peroxide cure is not that high (Section 16.9.3). This explains why up to 10 wt% dienes are incorporated in the EPM chain; it is to allow sulfur vulcanization and to enhance the peroxide curing efficiency. The diene should have two unsaturated bonds which have quite different affinities for polymerization, in order to prevent gelation during polymerization. 5-Ethylidene-2-norbornene (ENB) and dicyclopentadiene (DCPD) fulfill this requirement, with the endocyclic norbornene unsaturation being the most reactive for polymerization (ring tension!), and upon incorporation they yield EPDM with pendent, unsaturated bonds that are susceptible to sulfur and peroxide crosslinking. 5-Vinylidene-2-norbornene (VNB) is sometimes used because the exocyclic vinyl unsaturation participates to some extent in polymerization, yielding LCB EPDM with enhanced processability. Recently, metallocene catalysts have been developed and are applied commercially, which have a much higher activity than the traditional catalysts and also have enhanced affinity for polymerization of higher α-olefins like 1-butene, 1-hexene and 1-octene. The ethene/α-olefin copolymers thus produced are a new class of materials, namely plastomers, bridging the gap between rigid polyethylene and elastic EP(D)M.

Traditionally, Ziegler–Natta EP(D)M polymerization is performed in a solution process using a volatile hydrocarbon as the solvent. The ethylene and propene monomers are cooled prior to the exothermic polymerization, to avoid heating to temperatures that are so high that the catalyst activity is significantly reduced. All the monomers and the solvent have to be purified: polar moieties especially have to be removed, since they kill the catalyst. After polymerization the unreacted monomers are recycled and the polymer is recovered, a process which includes removal of traces of unreacted monomers and residual solvent, deactivation and washing of the catalysts, and rubber crumb formation. The activity of the metallocene catalysts is so high that catalyst removal is not necessary. Finally, gas-phase technology has been developed, and has recently been applied commercially, by which the polymerization is performed in a fluidized-bed reactor with carbon black as fluidizing agent.

16.9.3
Crosslinking

Most EPDM applications involve some sort of crosslinking [15], about 80% via sulfur vulcanization. Sulfur vulcanizates have a relatively low thermal stability, which explains the slow transition to peroxide cure in critical applications. The tensile and dynamic properties of sulfur-vulcanized EPDM are superior, whereas the elastic recovery of peroxide-cured EPDM is superior.

16.9.3.1 Sulfur Vulcanization

Sulfur vulcanization is the traditional crosslinking method for unsaturated elastomers. It was developed by Charles Goodyear and Thomas Hancock in the 1840s. Since then the sulfur vulcanization recipes have been further developed and optimized and today sulfur is always used in a cocktail with accelerators (mainly sulfur- and nitrogen-containing chemicals) and activators (zinc oxide, stearic acid). EPDM is an apolar rubber with relatively little unsaturation in comparison to the polydiene rubbers, and it has low solubility for the polar accelerators. As a result, the vulcanization mixture may consist of up to ten ingredients, which allows for fine-tuning of the crosslinking rate, the crosslink density, and the type of sulfur crosslink. The last two determine the physical properties of sulfur-vulcanized EPDM. In Scheme 16.37 a generally accepted mechanism for sulfur vulcanization of EPDM is presented. First, sulfur reacts with the accelerators and the activators, yielding a so-called active sulfurating species (its precise structure has still to be elucidated). Next, the crosslink precursor is formed by substitution of one of the allylic hydrogen atoms, yielding an alkenyl sulfide. Note that the unsaturation of EPDM is essential for activating the allylic position, but is not consumed during vulcanization. Next, the crosslink precursor reacts with a second EPDM chain, yielding the initial crosslink with a relatively high number of sulfur atoms in the crosslink, which upon further heating converts into shorter crosslinks. In the case of ENB as the diene, about 50 crosslink structures can be formed, because of the variation in the length of the sulfur bridge (typically one to five sulfur atoms), the three different allylic sites (C3(exo), C3(endo) and C9) and the isomerism at C8/C9 (*Entgegen* versus *Zusammen*). For normal recipes about 25–75% of the ENB units are involved in crosslink formation. Zinc salts act as "catalysts" for all three reaction steps. There is still a debate on the exact nature of the sulfur vulcanization reactions (ionic versus concerted). Oxidation may occur as a side reaction. Side reactions that are frequently observed during sulfur vulcanization of polydiene rubbers, such as cis–trans isomerization, allylic rearrangement, and/or the formation

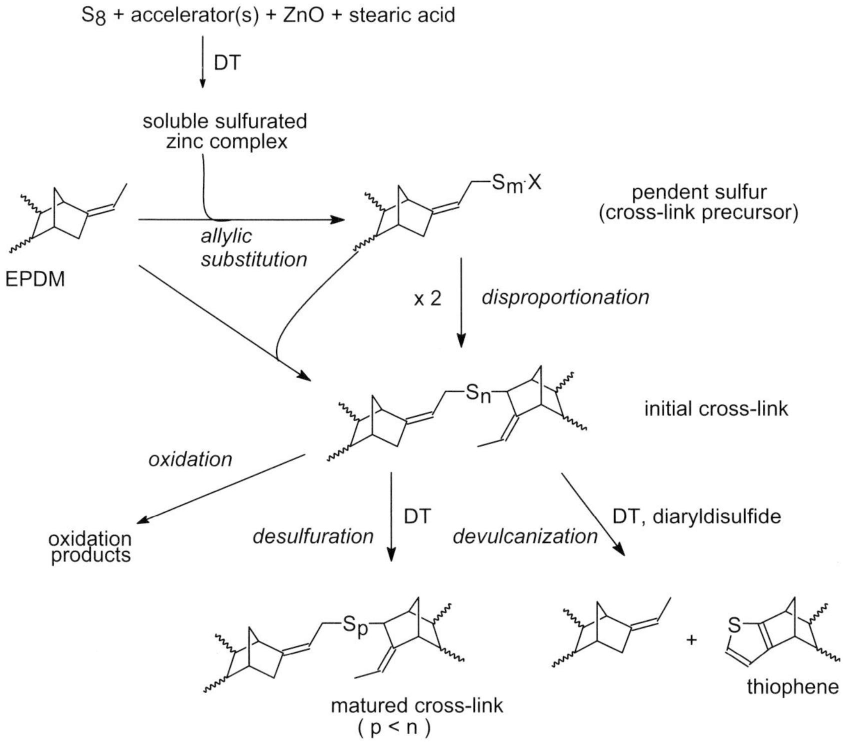

Scheme 16.37. Products and generally accepted mechanism for accelerated sulfur vulcanization of EPDM (X = accelerator residue; sulfur substitution at C3(exo) and C9 of ENB is just given as an example).

of conjugated dienes/trienes, do not occur for ENB–EPDM, because of the stability of the tris-substituted unsaturation in ENB and its isolation from other ENB units.

16.9.3.2 **Peroxide Curing**

The mechanism of peroxide crosslinking is shown in Scheme 16.38. Upon heating, the peroxide decomposes into the primary alkoxy radicals, which may further react to secondary radicals. These radicals abstract hydrogen atoms from the EPM chain. In the case of EPM, crosslinks are formed by combination of two macroradicals, whereas in the case of EPDM crosslinks are formed via combination, but also via addition of the macro-radical to the pendent unsaturation of a second EPDM chain. The formation of C–C bonds explains the higher thermal stability of peroxide-cured EPDM in comparison with sulfur-vulcanized EPDM with its labile S–S crosslinks. The reactivity for peroxide cure increases in the series ENB $\sim$ DCPD $<$ VNB, because of the decreased steric hindrance at the unsaturation. The efficiency of peroxide curing is enhanced by the addition of co-agents, that is, chemicals with two or more unsaturated bonds, which are actually built

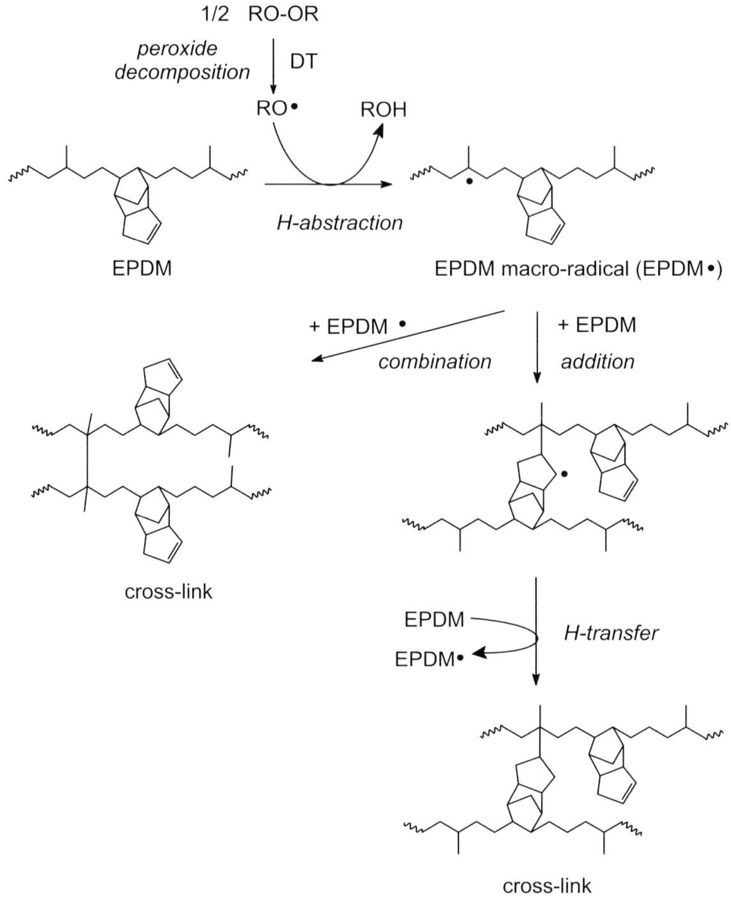

Scheme 16.38. Products and mechanism for peroxide cure of EP(D)M (DCPD is just used as an example; the formation of tertiary free radicals is chosen arbitrarily).

into the elastic network, whereas the peroxide "only" initiates the crosslinking reaction. Note that in contrast to sulfur vulcanization, the EPDM unsaturation is consumed during peroxide cure.

16.9.3.3 Processing

EPDM is first mixed with fillers and oil on two-roll mills or batch kneaders, and next the crosslinking chemicals are added at moderate temperatures. Then the EPDM compound may be shaped and crosslinked in a hot ($> 160°C$) press or calendered into a foil, and cured in a steam autoclave. EPDM strips and profiles may be continuously extruded and subsequently crosslinked in a hot-air oven, a salt bath, or an ultrahigh-frequency radiation unit.

16.9.4
Properties and Applications

In Section 16.9.1 it was explained that EPDM is a very versatile polymer, because of the large number of structural combinations. A broad molecular weight distribution (MWD) at a given viscosity is beneficial for processability, but detrimental for network properties (a large number of dangling ends). LCB has been introduced to enhance processability. The combination of a narrow MWD with LCB yields superior materials. At ethylene levels above 55 wt%, the average length of the ethylene sequences is sufficiently large to give rise to diffuse crystalline regions, which give higher hardness, modulus, and tensile strength to EPDM. At lower ethylene levels, EPDM is a fully amorphous rubber. The type and level of diene determines the crosslinking rate and density (Section 16.9.3). The latter in its turn determines the tensile, elastic, and dynamic properties. It should be noted that EPDM, like most rubbers, is hardly used as such, but always in compounds with fillers and extender oil, because of performance improvement and cost reduction. The addition of large amounts (up to 150 phr) of reinforcing filler, mainly carbon black, results in greatly improved physical properties and UV stability at the cost of increased compound viscosity. To maintain good processability large amounts (up to 150 phr) of extender oil are added in addition to the filler.

16.9.4.1 Advantages and Disadvantages

The main advantage of EPDM over the polydiene rubbers, such as NR, BR, SBR, and NBR, is the saturated character of the main chain. As a result, EPDM has a relatively high resistance against oxidation, ozonolysis, heat, and UV light. This makes EPDM especially suitable for outdoor and high-heat applications, such as roof sheeting, window profiles, automotive seals (for both doors and windows), radiator hoses, cable and wire insulation, and all sorts of technical items like seals and tubes. The main disadvantages of EPDM are also related to the structure of the main chain: actually being a hydrocarbon results in a relatively low oil resistance and poor adhesion to polar substrates (metal, glass, and polar polymers). Most EPDM applications require elasticity and thus crosslinking. However, EPM is also used without crosslinking, as an impact modifier for crystalline thermoplastics such as polypropylene (PP) and polyamides, and as an oil additive.

16.9.4.2 Thermoplastic Vulcanizates

We started this chapter with a description of the fundamental differences between thermoplastic and thermoset materials; we will end it with an example of a subtle blend of their respective properties. A disadvantage of the three-dimensional network of EPDM, but actually of all thermoset materials, is the lack of recyclability. Crosslinked EPDM, both the waste from production and after use, cannot be processed in the melt again like thermoplastics. Reclaiming technologies have been developed for vulcanized rubber, degrading part of the network via high-temperature and shear treatment, but these technologies are less effective for EPDM vulcanizates, probably because the EPDM chains are so stable. A breakthrough in this respect has been the development of thermoplastic vulcanizates (TPVs), which combine the elastic properties of thermoset crosslinked rubbers with the melt processability of thermoplastics. The most important TPVs from a commercial perspective are based on blends of EPDM and PP. Dynamic vulcanization of EPDM/PP blends results in a crosslinked EPDM phase, which is finely dispersed in a PP matrix. PP being the matrix in TPVs makes them melt-processable. At high EPDM loadings the thin PP layers, surrounding the crosslinked EPDM particles, act as a sort of glue and transfer the macroscopic stress to the rubber particles. Resol resins, that is, *p*-alkylphenol–formaldehyde condensates produced at high formaldehyde/phenol ratios and at high pH (Section 16.2.2) are used as the main crosslinking agent activated by acids for the production of TPVs. In addition to being melt-processable, TPVs have other advantages over thermoset crosslinked EPDM, such as higher oil resistance, colorability (no carbon black reinforcement), and co-extrusion with polyolefin thermoplastic parts. The volume of TPVs is growing above the average for the rubber market, and they are partly replacing thermoset EPDM in technical goods and sealing systems.

Notation

M_n	(number-averaged) molecular weight
M_w	weight-averaged molecular weight
T_g	glass transition temperature

Acronyms and Abbreviations

1K	one-component
2K	two-component
BR	polybutadiene = butadiene rubber
BMC	bulk molding compound
DCPD	dicyclopentadiene
DBTDL	dibutyltin dilaurate
DGEBPA	bisphenol-A diglycidyl ether
EAN	elastically active network chain
ECH	epichlorohydrin
ENB	5-ethylidene-2-norbornene
EPDM	Ethene–propene–diene terpolymer
EPM	ethene/propene co-polymer
HEA	2-hydroxyethyl acrylate
HDT	heat deflection temperature
HMTA	hexamethylene tetramine
IPDI	isophorone diisocyanate
LCB	long-chain branching
LSC	low-styrene content
LSE	low-styrene emission

MF	melamine–formaldehyde
MPa	10^6 Pascal
MWD	molecular weight distribution ($= M_w/M_n$)
NBR	acrylonitrile–butadiene copolymer = nitrile rubber
NR	natural rubber
PF	phenol–formaldehyde
phr	parts per hundred
PP	polypropylene
Sh	Shore (hardness degree classes)
SBR	styrene–butadiene copolymer
SMC	sheet molding compound
TGIC	triglycidyl isocyanurate
TPV	thermoplastic vulcanizate
UF	urea–formaldehyde
UP	unsaturated polyester
UV	ultraviolet light
VNB	5-vinylidene-2-norbornene

References

1 a] D. Stoye, W. Freitag, *Resins for Coatings*, Hanser Publishers, Munich, **1996**. b] Z. W. Wicks Jr., F. N. Jones, S. P. Pappas, *Organic Coatings Science and Technology*, Wiley-Interscience, New York, **1992**, Volumes I, II. c] T. Brock, M. Groteklaes, P. Mischke, *European Coatings Handbook*, Vincentz Verlag, Hannover, **2000**. d] W. F. Gum, W. Riese, H. Ulrich, *Reaction Polymers: Polyurethanes, Epoxies, Unsaturated Polyesters, Phenolics, Special Monomers and Additives: Chemistry, Technology and Applications*, Hanser Gardner Publications, Munich, **2000**. e] S. H. Goodman, *Handbook of Thermoset Plastics*, Noyes Publications, Norwich, **1999** (2nd edition).

2 A. Knop, L. A. Pilato, *Phenolic Resins*, Springer Verlag, Berlin, **1990**.

3 a] M. Dunky, P. Niemz, *Holzwerkstoffe und Leime*, Springer Verlag, Berlin, **2002**. b] H. Diem, M. Gunther, "Amino resins", in *Ullmann's Encyclopedia of Industrial Chemistry*, Wiley-VCH, Weinheim, **1999**.

4 a] S. Tohmura, *Journal of Wood Science*, **2001**, *47*, 451–457. b] J. J. H. Nusselder, "Co-condensation of Melamine, urea and formaldehyde", in *Proceedings of the 1998 TAPPI Plastic Laminates Symposium, Atlanta, Georgia*, Tappi Press, Atlanta.

5 a] K. Holmberg, *Progress in Organic Coatings*, **1992**, *20*, 325. b] A. Hofland, *Journal of Coating Technology*, **1995**, *67*(*848*), 113. c] G. Hardeman, J. Beetsma, in *14th International Conference, Coatings Community and Care, Copenhagen*, **1994**, paper 17.

6 *Ullmans Encyklopädie der Technischen Chemie, Vol. 9*, Wiley-VCH, Weinheim, 6th ed. **2004**.

7 T. Misev, *Powder Coatings, Chemistry and Technology*, John Wiley & Sons, New York, **1992**.

8 R. Vieweg, L. Goerden, "Polyester", in *Kunststoff-Handbuch, Vol. VIII*, Carl Hanser Verlag, Munich, **1973**.

9 a] B. T. Åström, *Manufacturing of Polymer Composites*, Chapman & Hall, London, UK, **1997**. b] G. Akovali, *New Handbook of Composite Fabrication*, Rapra Technology, Shrewsburg, UK, **2001**. c] D. Hull, T. W. Clyne,

D. R. Clarke, S. Suresh, I. M. Ward, *An Introduction to Composite Materials*, Cambridge University Press, Cambridge, UK, **1996**.

10 a] A. K. Kaw, *Mechanics of Composite Materials*, CRC Press, Boca Raton, FL, **1997**. b] F. L. Matthews and R. D. Rawlings, *Composite Materials: Engineering and Science*, CRC Press, **1999**.

11 a] Sita Technology Ltd., *Chemistry & Technology of UV and EB Formulations for Coatings, Inks and Paints*, Ed. P. K. T. Oldring, 8 volumes, Wiley, London, **1998**. b] J. P. Fouassier, J. F. Rabek, *Radiation Curing in Polymer Science and Technology*, 4 volumes, Elsevier, London, **1993**.

12 a] J. P. Fouassier, *Photoinitiation, Photopolymerization and Photocuring*, Hanser, Munich, **1995**. b] K. D. Belfield, J. V. Crivello, *Photoinitiated Polymerization*, ACS Symposium Series Vol. 847, ACS, Washington DC, **2003**.

13 W. Hofmann, *Rubber Technology Handbook*, Hanser Publishers, Munich, **1989**.

14 a] J. W. M. Noordermeer, "Ethylene-propylene polymers", in *Kirk-Othmer Encyclopedia of Chemical Technology*, Wiley InterScience, New York, 5th ed. **2004**. b] F. P. Baldwin, G. Ver Strate, "Polyolefin elastomers based on ethylene and propylene", *Rubber Chemistry and Technology* **1972**, *45*, 709.

15 M. van Duin, "Chemistry of EPDM crosslinking", *Kautschuk und Gummi Kunststoffe*, **2002**, *55*, 150.

17 Fibers[1]

J. A. Juijn

17.1 Introduction

17.1.1 A Fiber World

Look around and see the fiber world! Clothes in a wide variety of shapes and colors, carpets and curtains in your house, sheets and blankets on your bed. Step in your car and see the safety belts, and hopefully never see the airbags. Rely on the strong cords reinforcing your tires, timing belt, V-belts. Enjoy many fiber-reinforced composites in your leisure activities: tennis rackets, golf clubs, skis, the frame of your racing bike. Go sailing and use those novel lightweight sails and strong ropes. Or maybe you're a military man and wear a helmet and bulletproof vest based on new advanced fibers.

Fiber production is a large-volume business: around 60 million metric tons per year. About 40% of this is natural fiber: cotton and wool; the remaining part is "man-made", synthetic fiber: polyester, polyamide, cellulose, acrylics, and so forth. More than 90% is applied in textiles or carpets. Only about 5% is used in industrial applications. And the new, "advanced", "high-modulus" fibers? Technically and commercially very interesting, but we are talking of no more than about 0.2% of the fiber capacity!

You may become curious and study some yarns with a magnifying glass and discover that single fibers are as thin as 10–30 μm. With a pair of tweezers you unravel the structure and pull out fibers a few centimeters in length or discover that in some cases the fibers have infinite length. You may test yarns by pulling them between your hands – and cut your fingers when it is an aramid yarn!

Gradually, you start realizing that fibers are a completely different class of polymer materials: different in shape, molecular structure, and physical and mechanical properties.

1) The symbols used in this chapter are listed at the end of the text, under "Notation".

Handbook of Polymer Reaction Engineering. Edited by T. Meyer, J. Keurentjes

ISBN: 3-527-31014-2

17.1.2
Scope of this Chapter

This chapter has been written for polymer engineers, not for fiber specialists. A comprehensive impression of fiber technology can be obtained from relevant sections in an encyclopedia [1] or more specific textbooks [2–5], but this would involve reading hundreds of pages of fairly detailed text. This chapter had to be much shorter but should nevertheless give a broad survey of the fiber field. Therefore, clear choices had to be made. Natural fibers will almost completely be neglected. Man-made fibers are still in part based on natural polymer (cellulose), but fully synthetic fibers have become more important. The choice was easy: we will pay more attention to fully synthetic fibers. Many of those are spun from solution but melt-spun fibers are growing more rapidly, especially polyester. Therefore, more attention will be given here to melt spinning than to "dry" and "wet" spinning. The advantage of this approach is that most polymer engineers are already familiar with melt processing and will more easily recognize the details of a melt spinning process.

The novel high-modulus, high-strength fibers, especially aramid and gel-spun polyethylene, deserve a special status. These yarns are spun from solution again, and we will discuss the specific modifications of the old wet spinning process that were required to produce "yarns stronger than steel".

This approach has led to the following arrangement of this chapter. It starts in Section 17.2 with an introduction to fiber terminology, because this will be new for most polymer engineers. In Section 17.3 we see which properties are required to make a polymer suitable as a fiber material, and how simple the selection of a spinning process seems to be. Melt spinning is discussed in some detail in Section 17.4, with attention to typical machine parts, rheology, and orientation during spinning and drawing. Examples of simple process calculations are given. The section includes the large melt-spun fiber materials polyester, polyamide, and polypropylene. Section 17.5, which is much shorter, and inevitably more superficial, contains accounts of cellulose (rayon), cellulose acetate, acrylics, and poly(vinyl alcohol), the large dry- or wet-spun fibers. Then in Section 17.6 we discuss the typical differences in process limitations between melt spinning and spinning from solution. The chapter concludes with Section 17.7 on high-modulus, high-strength fibers: aramid and gel-spun polyethylene, but including carbon fibers.

Finally, for those who have read this chapter as an appetizer, a full menu is provided in the References.

17.2
Fiber Terminology

17.2.1
Definitions: Fibers, Filaments, Spinning

Until the end of the 19th century all yarns were based on natural fibers, such as cotton and wool; all these fibers are thin (10–100 μm) and short (2–30 cm).

For continuity in a yarn the fibers must be twisted. This is the process we call "spinning" – producing yarn from fibers, for example on a spinning wheel – but industrial equipment has been developed for spinning on a larger scale. More than 50% of the world's yarn production is based on fibers. Cotton is still fiber material number one, being used in pure cotton yarns but also in blends with synthetic fibers.

One natural fiber has a much greater length: silk. About 1000 m of fiber quality can be unwound from one cocoon, and such filaments are combined into a yarn. This introduces the word "filament" – a fiber of (almost) infinite length. And we have met the first spinning machine, the silkworm. Please note that "spinning" has a second meaning here: the production of continuous filaments; all synthetic fibers are spun as continuous filaments.

A yarn spun from fibers is called a "fiber yarn" (Figure 17.1a), whereas a yarn containing endless filaments is called a "filament yarn" (Figure 17.1b). It seems logical that all synthetic yarns would be filament yarns, but this is not the case. Filaments are often cut into (short) staple fibers at the end of their spinning process, and then spun again into a fiber yarn, either as a blend with cotton or wool, or in a fully synthetic product. The reason is that the fibrous character of a yarn is preferred in many textile applications, and that it is difficult to imitate these tactile properties by treatment of filament yarns.

Finally, we must introduce the term "cord" – a construction of two or more twisted yarn ends (Figure 17.1c). For example, staple fibers are first twisted into a thin yarn and two or three yarn ends are then twisted – in the opposite direction – to form a corded yarn. This is a common construction for textile knitware or fabrics, and carpet yarns. Cords are also used in industrial applications. Filament yarns are then twisted to high levels (several hundred turns per meter) and twisted yarns are combined in a second twisting step – again in the opposite direction – to form twofold or threefold cords. A well-known example is tire cord.

Strictly speaking, one should distinguish between "fiber" and "filament": a fiber has a limited length and a filament is essentially endless. To avoid misunderstanding, synthetic fibers with a limited length are in most cases called "staple fibers". It

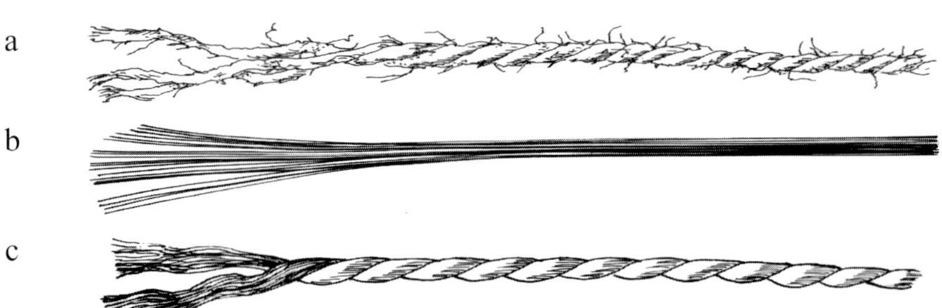

Fig. 17.1. (a) Fiber yarn (the construction shown is in fact a loosely twisted cord); (b) filament yarn; (c) industrial cord.

is common practice, however, to use "fiber" also as a general term for fiber, filament, yarn, and cord materials.

17.2.2
Synthetic Yarns

Synthetic cellulose yarns were developed between 1880 and 1910, first from a nitrocellulose solution, and later as "copper rayon" and "viscose rayon". The cellulosics are often called half-synthetic because the raw material is a natural polymer. The most important fully synthetic yarns were developed between 1935 and 1942 – polyamides (PA66, PA6), polyester (PET), and acrylic yarns (PAN copolymers). Another half-century later, many high-performance fibers were introduced, for example aramid (PPTA), gel-spun polyethylene, and carbon fiber.

All synthetic yarn processes are continuous, so endless filaments are formed. For filament yarns, the spinneret contains as many holes as the number of filaments required for that particular type of yarn. For synthetic staple fiber production, the spinneret contains thousands of holes and the filaments are cut into fibers at the end of the process.

17.2.3
Titer: Tex and Denier

It is fairly difficult and inaccurate to measure the diameter of fibers or filaments. It is much easier to weigh a certain length of filament or yarn. We thus determine a linear density, or titer. The tex unit is officially included in the SI system, and has become the standard in Europe. The older unit denier stems from the silk industry and is still common in the USA and in many Asian countries. The definitions are:

tex = grams per 1000 meter (filament or yarn).

denier = grams per 9000 meter (filament or yarn).

It is fairly common practice to use decitex (dtex) instead of tex (dtex = grams per 10,000 meter). Numbers in dtex and denier differ by about 10%; for accurate relationships, see Table 17.1.

The diameter of a filament can be calculated from the filament titer if the density

Tab. 17.1. Conversion factors for denier, tex, and dtex.

Multiply ↓ to obtain ⟶	*denier*	*tex*	*dtex*
denier	1	1/9	10/9
tex	9	1	10
dtex	0.9	0.1	1

is known. Using the definition of dtex (g 10 000 m^{-1}) and expressing density in g cm^{-3}, we can calculate the diameter D in μm from Eq. (1).

$$D = 10\sqrt{\frac{4\ \text{dtex}}{\pi\rho}} \tag{1}$$

For example, a polyester filament ($\rho = 1.38$ g cm^{-3}) of 10 dtex has a diameter of 30.4 μm, and a filament of 1 dtex has a diameter of 9.6 μm.

For yarn titers it is useful to specify the number of filaments. For example, an aramid yarn type could be indicated as 1500 denier f 1000, which means that the yarn contains 1000 filaments, each of 1.5 denier. In dtex, this yarn type would be 1670 dtex f 1000, a filament titer of 1.67 dtex.

17.2.4 Tenacity and Modulus: g $denier^{-1}$, N tex^{-1}, or GPa

After having introduced the denier and tex units the next logical step is to express breaking strength (tenacity) and modulus in g $denier^{-1}$ or N tex^{-1} instead of in Pascals (1 Pa = 1 N m^{-2}), MPa, or GPa. Fiber engineers often deviate from using the basic SI units and convert N tex^{-1} to cN tex^{-1} or mN tex^{-1}, or even cN $dtex^{-1}$: 1 N tex^{-1} = 100 cN tex^{-1} = 1000 mN tex^{-1} = 10 cN $dtex^{-1}$.

The unit g $denier^{-1}$ (g den^{-1}, g d^{-1}, gpd) stands for gram-force per denier. In conversion factors to Newtons, the gravity constant therefore appears: 1 gf = $9.806\,650 \times 10^{-3}$ N. In combination with the conversion factor from denier to tex, this results in:

$$1\ \text{g denier}^{-1} = 8.826 * 10^{-2}\ \text{N tex}^{-1}, \quad \text{or} \quad 1\ \text{N tex}^{-1} = 11.33\ \text{g denier}^{-1}$$

For conversion of g $denier^{-1}$ or N tex^{-1} to Pa or GPa one needs to know the density ρ of the yarn. The main conversion factors are given in Table 17.2.

Differences in densities can be large, especially when we compare organic fibers (1–1.5 g cm^{-3}) with glass fibers (2.5 g cm^{-3}) or steel cord (7.8 g cm^{-3}). Therefore, aramid producers may argue that their product is five times stronger than steel (in g $denier^{-1}$ or N tex^{-1}), while steel cord producers can rightfully respond that steel has the same strength (in GPa).

Tab. 17.2. Conversion factors for g $denier^{-1}$, N tex^{-1} and GPa (express ρ in g cm^{-3}).

Multiply ↓ to obtain →	*g denier^{-1}*	*N tex^{-1}*	*GPa*
g $denier^{-1}$	1	8.826×10^{-2}	$8.826 \times 10^{-2}\rho$
N tex^{-1}	11.33	1	ρ
GPa	$11.33/\rho$	$1/\rho$	1

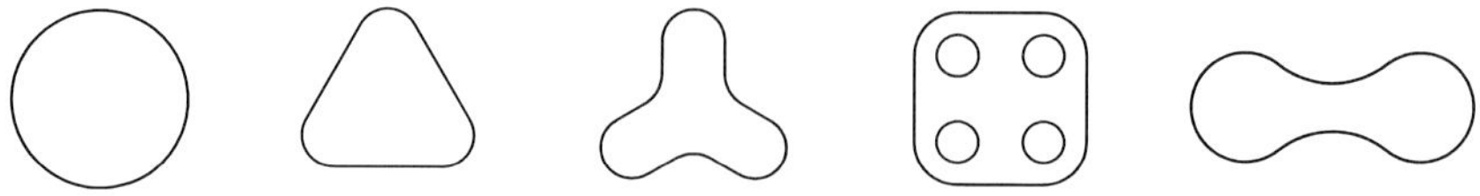

Fig. 17.2. Examples of cross-sections of fibers and filaments.

17.2.5
Yarn Appearance

Most natural fibers have a non-round cross-section and low luster (Figure 17.2); silk is the exception. Moreover, yarns spun from these fibers have a high volume (bulk) and fiber ends make them hairy. All these factors are appreciated for applications in textiles and carpets. Synthetic yarn is most easily spun from round spinning holes, the filament yarns have a high luster, they are "flat" (not textured), and hairiness is absent. This may be perfect for industrial yarn applications, but it is not desirable for most textile and all carpet yarns.

The cross section can be adapted by using profiled spinning holes, for example trilobes, triangles or hexagonal stars. Delustering can be achieved by the addition of fine titanium dioxide or another dulling agent, thus producing semi-dull, dull, or deep-dull yarns, as desired.

To give the yarns a voluminous appearance (bulk), a texturing process is required. For low-titer textile filament yarns this is often a false-twist operation: twisting by torsion, heat setting, and subsequently detwisting, in a continuous operation. For higher yarn titers the yarn may be stuffed into a hot chamber and then pulled out with simultaneous cooling: this is called stuffer box texturing, and is the common process for staple fiber and carpet yarns. An example of a textured carpet yarn is shown in Figure 17.3(a).

Blowing with air can be used as a texturing process, but also to give the yarn coherence. The yarn is led through a narrow channel or slit and air is blown perpendicularly onto it. At low yarn tension, air blowing creates filament loops. These loops give volume to the yarn and air blowing can therefore be regarded as a texturing process. The effect of air impingement on a yarn under tension is an intermingling of the filaments. This is called "interlacing" for improvement of the weavability of textile yarns. For industrial yarns the air impingement may be quite

a
b

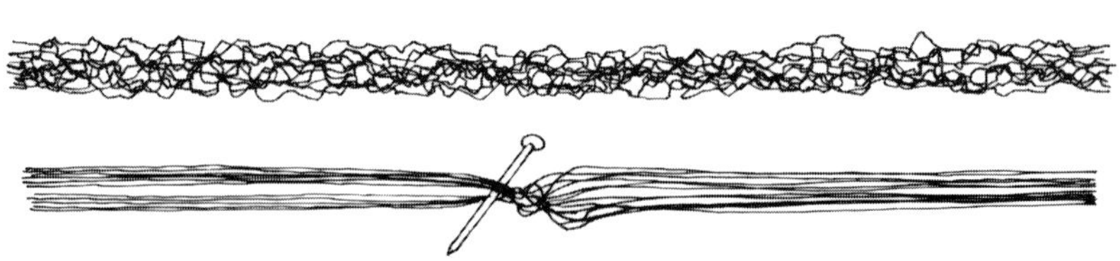

Fig. 17.3. Textured and tangled yarns: (a) textured filament carpet yarn; (b) a knot appears when a needle is moved through a tangled yarn.

vigorous, in a so-called "tangling jet". If a needle is pulled through a tangled yarn a "knot" is formed, typically after a pull-through length of about 10 cm. The effect is shown in Figure 17.3(b). Tangled yarn packages can be easily unwound, and the yarns are weavable; the tangling operation replaces a more expensive twisting after-treatment.

17.2.6
Textile, Carpet, and Industrial Yarns

Further aspects of yarn appearance are the yarn titer and the number of filaments: there may be thin or thick yarn bundles. This will obviously depend on the application.

Textiles and carpets are often based on fiber yarns – cotton, wool, or blends, for example cotton/polyester and wool/polyamide – but 100% synthetic fiber yarns also find wide application. Fiber yarns are hardly ever found in industrial applications.

Textile filament yarns have low tex values and a low number of filaments, although there is a trend toward lower filament titers, as in microfilaments (≤ 1 dtex per filament). Textile yarns with higher titers are found in home furnishing. Textile yarns often have non-round filaments and are often textured. Carpets can be produced from fiber yarns as well as from filament yarns. The filaments or fibers are profiled and very thick, typically 10–30 dtex. Carpet yarns are always textured. Industrial yarns, including high-performance yarns such as aramid and polyethylene, are always synthetic filament yarns. They have high dtex values and a filament thickness depending on the spinning process. Wet-spun yarns usually have thinner filaments than melt-spun yarns (this is also valid for textile yarns). Industrial yarns are flat (without texture) and have round filaments, with only a few exceptions. A survey is given in Table 17.3.

The mechanical properties of textile and industrial yarns differ considerably, even if we neglect the influence of texturing; see Figure 17.4. A textile yarn has a high elongation (20–50%) and a low tenacity (200–400 mN tex^{-1}). Industrial yarns (always flat) have an elongation below 20% and a tenacity of 600–900 mN tex^{-1}. High-modulus, high-strength fibers (HMHS), like aromatic polyamide yarns and

Tab. 17.3. Yarn composition in synthetic filament yarns.

	Yarn titer [dtex]	*Number of filaments*	*Filament titer [dtex]*
Textile yarns			
melt-spun	20–200	3–200	1–5
wet-spun	50–200	50–200	1–2
Carpet yarns	ca. 1500	ca. 50	10–30
Industrial yarns			
melt-spun	500–2000	100–400	3–10
wet-spun	500–2000	250–2000	1–2

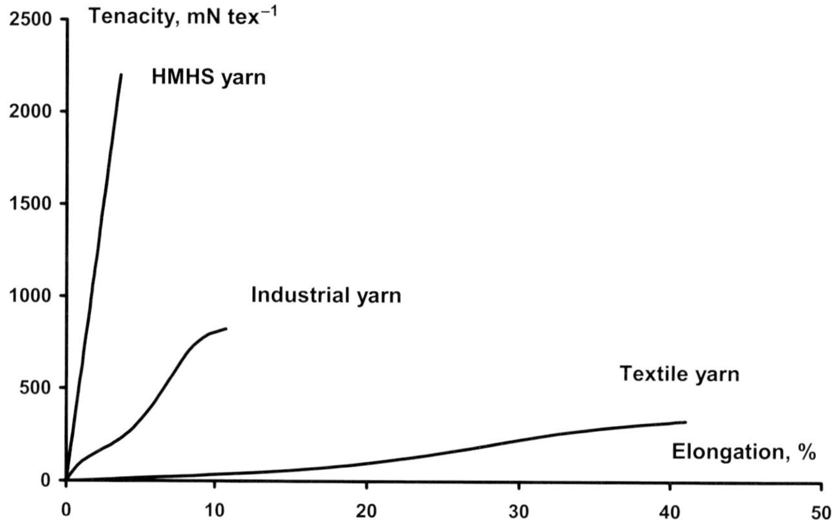

Fig. 17.4. Typical stress–strain curves of textile, industrial and high-modulus, high-strength yarns.

gel-spun polyethylene, have an even lower elongation (2–4%) and a much higher tenacity (2000–4000 mN tex^{-1}). The initial moduli are accordingly low for textile yarns, medium to high for industrial yarns, and very high for HMHS fibers. The stress–strain curve of a typical textile yarn shows an intermediate flat section, indicating that the orientation could have been improved further still. The curves of polyamide or polyester industrial yarns have the remnant of this intermediate section although the yarns have been drawn almost to the limit. The aramid and gel-spun polyethylene processes enable complete orientation. As a result, the stress–strain curves have become almost straight lines.

17.2.7
Physical Structure

Most fibers are semicrystalline. A few of them – aramid and gel-spun polyethylene – approach 100% crystallinity. Examples of the fiber structure of PET and PPTA are shown in Figure 17.5. All fibers have a uniaxial organization: properties in the direction of the axis are completely different from those in the cross-direction. This is an essential difference from other polymer materials.

The crystallization of polymers in a spinning or drawing process differs from the crystallization in quiescent melts, for example in extrusion or injection molding processes. The conditions in fiber production are highly anisotropic, and nucleation and growth rates are orders of magnitude higher. Therefore Avrami equations from the literature and values from DSC studies cannot be used to describe fiber crystallization.

A direct result of the anisotropic crystallization conditions is that the crystalline

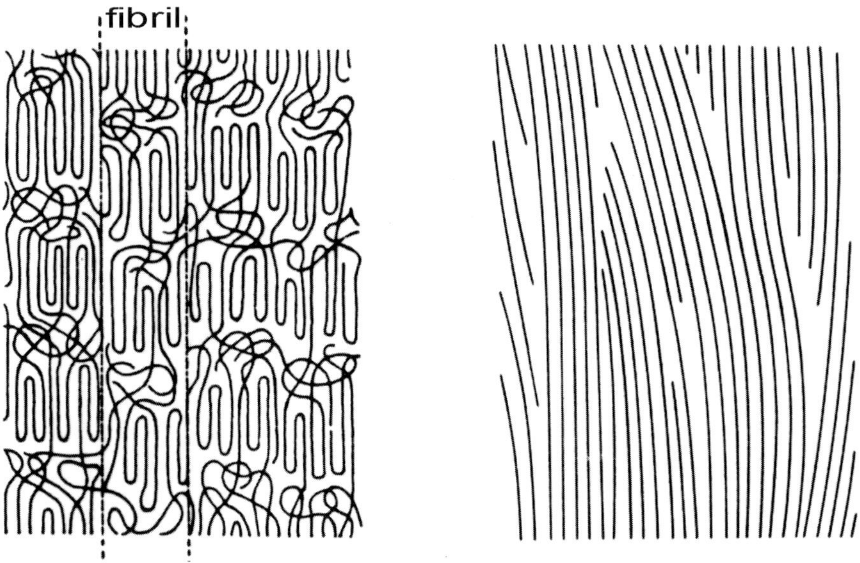

a b

Fig. 17.5. Physical structure of fibers: (a) semicrystalline structure of poly(ethylene terephthalate), PET; the broken lines are the borders of a fibril; (b) paracrystalline structure of poly(*p*-phenylene terephthalamide), PPTA.

part will become almost perfectly oriented in the axis direction. The remaining amorphous regions are more disordered, and will determine the fiber properties to a large extent. Fibers drawn to a high ratio will also have a high orientation in the amorphous regions between the crystallites. It is understandable that amorphous orientation will be moderate for most textile yarns and high for industrial yarns.

The total result of crystallization and orientation is a measurable overall orientation in the fiber direction, uniaxially. Fibers are always birefringent: the index of refraction in the axis direction ($n_{//}$) is larger than the index crosswise ($n_{\perp}$). The difference is the birefringence: $\Delta n = n_{//} - n_{\perp}$, and is a measure of fiber orientation. Orientation factors can be calculated from birefringence measurements ($f = 0$ for random orientation, $f = 1$ for complete orientation), and are often split into crystalline orientation ($f_c \approx 1$) and amorphous orientation ($0 < f_a < 1$).

An understanding of the physical structure is important to explain fiber properties. For example, diffusion of molecules is rapid through large amorphous regions and extremely slow in crystallites. Fibers with low crystallinity and a coarse structure will therefore dye fast and deep, but their chemical stability may be low because small molecules (oxygen, ozone, water, and suchlike) can rapidly diffuse into the filaments.

For industrial fibers the relationship between mechanical properties and physical structure is important. With almost perfectly oriented crystals, the orientation of the amorphous regions largely determines the fiber modulus and its tendency to shrink. A high crystallinity, implying a low amorphous content, will contribute to modulus and will reduce shrinkage. These factors combined give the fiber engineer the possibility of developing yarns with high modulus and yet low shrinkage.

A typical aspect of a fiber structure is fibril formation. Fibrils are an arrangement of many crystallites and amorphous regions in series, and are common in both natural and synthetic fibers (see Figure 17.5). Chains travel through many crystals within a fibril, but rarely transfer to neighboring fibrils. An effect of this structure is the tendency of many synthetic fibers, especially those with a high orientation, toward fibrillation. This axial splitting of fibers can be a disadvantage, but it can be applied advantageously, for example to produce the aramid "pulp" used as a replacement for fine asbestos fibers.

The word "paracrystalline" is used for aramid and gel-spun polyethylene. Amorphous regions are no longer present; rather, the discussion is about defects in the crystal regions. Fiber moduli approach theoretical crystal moduli and shrinkage is virtually absent. Indeed, this is a completely different class of materials.

17.3
Fiber Polymers: Choice of Spinning Process

17.3.1
Polymer Requirements

Fibers must have a reasonable thermal stability. This means that the melting and/or decomposition temperature must be high, preferably above 200°C. This will allow easy ironing of textiles, curing of reinforced rubber at 190°C, coating of fabrics with PVC around 190°C, and so on. All the large-volume fibers are semicrystalline, with crystalline contents of 30% or more. One can argue about the proper value of the crystallinity, which depends on the way it is measured. Sometimes there is discussion of whether a structure is really crystalline: for example, acrylics (PAN copolymers) would rather have a kind of "organized amorphous" or "defective crystalline" character.

Table 17.4 shows the well-known fiber polymers, and a few which seem unsuitable as fiber material, namely polystyrene and poly(vinyl chloride). The latter finds limited fiber application, but a special syndiotactic grade with some crystallinity is then used. Polypropylene has a fairly low melting point but is nevertheless a large fiber product, because the material is cheap and versatile. Polyethylene is even lower melting and is used only as a superstrong fiber at ambient temperature.

17.3.2
Selection of Spinning Process

On paper, the selection of a spinning process seems simple. If the polymer melt is thermally stable a melt-spinning process will be preferred, because this pro-

Tab. 17.4. Properties of fiber polymers.[a]

Polymer	T_g [°C]	T_m or T_d [°C]	Crystallinity [%]
Polyethylene	−100	130–140	>50 >80, gel-spun
Polypropylene	−20	170	>50
Poly(vinyl chloride)	85	200 (d)	<10
Polystyrene	100	275 (d)	0
Polyacrylonitrile	85	ca. 250 (d)	ca. 30
Poly(vinyl alcohol)	70/85	ca. 250 (d)	40–60
Cellulose	230	200–250 (d)	40–50
Polyamide 6	≤50	225	40–50
Polyamide 66	≤50	260	40–50
Poly(ethylene terephthalate)	75	255	30–40
Poly(*p*-phenylene terephthalamide)	–	ca. 400 (d)	ca. 100

[a] T_g = glass transition temperature, T_m = melting temperature, T_d = decomposition temperature.

cess is relatively simple and cheap. If a polymer is not melt-spinnable, a solvent must be found which can be used in a solution-spinning process. If the solvent can be easily evaporated, dry spinning can be applied. In many cases it will be necessary to spin into a bath containing a nonsolvent, which results in coagulation of the polymer solution. This is called wet spinning. In a single case – viscose rayon – a chemical reaction taking place in the spinning bath is a further complication.

The preference for melt spinning is very strong. Attempts have therefore been made to make solution-spun polymers melt-spinnable. For example, small amounts of solvent/plasticizer can make acrylics melt-processable. Cellulose triacetate can be melt-spun if the residence time is held very short. These attempts have not led to commercial processes, however.

If a choice can be made between dry and wet spinning, the trend is to favor wet spinning. This may seem surprising because wet spinning is technically complicated. Organic solvents used in dry spinning are often highly flammable, however, which considerably increases the processing costs.

New, advanced fiber types are often spun from solution. Aramid, for example, has to be spun from solution because the polymer does not melt, but the developed process strongly deviates from standard wet spinning. The new process is called air-gap spinning, or dry-jet wet spinning. In the case of polyethylene the choice was made to spin ultra-high molecular weight polymer from a dilute solution. This sounds economically unattractive, but the resulting fiber is much stronger than a melt-spun product. A new name, gel spinning, was introduced for this process.

A survey of the most important fiber polymers and the spinning processes applied is given in Table 17.5.

Tab. 17.5. The most important polymer fiber processes.

Polymer	*Spinning process*
Polyethylene	gel spinning (dilute solutions of UHMW polymer)
Polypropylene	melt spinning (or fibrillation of melt-extruded film)
Acrylic copolymers	dry spinning (e.g., DMF solvent)
	wet spinning (e.g., NaSCN/water solvent)
Poly(vinyl alcohol)	wet spinning or dry spinning (solvent: water)
Cellulose	wet spinning
	rayon: polymer is derivatized during dissolution in NaOH
	lyocell: direct dissolution in NMMO
Cellulose acetate	dry spinning
	acetate in acetone/water
	triacetate in DCM/methanol
Polyamide 6	melt spinning
Polyamide 66	melt spinning
Poly(ethylene terephthalate)	melt spinning
Poly(*p*-phenylene terephthalamide)	air-gap spinning (solvent H_2SO_4)

17.3.3 Spinnability

What makes a polymer melt- or solution-spinnable? A theoretical prediction is hardly possible [6]. Laboratory tests are in fact small spinning trials. This is the way most fiber producers will test newly developed polymers, or modifications and new grades of known polymers: on a small spinning machine, which may be a "one-hole" laboratory machine. The reason is that in the initial stage of development often only small amounts of material are available.

The first aspect of the test is whether the extrusion through the spinning holes proceeds smoothly, without excessive die swell, and without extrudate distortion. Modest problems may be solved by increasing the temperature in melt spinning or reducing the polymer concentration in solution spinning. In more severe cases one must probably conclude that the molecular weight is too high or the molecular weight distribution too broad. Elastic behavior is more problematic in spinning than in most other polymer shaping processes.

The second aspect of spinnability is whether the filaments will break under their own weight or if the spin-line stress is increased by imposing higher spinning speeds. Filament breaks during spinning indicate too low a melt strength. This problem can only be solved by increasing the molecular weight of the polymer and/or the concentration of the spinning solution.

Obviously, the use of a high molecular weight polymer grade will be preferred. It gives increased melt strength, and will usually give better yarn properties. Problems with die swell and melt fracture can make a compromise necessary, however.

For wet spinning there is an additional factor influencing spinnability. The thread formation process is complicated: it involves diffusion of solvents out of the filament and nonsolvent from the spinning bath into the filament, followed by gelation and/or coagulation. Unstable coagulation can make the spinning filaments weak. A polymer solution may therefore appear unspinnable until the proper composition of the spinning bath has been discovered. With tongue in cheek, one could say that melt spinning is for engineers while wet spinning is a technique for artists.

17.4 Melt Spinning

The most important melt-spun polymers are polyester (PET), polyamide (PA6 and PA66) and polypropylene (PP). A scheme for a melt-spinning machine is given in Figure 17.6. We will follow the process scheme, from extruder to winder.

17.4.1 Extrusion

Some large-scale processes (for polyester and polyamide 66) use "direct spinning", which means that polymerization and spinning are integrated. In most cases poly-

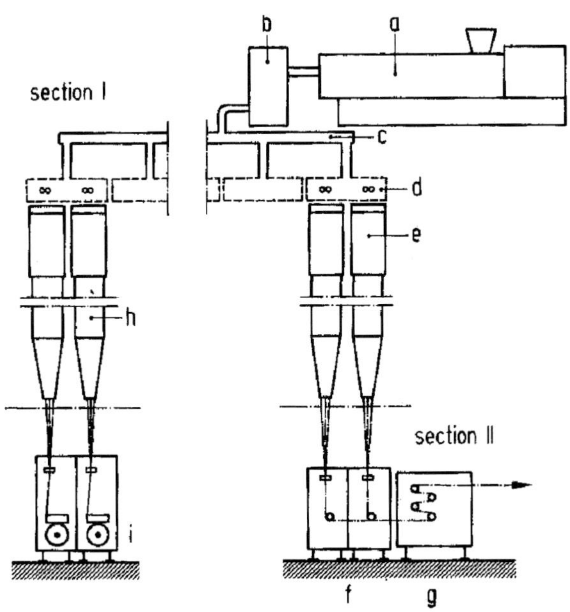

Fig. 17.6. Scheme for a melt spinning machine (reproduced from Ref. 5): a, extruder; b, central filter; c, manifold (polymer lines); d, spin-box with pumps and spinnerets; e, cooling; h, interfloor tube. In section I as-spun yarns are wound separately (i). Section II is a typical arrangement for staple fiber production: f, deflection of bundles; g, combination of bundles, which are laid down in containers.

mer chips are extruded, however. For melt spinning single-screw extruders are used, no different from those applied in other fields of polymer melt processing. Screw designs may vary considerably, and certainly depend on polymer type and grade, but there are no typical melt-spinning screws.

Twin-screw extruders are necessary for blending, compounding, or reactive extrusion, but these operations are not usually applied in melt spinning. Moreover, twin-screw extruders may not easily build up the high pressure necessary to pump the melt through a long distribution system (see Section 17.4.2). And, of course, twin-screw extruders are more expensive than single-screw extruders.

Complete de-aeration of the melt is an absolute requirement. Remnant undissolved gas will give small bubbles in the extruded filaments and cause breaks in the spin-line. Vacuum in the hopper would solve this problem but is technically complicated. Spinning under nitrogen is more common. A proper screw design, especially of the feed zone, and temperature control should ensure sufficient de-aeration.

The size of the extruders for commercial production varies between 90 and 150 mm screw diameter, with outputs of about 100–600 kg h^{-1}. This mass flow should feed many spinning positions. For textile yarns with a low titer, the output per spinneret can be as low as 1 kg h^{-1} and consequently an extruder then should feed at least about 100 spinning positions. For industrial yarns the output per bundle can easily go up to 50 kg h^{-1}, and one would see about ten spinning positions per extruder.

17.4.2
Polymer Lines and Spin-box

Each spinning position should receive polymer with the same time and temperature history because yarn processing and properties depend on molecular weight in particular. Residence times of a few minutes between head extruder and spinning plate are common, and some degradation will always occur. Therefore, polymer line systems (manifolds) have simple splits (into two or three) and the same length between the head of the extruder and all positions. An example is given in Figure 17.7.

It is obvious now that the examples in Section 17.4.1 were inaccurate. The textile machine would have 96 positions (3×2^5) rather than "about 100", and the industrial yarn machine eight positions (2^3).

Many producers insert static mixers in their polymer lines, for example before

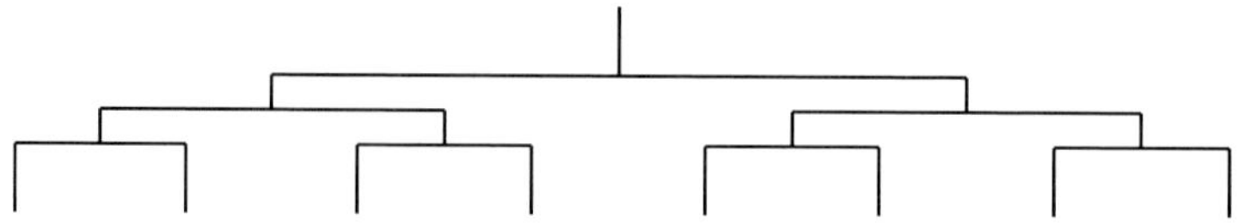

Fig. 17.7. A polymer line system for eight spinning positions.

each split in the system. The disadvantages are an additional pressure drop and fouling of the elements. There are special "redistribution" elements however, guiding the stagnant wall layer to the center of the line again, with only a modest pressure drop.

Fouling has already been mentioned: it is a very critical issue for spinning machines. Small particles of degraded polymer (gels) will cause filament breaks. Dead pockets must be prevented and bends in polymer lines are therefore as smooth as possible. A careful cleaning operation must be performed before a spinning machine is started. Not surprisingly, this is often carried out with high molecular weight polyolefins, which are highly viscous, very elastic, not spinnable, but effective as cleaning compounds. In some occasions (PA66) the distribution system can even be completely dismantled in order to be cleaned by burning in an oven at 500°C or higher.

Polymer lines can be heated electrically, but temperature control is more accurate when a thermal vapor fluid is used. The standard fluid is a mixture of diphenyl and diphenyl oxide, with a boiling point around 260°C. Double-walled lines are therefore almost standard.

The spin-box is a vapor-heated enclosure containing the spinning pumps and the spinning assemblies. Each spinning position has its own metering pump. The pump and spinning assembly are installed close to each other. The spinning assembly is inserted in a hole in the spin-box. Both top loading and bottom loading are possible.

17.4.3 Spinning Pumps

Each spinneret of a spinning machine should be fed with a constant volume flow of polymer melt because otherwise the titer of the resulting yarn bundle would vary. This is done with positive displacement pumps: that is, gear pumps with low tolerances. Their output should be independent of the pressure difference over the pump. A melt-spinning pump must ensure a constant output up to 400 or 500 bar. Most pumps have one driven gearwheel and one co-rotating wheel, but two, three or more co-rotating wheels are also possible. Having a combination of two (or more) spinning pumps on one driving axis is also a fairly common practice. All spinning pumps on a machine run at the same speed, and thus the linear density of all yarn bundles is the same.

The capacity of spinning pumps is indicated by their volume per revolution, usually in cm^3; values of 0.5–50 cm^3 are common. They are operated at 10–50 rpm. For example, for a low yarn titer the required volume flow could be 50 $cm^3\ min^{-1}$, which would be delivered by a 2 cm^3 spinning pump running at 25 rpm; or a heavy industrial yarn titer could require a 20 cm^3 pump at 30–40 rpm. Note that volume flows are adjusted while mass flows are prescribed. A recalculation is easy when the density of the melt is known, but in practice a mass flow check is always made.

Melt-spinning pumps can be constructed from normal steel. Tolerances are not too critical when melt viscosities are sufficiently high. Pumps are cleaned in hot ovens at ≥500°C.

17.4.4 Spinning Assembly

The spinning assembly can be round or rectangular. It contains a filter package on top of a supporting/distribution plate. The spinning plate is the bottom of the spinning assembly.

17.4.4.1 Filtration

For production of fibers, usually having diameters of 10–30 μm, very fine filtration is necessary. The rule of thumb is that particles of more than one-third of the eventual fiber diameter will cause interruptions of the spinning or drawing process and should therefore be removed. The particles that must be removed can be solid (catalyst remnants, dust) or gel-like (not fully melted or crosslinked polymer). Solid particles can more easily be removed than gel particles. Pigments or delustrants may be added to the spinning melt. Note that these must almost be submicronic and should easily pass any filter.

Filtration can be applied in all stages of the process: in a central filter in the polymer melt line, or at the end of the line, in the spinning assembly. Central filters are common in large-scale melt-spinning processes, for example, staple fiber production. Two parallel filters may be used to enable filter cleaning without interruption of the process. Rotating filters with continuous cleaning are also used.

Spinning assemblies always contain filters. In many cases it is the only place where filtration takes place, but there may also be a combination of prefiltration in a central filter and a last, and finest, filtration just above the spinning plate.

Filter packages are usually a stack of screens, going from coarse to fine in the streamline direction, on top of a support plate with holes for further distribution of the flow. Screens are plain-weave, or sometimes twill-weave, steel wire fabrics. Their fineness is indicated in mesh (threads per inch), openings per cm^2, or pore size in microns. For example, 325 mesh = 16 800 opening s $cm^{-2} \approx 40$ μm. 500 mesh (25 μm) filtration is common in melt spinning. For even finer filtration, steel nonwoven materials are used, with a filter fineness going down to 5 μm.

An alternative filtration method is a "sand" filter 1–5 cm thick, on top of supporting screens. Rather than sand, a sharp-edged stainless steel powder would nowadays be used. The alleged advantage of a sand filter is the breakdown of gel particles, which would pass through a relatively thin screen package.

17.4.4.2 Spinning Plate

The bottom of the assembly is the spinning plate: 5–30 mm thick. The number of holes varies enormously. A hosiery yarn may have one or three filaments, a carpet yarn 60 filaments, an industrial yarn 350 filaments. For low filament numbers per yarn bundle, two or more bundles can be spun from one plate. The number of holes can go up to 1000–10 000 for staple fiber production. The size of the spinning plates is chosen accordingly, and the plates can range, for example, from round 60 mm textile plates to rectangular plates 1 m wide for staple fiber production.

The spinning holes are relatively large: 200–500 μm round is common, <200 μm may be used for textile microtiter yarns, and >500 μm for some industrial yarns. The holes are large in comparison with the eventual diameter of the filaments. This implies that a considerable draft (see Section 17.4.13.3) is applied in the spin-line, reducing the filament diameter without however inducing much orientation.

It is impossible to manufacture 200–500 μm holes in a melt-spinning plate, which must be 5–30 mm thick to withstand the high spinning pressure. Moreover, such a long narrow hole would give too high a pressure drop. In practice there is always a "backhole" (pre-channel, counterbore), with a diameter of 2–3 mm. The actual spinning capillary is usually short, with an L/D ratio of only 1–2.

If possible, spinning holes are made by drilling or punching, because these are cheap operations. Very small and non-round holes, however, must be made by spark erosion, making use of an electrode in the form of the hole. The simplest shape of a hole – backhole + spinning capillary (see Figure 17.8a) – is rarely applied. At the least there is some smoothening of the entrance and especially of the transition region to avoid turbulence flow (Figure 17.8b). Holes may have a tapered section or even a trumpet form (Figure 17.8c). The backhole is always round, but the spinning capillary may be profiled. One can imagine how complicated it then is to make a smooth transition zone.

The holes are positioned 5–10 mm from each other. This spacing is necessary for the air-cooling process. Cold air must penetrate easily into the bundle, and some mobility of the filaments should be allowed without filaments immediately sticking as a result.

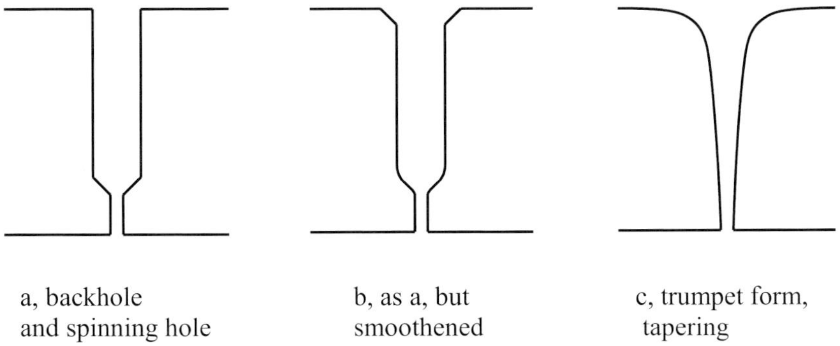

Fig. 17.8. Examples of spinning holes.

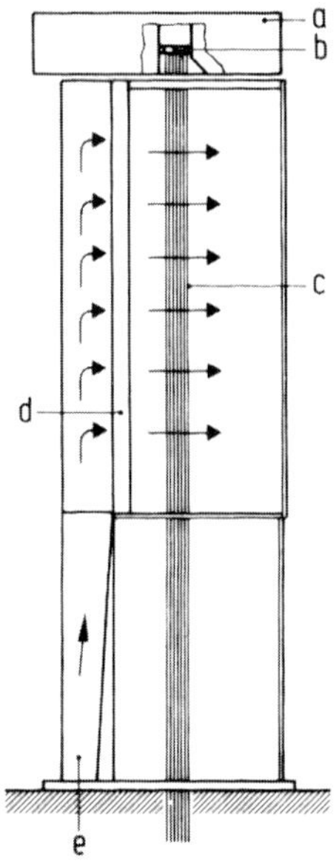

Fig. 17.9. Principle of a crossflow quenching system (reproduced from Ref. 5): a, spin-box; b, spinneret; c, filament bundle; d, filter; e, air supply.

17.4.5 Quenching

The standard quenching or cooling system is crossflow, in which cold air (at 15–70°C) flows from one side, perpendicular to the yarn bundle (see Figure 17.9).

The disadvantage of this system is that filaments close to the blow-box cool faster than those in the front of the yarn bundle (see Figure 17.10a). This can be minimized by adapting the drilling pattern and limiting the number of rows of holes in the cross-direction (Figure 17.10b). An alternative is radial cooling (Figure 17.10c), usually outside to inside, with the disadvantage, however, that a column of hot air will be enclosed in the bundle, again resulting in differences in cooling across the bundle.

Enforced cooling only takes place in the first 0.5–1.5 m, whereas the total height of the spin-line is 1.5–6.0 m. Cooling air speeds are in the order of 0.1–1 m s^{-1}.

A standard spinning machine with a cooling height of 6 m requires at least a four-story building, with chips handling on the fourth floor and extruders, spin-boxes, and cooling on the third floor (the spinning floor). On the second floor one would only see "interfloor tubes", or "chimneys", through which the bundles travel downward and cool further. The first floor is the winding floor.

At high spinning speeds the yarn bundle attracts its own cooling air, by self-suction. With only a few measures to avoid turbulence and ensure symmetric stability, this can be sufficient for textile yarns. For industrial yarns the initial cooling

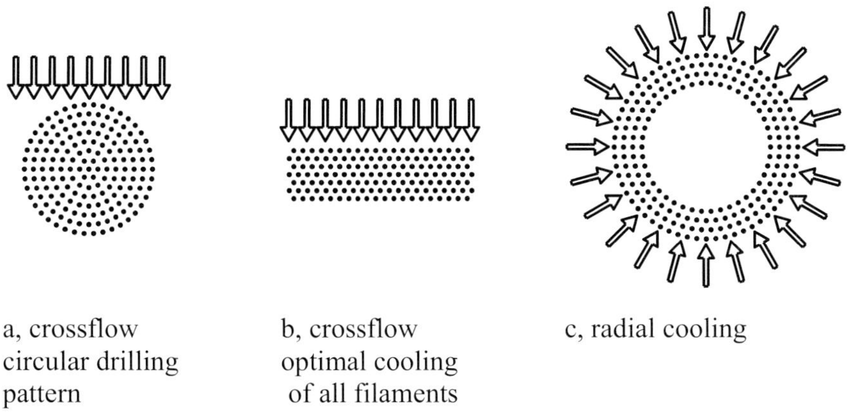

Fig. 17.10. Yarn cooling.

is always enforced, but the remaining necessary cooling air is attracted "automatically" by the bundle.

For highly viscous polyester and polyamide 66 yarns there may be *no* cooling in the first 0.1–0.5 m. On the contrary, an electrically heated ring or box, which keeps the filaments at the spinning temperature, would be positioned around the bundle. The technique is called retarded cooling, and the device is a "quench collar". In this way the spinning plate is kept at an even temperature and the orientation level is reduced as well as the variation in orientation between filaments. This results in applicability of higher draw ratios, and in higher tenacities of the yarns.

The cooling speed depends on the mass flow per hole and the temperature difference between the filament and the air. Of course, large bundles spun at high speeds require much cooling air; this is a simple energy balance calculation. The limiting step in the cooling process is the formation of a sublayer around the filaments. Removing or refreshing this sublayer is the key for a fast and even cooling process.

The filaments in a bundle should not touch each other before they are solid, below the glass transition temperature or below the crystallization temperature. Polypropylene and polyamide 66 always crystallize in the spin-line, polyamide 6 and polyester crystallize in the spin-line only at high speeds.

17.4.6 Finish

A yarn without spin finish cannot be processed; yarn–metal friction is too high, and static charge would build up. A finish is therefore applied before the yarn first touches guides, rolls, and so on. Finishes can have sophisticated compositions, but two components are always present: lubricants and antistatics. In water-based finishes emulsifiers are a third essential component. Finishes are specific for certain yarn types: a polyamide finish cannot be used on polyester or polypropylene; a finish for polyester textile cannot be used on a polyester tire yarn.

Finishes must be adapted to the polarity of the fiber polymer. They must spread over the surface within a very short time. Another criterion is that the finish should not interfere with later applications. For example, components which are necessary in a textile finish could ruin the rubber adhesion of a tire cord.

Knowledge of finishes used to be proprietary, with most fiber producers using their own finishes and keeping the compositions secret. Commercial producers have developed good general-purpose finishes, however, which have found wide application. Only a few larger fiber producers still develop their own finishes, especially for critical processes. One feature has remained: still, fiber producers do not reveal which finish is used or how it is applied.

Spin finishes often used to be solutions in white spirit, but for environmental reasons this is no longer the case. Emulsions in water, or finishes without a diluent – so-called neat oils, with components of a sufficiently low viscosity – are used instead.

Application of finish requires special applicators. Fairly old-fashioned is a finish roll, a porous ceramic roll which constantly takes up finish from a bath, which is then transferred to the yarn that just touches the surface of the roll (see Figure 17.11a). More modern applicators accurately supply the required amount of finish via a small gear pump, to "stift" or "block" applicators, through a hole or slit. The yarn is spread as a ribbon over the applicator and takes up the finish (Figure 17.11b,c). At higher yarn speeds the even distribution of finish becomes increasingly difficult. An applicator with multiple slits or a finish wheel (a roll with small holes under the yarn path) may then be applied.

Finishes of a different nature, so-called after-oils, may be applied at the end of the spinning and drawing process to improve the yarn for a specific application, for example to produce a wrinkle-free textile fabric, to make carpet yarns antistatic

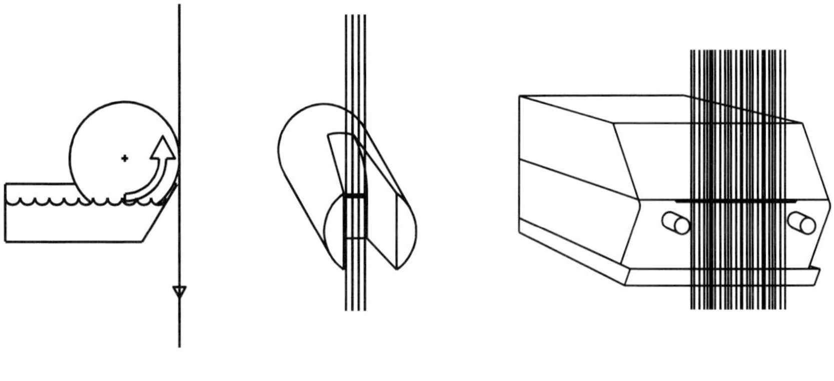

Fig. 17.11. Application of finish.

and/or non-soiling, or to enhance PVC or rubber adhesion for industrial yarns. There is a distinct overlap with other finish-like aftertreatments such as sizes for weaving and dips for tire cords.

17.4.7 Spinning Speed

Falling under its own weight a spinning bundle would reach a speed of 300–400 m min^{-1}. In practice, spinning speeds of at least 500 m min^{-1} are required to obtain a sufficiently high tension in the spin-line. Much higher speeds can be enforced, as long as the filaments do not break under the constantly increasing tension. Speeds of 4000–8000 m min^{-1} have become common practice, especially for polyester. This is called high-speed spinning (HSS), in contrast to LSS ($<$ 1000 m min^{-1}). The air friction can only build up modest forces in the filaments. Even at high speeds the molecular orientation level remains limited, and further drawing is always possible, or even necessary. For further information on HSS see Ref. 7 and Section 17.4.14.

17.4.8 Winding

An old-fashioned winder, operating at 500 m min^{-1}, was still a simple piece of equipment. Modern high-speed winders, operating at 6000 m min^{-1} or more, are "high-tech" devices, the area of specialists. In order to make packages with a good cylindrical shape, the traverse system of the yarn must be very rapid and smooth, especially at the reversal points. A second problem is how to handle the very great weights of the packages, for example, four packages of 30 kg each, wound on one spindle, at $>$5000 rpm! A third problem is that threading in and transfer to a new package must be fully automated; winders therefore have "revolver" systems which simultaneously end the winding of one package and start the next one.

For textile production the capacity of winders has been increased by winding eight or even 12 packages on one winder. For industrial yarns the standard number of packages is two to four.

Winders are used at the end of the spinning process, LSS or HSS, or at the end of an integrated process (spin–draw winding; see Section 17.4.9), always at high speed. Winders for separate drawing processes (see Section 17.4.9) usually operate at $\leq$1000 m min^{-1}. At these low speeds it is possible to wind and twist the yarn simultaneously. This is done for textile yarns. For industrial yarns twist is often applied as a separate aftertreatment.

17.4.9 Drawing

A yarn spun at low speed ($<$ 1000 m min^{-1}) still requires additional drawing. The elongation is still a few hundred percent and the tenacity is too low. A yarn spun at high speed (> 4000 m min^{-1}) may already have sufficient strength for textile or nonwoven applications, but is never strong enough for industrial use. If additional orientation is required, this can be enforced in the spin-line, in a narrow hot tube where drawing takes place. This is called hot-tube spinning (HTS) and is applied only for textile yarns. In most cases drawing is a separate operation. This can be on a separate machine, or combined with the spinning operation, but then in a separate step.

The drawing process is usually carried out in two steps: cold and hot drawing. Cold drawing is the so-called neck-drawing step. For a stable process the position of the neck should be fixed, by tension and/or temperature. Hot drawing is distributed over a longer distance, in a more homogeneous deformation process.

The simplest setup of a drawing machine is a drawing pin (with one yarn wrap around it) and a hotplate (the yarn loosely touches the surface) between two rolls (see Figure 17.12a). The pin is heated to about the glass transition temperature, while the hotplate is held at a temperature safely below the melting temperature. The speed ratio of the two rolls is the adjusted draw ratio (for example, 5.4×). The yarn necks on the pin and completes a 'natural' draw (for example 3×). The remaining factor (1.8×) is the draw ratio over the hotplate. One lets the as-spun yarn decide how and where it wants to draw, which is not necessarily the optimum situation.

More control is possible if the two steps are separated. As an example, drawing in hot gas (steam or air, Figure 17.12b) is depicted, with a set of pins for the cold drawing step and an oven for the hot drawing step. The scheme in Figure 17.12(c) shows that drawing is also possible on rolls only, provided that these are heated to

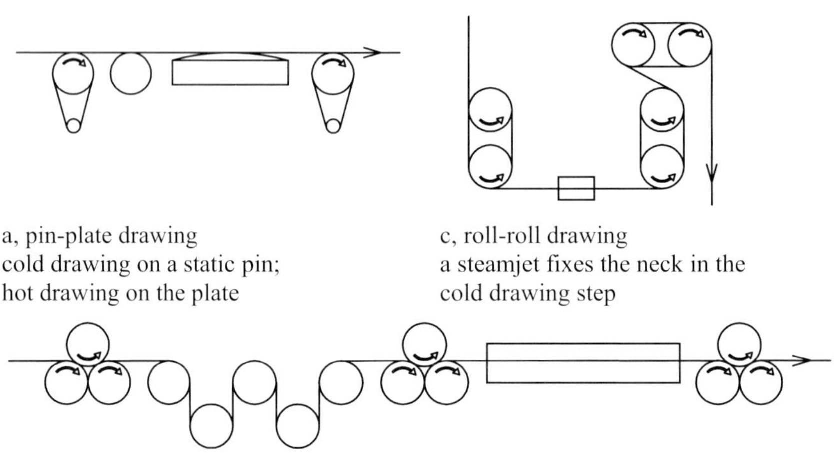

a, pin-plate drawing
cold drawing on a static pin;
hot drawing on the plate

c, roll-roll drawing
a steamjet fixes the neck in the
cold drawing step

b, hot gas drawing
three-roller sets determine the initial, intermediate and end speed
cold drawing on 5 static pins; hot drawing in a oven, with hot steam or air

Fig. 17.12. Drawing processes.

approximately the previously applied pin and plate temperatures, respectively. An alternative is to insert a steam jet between the rolls, which fixes the neck position.

Traditionally, spinning and drawing were two separate processes. A spinning spool was made and then unwound again to feed a drawing machine. Later, the two steps were combined. This technique is called spin–draw winding (SDW). The winding speeds on spin–draw winders are always high. For example, a minimum spinning speed of 500 m min^{-1} should be multiplied with a draw ratio (DR) of 5–6, resulting in speeds of 2500–3000 m min^{-1}. For better economics, speeds of $\geq$4000 m min^{-1} are preferred, which are achieved at spinning speeds of $\geq$700 m min^{-1}. Figure 17.12(c) shows a typical set of drawing rolls (godets) for an SDW machine.

Textile yarns are drawn to moderate tenacity and still high elongation. A (super)high-speed spinning process may fulfill the demands; the term fully oriented yarn (FOY) is used for the product. If a drawing process is still necessary it is often very simple, in one step, for example combined with a texturizing process. High-tenacity industrial yarns always require extensive drawing, in most cases in two steps, even if the spun yarn is already highly preoriented.

When higher draw ratios are applied the stress–strain curves become steeper: tenacity is increased and elongation is reduced (see Figure 17.13). We find the endpoints of a series of curves on an envelope, often described with empirical relationships. A popular one for polyester and nylons is $TE^{0.5}$ = constant (T = tenacity; E = elongation). A demand to raise this factor, which implies a higher breaking energy, cannot be fulfilled by adapting the draw ratio, but only by adapting a "quality" factor. In the example given this is an increase of the polymer molecular

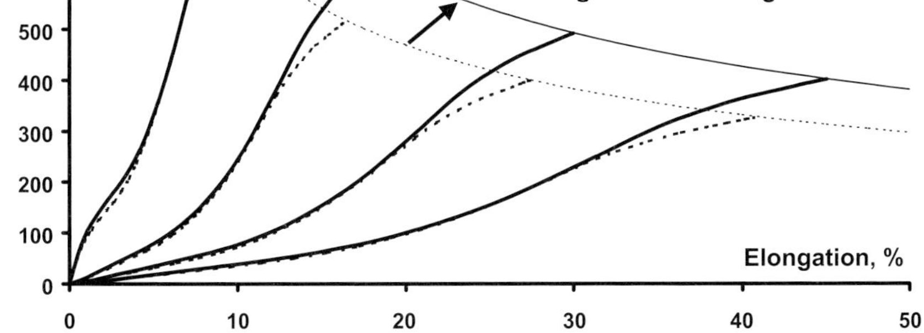

Fig. 17.13. Stress–strain curves of polyester yarns: influence of draw ratio and molecular weight.

weight. Figure 17.13 shows realistic values of *T* and *E* for polyester textile and industrial yarns, the latter being produced from a 50% higher molecular weight.

17.4.10
Relaxation and Stabilization

High orientation of the amorphous phase results in yarn shrinkage, for example in boiling water or in hot air. The shrinkage can be partly relieved in a stabilization step: a heat treatment at (almost) constant length ($DR = 1$). If the heat treatment is carried out at low tension the term "relaxation" is used ($DR < 1$). Almost zero-shrinkage yarns ("pre-shrunk") can thus be produced. The equipment is the same as for drawing: hotplates, hot ovens, or rolls, often installed as an additional step in the drawing process.

17.4.11
Process Integration

It is expensive to let operators handle intermediate products and it is therefore necessary to reduce the number of process steps or to integrate process steps, in order to remain competitive. The first trend has been to increase spinning speeds and get so much orientation that the drawing step can be omitted. This is the case for textile yarns spun at high speed, "fully oriented" and spun-bonded polyester or polypropylene nonwovens. A second trend is to combine spinning and drawing (SDW) and, if possible, to include stabilization and texturing. An example of the latter integration is spin–draw bulk winding (SDBW) of carpet yarns. A third possibility is to couple polymerization with yarn production, by direct spinning – for example, in a continuous polyester polymerization unit coupled with staple-fiber spinning or a series of spin–draw winders. This is the ultimate wish: for monomers going in and full yarn spools coming out.

17.4.12
Rheology

For the design of polymer lines, filter packs and spinning plates, rheological data are required to make the proper calculations. Fiber polymer melts have a fairly common rheological behavior: they are viscoelastic and shear thinning.

17.4.12.1 Shear Viscosity

Most rheological processes during spinning are determined by shear. It is important to understand the typical shear rates in spinning machines. We will use Eq. (2), for the apparent shear rate, to work out a few examples.

$$\dot{\gamma}_{app} = \frac{32\Phi_V}{\pi D^3} \quad (2)$$

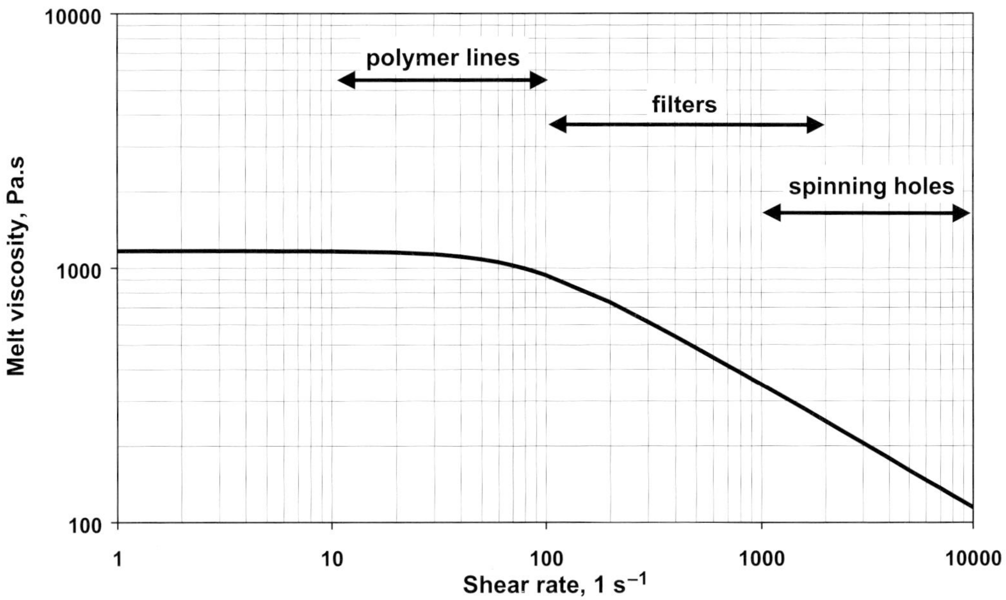

Fig. 17.14. Melt viscosity of polyester as a function of shear rate. $[\eta] = 0.9$, at 300°C and 100 bar. An indication of the shear rate domains in a spinning machine has been added.

For a polymer line with a diameter of 4 cm through which a mass flow of 360 kg h^{-1} ($\rho_{melt} = 1$ g cm^{-3}) is pumped, we calculate $\dot{\gamma} = 16$ s^{-1}, a low-shear condition. For a flow of 5 g min^{-1} through a 500 μm spinning hole, however, we calculate $\dot{\gamma} = 6790$ s^{-1}, a high-shear condition. In the filter package the pores can be small, but the total flow is divided over as large a surface as possible, in order to prevent a high pressure buildup. As a result, shear rates will be in the order of 100–2000 s^{-1}. Finally, in the metering pumps the shear rates between gearwheel and housing may be in the order of 10^5 s^{-1}. For melts with pronounced shear thinning, narrow tolerances may thus be required for a proper metering.

For appropriate calculations of the flow in spinning machines one thus needs a rheological curve over about four decades of shear rate, such as Figure 17.14. A zero-shear viscosity (η_0) or a melt flow index (MFI) gives insufficient insight.

The polymer molecular weight is the most important factor determining the viscosity of melts. The scaling rule is $\eta \sim M_w^{3.4}$. In many cases a solution viscosity (for example, intrinsic viscosity, $[\eta]$) is used as a measure for polymer molecular weight. Since for many flexible polymers $[\eta] \sim M_w^{0.7}$ we can derive another scaling rule, of melt viscosity as a function of solution viscosity: $\eta \sim [\eta]^{5.0}$. It is evident that a higher melt temperature will reduce melt viscosity, but only small variations in molecular weight can be compensated for by temperature adjustment.

17.4.12.2 Elasticity

Elastic behavior is the background of melt fracture just below the spinning plate, which in practice means the immediate interruption of a running process. Elasticity can be controlled by avoiding the use of high molecular weight polymer. For example, the common grades of polypropylene have a broad molecular weight distribution and it may be required to remove the high molecular weight tail by degradation, for example with a peroxide. The resulting "fiber grade" is indicated as "controlled rheology".

17.4.12.3 Elongational Viscosity

Elongational behavior is induced in the entrance of the spinning hole and in the transition region from backhole to actual capillary. In practice hardly any permanent orientation is built up in this way, however, because molecular relaxation is rapid. Spinning hole profiles are smoothened only to prevent the formation of vortices which would lead to extrudate distortion. Promoting orientation already in the spinning holes is not common for melt spinning. It could be beneficial for the orientation of melt-spun liquid-crystalline polymers, however, for example in the production of carbon fiber from pitch.

More important is the elongation in the spin-line, where uniaxial deformation of the material obviously takes place. Measurement of elongational viscosity on laboratory equipment is very complicated. Fiber engineers may include it in developing spinning models, thereby using the spinning machine itself as a "rheometer", but will usually keep the information they gather proprietary.

In situations of low elongational rate one can simply apply Trouton's ratio [Eq. (3), where η_E is the elongational viscosity and η_S is shear viscosity; the subscript S is often omitted].

$$\eta_E = 3\eta_S \tag{3}$$

At higher rates, and in contrast to shear thinning, strain hardening (increasing η_E) may occur, resulting in rapid orientation. In practice this is often induced or accompanied by cooling, solidification, and crystallization of the melt, making the analysis of elongational behavior even more complicated.

Catastrophic strain hardening, resulting in breaking filaments, is rare for melt-spun polymers. Buildup of fairly high orientation in the spin-line has become common, however, in modern high-speed spinning processes.

17.4.13 Process Calculations

Fiber engineers are notorious users of non-SI units: feet or yards instead of meters, minutes or hours instead of seconds, grams instead of kilograms, and denier or dtex instead of tex. Nevertheless, most process calculations are quite simple, without difficult conversions being necessary.

A few examples for a polyester spin–draw winding process will show this, and hopefully give an impression of how melt-spinning machines are designed.

The process concerns the production of an industrial yarn with a titer of 1670 dtex f 325. We assume that spinning holes are used with a diameter of 500 μm and an L/D ratio of 1.5; there are 325 holes per spinning plate. There are eight spinning bundles per extruder. The spinning speed is 800 m min^{-1} and the spun yarn is immediately drawn five times on the same machine (an integrated process) and then wound with a speed of 4000 m min^{-1}.

17.4.13.1 Mass Flow

By definition 1 dtex = 1 gram per 10,000 m. The output per bundle is then calculated from Eq. (4).

$$\{(\text{yarn titer in dtex})\ 10{,}000^{-1}\} * (\text{spinning speed in m min}^{-1})$$
$$= (1670/10{,}000) * 4000 = 668\ \text{g min}^{-1} \tag{4}$$

Note that the mass flow is constant in the process, from spinning pump to winder. No mass is lost, in contrast to dry or wet spinning. For example, at the end of the spin-line the speed is 800 m min^{-1}, five times lower than after drawing, but the undrawn yarn still has a five times higher titer. Also note that the volume flow is *not* completely constant, because the density increases (by about 20%) when the melt cools down, solidifies, and crystallizes.

Eight yarn bundles make $8 \times 668 = 5344$ g min^{-1} = 320.6 kg h^{-1}, which would require a 120 mm, maybe 150 mm, extruder. The unit produces about 7.6 ton day^{-1}, or about 2700 ton y^{-1}.

17.4.13.2 Volume Flow

The density of a polyester melt is approximately 1.18 g cm^{-3}. The volume flow per yarn bundle and spinning pump thus is $668/1.18 = 566$ cm^3 min^{-1}. This flow can be achieved with a 20 cm^3 pump at 28.3 rpm.

17.4.13.3 Extrusion Speed and Elongation in the Spin-line

A volume flow of 566 cm^3 min^{-1} is extruded through 325 holes, which is $566/325 = 1.74$ cm^3 min^{-1} per hole. The holes have a diameter of 500 μm (0.05 cm) and their cross-section is $(\pi/4) \times 0.05^2 = 0.001964$ cm^2. The extrusion speed through the holes is $(1.74\ \text{cm}^3\ \text{min}^{-1})/(0.001964\ \text{cm}^2) = 887\ \text{cm min}^{-1} = 8.87\ \text{m min}^{-1}$.

The spinning speed is 800 m min^{-1}, and the filaments are thus accelerated in the spin-line by a factor of $800/8.87 = 90$. In practice this is viscous flow rather than molecular orientation. Calling this a draw ratio is misleading; "draft", "draw-down" or "spin stretch factor" are better notions.

17.4.13.4 Pressure Drop over the Spinning Holes

For the calculation of shear rate it is essential to convert accurately to meters and seconds in order to obtain shear rate in reciprocal seconds (s^{-1}). The formula is given by Eq. (5), with $\Phi_v = 1.74\ cm^3\ min^{-1}$ and $D = 500\ \mu m$.

$$\dot{\gamma}_{app} = \frac{32\Phi_v}{\pi D^3} \tag{5}$$

The shear rate in the spinning holes in this case is $\dot{\gamma} = 2363\ s^{-1}$. In the rheology graph (Figure 17.14) we see that $\eta = 220$ Pa s at this shear rate.

We can now calculate the pressure drop over the spinning hole with Poiseuille's law [Eq. (6)].

$$\Delta P = \frac{128}{\pi}\Phi_v \eta \frac{L}{D^4} \tag{6}$$

The result is 3119335 Pa = 31 bar. The pressure drop over the much wider back-hole would add only a few bars.

This is a "reasonable" pressure drop: not too high, but also not too low, because this would lead to an uneven distribution of the flow over the holes, which would result in filament titer differences within a yarn bundle. For a lower melt viscosity (for example, 100 Pa s), narrower spinning holes (350 μm) would have been necessary.

The pressure drop over the filter package would add 30–50 bar, resulting in an initial pressure drop over the spinning assembly of 60–85 bar. The pressure drop over the filter does increase during the lifetime of the spinning assembly. The assembly would be taken out of the machine at 200–250 bar.

17.4.14
Polyester (Poly(ethylene terephthalate), PET)

Polyester and PET are almost synonyms. Other polyesters, such as poly(butylene terephthalate) (PBT), poly(trimethylene terephthalate) (PTT), and poly(ethylene naphthalate) (PEN), have hardly any significance as fiber materials.

17.4.14.1 PET Polymer

Textile-grade polyester has an intrinsic viscosity around 0.6 (degree of polymerization 100) and is spun at about 285°C, which is 30°C above the melting point. Grades for industrial yarns have an intrinsic viscosity of 0.8–0.9 (degree of polymerization 140–160) and must be spun at 300–310°C. The main degradation problem for PET is hydrolysis: each water molecule gives one chain break in the PET. There is no hydrolysis problem when a polymerization unit directly feeds the spinning machine. Handling of polymer chips requires closed systems, ultra-dry air or nitrogen, among similar precautions. Thermal degradation is unproblematic for textile yarn polyester, but becomes severe for industrial yarn polyester grades

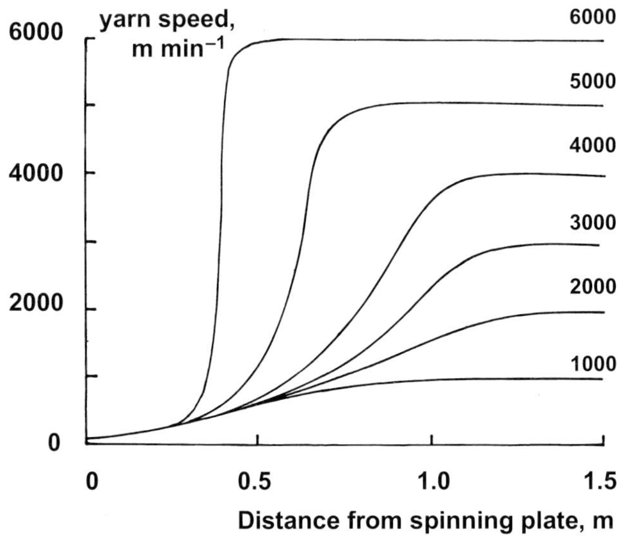

Fig. 17.15. Yarn speed curves for LSS and HSS of polyester.

spun above 300°C. Even at a short residence time in the machine (below 10 min) a total drop of 0.1 in the intrinsic viscosity must often be accepted.

17.4.14.2 **Spinning of PET**

PET is a slowly crystallizing polymer. It remains amorphous in the spin-line at speeds below 3000 m min^{-1}. At higher speeds nucleation takes place high in the spin-line, at high temperature, and these nuclei grow rapidly on their way down. The air drag forces deform the crystalline network in a "neck-like" fashion in the spin-line (see Figure 17.15). This necking is more pronounced for higher speeds, finer filaments, and higher molecular weights. These as-spun yarns have a low shrinkage even when the crystallinity is still low. The remaining draw ratio of such yarns is below 2.

17.4.14.3 **PET Staple Fiber**

The majority of polyester fiber production is for textile application, and most of this is staple fiber. Staple fiber is produced by direct spinning from continuous polymerization units based on pure terephthalic acid. The spinning plates have 2000–5000 holes, and the spinning speed is around 2000 m min^{-1}. The bundles are not wound but are laid down loosely in a container. Drawing (more than three times) and texturing is a second, separate step: spinning bundles are combined into a tow of several hundred thousand dtex and drawn, crimped, and heat-set collectively. Cotton-type fiber is the standard: 0.5–2 dtex, drawn to 20% elongation, cut to 32–40 mm in length. Wool-type fibers have a titer of 2–7 dtex; they are drawn to 40% elongation and cut to 40–120 mm in length. Staple fiber is delivered to the customer in large bales.

17.4.14.4 PET Textile Filament Yarns

Most textile filament yarns are also direct-spun. The output per bundle is small, and therefore hundreds of spinning positions must be fed from one polymerization unit, making the polymer melt-line system very complicated. Winders make four, six, eight, or even 12 packages simultaneously, and spinnerets and spinning pumps are clustered accordingly.

Partly oriented yarn (POY) is a large product; spun at about 3500 m min^{-1}, with a round cross-section, and draw-textured (DR $\sim$ 1.8, at about 1000 m min^{-1}) in a separate step. Fully oriented yarn (FOY) can be spun at $\geq$6000 m min^{-1}, or drawn in the spin-line (hot-tube spinning, HTS), or spin-drawn. Typical yarn titers are 30, 50, 76, 110, and 176 dtex, and the filament titers 2–3 dtex, although there is a trend toward values of 1 dtex or lower.

17.4.14.5 PET Industrial Yarns

Direct spinning is not very common for industrial yarns because there are only a few yarn types that would match the large capacity of polymerization units ($\geq$ 25 000 ton y^{-1}). A further complication is that the polymer must be condensed to a high molecular weight, but this can be achieved in deep-vacuum, thin-film "finishers". In most cases chips with a textile viscosity are solid-state postcondensed, at a relatively low temperature (about 230°C), which takes many hours but has the advantage that thermal degradation is minimized.

There is a limited field of application for low yarn titers, 200–550 dtex, in sewing yarns and fine fabrics. Most industrial yarns have titers of 1100–2200 dtex. The filament titer is usually around 5 dtex, but yarns for safety belts have coarser filaments (10–15 dtex) and modern tire yarns may have finer filaments (around 3 dtex).

Large spinning holes (350–800 μm) are used to handle the high melt viscosities.

The standard process for PET is spin–draw winding (SDW) at speeds of 4000–5000 m min^{-1}. This is the cheapest process, especially when two, three, or four bundles are combined on one set of spinning and drawing godets and one winder. The yarns are strong (700–850 mN tex^{-1}), but may have a fairly high shrinkage. For applications such as nets, ropes, cables, and most fabrics (safety belts, conveyor belts) this is an ideal combination. For some applications almost zero shrinkage can be required (PVC-coated fabrics) and a separate, slow, drawing and stabilization process may still be applied.

Tire cord is a different case. Polyester is the most important reinforcing material for radial tires. The cords run radially, from rim to rim, and their high modulus reduces the deformation of the rolling tire, and thus fuel consumption. The properties of the reinforcement in the tire depend on how well the modulus is retained during the dipping and curing processes. Therefore, tire yarns are always spun at high speeds (> 3000 m min^{-1}), which gives lower shrinkage and somewhat lower tenacity in the yarn. After dipping of the cords and curing of the rubber, these yarns offer the best strength/modulus combination. To reduce costs, even these yarns are spin-drawn, which implies winding speeds of at least 6000 m min^{-1} (high-speed spin-draw winding, HSSDW).

The most important applications of polyester industrial yarns have been mentioned above: tires, other rubber reinforcement, narrow and wide fabrics, nets, ropes and cables, and sewing yarns. It is not unusual for a company to list 20 different types of polyester yarn, each type being further divided into various yarn titers and twist levels.

17.4.15 Polyamide (PA6 and PA66)

The former difference in usage between the USA (PA66) and Europe (PA6) is still evident. PA66 seems to be the larger-volume product but PA6 is still large in South American countries, eastern Europe, and India. High-melting PA46 is gradually finding application in airbag fabrics. Other polyamides (PA11, PA12) are not important as fiber materials.

17.4.15.1 PA Polymer

Polyamide 6, melting point 225°C, is spun at 260–280°C (290°C); polyamide 66, melting point 265°C, is spun at 290–300°C (310°C). Thermal degradation is relatively unproblematic for polyamide 6, but is more severe for polyamide 66 because of the higher spinning temperature and the tendency of polyamide 66 to crosslink. Filtration of gel particles is an issue for PA66, not for PA6. PA66 polymer melt-lines must be burned out once a year because a crosslinked, charred polymer layer is built up at the hot walls. High-shear conditions in narrower polymer lines can help to remove the gel layer from the walls and prolongs the cleaning cycle.

Both polyamides evolve oligomer vapor upon extrusion from the spinning holes, and snow-like deposits are formed on cold spots in the top of the spinning machine. Measures such as steam injection, suction, and regular manual cleaning must be taken to keep these under control.

Polyamides are prone to oxidative degradation, resulting in yellowing. The nitrogen gas in the chips hoppers must therefore be completely free of oxygen.

The reaction of polyamide with water is an equilibrium. When the chips are too dry, the polymer will postcondense in the spinning machine. When they are too wet hydrolysis will take place. The equilibrium water content depends on the molecular weight: for a higher molecular weight, the water content must be lower. There is always some loss of molecular weight by thermal degradation. This can be compensated for by postcondensation when the water content is adjusted to slightly below the equilibrium level.

17.4.15.2 PA Spinning

Polyamides crystallize faster than polyester. At low spinning speeds polyamide 6 does not crystallize in the spin-line but fairly rapidly on the spinning spool, thereby also attracting water. As a result, the packages increase in volume ("grow") and can be unwound only with difficulty when the storage conditions are not kept very constant. At speeds of 1000–4000 m min^{-1} no proper spinning spools can be made. Above 4000 m min^{-1} the yarns crystallize sufficiently in the spin-line and stable packages are built. Polyamide 66 crystallizes faster than PA6. As-spun yarns are always crystalline, even at low spinning speeds. Complete melting of the crystalline chips is essential because remnant crystal nuclei can even cause spherulitic crystallization.

17.4.15.3 **PA Staple Fiber**

Staple fiber is a large product, especially for applications in carpets. The production resembles that for polyester staple. The spinning speed is about 2000 m min^{-1}, which is not problematic because the as-spun yarn is not wound but laid down in containers. The filament titer may be low (4 dtex, for velour carpet) but is usually around 20 dtex. The cross-section is profiled: it may be trilobal or a square with holes.

17.4.15.4 **PA Textile Filament Yarns**

The yarn counts are lower than for polyester because the main applications are as hosiery yarn (for example "20 denier": 22 dtex f 1 or 22 dtex f 5) or in ladies' underwear (typical yarn titers 44 dtex f 10, 78 dtex f 28, or 78 dtex f 60). The processes are, for POY, spinning at >4000 m min^{-1} followed by a draw-texturing treatment (at approximately 1000 m min^{-1}), or SDW at >5000 m min^{-1}. The elongation of nylon textile yarns is 45–50%.

17.4.15.5 **PA Industrial Yarns**

Polyamide industrial grades have much lower melt viscosities than polyester, and are therefore spun through smaller spinning holes (250–400 μm). Polyamide yarns can be drawn to elongations slightly below 20%, but subsequent stabilization to reduce shrinkage may add a few percent elongation. Tenacities are at least as good as for polyester, but the modulus is much lower.

Polyamide is applied in fabrics, especially in airbags, where PA66 has an important advantage over other fiber materials. Airbags are blown up with hot gas and PA66 does not melt because it has a high specific heat and a high melting point. In this respect, PA46 is even better. The high tenacity and breaking energy of polyamide yarns make them suitable for application in fishing nets, ropes, and cables, but the competition with cheaper polyester is fierce. For rubber reinforcement PA6 has the disadvantage of its low melting point. Nevertheless it is still widely applied in India and South America. PA66 is used on a large scale: in conveyor belts, rubber hoses, and tires. In radial tires PA66 cannot be used in the tire walls because its modulus is too low, but it is present as a cap ply around the steel belt. In old-fashioned bias-belted tires, for bumpy roads, polyamide is the perfect reinforcement. Aircraft tires also have a "diagonal" construction, and usually contain many layers of nylon cords.

Spin–draw winding (SDW) is the standard process for the polyamides: spinning speeds are 500–1000 m min^{-1}, winding speeds 3000–5000 m min^{-1}. Drawing usually takes place in two steps, cold and hot drawing, but one-step drawing is possible for low yarn counts. For better economics it is essential to process two, three, or four bundles on one set of drawing godets, and to wind them on one winder.

Yarn titers are 200–600 dtex for airbags and other fine fabrics, and 1100–2200 dtex for most other applications. Filament titers are around 5 dtex. Tenacities are between 750 and 850 mN tex^{-1}, and elongations between 18 and 25%.

17.4.16
Polypropylene (PP)

17.4.16.1 PP Polymer

Polypropylene is an addition polymer with a fairly broad molecular weight distribution. The strong viscoelastic effects make spinning through small holes, or slits in profiled holes, difficult. PP fiber grades are therefore made by cracking the high molecular weight tail by oxidative degradation, for example by the addition of peroxide in a twin-screw extruder.

Polypropylene is full of tertiary carbon atoms which give oxidative and light stability problems. Stabilizer packages are therefore always included in fibers, with their small diameters and high specific surface. The polymer is hydrophobic. Special treatments or additions are required to make polypropylene dyeable.

Polypropylene melts at about 175°C but is spun relatively hot, ≥75°C above its melting point. Polypropylene is often spun from large and long spinning holes (800–1500 μm, $L/D = 3$–5). It is evident that the low melting point is a limit for industrial applications. A further disadvantage of polypropylene is its low creep resistance.

17.4.16.2 PP Spinning

Polypropylene crystallizes fast, always in the spin-line, even at low speeds.

17.4.16.3 PP Staple Fiber

This is probably the largest-volume spun product, for application in carpets. The usual machines resemble those for polyester and polyamide, with the spinning speed around 1000 m min^{-1}, collection of as-spun yarn in a container, and draw-texturing a tow of several hundred thousand dtex. "Short spinning" is an alternative: it involves a one-floor machine, holes at a very short distance from each other, cooling over a short distance, and the spinning speed so low (~100 m min^{-1}) that drawing and texturing can be included on the same machine.

17.4.16.4 PP Split Fiber

Much polypropylene fiber is not spun, but produced from film. A film is blown, or cast on a chill roll, drawn 6–10 times and then fibrillated. There are numerous ways of slitting, fibrillating, and cutting. Split fiber finds applications in twines and ropes (as a replacement for sisal), cheap fabrics for bags and tarpaulins, and carpet backing (replacing jute).

17.4.16.5 PP Filament Yarns

The production machines can be normal spin–draw winders, as for the other melt-spun fibers. But here also, "compact" machines have been developed, for example for bulked continuous filament yarn (BCF, for carpet) with all the process steps on one machine, at an end speed of ≤ 1000 m min^{-1}.

Polypropylene is small in textile applications (in sportswear) but it has a reasonable position in high-tenacity yarns, in low-temperature applications such as ropes, cables, and geotextiles. It should be added that polypropylene can be drawn to high ratios (see Section 17.7.2). This results in very good tenacities, but the helix configuration of the isotactic chain in the crystals severely limits the modulus.

17.5 Solution Spinning

17.5.1 Preparation of Spinning Dope

If a polymer cannot be melt-spun, a solution must be made which can be spun. Chemical ractions may be involved, for example the xanthogenation or acetylation of cellulose. In general, the spinning dope is prepared in large vessels, not in extruders. Filter presses for very fine filtration are included in the equipment setup and de-aeration is carried out in a storage tank just before pumping the solution to the spinnerets.

Solution spinning implies the handling of large quantities of solvent. At a polymer concentration of 20% the total mass flow for dry spinning is five times higher than the polymer mass flow. For wet spinning there is a very large additional flow of nonsolvent from the spinning bath.

17.5.2 Dry Spinning

The principle of dry spinning is shown in Figure 17.16. Dry spinning is not widely applied; cellulose acetate fibers are dry-spun, and whereas acrylics and polyvinyl alcohol can be dry spun, wet spinning is preferred.

The gas in the column is preferably hot nitrogen when flammable organic solvents are used. If the solvent is water (for poly(vinyl alcohol)), hot air can be used. The column height is limited to about 5 m and the residence time of the running filaments in the column is about 1 s, during which most of the solvent must be removed from the filaments. Complete removal is not necessary because at a high polymer concentration the filaments solidify by gelation and can then be handled. Removal of solvent can be completed in later process steps, for example during additional drawing.

Control of gas flows in the column, avoiding turbulence or sticking of filaments, is important. Therefore, the hot gas flow would often be downward rather than in

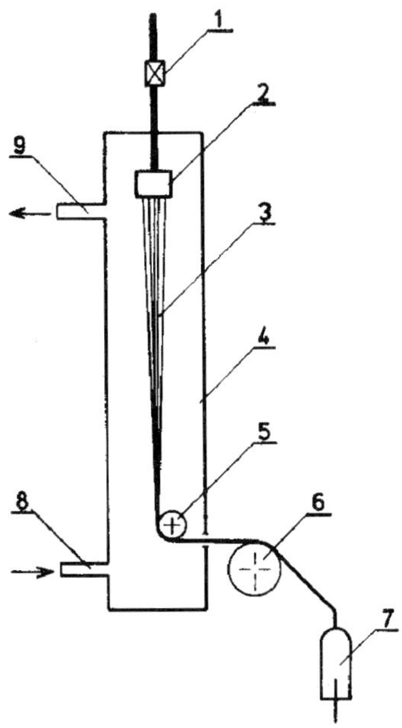

Fig. 17.16. Scheme for dry spinning (reproduced from Ref. 6): 1, metering pump; 2, spinneret; 3, spinning bundle; 4, drying column; 5, 6, 7, take-up system; 8, 9, inlet and outlet of drying gas.

countercurrent, and spinning holes must be a few millimeters apart, as in melt spinning.

Evaporation of solvent is a relatively slow process, slower than cooling in melt spinning. It is therefore essential that the spinning filaments are thin. The common approach is to use small spinning holes (50–100 μm) and apply a "low draft" in the spin-line. For example, both the extrusion speed from the holes and the exit speed from the dry-spinning column would be around 300 m min^{-1}. Even a draft below 1 is possible, which means that the die swell effect is not completely undone.

Another approach is a "high-draft" process, spinning from large holes (500 μm) and taking up the yarn at relatively high speeds. For example, the extrusion speed through the holes could be 20–50 m min^{-1} and the exit speed around 500 m min^{-1}, which implies a draft of 10–25. The draft in the spin-line is viscous flow and only limited orientation is built up, a situation completely comparable with melt spinning.

17.5.2.1 **Cellulose Acetate**

There are two types of cellulose acetate fibers. When all three hydroxyl groups per cellulose unit (see Section 17.5.3.1) are acetalized, cellulose triacetate is obtained. Cellulose triacetate is spun from dichloromethane (DCM) plus 5–15% methanol or ethanol. Triacetate is a small product, mainly in the form of yarns with low titers for textile applications.

Cellulose acetate contains approximately 2.5 acetate groups per cellulose unit, and is actually produced by partially hydrolyzing the triacetate. The polymer is spun from a solution in acetone/water (approximately 95:5). With a large excess of drying air it is possible to remain below explosion limits. Nevertheless, all operations are carried out in closed equipment.

"Secondary" acetate has a remarkably poor crystallinity, in fact too low to survive in a strong interfiber competition for textile applications. The main application of cellulose acetate is as tow in cigarette filters.

17.5.2.2 Acrylics

Most acrylic fibers are wet-spun, but dry spinning is also applied. The most common solvent is dimethylformamide, DMF. The polymerization of acrylics can also be carried out in DMF and the polymerization solution can then be directly spun. The boiling point of DMF is 153°C, making complete removal of solvent in the spinning column almost impossible. Most dry-spun acrylic production is staple fiber, and the remaining solvent is then removed during tow processing.

Further information on acrylic fibers is given in Section 17.5.3.2.

17.5.2.3 Poly(vinyl alcohol)

The preferred process for polyvinyl alcohol is wet spinning, but dry spinning is also possible. The solvent is water and the polymer concentration is so high (25–50%) that solid chips can be produced at room temperature. These can be processed in a single-screw extruder, as for melt spinning. PVA dry spinning can be a low-draft or high-draft process. The low-draft as-spun yarn has a better drawability. Very high draw ratios are possible for PVA, resulting in high tenacity and modulus.

For more information see Section 17.5.3.3.

17.5.3 Wet Spinning

The principle of wet spinning is shown in Figure 17.17. Wet spinning is applied for two large-scale fiber products – viscose rayon and acrylics – and one smaller product – poly(vinyl alcohol).

The critical factor for each wet spinning process is how the coagulation process in the filaments proceeds. In the ternary diagram shown in Figure 17.18 the three corners represent polymer (P), solvent (S), and non-solvent (N, from the spinning bath). SD is the spinning dope composition on the line P–S. The hatched area is where phase separation takes place.

The route from SD into the separation region will determine how the coagulation will proceed. (Note that a path on the line SP represents dry spinning: solvent is removed from the fiber and the polymer concentration is increased; there is no coagulation but there is gelation of the system.) If the amount of nonsolvent enter-

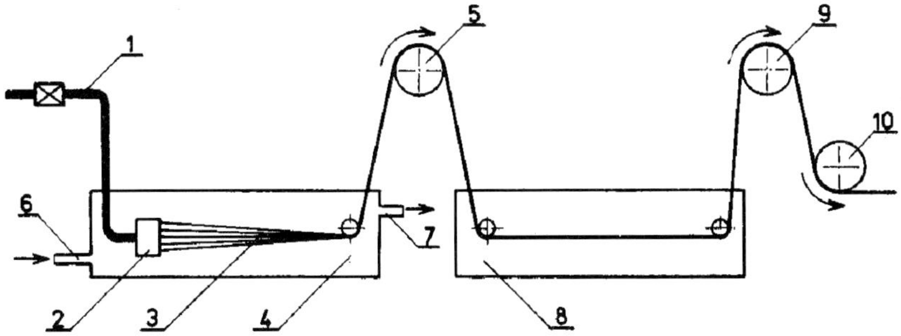

Fig. 17.17. Scheme for wet spinning (reproduced from Ref. 6): 1, metering pump for spinning dope; 2, spinneret; 3, spinning bundle; 4, spinning or coagulation bath; 5, godet; 6, 7, inlet and outlet of spinning bath; 8, drawing bath; 9, 10, take-up system.

ing the fiber is relatively small in comparison with the flow of solvent out of the fiber (path 1), phase separation will occur at high polymer concentrations. A homogeneous, dense structure will be formed; the continuous phase is polymer-rich. This takes a long time, however, because diffusion of solvent will become slower at higher polymer concentrations. The opposite occurs along path 2: nonsolvent enters relatively rapidly into the fiber and coagulation occurs at low polymer concentrations. A heterogeneous, porous structure is formed; the continuous phase has a low polymer concentration and the fiber will be very weak. Path 1 is preferred, but fiber formation is relatively slow then.

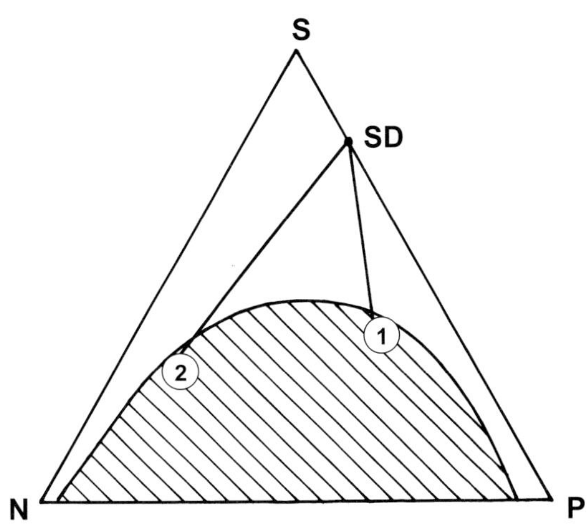

Fig. 17.18. Ternary diagram for wet spinning: P = polymer, S = solvent, N = nonsolvent, SD = spinning dope. 1, 2, coagulation paths (see text).

It is noteworthy that "coagulation" is the general term for the solidification process in wet spinning, regardless of whether the solidification process is heterogeneous (coagulation) or homogeneous (gelation). Even aramid spinners use the word coagulation (see Section 17.7.1.1).

In any case, solvent diffusion and coagulation will be more rapid in the skin than in the core of filaments. Skin formation is an inherent problem for wet spinning (and also for dry spinning). The main effect is that the filaments collapse: the original round cross-section is transferred into a dog bone shape. Further information is available in Ref. 6, Chapter 4.

17.5.3.1 **Viscose Rayon**

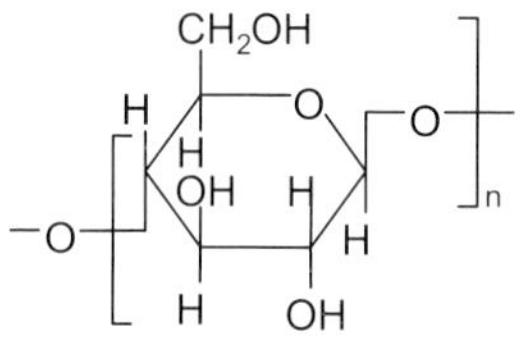

(**1**) Cellulose

It is amazing to see how complicated the technology is for this *first* large-scale man-made fiber. The complications are caused by the insolubility of cellulose (**1**). Chemical modification was necessary to make the polymer spinnable. The first attempt was nitration of the hydroxyl groups, but cellulose nitrate proved more useful as gunpowder than as a basis for fiber manufacure. Xanthogenation proved more successful, although the first attempts were aiming at producing carbon filaments for electric lamps, rather than textile fibers.

In the modern process (Figure 17.19), sheets of wood pulp cellulose are swollen in concentrated alkali and alkali cellulose is formed [Eq. (7), writing CellOH for a cellulose hydroxyl group].

$$\text{CellOH} + \text{NaOH} \rightarrow \text{CellONa} + H_2O \tag{7}$$

The sheets are shredded and the "white crumbs" are aged, which implies depolymerization to a molecular weight which is suitable for textile applications or tire cord. The material is then treated in closed equipment (batchwise or continuous) with carbon disulfide in order to xanthate the cellulose [Eq. (8)].

$$\text{CellONa} + CS_2 \rightarrow \text{CellOCS}_2\text{Na} \tag{8}$$

Byproducts give an orange-yellow color: "yellow crumbs" are formed. The degree of xanthation (of cellulose) is low: fewer than 0.5 hydroxyl groups need be derivatized to accomplish dissolution in dilute alkali. The polymer concentration for a textile yarn would be around 9% and the alkali concentration 5–6%. For a tire cord the polymer/alkali ratio is lower, both concentrations being 7%, for example. The

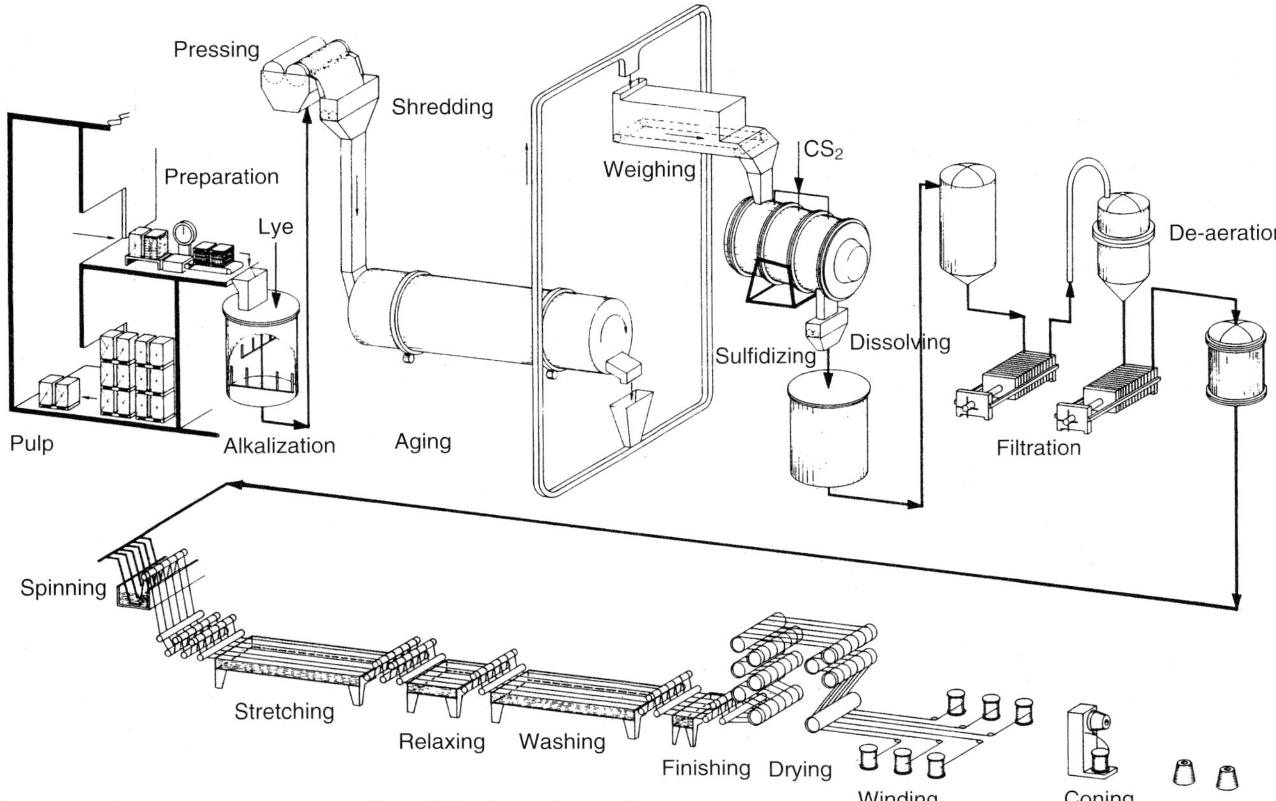

Fig. 17.19. Viscose rayon process scheme (Enka tire cord process).

spinning solution is then ripened, which implies a redistribution and slight reduction of xanthate groups. The eventual degree of xanthation would be below 0.3 for a textile yarn and close to 0.4 for a tire cord. After fine filtration and de-aeration the viscose spinning dope is spun into an acid bath, where the cellulose is regenerated [Eq. (9)].

$$2\ \mathrm{CellOCS_2Na} + \mathrm{H_2SO_4} \rightarrow 2\ \mathrm{CellOH} + 2\ \mathrm{CS_2} + \mathrm{Na_2SO_4} \tag{9}$$

The spinning bath for viscose contains sulfuric acid (at about 10% concentration) for the decomposition of the xanthate and neutralization of the alkali. Sodium sulfate is formed anyway, but is also dissolved in large quantities (about 20%) in the spinning bath to control the coagulation process. A further addition is zinc sulfate ($<$ 3%), again to control coagulation.

The spinning solutions – caustic plus acid – are highly corrosive. The spinnerets are made of gold/platinum, round, with diameters of a few centimeters only. The capillaries are small (50–75 μm) and close to each other ($<$ 1 mm). The number of holes must be large because the standard filament titer of rayon yarns is around 1.7 dtex. The freshly spun filaments are very weak and a tube is often placed around the spinning bundle.

Drawing is always carried out in a combined process, in most cases at least partly before the complete decomposition of the xanthate groups. Draw ratios may be low for rayon staple fiber. Tire cord is drawn to an elongation of 12–13% and a tenacity of about 500 mN tex^{-1}. Note that these are conditioned values; in a wet state the high-modulus character is lost; the tenacity becomes lower (400 mN tex^{-1}) and the elongation much higher (25%).

A few alternatives for the derivatization of cellulose have been found recently: direct dissolution of cellulose has been developed. For textile filament yarns, dissolution in *N*-methylmorpholine oxide (NMMO) is possible. The process is applied by Courtaulds and Lenzing. A solution in formic acid/phosphoric acid was found to have lyotropic behavior and tire yarns with very interesting properties could be produced (patented by Michelin). It even proved possible to use phosphoric acid alone (patented by AKZO), but the process was never commercialized.

Yarns spun from solution have a different crystalline structure (cellulose II) than natural cellulose (cellulose I). The difference is that cellulose I has two *inter*molecular hydrogen bonds formed parallel to the glucosidic bond, whereas cellulose II has only one parallel hydrogen bond. The main effect is a large difference in crystal modulus: 130–180 GPa for cellulose I, 60–90 GPa for cellulose II. All attempts to produce man-made fibers with a cellulose I structure, and hence an even higher modulus, have remained unsuccessful, however.

Applications of viscose rayon The main application of rayon is as staple fiber. Spinning is carried out from clusters of spinnerets and bundles are processed collectively as a tow. Rayon staple fiber is applied unblended, but is also used in blends with cotton and/or polyester, in outerwear. Filament yarns have become a minor product; one may still find them applied in lining fabrics. Rayon tire cord has survived the strong competition with polyester, at least in Europe. Rayon has an unproblematic adhesion to rubber, a high modulus, and a perfect fatigue behavior, and therefore remains the ideal reinforcement for high-speed radial tires.

17.5.3.2 **Acrylics**

The crystallinity of acrylic fibers is rather ill-defined. Polyacrylonitrile (PAN) is known to be atactic, but the very strong polar interactions of the nitrile groups give PAN a definite crystalline behavior. In water a melting endotherm slightly below 200°C is measured and an extrapolated dry melting point could be as high as 320°C, above the decomposition temperature. This finding of melting point depression by water has led to suggestions that PAN could be melt-spinnable when water, or another plasticizer, is added to the polymer. So far, the technique has been considered too complicated for technical realization, however.

In practice, the crystalline character of pure PAN is a problem for dissolution, drawing, and dyeing in various stages of acrylic processing. Therefore comonomers (in acrylics) are added to reduce the crystallinity. Yarns containing $\geq$ 85% acrylonitrile are called "acrylics". Comonomers that are often used are methyl acrylate and vinyl acetate. Yarns with less than 85% acrylonitrile are called "modacrylics". In this case halogen-containing comonomers are often used, such as vinyl chloride, vinylidene chloride, and vinyl bromide, obviously to improve the flame resistance of the yarn.

For dissolution of acrylics highly polar solvents are required to disrupt the intermolecular bonds between the nitrile groups. Frequently used organic solvents are dimethylformamide (DMF) and dimethylacetamide (DMAc). The polymer concentration is about 20%. The spinning bath is water, in most cases mixed with the organic solvent being used, in order to slow down the coagulation and precipitation. An alternative is the use of concentrated solutions (50%) of sodium thiocyanate (NaSCN) in water. The spinning bath then is a dilute solution of NaSCN in water.

Acrylic solutions have the tendency to gel with time, a thermoreversible process. For spinning it is an advantage to have gelation rather than precipitation of the filaments. As a result of the gelation, as-spun acrylic filaments maintain the cross-section of the spinning holes, but are very porous. A density of 0.4–0.5 g cm^{-3} is usual, while the density of a drawn and dried fiber is 1.17 g cm^{-3}. The voids are very small (0.1–1 μm), and collapse during drawing and drying.

Most acrylic production is for staple fiber. Huge spinning plates with up to 60 000 holes are used, and tow drawing and crimping are included in the process. Drawing takes place in a hot-water bath, which is possible because the wet glass transition temperature is about 75°C. The above-mentioned porosity of acrylics is used to control luster. Dried yarns are very lustrous but can be made dull again by a hot wet treatment.

Applications of acrylics The largest application is in fiber yarns for clothing (especially sweaters), home furnishings, covers, and blankets. The filament titers vary between 1.3 and 17 dtex, the higher titers being for carpets. Tenacities are modest (200–300 mN tex^{-1}), and elongations are between 30 and 60%. The tenacity may seem low, but acrylics are still clearly stronger than wool.

Filament textile acrylics seem to have lost the competition with polyester and nylon.

There is an interesting industrial application in outdoor fabrics, based on the perfect UV stability of acrylics, which are used for awnings, tents, automobile upholstery, and outdoor furniture. These products are often pigment-dyed, for better lightfastness. A further step is when hot drawing is applied. Yarns with low comonomer levels can be drawn to ratios of 10 or more. Such yarns are used as precursors for carbon fiber production (see Section 17.7.3), but may also be applied as such, for asbestos replacement.

17.5.3.3 Poly(vinyl alcohol)

PVA is produced by hydrolyzing poly(vinyl acetate). PVA fiber is an almost exclusively Japanese product; some production takes place in Korea and China, based on Japanese technology. The fiber has Vinylon as a general name and Kuraray is the largest producer (Kuralon). Poly(vinyl alcohol) is water-soluble, which makes the choice of a solvent easy. And, of course, it has proved possible to make the fibers water-insoluble.

Wet spinning solutions contain about 15% polymer and are spun above 70°C, because otherwise premature gelation would take place. The spinning bath (for PVA) is an almost saturated solution of sodium sulfate in water, at 40–50°C. Note that this is not the usual solvent–nonsolvent situation. Diffusion of water out of the filament is rapid, diffusion of the salt into the filaments relatively slow. The polymer concentration increases and solidificaton is by gelation. The solidification is slow, however – much slower than for viscose rayon, for example. Therefore, the spinning speeds are low. Fairly exceptional are the vertical spinning machines, spinning upward, with the spinning bath moving in co-current flow in a tube around the yarn bundle (see Figure 17.20). The spinning holes are small (100–150 μm).

Hot drawing of PVA, at 210–240°C, takes place after drying. Draw ratios (> 10) and tenacities achieved (> 900 mN tex^{-1}) are high, and the resulting yarns are highly crystalline (40–50%) and no longer water-soluble. For staple fiber the draw ratios may be lower, and a heat treatment for 0.5–3 min at 210–230°C is given to further crystallize the fibers.

For complete insolubility in water, the hydroxyl groups can be made to react with formaldehyde to a maximum degree of 85%. This takes 10–20 min, however, and only seems interesting in tow processing.

An alternative spinning bath for PVA contains alkali (NaOH, ≥20%), which penetrates more rapidly into the filaments. The polymer concentration in the spinning dope is somewhat higher (18%) and higher molecular weights are used. The filaments solidify more homogeneously and remain round. This process allows even higher draw ratios than spinning in a sulfate bath, for example DR > 15 and tenacity > 1300 mN tex^{-1}. For further improvement of water insolu-

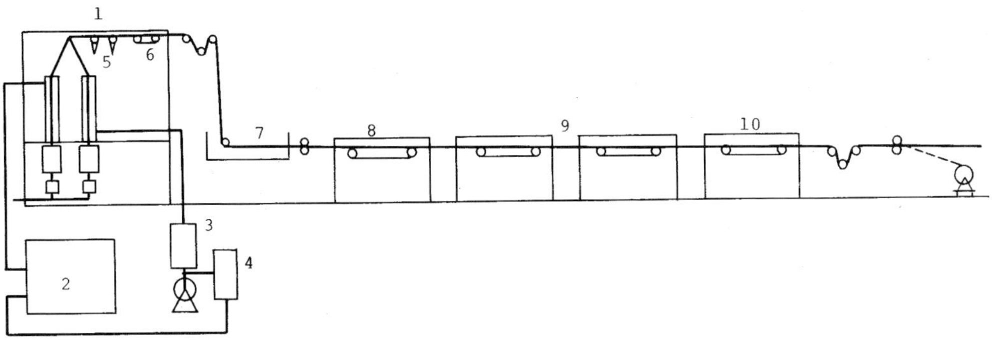

Fig. 17.20. Spinning machine for PVA (reproduced from Ref. 2c): 1, vertical spinning machine, spinnerets at the bottom; 2, 3, 4, spinning bath circulation; 5, 6, godets; 7, washing; 8, drying; 9, drawing; 10, heat treatment; 11, winder.

bility, small amounts of boric acid may be added to the spinning solution, inducing crosslinking between chains. This seems to be the preferred route to high-tenacity filament yarns. The term "gel spinning" is often used for spinning PVA according to this route (see Section 17.7.2).

Applications of PVA The largest application of PVA fibers is in paper and non-woven fabrics, where a fraction of water-*soluble* fiber is often used as a binder. Further applications are in twines, ropes, fabrics (tatami mats), and tarpaulins. "Gel-spun", very strong, PVA fibers have become an important replacement for asbestos in cement reinforcement.

PVA filament yarns are used almost exclusively in industrial applications where high strength is important, such as in fishing nets, ropes and cables, reinforcement of (high-pressure) hoses, and conveyor and transmission belts. PVA is not suitable as a tire cord, however, because of its inadequate fatigue behavior.

17.6 Comparison of Melt and Solution Spinning

Having given a rough technical outline of melt and solution spinning, we can now make a comparison between the speed spinning controlling mechanisms for the two processes: cooling for melt spinning, and diffusion of solvents for solution spinning. We first make a simple analysis of the fiber formation processes, neglecting sublayer effects outside the filaments. An example of the situation is shown for cooling (melt spinning) in Figure 17.21 (at the top).

Initially, the temperature will be flat throughout the filament (at the spinning temperature, for example 300°C) and the surrounding cooling air temperature is assumed to be constant (20°C). Gradually a temperature profile will be formed in

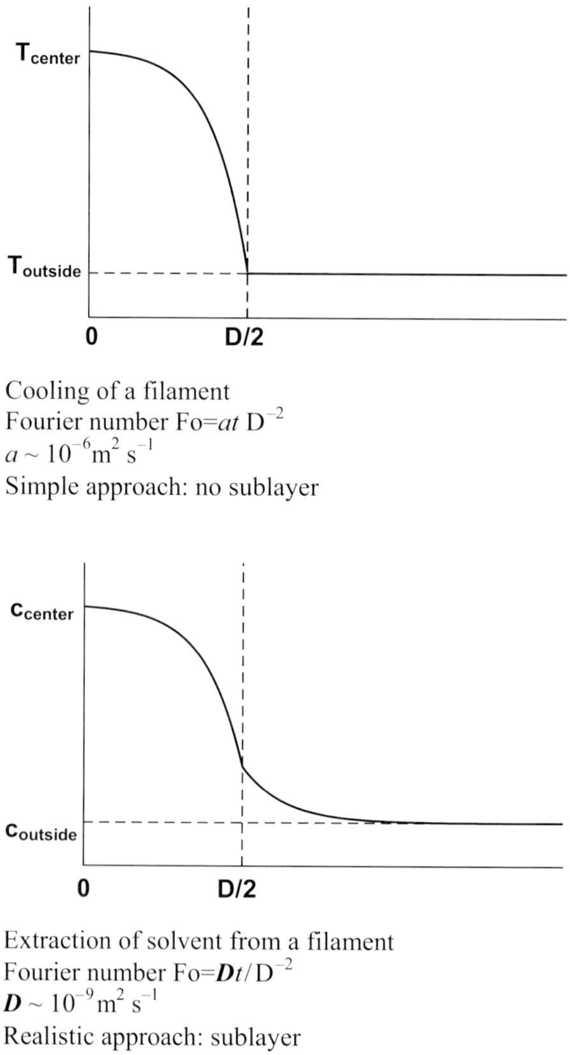

Fig. 17.21. Comparison of melt and solution spinning.

the filament; the surrounding air remains at 20°C. This situation is shown in the figure. For solidification of the filaments we would like to have a temperature in the center of, say, 100°C. The initial temperature difference (300 – 20 = 280°C) must be reduced to 100 – 20 = 80°C. In relative terms, which are used in a Fourier analysis of such a cooling process, the desired "dimensionless temperature" is $80/280 \approx 0.3$.

For solution spinning, the analysis proceeds similarly, but now in terms of concentrations instead of temperatures. For example, we assume an initial solvent concentration of 0.8 (polymer concentration 20%) and solidification at $c = 0.3$

(70% polymer concentration). We assume the spinning bath consists of pure non-solvent; hence the solvent concentration $c = 0$ around the filaments. The desired "dimensionless concentration" then is: $0.3/0.8 \approx 0.375$, not very different from the value for the dimensionless temperature in the cooling process.

We proceed to the Fourier analysis for the cooling and extraction/diffusion processes. The dimensionless Fourier number for cooling is $\mathrm{Fo} = at/D^2$, with a = thermal diffusivity ($a = \lambda/\rho C_\mathrm{p}$), t = time required for cooling and D = diameter of a filament. The thermal diffusivity is of the order of 10^{-6} m^2 s^{-1}. The Fourier number for diffusion processes is $\mathrm{Fo} = \boldsymbol{D}t/D^2$, with $\boldsymbol{D}$ = diffusion coefficient, t = time and D = diameter. Diffusion coefficients in dilute systems are of the order of 10^{-9} m^2 s^{-1}.

The difference by a factor of 1000 between the values of thermal diffusivity and diffusion coefficient can only be compensated for by adapting the values of t/D^2. As a first approximation, filaments in solution spinning should, for example, be made ten times thinner and process times ten times longer than in melt spinning.

For cooling we had a dimensionless temperature of 0.3. A Fourier plot then shows a value of $\mathrm{Fo} = at/D^2 = 0.1$. If we have an average diameter in the cooling process of 250 μm (2.5×10^{-4} m) we can calculate the necessary cooling time: $t \approx 6$ ms. For wet spinning we would find $t \approx 50$ ms. Experienced spinners will say that the typical difference between melt and wet spinning is correct, but that both answers are one order of magnitude wrong. Both melt spinning and wet spinning are slower by a factor of about 10 than calculated with our model, which is *too* simple.

The mistake that we have made is obvious: we have neglected the speed limitation in the sublayer, or, even worse, in sublayers that touch each other in a bundle of filaments. Figure 17.21 (at the bottom) is a graph showing a profile in the filament and in the sublayer around the filament. The example is for concentration as the parameter, for solution spinning, but the same graph can be used for cooling by using T instead of c.

Exact calculations become much more complicated, but a few qualitative statements can be made. In sublayers we are dealing with diffusion: of molecules in an air layer (cooling process *and* dry spinning) or of solvent molecules in a liquid layer (wet spinning). The mobility of molecules in a gas is much greater than in a liquid. For the three types of processes we thus have:

- melt spinning: fast conduction in the filament, fast diffusion in the gas sublayer;
- dry spinning: slow diffusion in the filament, fast diffusion in the gas sublayer;
- wet spinning: slow diffusion in the filament, slow diffusion in the liquid sublayer.

After all, the consequences for filament titers and diameters and process speeds do not differ very much from our first rough estimate. Melt-spun filaments may have a titer of 1–5 dtex, but 20–30 dtex is still possible. Their eventual diameters range from 10 to 50 μm, but they are spun from much larger holes. Dry- and wet-spun filaments often have a titer of 1.7 dtex (13 μm), and are usually spun from holes

Tab. 17.6. Typical process data for melt and solution spinning.

Process	*Typical spinning hole size [μm]*	*Average speed in spin-line [m min^{-1} (m s^{-1})]*	*Height of spin-line or length of bath [m]*	*Residence time in spin-line [s]*
Melt spinning	250–500	3000 (50)	5	0.1
Dry spinning	50–100	300 (5)	5	1
Wet spinning	50–75	30 (0.5)	0.5	1

smaller than 100 μm. Wet spinning speeds are limited to a maximum of 300–400 m min^{-1}, whereas melt-spun yarns can be wound at speeds up to 8000 m min^{-1}. Dry-spinning speeds are in between, at 500–1000 m min^{-1}. Typical data are shown in Table 17.6.

Not surprisingly, there are always exceptions to the rule. For example, acrylic carpet fibers up to 17 dtex are wet-spun, evidently from much larger holes and at lower speeds than are indicated in Table 17.6!

17.7 High-modulus, High-strength Fibers

In the last three decades of the 20th century many advanced fibers were developed (see the surveys in Refs. 8 and 9). Carbon fiber may be described as an inorganic fiber, but is produced by aftertreatment of organic fibers, usually acrylic yarns. High-modulus, high-strength fibers were developed from stiff-chain polymers showing liquid-crystalline behavior in solution or melt. Aramid yarns spun from solution became an important product. How flexible-chain polymers can be superdrawn was discovered, and gel-spinning of polyethylene was developed, adding a valuable product to the spectrum of high-tenacity fibers.

17.7.1 Air-gap Spinning

17.7.1.1 Aramids

Aramid yarns (Kevlar of DuPont, Twaron of Teijin–Twaron) are produced from poly(*p*-phenylene terephthalamide), PPTA (**2**), which is specially developed for fiber spinning and not used in any other application. DuPont had experience with poly(*m*-phenylene isophthalamide) in a fiber product called Nomex for high-temperature applications. The polymer is produced in dimethylacetamide and the solution is dry-spun. This cannot be done with the stiff-chain para–para analogue PPTA. The polymer does not dissolve in organic solvents. A special polymerization route had to be developed, and the discovery of lyotropic behavior of concentrated solutions in sulfuric acid then led the way to the production of a magnificent new fiber material.

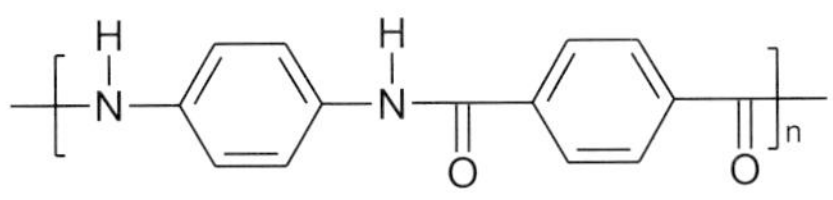

(**2**) PPTA (Kevlar, Twaron)

The polymerization of PPTA starts at low temperature (−10°C) with a solution of *p*-phenylene diamine (PPD) in a mixture of *N*-methylpyrrolidone (NMP) and calcium chloride [10]. The second monomer, terephthaloyl dichloride, TPC, is injected as rapidly as possible. A fast condensation reaction takes place, hydrochloric acid being the condensate. Much heat is evolved and a high cooling capacity is required to keep the temperature below 50°C. An oligomer is formed that becomes insoluble in the reaction medium at a degree of polymerization of about 10. In the rubbery mass further condensation must be enforced by vigorous mixing with a powerful stirrer. After neutralization, washing, and drying a yellow polymer, with a degree of polymerization of 70–100, is obtained in an irregular powder form.

Dissolution of the polymer powder in water-free sulfuric acid can be achieved by freezing the acid and blending it with the powder. This "dry blend" melts at 60–70°C. An alternative is direct dissolution in a kneader.

The key invention for aramid spinning was the discovery of the liquid-crystalline behavior (lyotropy) of PPTA–sulfuric acid systems at polymer concentrations above 10%. Morgan (Monsanto) was the first to describe spinning of fully aromatic polymers from concentrated solutions (in organic solvents), but did not report their liquid-crystalline behavior [11]. Kwolek (DuPont) patented "optically anisotropic" solutions of PPTA in sulfuric acid [12] and Blades (DuPont) further specified the use of an air gap in the spinning process [13].

An example of the liquid-crystalline behavior is shown in Figure 17.22, in which the viscosity–concentration relationship of a solution in a shear field is plotted. Domains of oriented chains are formed and these domains are easily oriented in a flow field ("like tree-trunks in a river"). An elongational deformation is much more effective for achieving a high orientation. In Figure 17.23 the elongational phenomena during spinning, and their effect on molecular orientation, are shown.

The polymer solution is spun at about 85°C. The polymer concentration is about 20%. Domains flow into the spinning hole and orient. The spinning holes are shaped to promote elongational flow: they have a conical entrance and a tapered capillary. In this way almost complete orientation is achieved. The filaments are quenched in cold water, which should not touch the spinneret because the spinning solution would then immediately freeze in. This makes an air gap between the spinneret surface and spinning bath necessary. An air gap is more often applied in other wet spinning processes, but in this case there is an additional very important function. A draft of 5–15 times is applied over the air gap of about 1 cm. The solidification in the cold bath is immediate: the end speed of the process is already achieved at 1 mm below the water surface. The elongational rate in the air gap is comparable with the situation of a drawing neck in cold drawing of

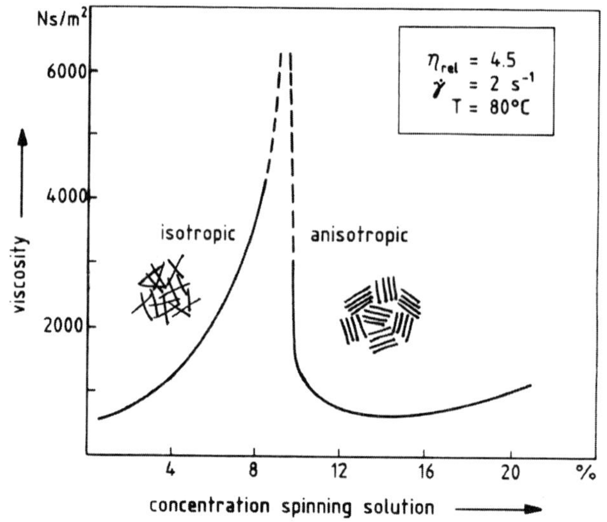

Fig. 17.22. Lyotropic behavior of PPTA in sulfuric acid.

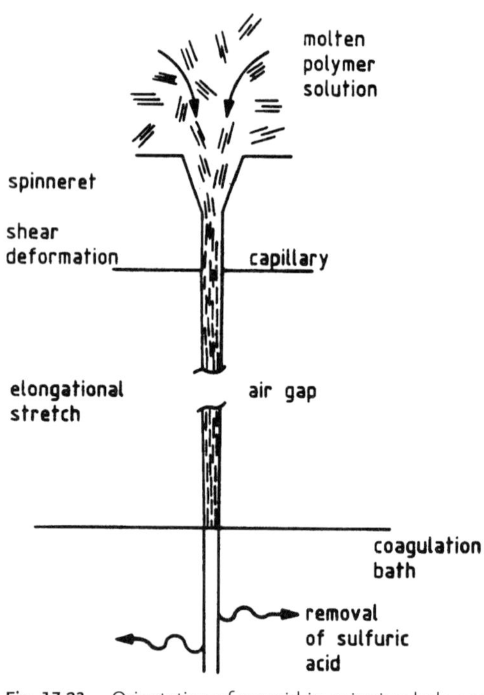

Fig. 17.23. Orientation of aramid in spinning hole and air gap.

flexible-chain polymers. One can imagine how beneficial this draw over the air gap is, to complete the molecular orientation in the filaments.

Solidification of the aramid filaments is called "coagulation" but is in fact a simple freezing-in of the solution. It is not a speed-limiting step. The subsequent removal of sulfuric acid from the solid filaments in the washing process is diffusion-controlled, however, and hence relatively slow.

Thus, the rest of the spinning process is careful washing, neutralization, further washing, drying, and winding. The standard titer of aramid yarns is 1670 dtex f 1000 (1.7 dtex filament titer). The elongation at break is around 3%, the tenacity 2100 mN tex^{-1}, and the modulus $\geq$ 50 N tex^{-1} (75 GPa).

Filaments of 1.7 dtex and density 1.45 g cm^{-3} have a diameter of 12 μm. Assuming a draft over the air gap of 5–15 and a polymer concentration of 20%, and neglecting changes in density, we can make an estimate of the spinning hole diameter by means of Eq. (10); hence $D_{hole} = 60\text{–}100$ μm.

$$D_{hole} = D_{filament}(DR/c)^{0.5} \tag{10}$$

Fibers with an even higher modulus are produced by a post-treatment of PPTA fibers above 400°C under high tension, reducing the elongation to 1.5–2% and enhancing the modulus to above 100 GPa. Special fibers for ballistics may be spun through smaller holes, and have finer filaments, for example <1 dtex.

The aramid process as described is a wet spinning process, but a fairly strange one. Extrusion, polymer lines, and filtration are like those in melt spinning. The heart of the spinning process, spinneret–air gap–spinning bath or "coagulator", is very specific for aramid, but would also be found on machines for other lyotropic systems. The section after the coagulation again deviates from a normal wet spinning process. Since the filaments are not deformable, high tensions can be applied but drawing is not necessary.

The construction material for PPTA in the "melt-spinning section" can be a normal steel because 100% sulfuric acid is not very corrosive. In places with high shear (extruder screw, spinning pumps), nitrated steel, stainless steel, or even ceramic inserts will be used, however. As soon as the acid becomes diluted, corrosion becomes very severe. Spinning plates are therefore made of gold/platinum or with capillary inserts of this material. The construction materials for the spinning bath and washing sections are stainless steel, PVC, PP, or glass.

Technora There is one aramid product with a deviating composition. Teijin's Technora (**3**) is produced from terephthaloyl dichloride (TPC) and a 50:50 mixture of two aromatic diamines: PPD (as in PPTA) and 3,4′-diaminodiphenyl ether. This "kinked" monomer has the advantage that the polymer remains dissolved in the polymerization solvent ($NMP/CaCl_2$). The solution can be spun directly, but is not liquid-crystalline! After washing and drying the yarn is drawn at high temperature (> 500°C) to a high ratio (> 10). In fact Technora is an example of superdrawability (see Section 17.7.2), which is surprising for a fully aromatic polymer, but the high temperature makes the chains drawable.

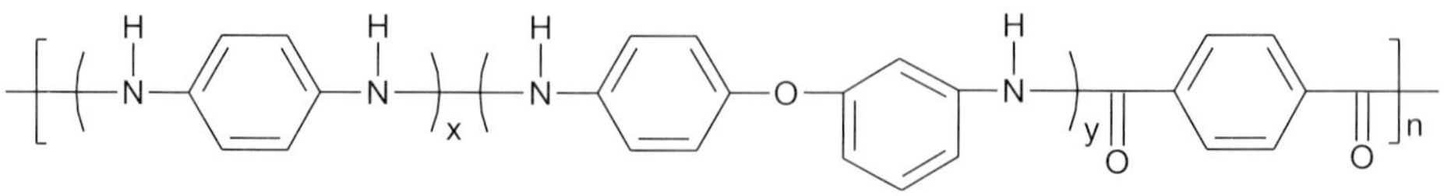

(**3**) 'Technora' ($x/y = 0.5/0.5$)

Applications of aramids Large fields of application for aramid yarns are: antiballistics (bulletproof vests, armored plates); reinforcement of rubber (high-pressure hoses, conveyor and transmission belts; limited application in automobile tires because the material is expensive, but wide application in bicycle tires, on account of its puncture resistance); fiber-reinforced composites (aircraft interiors, sports goods); reinforcement of optical cables; ropes and cables; protective clothing (fireworkers, metalworkers, butchers). Maybe the largest application of aramid fiber, however, is asbestos replacement, in clutches, brakes, and gaskets. To this end filaments are cut into short fibers and fibrillated in a mill. The resulting pulp can be blended with the resins and minerals to produce the clutch faces and brake linings. For replacement of asbestos in gaskets filament yarns can be applied, in the form of thick cords or fabrics.

17.7.1.2 Other Liquid-crystalline Polymers

Higher moduli and strength than in aramid can be reached with polymers having even stiffer chains. We may call them semi-ladder polymers. Examples are the polyazoles polybenzothiazole (PBT, **4**) and polybenzoxazole (PBO, **5**); the latter yarn is produced on small scale by Toyobo (Zylon). An other example is "M5" (**6**), a semi-ladder polymer with higher polarity, yet having a very high modulus; the product is developed by Magellan Systems International, in the USA. PBO and M5 are polymerized in polyphosphoric acid and air-gap spun from this solution.

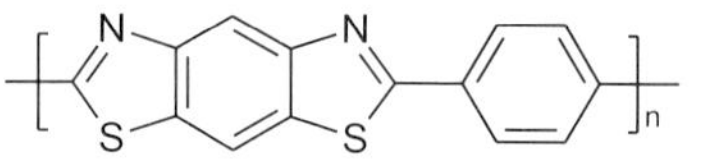

(**4**) PBT

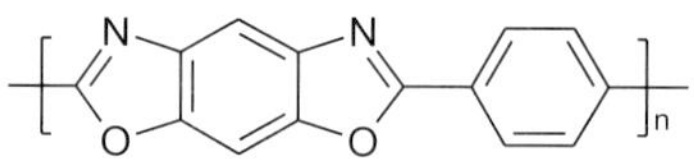

(**5**) PBO

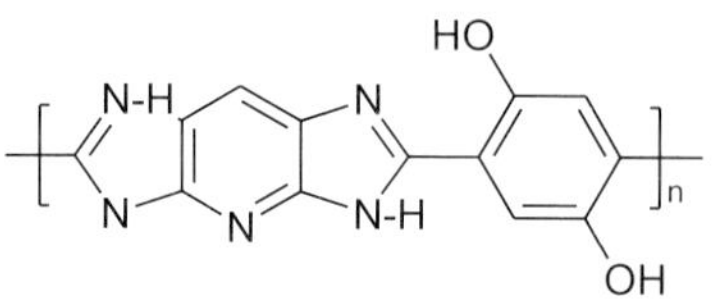

(**6**) 'M5'

17.7.2 Gel Spinning

17.7.2.1 Theory

Some flexible polymers can be drawn to (much) higher ratios than are achieved in a standard process. In the 1970s many publications were issued on the superdrawability of, especially, polyethylene. Ward and co-workers (University of Leeds, UK) showed [14] that polyethylene can be drawn 30 times, and that the modulus increased linearly with the draw ratio, up to values of 60 GPa, but the tenacities remained modest, ≤1 GPa. The undrawn filaments or films (strips) were melt-spun. Too high a molecular weight gave spinnability problems, and, moreover, lower molecular weights were found to be more easily drawable. *Careful* drawing seemed to be the key to high drawablity, in the sense that the drawing speed had to be low and/or the drawing path long. In other words, one should have a low elongational rate, $\dot{\varepsilon} = dv/dx$. For comparison, elongational rates in common drawing processes of melt-spun yarns are in the order of 50–500 s^{-1}. For superdrawing, values of 0.1–1 s^{-1} are preferred, for example a homogeneous drawing on a hotplate of 1 m at a speed of 1 m s^{-1} (60 m min^{-1}).

Superdrawability also proved applicable to polypropylene and slightly more polar polymers, such as polyoxymethylene (POM) [15].

At the same time Pennings (DSM, later of the University of Groningen, The Netherlands) studied the fiber formation from dilute solutions of high molecular weight polyethylene. He started with fibers formed in a Couette device: these were stirring-induced fiber crystals, with the famous "shish-kebab" structure [16]. The best properties were obtained when fibers were slowly withdrawn from a gel layer on a rotor. The molecular weight was above 10^6, the polymer concentration about 1%, the fiber growth rate below 1 m min^{-1}, the moduli around 125 GPa [17], and tenacities far above 1 GPa.

Smith and Lemstra (DSM, The Netherlands) combined the ideas of superdrawability and working with dilute solutions of (ultra-)high molecular weight polyethylene, and added the essential function of a continuous spinning process [18]. Gel spinning was born.

To explain the concept of gel spinning and superdrawing, it is necessary to introduce the notion of entanglement of chains. If two chains form a loop, the disentangling of one of those chains depends on the length between the entanglement point and the chain end. A short chain end will fairly rapidly diffuse through the network of neighboring polymer chains and slip through the entanglement point, without chain break. The "reptation time" necessary for disentanglement scales with $(MW)^3$, according to De Gennes [19]. For a polyethylene chain end with a molecular weight of 100 000 (approximately 4000 monomer units), the time would already be in the order of one second, a very high value for a drawing process.

The consequences for gel-spinning are:

- A high molecular weight is required for an eventually high draw ratio. Draw ratio scales as $DR \sim MW^{0.5}$. In gel spinning the molecular weight is higher than in

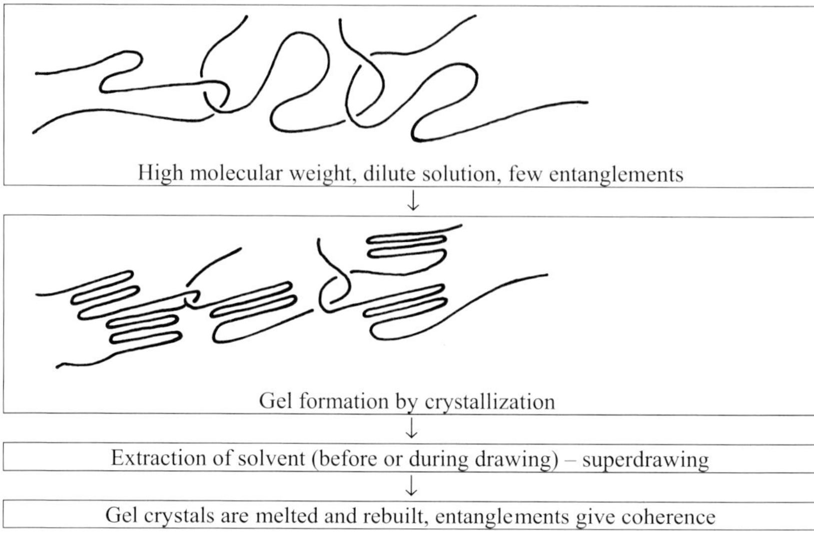

Fig. 17.24. Scheme for gel spinning.

superdrawing of melt-spun polyethylene by a factor of about 10. One may thus expect a draw ratio of $30(10)^{0.5} \approx 100$, which is close to reality.

- The number of entanglements must be reduced, but a few should remain to give the material coherence. This is only possible by working from a dilute solution. Concentrations of 5–10% seem realistic; for ultra-high modulus products an ultra-high molecular weight ($\gg 10^6$), and an even lower concentration (2–5%), will be selected to keep the material spinnable.

Figure 17.24 depicts a scheme for the gel spinning process.

17.7.2.2 **Gel Spinning of Polyethylene**

Gel-spun polyethylene is produced by DSM in cooperation with Toyobo (Dyneema) and Honeywell, formerly Allied Signal (Spectra). From the relevant patents it is fairly obvious that different solvents are used: a volatile solvent in the Dyneema process and a nonvolatile solvent in the Spectra process.

A nonvolatile solvent must be completely removed before drawing, by extraction with a volatile solvent. The dried filaments have a "xerogel" structure, a dry, porous structure which collapses during drawing. A volatile solvent can be partly removed in the spinning step, but may well be further removed by evaporation and squeezing out in the drawing step.

Dyneema yarns have filament titers of 1–2 dtex, Spectra filaments are coarser (4–11 dtex). Filaments of 2 dtex with a density of 0.97 g cm^{-3} have a diameter of 16 μm. How is this diameter produced? If we assume a draw ratio of 100 and a concentration of 10% we can calculate the diameter of the undrawn gel filaments from Eq. (10) to be $D_{\text{gel, undrawn}} = D_{\text{drawn}}(\text{DR}/c) = 16(100/0.1)^{0.5} = 506$ μm. We

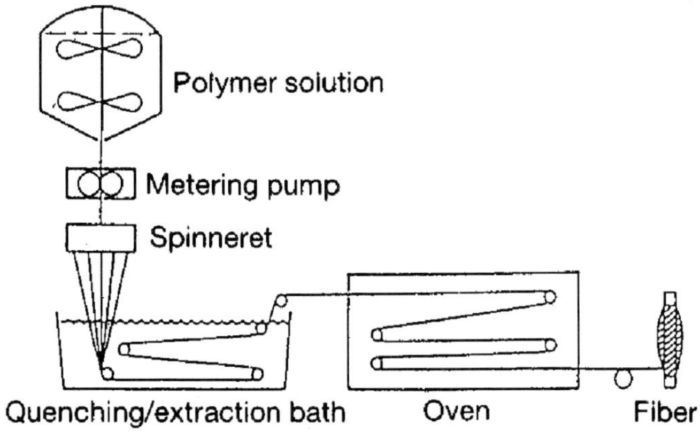

Fig. 17.25. Gel spinning process (reproduced from Ref. 8).

may further take into account that the density of the spinning solution is lower than that of the drawn filaments (~0.8 g cm^{-3} versus 1.0 g cm^{-3}), and that there may be some draft in the spin-line. A realistic estimate of the spinning hole diameter will thus be about 1000 μm, greater than in most melt spinning processes!

The spinning solution extruded from the large holes is above 100°C, in order to keep the polymer in solution and to keep the viscosity acceptable. Fiber formation for gel spinning by quenching is relatively rapid: solidification is by crystallization, rather than by evaporation or extraction of solvent. Complete removal of solvent from the thick filaments proceeds much more slowly. The quench liquid can simply be water and the spinneret cannot be immersed in the water bath because the spinning solution would then freeze in. An air gap must be used, as in aramid spinning, but without the function of increasing the molecular orientation.

The extraction step depends on the solvent used. The nonvolatile solvent (high-boiling, and acting as a plasticizer) must be removed completely before drawing. One can imagine that most of a volatile solvent would be removed in the initial stages of the drawing process (at 100–130°C): at least, this is how DSM (and Toyobo) illustrate it in a simplified process scheme presented in articles and folders (see Figure 17.25).

The process scheme for gel-spun PE shows a stirred vessel for making up the spinning solution, but the use of a kneader in a continuous operation should be possible. It also suggests the integration of spinning and drawing, but this is improbable. As discussed in Section 17.7.2.1. the superdrawing process must be "careful" and slow. Even for very long drawing ovens, an end speed of 200 m min^{-1} seems a limit. (As an example, at this speed, 3.3 m s^{-1}, and an oven length of 33 m one would have an elongational rate ($\dot{\varepsilon} = dv/dx$) of 0.1 s^{-1}). The spinning speed would then have to be about 2–3 m min^{-1}. This is technically difficult and economically unattractive. The output per hole would be as low as 0.4 g solution min^{-1} (0.04 g polymer per minute). A ten times higher output is technically feasible. The technical implication then is that spinning and drawing must be separate steps; there should be ten times more drawing than spinning positions. This can be realized if a collective drawing process is used, such as a warp of yarns drawn in a wide oven.

Applications of gel-spun PE Many applications overlap with those of aramid. Polyethylene has the advantage of its lower density and often a higher tenacity. This is important in ballistics and protective clothing, where Dyneema and Spectra find wide application. A special feature in armored plates is the Spectra Shield or Dyneema UD construction: arrangements of filaments in unidirectional layers bonded by thermoplastic layers in between; the UD layers are arranged crosswise. This construction is more effective than woven structures in ballistic protection.

A second large application, especially of Dyneema, is in marine twine, ropes, and nets, where it replaces polyamide and polyester yarns. Outstanding tenacity, abrasion resistance, UV stability, and low density (floating in water) give it a competitive edge, even at about ten times the selling price of the melt-spun yarns!

Gel-spun polyethylene is too low-melting (about 140°C) to be applied in rubber reinforcement. In composites the curing temperature should not exceed 130°C, and a surface treatment is required for sufficient adhesion. Finally, performance under constant load is restricted because of creep limitations.

17.7.2.3 Other Gel-spun or Superdrawn Fibers

The ultra-high draw ratios for polyethylene are not found for any other polymer, which may not be surprising because the interchain interactions in polyethylene are so low. This makes molecular rearrangements in a drawing process easy. This "polarity concept" was worked out by Smook and co-workers [20]. They took values of maximum draw ratios from the literature and their own experience. For the value of the polarity of a polymer the cohesive energy was taken, which is the energy required to make a polymer chain completely free from interactions with its neighbors. This factor was made dimensionless by taking the ratio with RT_d, T_d (K) being the drawing temperature, and R the gas constant. The best relationship was found to be Eq. (11), where λ represents the draw ratio DR; this is plotted in Figure 17.26.

$$\ln \lambda_{max} = 360 \exp(-E_{coh}/RT_d)^{0.5} \quad (11)$$

Note the maximum value for polyethylene (PE) at $\ln \lambda = 4.6$, implying a draw ratio of 100. The polar polymers, polyester and polyamide(s), which are not superdrawable and not candidates for a gel spinning process, are at the other extreme. There is an interesting group of polymers with intermediate polarity and drawability, however: polypropylene (PP, DR = 47.5), poly(vinyl alcohol) (PVA, DR = 30), polyacrylonitrile (PAN, DR = 28), poly-L-lactic acid (PLLA, DR = 20).

The very good drawability of polypropylene was mentioned when discussing that polymer; it is not gel-spun, but-melt spun fibers are also highly drawable. Poly(vinyl alcohol) is wet-spun, and special versions of this process, at Kuraray,

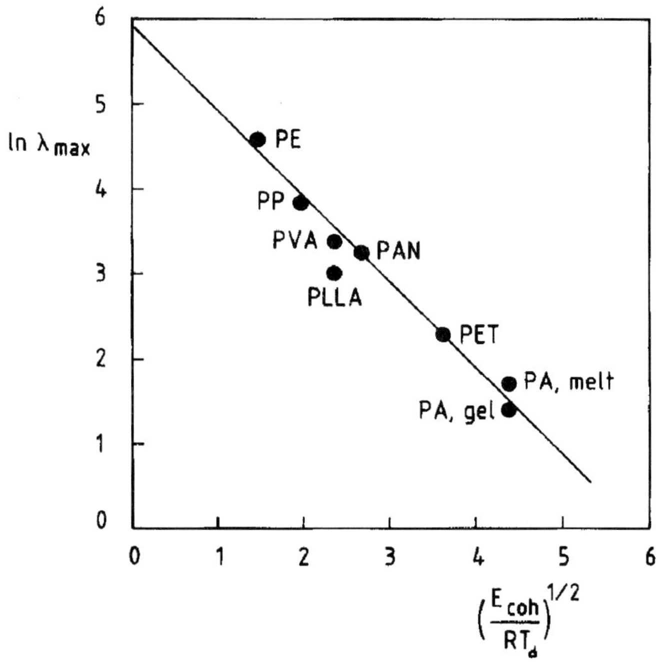

Fig. 17.26. (Super)drawability of flexible-chain polymers (reproduced from Ref. 20).

with high molecular weight polymer, are indeed gel-spun; the fibers replace asbestos in cement reinforcement. Polyacrylonitrile, or a copolymer with a low comonomer level, is drawn 10–15 times in the production of precursor for carbon fiber.

17.7.3 Carbon Fiber

Carbon fibers are not directly spun but are the product of a complicated aftertreatment. Nowadays, most carbon fibers (90%) are produced from an acrylic precursor. Cellulose rayon is no longer applied as a precursor. Production from pitch has been developed, but is still a small-volume business.

17.7.3.1 Carbon Fiber from PAN

Most textile acrylics contain 10–15% comonomers. For carbon fiber precursors lower comonomer levels are used (about 5%); comonomers are selected that promote the reactions in the aftertreatment (methyl acrylate, itaconic acid). Wet spinning is preferred because the cross-section can be controlled better then. In dry spinning skin formation can hardly be prevented and eventually the cross-section collapses into a "dog bone" shape, which is not desirable in carbon fiber applications. Precursor filaments are drawn to much higher draw ratios ($\geq 10\times$) than textile yarns. Precursor filaments must be free from solid particles; sharp filtration of spinning solutions is therefore required. Textbooks even mention spinning under "clean room" conditions.

The aftertreatment proceeds in three steps: cyclization, oxidation, and carbonization/graphitization. All steps are carried out under tension. In the cyclization step (around 200°C) N-containing rings are formed, in a ladder structure: PAN, "Orlon", is transferred into "Black Orlon". This first step is still without loss of material. In the second step this structure is made unmeltable by careful oxidation and stabilization, at temperatures between 200 and 300°C. Some hydrogen is removed, carbonyl groups and double bonds are formed, and the aromatic character is increased. The third step is a pyrolysis reaction carried out in an inert atmosphere, at temperatures which are gradually increased from 400 to 1700°C and further to 2800°C. Below 1000°C most volatile products are formed, such as H_2O, HCN, NH_3, CO, CO_2, N_2, and so on. Note that the nitrogen content of PAN is about 28%, while the weight loss of PAN to carbon fiber is about 50%.

The term "graphitization" is not really correct because a true graphite structure is not formed. The graphite layers, condensed ring systems, are indeed there, but the layers are not strictly coordinated. This "misfit" in the coordination of layers is called turbostratic. Nevertheless, carbon fibers are often referred to as graphite fibers, especially in the USA.

17.7.3.2 Carbon Fiber from Pitch

Pitches can be produced from petroleum and coal tar, preferably from the first feedstock. A useful pitch should have a mesophase character, which means that it must contain ordered, liquid-crystalline domains. The mesophase content can be increased by solvent extraction. A pitch would typically have a molecular weight of 1000 (20 aromatic rings) and a melting point of 300°C.

Pitch can be spun at about 325°C and must then be aftertreated. The cyclization step necessary for a PAN precursor can be omitted in this case. Stabilization by air oxidation and carbonization/graphitization proceeds as for a PAN precursor, but the amount of volatiles is much lower because the pitch does not contain nitrogen.

17.7.3.3 Applications of Carbon Fibers

Carbon fibers are used almost exclusively in composites. Two main types are offered: one with high strength (1.9–3.9 N tex^{-1} tenacity, 140 N tex^{-1} modulus) and one with a high-modulus (2.2 N tex^{-1} tenacity, 280 N tex^{-1} modulus). Carbon fiber moduli are much higher than those of aramid and gel-spun polyethylene.

17.7.4 Other Advanced Fibers

One class of HMHS fibers has not been mentioned: liquid-crystalline polyesters, thermotropic polymers, melt-spun. In the 1970s and 1980s many compositions were studied, in most cases fully aromatic polyesters, in one case PET enriched with large quantities of *p*-hydroxybenzoic acid (pHBA). Only one product "sur-

Tab. 17.7. Survey of properties of HMHS fibers (PET, PA, glass and steel are included for comparison).

Fiber	*Density* [g cm^{-3}]	*Elongation* [%]	*Tenacity* [N tex^{-1}]	[GPa]	*Modulus* [N tex^{-1}]	[GPa]
Aramid (standard)	1.45	3.5	2.1	3.0	50–60	72–87
Aramid (HM)	1.45	2.5 (1.5)	2.1	3.0	75 (115)	109 (167)
Technora	1.45	4.4	2.2	3.2	50	72
Vectran	1.40	≥3.3	2.0–2.5	2.8–3.5	46–62	65–87
PBO	1.56	1.5	3.7	5.8	180	280
PE (standard)	0.97	3.5	2.6–2.8	2.6–2.8	90	90
PE (HM)	0.97	3–3.5	ca. 3.5	ca. 3.5	120	120
C-fiber, HS	1.8	1.5–2.0	1.9–3.9	3.5–7.0	ca. 140	ca. 250
C-fiber, HM	1.8	0.7	2.2	3.9	ca. 280	ca. 500
C-fiber, pitch	2.1	0.2–2.0	0.5–1.0	1.0–2.0	19–380	40–800
Polyester	1.38	10–15	ca. 0.8	1.1	ca. 10	ca. 14
Polyamide	1.18	18–25	ca. 0.8	0.95	ca. 5	ca. 6
E-glass	2.55	1.8–3.2	0.6–1.2	1.5–3.0	28	72
S-glass	2.5	4.0	1.2	3.5	35	87
Steel	7.8	–	≤0.27	≤2.1	27	210

vived" and became a small commercial product: Vectra(n) polymer was developed by Celanese; the fiber is now produced by Kuraray. The polymer has a 75:25 composition of pHBA and HNA, the latter being the code for 2,6-hydroxynaphthoic acid. Mechanical properties are on the level of a standard-type aramid.

There are many polymers with high-temperature resistant fibers, most of them with textile properties (low tenacity, high elongation) (see Table 17.7). Their application is in insulation, hot gas filtration, and suchlike. Large products are meta–meta aramid (Nomex, Teijin-Conex) and polybenzimidazole (PBI). Smaller products are poly(phenylene sulfide) (PPS), several aromatic polyketones (for example poly(ether ether ketone), PEEK) and aromatic polysulfones; see Reference 9, Chapters 8 and 9.

Notation

Symbol	*Name*	*Unit*
a	thermal diffusivity	$m^2\ s^{-1}$
c	concentration (of spinning solutions)	– (or %)
C_p	specific heat	kJ $°C^{-1}$ kg^{-1}
$\dot{\gamma}$	shear rate	s^{-1}
D	diameter (of spinning holes)	μm
$\boldsymbol{D}$	diffusion coefficient	$m^2\ s^{-1}$
DR	draw ratio	–

E	elongation at break (in $TE^{0.5}$)	%
E_{coh}	cohesive energy	$kJ\ kmol^{-1}$
$\dot{\varepsilon}$	elongational rate	s^{-1}
f	overall orientation factor ($0 < f < 1$)	–
f_a	amorphous orientation factor ($0 < f_a < 1$)	–
f_c	crystalline orientation factor ($0 < f_c < 1$)	–
Φ_v	volume flow	$m^3\ s^{-1}$ or $cm^3\ min^{-1}$
Φ_m	mass flow	$kg\ s^{-1}$ or $g\ min^{-1}$
η	melt viscosity	Pa s
$[\eta]$	intrinsic viscosity	$dL\ g^{-1}$
L	length (of spinning holes)	µm
λ	thermal conductivity	$kJ\ m^{-1}\ °C^{-1}\ s^{-1}$
λ	draw ratio (DR)	–
n	index of refraction	–
Δn	birefringence	–
P	pressure	Pa or bar
ΔP	pressure drop	Pa or bar
ρ	density	$kg\ m^{-3}$ or $g\ cm^{-3}$
R	gas constant	$8.3143\ kJ\ K^{-1}\ kmol^{-1}$
T	tenacity at break (in $TE^{0.5}$)	$mN\ tex^{-1}$
T	temperature	°C
T_d	drawing temperature (in Figure 17.25)	K
T_d	decomposition temperature	°C
T_g	glass transition temperature	°C
T_m	melting temperature	°C

Acronyms

BCF	bulked continuous filament
DCM	dichloromethane
DMAc	dimethylacetamide
DMF	dimethylformamide
DR	draw ratio
FOY	fully oriented yarn
HMHS	high-modulus, high-strength fibers
HNA	2,6-hydroxynaphthoic acid
HSS	high-speed spinning
HSSDW	high-speed spin-draw winding
HTS	hot-tube spinning
LSS	low-speed spinning
MFI	melt flow index
NMMO	*N*-methylmorpholine oxide
NMP	*N*-methylpyrrolidone
PA	polyamide

PA66, PA6	polyamide 66, polyamide 6
PAN	polyacrylonitrile
PBI	polybenzimidazole
PBO	polybenzoxazole
PBT	poly(butylene terephthalate); polybenzothiazole
PEN	poly(ethylene naphthalate)
PET	poly(ethylene terephthalate)
pHBA	*p*-hydroxybenzoic acid
PLLA	poly-L-lactic acid
POM	polyoxymethylene
POY	partly oriented yarn
PP	polypropylene
PPD	*p*-phenylene diamine
PPS	poly(phenylene sulfide)
PEEK	poly(ether ether ketone)
PPTA	poly(*p*-phenylene terephthalamide)
PTT	poly(trimethylene terephthalate)
PVA	poly(vinyl acetate); poly(vinyl alcohol)
SDW	spin–draw winding
SDBW	spin–draw bulk winding
TPC	terephthaloyl dichloride

Acknowledgments

The information in this chapter was gathered during 25 years service at Akzo Fibers Research and recently updated, extended, commented upon, and improved by former colleagues now working for Acordis and Teijin–Twaron.

References

1 *Ullmann's Encyclopedia of Industrial Chemistry*, 5th Edition, VCH Verlag, Weinheim, **1987**: fibers, *A10*, pp. 451–655, *A11*, pp. 1–84; high-performance fibers, *A13*, pp. 1–23; cellulose and cellulose esters, *A5*, pp. 375–459; silk, *A24*, pp. 95–106; wool, *A28*, pp. 395–421.

2 *International Fiber Science and Technology Series*, Marcel Dekker, New York: (a) Volume 7, M. Lewin, E. M. Pearce (eds.), *Fiber Chemistry*, **1985** (polyester, polyamide, acrylic, polypropylene, polyvinyl alcohol, wool, silk, cotton, rayon, cellulose acetate); (b) Volume 4, H. A. Krässig, J. Lenz, H. F. Mark (eds.), *Fiber Technology*, **1984** (from film to fiber); (c) Volume 6, I. Sakurada (ed.), *Polyvinyl Alcohol Fibers*, **1985**; (d) Volume 10, J.-P. Donnet, R. C. Bansal (eds.), *Carbon Fibers*, 2nd Edition, **1990**.

3 B. von Falkai (ed.), *Synthesefasern*, Verlag Chemie, Weinheim, **1981**.

4 F. Fourné, *Synthetische Fasern*, Carl Hanser Verlag, Münich, **1995**.

5 M. Mayer, Chapter 17: Fiber Extrusion, in F. Hensen (ed.), *Plastics Extrusion Technology*, Hanser Publishers, Münich, **1981**.

6 A. Ziabicki, *Fundamentals of Fibre Formation*, John Wiley, New York, **1976**.

7 A. Ziabicki, H. Kawai, *High-Speed Spinning*, John Wiley, New York, **1985**.

8 T. Nakajima (ed.), *Advanced Fiber Spinning Technology*, Woodhead Publishing, Cambridge, **1994**.

9 J. W. S. Hearle (ed.), *High-Performance Fibres*, Woodhead Publishing, Cambridge, **2001**.

10 T. J. Veerman, L. Vollbracht, US patent 4 308 374, **1981** (AKZO).

11 H. S. Morgan, US patent 3 414 645, **1966** (Monsanto).

12 S. L. Kwolek, Netherlands patent 6 908 984, **1969** (DuPont).

13 H. L. Blades, Netherlands patent 7 205 836, **1971** (DuPont).

14 G. Capaccio, T. J. Chapman, I. M. Ward, *Polymer* **1975**, *16*, 469 (and numerous other publications).

15 E. S. Clark, L. S. Scott, *Polym. Eng. Sci.* **1974**, *14(10)*, 682.

16 A. J. Pennings, J. M. A. A. van der Mark, H. C. Booij, *Kolloid-Z.* **1970**, *237*, 336.

17 A. Zwijnenburg, Thesis, University of Groningen, **1978**.

18 P. Smith, P. J. Lemstra, UK patent 2 051 661, **1979** (Stamicarbon).

19 P.-G. de Gennes, *Scaling Concepts in Polymer Physics*, Cornell University Press, **1979**.

20 J. Smook, G. J. H. Vos, H. L. Doppert, *J. Applied Polym. Sci.* **1990**, *41*, 105–116.

18
Removal of Monomers and VOCs from Polymers[1]

María J. Barandiaran and José M. Asua

18.1
Introduction

The polymerization of monomers rarely proceeds to completion and there is inevitably a level of unreacted monomer remaining in the polymer. The presence of residual monomer is undesirable, particularly when the monomers are toxic, which is the case for vinyl chloride and acrylonitrile. Monomers such as acrylates and methacrylates can also have strong (and offensive) odors, and the residual volatile components may form an explosive mixture during transportation and storage. Contamination with unreacted monomers is a problem particularly when the polymer is used for food packaging, as well as in biomedical applications and in interior paints. Moreover, polymers may contain small amounts of nonpolymerizable compounds that result from impurities in the raw materials or by-products formed from side reactions during the polymerization process. The presence of such nonpolymerizable compounds results in an increase in the level of the residual volatile organic compounds (VOCs) in the polymer and can adversely affect its ultimate properties. In addition, many polymers are produced in the presence of organic solvents, which are considered VOCs and must be eliminated before use.

The increasingly strict environmental regulations [1, 2] and the higher market sensitivity to environmental and health issues are pushing the polymer manufacturing industry to reduce the amount of monomers and VOCs in polymers. Reduction of VOC emission can also lead to better workplace conditions, reduce risks of fire, reduce nuisance, and lead to economic savings. The industrial importance of VOC removal is reflected in the large number of patents that can be found in the literature. The open literature, however, is scarce. The review in 2002 by Araujo et al. [3], which summarized the principal methods employed for reducing residual monomer content, and the books edited by J. A. Biesenberger [4] and R. J. Albalak [5] reviewing the various aspects of polymer devolatilization, must be highlighted.

Several definitions of VOC have been offered. The European Union (EU) in Di-

1) The symbols used in this chapter are listed at the end of the text, under "Notation".

Handbook of Polymer Reaction Engineering. Edited by T. Meyer, J. Keurentjes

ISBN: 3-527-31014-2

rective 1999/13/EC [6] defines a VOC as any organic compound with a vapor pressure equal or greater than 10 Pa at 293.15 K. The U.S. Environmental Protection Agency (EPA) [7] defines the term VOC as any organic compound that participates in atmospheric photochemical reactions. Those two definitions can lead to contradictory situations. This is the case for acetone, an organic solvent common in polymer production, that is considered to be a VOC following the EU definition, while it is exempted by the EPA's criterion.

The understanding of the origin of residual monomers and VOCs will help to design strategies to polymerize up to very high monomer conversions (the higher the monomer conversion, the less residual monomer) while simultaneously avoiding the formation of new VOCs by side reactions. However, economic constraints that limit the polymerization time as well as intrinsic characteristics of the polymerization technique (such as solution polymerization) always yield systems containing VOCs. Therefore, post-treatments are necessary in order to fulfill the legislation and market requirements. Both the polymerization strategy and the type of post-treatment strongly depend on the polymerization process used for the manufacture of the polymer. The problems associated with monomer and VOC removal in highly viscous polymers produced by mass and solution polymerizations are different from those encountered in low-viscosity, heterogeneous aqueous dispersion systems. Therefore, this chapter will be organized according to the different polymerization processes. It is focused on polymers produced by chain polymerization. In polycondensation, the problem concerns only very specific cases, such as the removal of ethylene glycol in some poly(ethylene terephthalate) routes, or the removal of phenol in phenolic resins; the problems associated with these removals are similar to those found in the treatment of polymer melts. Section 18.2 is devoted to monomer and VOC removal from polymer melts and solutions. Monomer removal in polyolefins is described in Section 18.3, and finally the treatments for VOC removal in waterborne dispersions is discussed (see Section 18.4).

18.2 Polymer Melts and Solutions

Polymers produced by mass and solution polymerization, including the polycondensation processes, represent about 45% of the total synthetic polymer production [8]. In a bulk process, polymerization at high conversions is complicated by the effect of viscosity on initiator efficiency, monomer diffusion, and heat transfer. The efficiency of the initiator radicals can be dramatically lowered due to the cage effect resulting from the low mobility of the radicals, leading to a decrease in the polymerization rate. On the other hand, in polymerizations taking place at temperatures below the glass transition temperature of the polymer, the mobility of the monomer molecules is restricted, causing a severe decrease in the polymerization rate (glass effect).

The use of higher temperatures would reduce the viscosity of the system, thus avoiding some of the drawbacks associated with high viscosities. However, high temperatures have deleterious effects on the properties of the polymer, such as a decrease in the molecular weight and degradation of the polymer. Furthermore, some free-radical polymerizations suffer a substantial depropagation at temperatures above a certain level (ceiling point). This phenomenon is quite well known for methyl methacrylate. Biesenberger et al. [9] showed that sustained exposures in a vented extruder at temperatures higher than 250°C caused poly(methyl methacrylate) to depolymerize to methyl methacrylate.

On the other hand, in spite of the much lower viscosities of solution polymerization, the economically achievable conversion is limited because, due to the low monomer concentration, the polymerization rate is extremely low at high conversions, and consequently unaffordable process times are required.

18.2.1 Devolatilization

Polymer devolatilization is a separation process in which residual VOCs are removed from the polymer matrix by application of a reduced pressure (lower than the equilibrium partial pressure of the volatile component), and/or heat. Stripping agents are also commonly used.

18.2.1.1 Fundamentals

Devolatilization of molten and solution polymers generally involves first the transport by diffusion of volatiles to a polymer/vapor interface and then the transport of volatiles to a gas stream through the boundary layer.

The rate of removal of the volatiles through the interface can be expressed by Eq. (1), where A is the interfacial area and N_{VOC} is the flux of the VOC through the interface given by Eq. (2).

$$\frac{dVOC}{dt} = -N_{VOC}A \tag{1}$$

$$N_{VOC} = -D\nabla c|_{interface} \tag{2}$$

Equation (2) requires the concentration profile of the VOC in the polymer melt that can be calculated from the material balance in the polymer melt. Assuming that the convective flux is negligible, Eq. (3) expresses this balance, where D is the diffusion coefficient of the volatile in the polymer matrix.

$$\frac{\partial c}{\partial t} = \nabla \bullet D\nabla c \tag{3}$$

The diffusion coefficient depends on both the temperature of the system and the concentration of volatiles, and can be estimated through the Vrentas–Duda free-volume theory [10].

The boundary conditions representing the diffusion process are:

$$t = 0 \quad c = c_0$$

$$t > 0 \quad c_{\text{interface}} = c_e$$

c_0 being the initial interface volatile concentration. In the devolatilization of polymer solutions and polymer melts, the diffusion of the volatiles through the polymer is usually the rate-controlling part of the process [4]. Therefore, the volatile concentration at the interface, c_e, is in equilibrium with the concentration of the volatile in the gas phase. The equilibrium concentration is related to the partial pressure of the volatile in the gas phase, P_1, by means of Henry's law, Eq. (4), where H, the Henry's law constant, depends on the temperature, pressure, and nature of the volatile.

$$c_e = \frac{P_1}{H} \tag{4}$$

The interfacial area between the polymer and the gas phase is one of the key parameters in the devolatilization processes, because the rate of VOC removal is proportional to it. The devolatilizers are designed to maximize this interface area. For free-bubble devolatilization, the area of the polymer film that is exposed to devolatilization can be calculated, taking into account the geometry of the equipment. Based on those principles, several models to describe the devolatilization in single-screw extruders [11–14] and twin-screw extruders [15, 16] have been presented.

However, bubbles are produced in most of the devolatilization processes. Bubbles can be formed when the polymer melt is exposed to a pressure lower than the partial pressure of the monomer or the solvent in equilibrium with the polymeric solution. Those bubbles are composed of monomer and solvent. After formation, the bubbles grow by diffusion of monomer or solvent from the polymeric solution, and ultimately they separate from the polymer melt, releasing the monomer and solvent.

Bubbles can also be formed when stripping agents are introduced into the polymer melt/solution under pressure. If the vapor pressure of the stripping agent is higher than the equilibrium partial pressure of the volatile, gas bubbles composed mainly of the stripping agent will be produced. In this case, devolatilization involves diffusion of the volatiles to the surface of these bubbles. The area of the bubbles is determined by the interplay between bubble nucleation, growth, coalescence, and rupture. Under these circumstances the determination of the interface area is not straightforward. Direct visual observation of foaming in different types of devolatilizers has been reported by numerous authors [17–21].

Several models for foam-based devolatilization are available. The fundamentals are reviewed by Lee [22]. Bubble nucleation during devolatilization of polymer melts has been explained by homogeneous nucleation [23, 24], heterogeneous nucleation [25–27], and a mixed-mode nucleation [28]. Yarin et al. [29] considered a secondary nucleation.

Models that account for the growth of the bubbles, assuming either a single spherical bubble growing in an infinite sea of liquid [30, 31] or many spherical bubbles growing together with a thin shell of liquid surrounding them [32], have been developed. More complete models consider bubble motion, coalescence, and rupture [26, 33, 34].

18.2.1.2 Implementation of Devolatilization

The foregoing equations show that devolatilization may be accelerated by acting on the diffusion coefficient, the thermodynamic equilibrium, or the interfacial area. The diffusion coefficient depends on the temperature of the system and on the concentration of volatiles. An increase in the temperature results in an increase in diffusivity and a decrease in viscosity, both beneficial for devolatilization. However, many polymers are thermally sensitive, so there may be a practical upper limit on the temperature to which the polymer may be exposed, as higher temperatures may degrade the polymer. Heat stabilizers have been used to enhance the thermal stabilization [35].

On the other hand, the diffusion coefficient through the polymer strongly decreases as the concentration of the volatile approaches to zero. This drawback can be overcome by introducing an inert substance (usually water), which increases the free volume of the polymer, enhancing the molecular diffusion of the volatile to be removed [36–41]. Furthermore, the inert reduces the partial pressure of the volatile in the gas phase, increasing the driving force for devolatilization. Adding an immiscible liquid to the melt presents the additional advantage that the total vapor pressure of the system increases, and therefore the temperature and volatile concentration needed to produce bubbles in the polymer solution is lower.

Recently the use of supercritical fluids in processing polymers has received much attention, because supercritical devolatilization has the potential for producing high-purity products with lower energy [42–44]. The high solubility of the volatiles in the supercritical fluid enhances the driving force for devolatilization. When the polymer is exposed to the supercritical fluid, it is swollen, and the free volume in the polymer is increased so that diffusion of the volatiles out of the polymer is enhanced.

18.2.1.3 Equipment

There exists a wide variety of devolatilization equipment. According to Biesenberger [4], the processes can be classified into two main categories: non-rotating or still equipment, and rotating equipment, in which devolatilization is enhanced by mechanical agitation.

Nonrotating equipment, such as the falling-strand [5, 29, 45, 46] (Figure 18.1) and falling-film [5, 47] devolatilizers, include mainly flash evaporators and other equipment with specific configurations of the flash chamber, aimed at increasing the exposed interfacial area between the gas and the polymeric phase. In this equipment, the polymer melt/solution is preheated before entering the flash chamber, where the conditions of pressure and temperature are such that the volatiles boil. Vaporization of volatiles is prompted by continuous removal of their vapor through the vacuum outlet port of the flash tank.

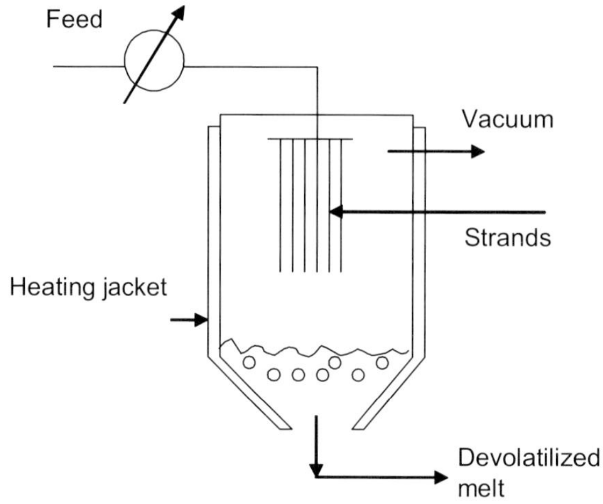

Fig. 18.1. The falling-strand devolatilizer.

In those processes three different regimes can be distinguished [48]. Each of them has a different rate-limiting mechanism, which depends mainly on the difference between the equilibrium partial pressure of the volatile in equilibrium with the melt (P_1) and the total pressure within the vacuum chamber (P). This difference ($P_1 - P$) is often called the "degree of superheat", SH.

The first regime, termed "free boiling", occurs when SH is large and the liquid viscosity is low. This usually happens under volatile-rich conditions. Vapor bubbles initiate and grow very fast, causing convective mixing that enhances mass transfer to the vapor phase. Under this regime, the temperature of the melt can rapidly drop if there is not an external energy input, since the latent heat of evaporation is provided by the sensible heat of the devolatilization mass. For example, under adiabatic conditions an MMA–PMMA melt at 250°C will drop approximately 15°C for every 10 wt.% of the melt that vaporizes. As the melt temperature drops, P_1 will drop and the degree of superheat will diminish. On the other hand, melt viscosity will rise, due to both the removal of volatiles and the temperature drop. These effects cause bubble growth and movement to slow down and the free-boiling regime gradually fades out.

The second regime is termed "bubble growth" and the rate-controlling mechanism is the bubble initiation and growth. The volatile removal rate is slower than in the previous case, and consequently cool-down is reduced. Melt viscosity increases due to depletion of the volatile.

In the third regime, where the viscosity of the melt is very high and the degree of superheat has been reduced due to depletion of volatiles and temperature drop, bubbles are hardly formed and existing bubbles grow very slowly. Under these circumstances, the rate of volatile loss is very slow, as it is controlled by the molecular diffusion at the melt/vapor interface. Therefore, for a very low level of volatiles, flash devolatilizers are not efficient.

Nonrotating equipment is efficient for low-viscosity polymer solutions containing large amounts of volatiles, and particularly useful when dealing with shear-sensitive polymers. However, care must be taken to ensure that they do not have stagnant areas where the polymer could degrade. Excessive foam expansion and polymer entrainment in the volatile vapor stream could be serious problem. Those systems are not operable with solutions that upon flashing retain enough solvent to form a sticky mass, such as in the case of rubbery polymer solutions, because the equipment has little self-cleaning and can become fouled up [49]. On the other hand, low volatile concentrations can hardly be achieved with those equipment, since very long residence times would be required to overcome their poor mass-transport efficiency, which would increase thermal degradation of the polymer. Multi-slit devolatilizers [50–52] were designed to overcome the problem of long residence times of the polymer at high temperatures. In this process, the polymeric solution is fed through an assembly of heated slits which discharge into a flash chamber where vacuum is maintained.

Rotating equipment is used for devolatilization of highly viscous polymer melts/solutions. The rotating parts serve not only to spread and move the polymer forward along the devolatilizer but also to generate and renew the polymer/gas phase interface. Typical devices are wiped-film evaporators and screw extruders.

The wiped-film evaporators (Figure 18.2) employ a rotor with an array of blades attached to the rotor shaft which transport the polymer melt through the evaporator, forming films. As the product goes down along the wall, bow waves developed by the rotor blades generate a highly turbulent flow, resulting in optimum heat and mass transfer. The melt and the vapor flow either in the cocurrent fashion

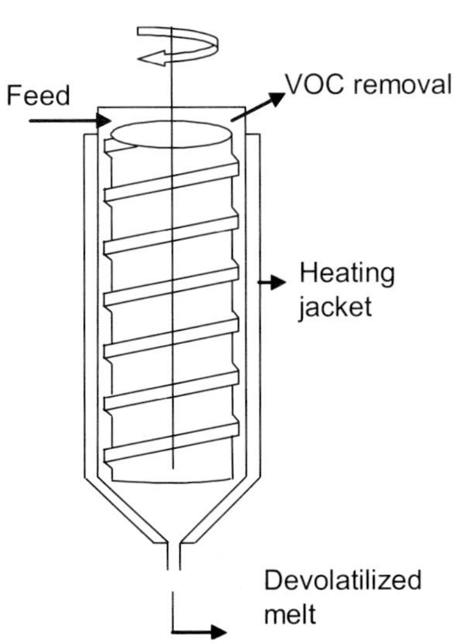

Fig. 18.2. The wiped-film evaporator.

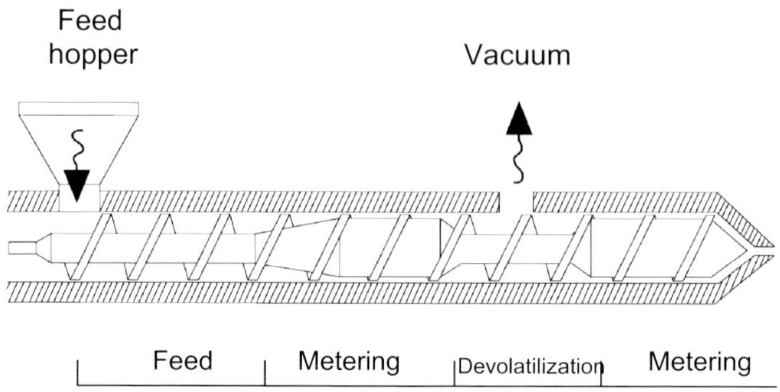

Fig. 18.3. A single-screw extruder.

(mainly for removing highly volatile contents) or in countercurrent (for removing low-volatility contents). The combination of the short residence time, narrow residence time distribution, high turbulence, and rapid surface renewal permits the wiped-film evaporator to successfully handle heat-sensitive, viscous (up to 10^4 Pa s) and fouling-type fluids. This equipment is also used as a final reactor in polycondensation reactions, where the low molecular weight species produced in the reaction must be removed to achieve high molecular weights.

The screw extruders used for devolatilization (Figure 18.3) present a vent zone with a deep channel, where the polymer is exposed to vacuum. Due to the clearance between the barrel and the screw flight, a thin layer of polymer is deposited in the barrel as the screw flight rotates. Evaporation may occur either at the surface of the rotating melt pool or at the surface of the deposit films. Bubbles can be generated inside the melt pool. Due to high viscosity, they hardly diffuse through the pool. However, as the pool rotates, some of them reach the exposed pool surface, increasing the devolatilization efficiency.

Both single- and twin-screw extruders have been used for devolatilization. Counter-rotating, non-intermeshing twin-screw extruders have been widely used in the isolation of polymers from solutions, emulsions, and suspensions, and in the devolatilization-driven reactive extrusion processes [53]. Co-rotating, intermeshing twin-screw extruders have been also used as bulk and solution finishing devolatilizers [54]. One of the main advantages of the twin screw over the single screws is its ability to offer additional surface removal for film exposure and the excellent interchange of pool and film polymer. On the other hand, the area under the vent of a single extruder is vulnerable to stagnation, while the twin-screw geometry reduces this possibility. However, design of twin-screw extruders is more complicated and they are more expensive than the single-screw extruders.

Although selection of the devolatilization system is not an easy issue, Figure 18.4 presents a guide based on the heat sensitivity, viscosity, concentration, and economics.

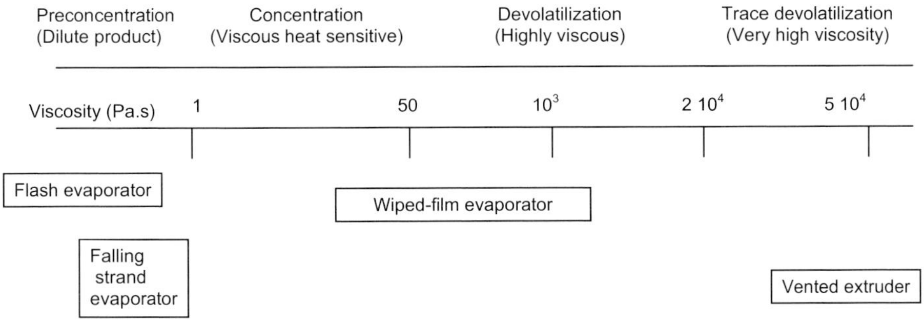

Fig. 18.4. Criteria for choosing the devolatilizer equipment.

18.3 Polyolefins

Polyolefins are nowadays one of the most important groups of commercial polymers, with a consumption of about 100 million tons per year, which represents more than 40% of the total word production of polymers [55]. Because the main monomers involved in the production, ethylene and propylene, have very high vapor pressure, these polymers are produced under pressure. Therefore, the residual monomers are easily removed by depressurization and recirculated to the polymerization zone. However, in the slurry and solution processes the polymerization medium is a solvent. In those cases, the polyolefin solution or slurry leaving the reactor is discharged into a vessel at lower pressure to flash off the solvent and unreacted monomers. The slurry from the polymerization reactor can also pass directly to a centrifuge where most of the volatiles are removed and recycled to the reactor. Further removal of solvent residues is achieved by passing a heated stream of inert gas (mainly nitrogen) through the bed of pellets.

In the gas-phase processes, the polymer is discharged into a cyclone separator, from which residual monomers are recovered and recompressed.

18.4 Waterborne Dispersions

Probably the main step to eliminate VOC in the polymer industry has been the substitution of solvent-based systems by waterborne products. Nowadays, about 5% of the polymers are produced by aqueous dispersion processes, mainly by suspension and emulsion polymerization [8]. The water-based polymers, however, are not totally free of monomer and VOCs, since the reaction does not reach completion and VOCs arising from impurities in the raw materials or from by-side reactions can be present. There are two main ways to reduce the residual monomer content in waterborne polymers: post-polymerization or/and devolatilization.

In order to develop efficient post-polymerization and devolatilization operations, it is critical to know where the residual monomer and VOCs are located, that is, if they are mainly in the polymer particles or in the aqueous phase. Table 18.1 (in Section 18.4.2.1) summarizes the partition coefficients (defined as the ratio between the concentrations of monomer/VOC in the polymer particles and in the aqueous phase) measured for several monomers and VOCs in latexes. It can be seen that the monomers are mainly located in the polymer particles. On the other hand, water-soluble VOCs such as acetaldehyde and *tert*-butanol are mainly in the aqueous phase. It is worth pointing out that the partition coefficient decreases (that is, the monomer partitioning shifts toward the aqueous phase) as the average monomer concentration in the system decreases [56, 57].

18.4.1 Post-polymerization

Post-polymerization consists of adding, after the end of the main polymerization process, fresh radical-generating systems to polymerize the residual monomer. Even though residual monomers may be post-polymerized using radiation [58, 59], post-polymerization is mainly carried out using initiators.

In suspension polymerization, the initiator must be oil-soluble. Those initiators, in order to be fed into the reactor, are commonly dissolved in an organic solvent, which increases the VOC content. On the other hand, the most important monomers (styrene, methyl methacrylate, and vinyl chloride) polymerized in suspension yield polymers with a high glass transition temperature (T_g). The polymerization temperature is usually below the T_g, and hence the polymerization rate decreases sharply at high conversions due to the glass effect. Increasing the temperature of the reactor above the T_g could be a way to improve the monomer conversion. However, this would imply temperatures higher than the boiling point of the water, and in addition, long exposures to high temperatures could degrade the polymer. Therefore, post-polymerization does not seem a promising way to reduce the residual monomer in suspension polymers.

In emulsion polymerization, any kind of free-radical initiation system may, in principle, be used for post-polymerization. However, some alternatives seem to be preferable. Thus, although both thermal and redox initiation systems may be used, and attempts to enhance the activity of thermal initiators by using ultrasound have been reported [60], redox systems are preferred because these systems generate a higher flux of radicals, in particular under mild conditions, leading to shorter post-polymerization times. A wide variety of oxidants and reductants can be combined in different redox systems [61, 62]. Ideally, radicals should have easy access to the place where the monomer is located. Because the residual monomer is mainly located in the polymer particles [57], oil-soluble initiators seem promising. However, as explained above, their addition would increase the VOC content. Consequently, water-soluble redox systems are more convenient. Ilundain et al. [63] demonstrated that independently of the water solubility of the monomer, water-soluble redox initiators yielding hydrophobic radicals (such as organic hydroperoxides) were advantageous for monomer removal by post-polymerization. The reason is that the hydrophobic radicals can enter directly into the polymer particles whereas the hydrophilic radicals must undergo a number of propagation steps before becoming hydrophobic enough to be able to enter into the polymer particle. These growth processes take rather a long time, because under post-polymerization conditions the monomer concentration in the aqueous phase is very low, and the oligoradicals suffer bimolecular termination leading to low initiator efficiency.

Emulsion polymerization is commonly used to produce film-forming polymers, and hence it is generally carried out at temperatures above the T_g of the polymers. Therefore, propagation does not become diffusion-controlled [64], and the kinetics of post-polymerization is not influenced by the reaction temperature differently than the emulsion polymerization stage [65].

The main advantages of post-polymerization treatments are that they can be carried out either in the polymerization reactor or in the storage tank, and no additional equipment is needed. However, only the polymerizable residual volatiles can be eliminated, and in some cases new VOCs are produced from secondary reactions. Thus, formaldehyde is formed when sodium sulfoxylate formaldehyde is used as the reductant [66] and acetone and *tert*-butanol are formed when *tert*-butyl hydroperoxide is used as the oxidant [67]. In addition, inorganic water-soluble initiators may be deleterious to both stability and water sensitivity of the film formed with the latexes.

It is worth mentioning that some initiator systems may modify the polymer microstructure [68, 69], which can be either a problem or an opportunity to extend the range of properties achievable with a given aqueous dispersion of polymers.

18.4.1.1 Modeling Post-polymerization

The processes involved in post-polymerization can be summarized as follows. The radicals that are generated in the aqueous phase must add some monomeric units in order to become hydrophobic enough to enter into the particles. This critical value is referred as z, and is a function of both the monomer hydrophobicity [70] and the type of radical formed from the initiator. The radicals may suffer bimolecular termination in the aqueous phase. On the other hand, the radicals inside the particles can propagate, terminate, or desorb. Under post-polymerization conditions, however, the radical desorption rate is negligible because, due to the low monomer content in the particles, the probability of generating small free-radical species by chain transfer to monomer is very low. A mathematical model based on the mechanisms summarized above is available [71]. Optimal strategies aiming at achieving maximum monomer removal with minimum formation of VOCs have been reported [71].

18.4.2 Devolatilization

Devolatilization of aqueous polymer dispersions is usually carried out using a stripping agent (steam and nitrogen are the most commonly used; air can also be

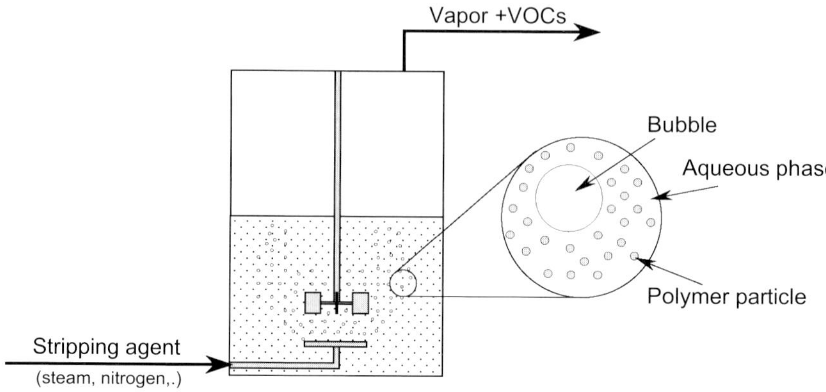

Fig. 18.5. The stripping of latexes.

used, but explosive vapor mixtures can be produced). Figure 18.5 summarizes the devolatilization process in a stirred tank using a stripping agent, which is usually fed in continuously until the volatile content is reduced to the desired level.

18.4.2.1 Modeling

From a microscopic point of view, devolatilization of aqueous-phase dispersions is a mass-transfer process, which involves the following steps: (1) diffusion of the VOCs to the particle surface; (2) transfer from the polymer surface to the aqueous phase; (3) diffusion through the aqueous phase; and (4) transfer from the aqueous phase to the gas phase (Figure 18.6) [72].

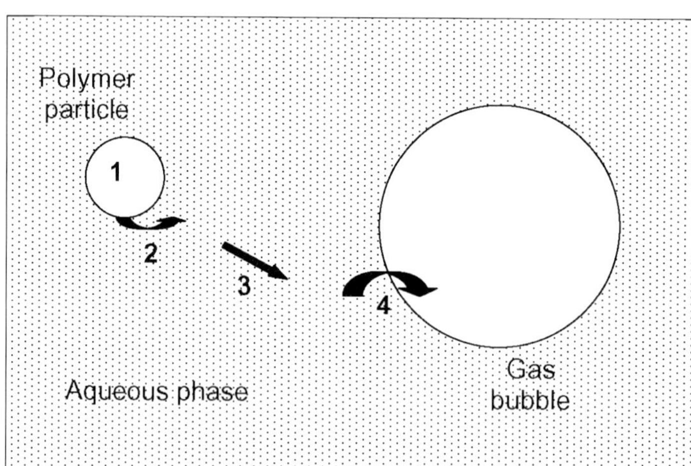

Fig. 18.6. Processes involved in the devolatilization of waterborne dispersions.

Assuming that the contact between the particles and the bubbles is negligible, and that the flow of the gas phase may be described by a well-mixed continuous system, the overall mass balances of the VOC in the different phases are given by Eqs. (5)–(7), where V_i is the volume of the phase i (w = water phase, p = polymer particles, g = gas phase); C_i is the concentration of the VOC in the phase i; $C_{i,ej}$ is the concentration of the VOC in the phase i that would be in equilibrium with the actual concentration of the VOC in phase j; K_{pw} represents the overall mass-transfer coefficient between the polymeric phase and the water phase, K_{wg} is the mass-transfer coefficient between the aqueous and the gas phase; A_{pw} and A_{wv} are the interfacial areas between the polymer particles and the aqueous phase and between the aqueous phase and the gas phase, respectively; y is the molar fraction of the VOC in the effluent; and G is the molar flow rate of the stripping gas.

Polymer particles:

$$\frac{d}{dt}(V_p C_p) = -K_{pw}A_{pw}(C_p - C_{p,ew}) \tag{5}$$

Aqueous phase:

$$\frac{d}{dt}(V_w C_w) = K_{pw}A_{pw}(C_p - C_{p,ew}) - K_{wg}A_{wg}(C_w - C_{w,eg}) \tag{6}$$

Gas phase:

$$\frac{d}{dt}(V_g C_g) = K_{wg}A_{wg}(C_w - C_{w,eg}) - Gy \tag{7}$$

In the devolatilization of polymers in aqueous dispersion, the equilibria are expressed in terms of the partition coefficient of the different VOCs between the polymer particles and the aqueous phase, k_w^p, and the Henry's law constant (H) for the partitioning of VOCs between the aqueous phase and the gas phase. Therefore, the equilibrium concentrations can be calculated from Eqs. (8) and (9).

$$C_{p,ew} = k_w^p C_w \tag{8}$$

$$C_{w,eg} = \frac{P_T y}{H} \tag{9}$$

Tables 18.1 and 18.2 present the values of those coefficients under devolatilization conditions (residual VOC content lower than 5000 ppm) for some monomers commonly used in emulsion polymerization and for the main VOCs found in those latexes after polymerization.

The polymer particle/aqueous phase interfacial area can be obtained from the particle size measurements, but the total transfer area between the aqueous and

Tab. 18.1. Polymer particle/aqueous phase coefficients under devolatilization conditions calculated by R. Salazar [100].

Monomer	Polymer system	T [°C]	k_w^p	Monomer content of particles (50 wt.% solids) [%]
VAc	poly(vinyl acetate)	65	10.5	91.3
VAc	VAc/BA/AA	65	10.9	88.2
BA	BA/S/AA	65	240	99.6
n-Butanol	VAc/BA/AA	65	2.5	71.4
Acetaldehyde	VAc/BA/AA	65	0.5	33.3
tert-Butanol	VAc/BA/AA	65	0.5	33.3

gas phases is difficult to measure. The degree of dispersion of the gas in a liquid is a function of several factors such as the geometry of the contact equipment, the agitation device and speed, the viscosity of the medium, and the gas flow rate. In addition, the physical-chemical properties of both phases, derived by the presence of surfactants or electrolytes in the medium, affect the interfacial area. The determination of the mass-transfer coefficients is not straightforward either. For the water/gas phase, the uncertainty of the estimation of both the interfacial area and the mass-transfer coefficient is overcome through the evaluation of the volumetric mass-transfer coefficient, $K_{wg}A_{wg}$. Estimated values from different authors are presented in Table 18.3.

The polymer particles/aqueous phase mass-transfer coefficient can be determined through the Frössling equation [73], where *Sh* is the Sherwood number, *Re* the Reynolds number, and *Sc* the Schmidt number.

$$Sh = 2.0 + aRe^b Sc^c \tag{10}$$

It is worth pointing out that few studies took into account the pseudoplastic behavior of the polymeric dispersions [74, 75]. There is little and sometimes contradictory information about the presence of the additional gas phase on the solid–liquid mass-transfer coefficients. Sanger et al. [76] observed that in a bubble column, K_{pw} increased upon increasing the gas flow rate. Grisafi et al. [77], observed just the opposite in a stirred tank; this was attributed to reduction of the effective power

Tab. 18.2. Henry's law constants at 65°C calculated by R. Salazar [100].

Compound	H [k Pa]
VAc	26
Butyl acrylate	29
Acetone	3
n-Butanol	1.5

Tab. 18.3. Estimated mass transfer volumetric coefficients.

Equation	*Estimated $K_{vw}A_{vw}$ [$m^3\ s^{-1}$]*	*Reference*
$\mathrm{K_{wg}}A_{wg} = 0.02218(P_g/V_L)^{0.5}u_s^{0.6}$	4.4×10^{-6}	118
$\frac{\mathrm{K_{wg}}A_{wg}d_i}{D} = 21.2\left(\frac{d_i^2 N\rho}{\mu_e}\right)^{1.11}\left(\frac{\mu_e}{\rho D}\right)^{0.5}\left(\frac{d_i u_s}{\sigma}\right)^{0.447}\left(\frac{\mu_g}{\mu_e}\right)^{0.694}$	0.14×10^{-6}	119
$\frac{\mathrm{K_{wg}}A_{wg}d_i^2}{D} = 1.41 \times 10^{-3}\left(\frac{d_i^2 N\rho}{\mu}\right)^{0.67}\left(\frac{\mu}{\rho D}\right)^{0.5}\left(\frac{\rho N^2 d_i^3}{\sigma}\right)^{1.29}$	0.91×10^{-6}	120
$\frac{\mathrm{K_{wg}}A_{vg}d_i^2}{D} = 0.06\left(\frac{d_i^2 N\rho}{\mu_e}\right)^{1.65}\left(\frac{\mu}{\rho D}\right)^{0.5}\left(\frac{N^2 d_i}{g}\right)^{0.19}\left(\frac{\mu u_s}{\sigma}\right)^{0.6}\left(\frac{N d_i}{u_s}\right)^{0.33}F$ $F = (1 + 2.0(\lambda N)^{0.5})^{-0.67}$	0.83×10^{-6}	121
$\mathrm{K_{wg}}A_{wg} = 8.48 \times 10^{-8}u_G^{0.56\pm0.08}N^{2.7\pm0.2} \quad N > 30$ rpm	5.3×10^{-6}	122

per unit volume used to mix the dispersion. Kim et al. [78, 79] measured experimentally the K_{pw} of vinyl acetate diffusion into poly(vinyl acetate) and polystyrene latex particles using a vapor-phase addition method. Values in the range of 5×10^{-7}–7×10^{-7} cm min^{-1} were obtained for diffusion of vinyl acetate into 160 nm PS latex particles.

18.4.2.2 Rate-limiting Steps

For suspension polymers, which are commonly high-T_g polymers and present small polymer particle/aqueous phase interfacial areas, the diffusion through the particle and the transfer from the particle surface to the aqueous phase are often the rate-limiting steps [80]. The ease of removal depends mainly on the polymer particle size, particle size distribution (the presence of large particles increases the diffusion time through the particle largely), and the presence of "glassy" particles. High temperatures speed up devolatilization.

For emulsion polymers, the monomer devolatilization rate is often limited by the mass transfer through the interface between the aqueous phase and the gas phase [57]. For these systems, diffusion in the polymer particles is fast because the T_g of the polymer is low and the particle size is small. In addition, the mass transfer from the particles to the aqueous phase is fast because of the huge interfacial area. For water-soluble VOCs such as acetaldehyde and *tert*-butanol, the rate-limiting step is also the mass transfer through the liquid/gas interface. Therefore, all the process variables that increase the interfacial area between the aqueous phase and the gas phase, such as agitation, geometry of the sparger, or gas flow rate, would improve devolatilization [81].

However, the stripping of emulsion polymers may cause a substantial fraction of the surfactant added to the system to desorb from the polymer particles to form foam, thus leaving the system susceptible to coagulation. Furthermore, the mechanical agitation and the creation of large gas/liquid interfacial areas promote coagulation. In addition, the solids content of the dispersion can vary due to either the evaporation of some water or the condensation of steam.

18.4.2.3 Devolatilization under Equilibrium Conditions

Under some circumstances, such as when the bubble size is very small and low gas flow rates are used, the concentrations of the VOCs in all phases are at thermodynamic equilibrium. In this case, the removal kinetics of the VOCs is given by Eq. (11) [82–84], where t is the removal time and k is the apparent removal rate constant, given by Eq. (12) [84]. M_w is the water molecular weight, E is the mass of latex in the devolatilizer, and X is the mass fraction of polymer in the latex.

$$\frac{C}{C_0} = e^{-kt} \tag{11}$$

$$k = \frac{GM_w(H/P_T)}{E\left[X\left(k_w^p \dfrac{\rho_w}{\rho_p} - 1\right) + 1\right]} \tag{12}$$

This prediction was found to be valid for different processes, such as the stripping of styrene–butadiene latexes when a constant steam sparge rate was used [83, 85], and acrylic copolymers at different solids contents and steam sparge rates [84].

Equations (11) and (12) show that devolatilization is strongly affected by the thermodynamic equilibria of the VOCs between different phases. High values of the polymer particles/aqueous-phase partition coefficient imply that the concentration in the aqueous phase will be low and hence it will be difficult to remove the VOC from the particles. Similarly, a low value of the Henry's law constant means that the concentration of VOCs in the gas phase is low and hence, devolatilization will be difficult. Figure 18.7 shows the kinetics of devolatilization of vinyl acetate, acetaldehyde, and n-butanol in a VAc/BA/AA latex, and that of BA in a BA/S/AA latex by stripping in laboratory-scale equipment, under equilibrium conditions. It can be seen that the devolatilization of BA was slow due to the high affinity (high k_w^p) of BA to the polymer particles. The removal of n-butanol was also very slow because of its high solubility in the aqueous phase and low vapor pressure (a low value of the Henry's law constant).

18.4.2.4 Equipment

Although flash devolatilization has been applied for removal of high-volatility monomers, such as vinyl chloride from poly(vinyl chloride) [86–88] or butadiene from polybutadiene [89], most of the processes described in the literature include the use of a stripping inert gas (usually steam) to improve the devolatilization efficiency.

Several types of equipment have been proposed to allow fast devolatilization without affecting the stability of the dispersion and avoiding foaming. Englund [83] reviewed the different latex strippers used in commercial practice. Batch,

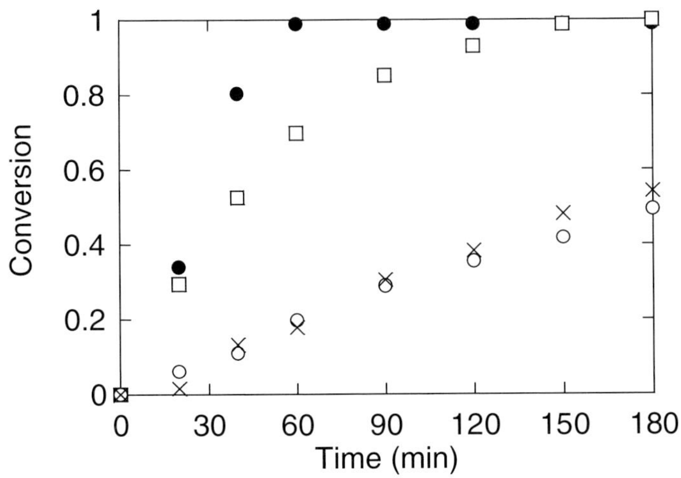

Fig. 18.7. Kinetics of devolatilization of vinyl acetate, acetaldehyde, and *n*-butanol in a VAc/BA/AA latex, and that of BA in a BA/S/AA latex: ●, acetaldehyde; □, VAc; ○, *n*-butanol removal rate by devolatilization from a VAc/BA/AA latex; ×, BA removal rate from BA/S/AA latex, under the same devolatilization conditions [100].

semibatch, and continuous tanks, using countercurrent or cross-flow gas [90–93], packed or perforated columns in countercurrent flow [94–98], and tubular devolatilizers [99] have been used.

Continuous steam stripping using countercurrent flow of an aqueous dispersion and steam requires less steam than batch and cross-flow systems, although the equipment needs considerable space. On the other hand, the column continuous countercurrent stripper requires the least steam of all the processes, and little floor space. This type of stripper is attractive because the short residence time allows short exposure to high temperatures, decreasing the risk of degradation. They are used extensively in suspension processes, such as the production of poly(vinyl chloride). Those stripping systems, however, are difficult to apply to latexes, because of severe foam formation, and the risk of coagulation due to the shear and the relatively high viscosity of the latex. The low T_g of most of the latexes is an additional drawback for using column stripping systems. In these cases, tank reactors are much more convenient.

The undesirable foaming can be suppressed by lowering the temperature, but this reduces the recovery of the low boiling point substances. Chemical defoaming agents may be used, but they may accelerate thermal degradation of the polymer when it is processed at an elevated temperature (as occurs with the polyvinyl chloride) and their use adds contaminants to the latex. One way to control foam formation is by a sudden increase of the pressure. This method has been proven to be very efficient in the stripping of acrylic latexes in tank reactors [100]. Many efforts have been devoted to the design of the devolatilizer to overcome this problem. The injection of the steam at a point in the autoclave such that it passes through the dispersion uncondensed has been found very convenient [101]. In other cases, the gas to be drawn off is passed through porous filter elements [102] or through a condenser adapted to the top of the packed tower [94]. Other authors [90] claimed to feed the aqueous latex as a spray to avoid the contact of the droplets of spray with the side wall of the chamber, and therefore to avoid foaming, which otherwise could cause flooding of the descending latex.

18.4.3
Combined Processes

Devolatilization is an adequate technique to remove both residual monomer and nonpolymerizable volatiles. However, this method is of limited efficiency for the removal of very hydrophobic monomers, for which high flow rates of the stripping gas should be used to appreciably reduce the monomer content, with the corresponding increase in cost, and risk of foaming and coagulation. Under these circumstances, combination of post-polymerization and devolatilization is an attractive alternative.

Some authors have proposed treating the dispersion initially by post-polymerization followed by stripping devolatilization [103–105]. Taylor [106] claimed a treatment in which devolatilization and post-polymerization proceeded concurrently. Salazar et al. [57] have compared the performance of concurrent combined strategies for removal of monomer and VOCs with those of other strategies including post-polymerization, devolatilization, and post-polymerization followed by devolatilization. The results obtained for the removal of BA and the VOCs present in BA/S/AA commercial latex are shown in Figure 18.8. It can be

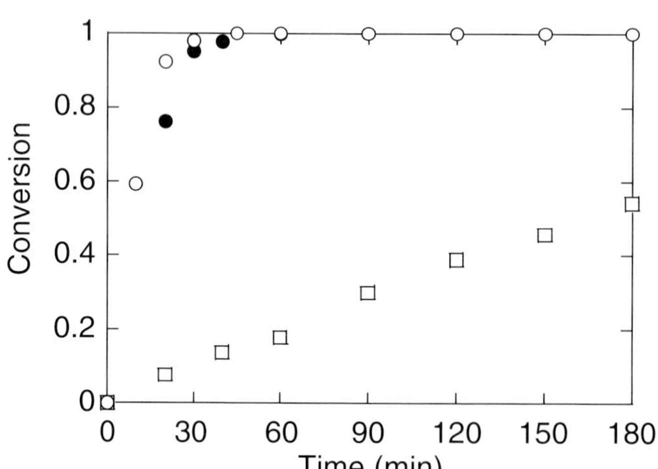

Fig. 18.8. Influence of the strategy on the efficiency of removal of BA from a BA/S/AA commercial latex: □, devolatilization; ●, post-polymerization; ○, simultaneous post-polymerization and devolatilization.

observed that the simultaneous process resulted in the highest efficiency of removal for monomer. For the nonpolymerizable VOCs, almost the same efficiency was obtained in all the strategies using devolatilization.

18.4.4 Alternative Processes

In addition to those conventional techniques widely applied in the industry, other alternatives that try to overcome the problems associated with the use of chemicals in post-polymerization or the high energy consumption associated with devolatilization have been implemented. The modification of the unreacted monomer into a harmless compound, or into a product easier to devolatilize, has been applied to reduce the VOC content. Ozone was found advantageous for vinylic monomer, because it can easily break the double bond and is innocuous [107, 108]. On the other hand, it could also attack the double bond of the polymer chain, causing the degradation of the polymer. Biological degradation through the use of peroxide-generating enzymes was also shown to be able to reduce residual monomer in latex [109, 110]. This process, however, is very slow.

Adsorbents, such as active carbon particles [111, 112], zeolites [111, 112], polymeric adsorbents [111], or ion exchangers [111], have been proven to be useful for VOC reduction in latexes. This treatment, however, requires additional equipment and the regeneration of the adsorbent.

Extraction of the VOCs with suitable solvents has been also extensively investigated. Treatment with organic solvents was found valid for particular cases, such as the removal of high-boiling organic compounds [113] or preparation of polymeric hydrogels [114]. Application of supercritical fluids to cleaning processes was presented in 1998 as an alternative to monomer removal in the near future [115]. Kemmere et al. [116, 117] studied the potential of a post-polymerization process based on supercritical carbon dioxide to remove methyl methacrylate from poly(methyl methacrylate). This process is attractive since CO_2 is very abundant, relatively inexpensive, and environmentally benign on this scale of use.

18.5 Summary

For polymer melts and solutions, the diffusion of the VOCs from the polymer to the polymer/vapor interface is the rate-limiting step. This limitation is more pronounced for highly viscous melts. Therefore, the devolatilizers must be designed to increase both the diffusion coefficient of the volatiles and the polymer/vapor interfacial area. This is achieved by using stripping agents, and equipment with rotating parts that generate and renew the interfacial area. Vented screw extruders are recommended for devolatilizing highly viscous systems. Wiped-film evaporators can handle polymer melts with viscosities up to 10^4 Pa s, and flash and falling-strand devolatilizers are recommended for low-viscosity polymers (viscosities up to 1 Pa s).

Residual monomer present in polyolefins is easily removed by depressurization. The solvent used in the slurry and solution processes is commonly eliminated by flash evaporation and subsequent contact with a heated inert gas stream.

There are two main ways to reduce the VOC content in waterborne dispersed polymers: post-polymerization and devolatilization. Post-polymerization can only be applied to emulsion polymers. Water-soluble redox initiators yielding hydrophobic radicals have been found to be advantageous for monomer removal by post-polymerization. In suspension polymers the devolatilization is limited by both the diffusion through the particle (because of the usually high T_g of the polymer) and the mass transfer from the particle surface to the aqueous phase (because of the large particle size that results in a relatively small interfacial area). Therefore, the temperature and the polymer particle size play an important role in the devolatilization efficiency.

In the devolatilization of emulsion polymers, the mass transfer between the aqueous phase and the gas phase is frequently the rate-determining step. Consequently, the agitation, the gas flow rate, and other process variables or design characteristics that increase the interfacial area between the aqueous phase and the gas phase will improve the devolatilization. However, care must be taken to avoid foaming and coagulation. Column continuous countercurrent strippers are often used in devolatilization of suspension polymers, while stripping in tank reactors is more convenient for low-T_g film-forming latexes.

Notation

A	interfacial area [m^2]
c	concentration of the VOC [mol m^{-3}]
c_e	volatile concentration at the interface [mol m^{-3}]
D	diffusion coefficient [m^2 s^{-1}]
d_i	impeller diameter [m]
E	mass of latex in the devolatilizer [kg]
G	molar flow rate of the stripping gas [mol s^{-1}]
g	gravitational constant [m s^{-2}]
H	Henry's law constant [kPa]
K	overall mass transfer coefficient [m s^{-1}]
k_w^p	partition coefficient between the polymer particles and the aqueous phase
M_w	water molecular weight [kg mol^{-1}]
N	agitation speed [s^{-1}]
N_{VOC}	flux of the VOC through the interface [mol m^{-2} s^{-1}]
P_1	partial pressure of the volatile in the gas phase [kPa]
P_g	power applied to the system [W]
Re	Reynolds number
Sc	Schmidt number

SH	degree of superheat
Sh	Sherwood number
T_g	glass transition temperature [°C]
u_g	linear gas velocity [m s^{-1}]
u_s	superficial gas velocity [m s^{-1}]
V	volume [m^3]
X	mass fraction of polymer in the latex
y	molar fraction of the VOC in the effluent

Greek

ρ	density [kg m^{-3}]
μ	viscosity [Pa s]
μ_e	effective viscosity [Pa s]
σ	interfacial tension [Nm s^{-1}]

Subscripts

g	gas phase
p	polymer particle
w	water phase

Acronyms

AA	acrylic acid
BA	butyl acrylate
S	styrene
VAc	vinyl acetate
VOC	volatile organic compound

References

1 Protocol of Montreal, September **1997**.

2 Protocol of Kyoto, December **1997**.

3 P. H. H. Araújo, C. Sayer, J. G. R. Poço, R. Giudici, *Polym React. Eng.* **2002**, *42*, 1442–1468.

4 J. A. Biesenberger (Ed.), *Devolatilization of Polymers*, Hanser Publishers, Munich, **1983**.

5 R. J. Albalak (Ed.), *Polymer Devolatilization*, Marcel Dekker, New York, **1996**.

6 European Directive 1999/13/EC.

7 J. Brezinski, *Paint and Coating Testing Manual*, 14th Ed., Ed. J. Koleske, American Society for Testing and Materials, Philadelphia, **1995**.

8 http://www.apme.org.

9 J. A. Biesenberger, S. K. Dey, J. Brizzolara, *Polym. Eng. Sci.* **1990**, *30*, 1493–1499.

10 J. L. Duda, J. S. Vrentas, S. T. Yu, H. T. Liu, *AIChE J.* **1982**, *28*, 279–285.

11 G. A. Latinen, *ACS Adv. Chem. Ser.* **1962**, *34*, 235–245.

12 R. W. Coughlin, G. P. Canevari, *AIChE J.* **1969**, *15*, 560–563.
13 G. W. Roberts, *AIChE J.* **1970**, *16*, 878–881.
14 J. A. Biesenberger, G. Kessidis, *Polym. Eng. Sci.* **1982**, *22*, 832–835.
15 G. P. Collins, C. D. Denson, G. Astarita, *AIChE J.* **1985**, *31*, 1288–1295.
16 R. H. Secor, *Polym. Eng. Sci.* **1986**, *26*, 647–652.
17 P. S. Mehta, L. N. Valsamis, Z. Tadmor, *Polym. Process Eng.* **1984**, *2*, 103–128.
18 J. A. Biesenberger, S.-T. Lee, *Polym. Eng. Sci.* **1986**, *26*, 982–988.
19 J. A. Biesenberger, S.-T. Lee, *Polym. Eng. Sci.* **1987**, *27*, 510–517.
20 R. J. Albalak, A. Tukachinsky, T. Chechink, Y. Talmon, Z. Tadmor, in *Polymer Devolatilization*, Ed. R. J. Albalak, Marcel Dekker, New York, **1996**.
21 A. L. Yarin, D. Lastochkin, Y. Talmon, Z. Tadmor, *AIChE J.* **1999**, *45*, 2590–2605.
22 S.-T. Lee, in *Polymer Devolatilization*, Ed. R. J. Albalak, Marcel Dekker, New York, **1996**.
23 J. H. Han, C. D. Han, *J. Polym. Sci.: Part B: Polym. Phys.* **1990**, *28*, 711–741.
24 J. H. Han, C. D. Han, *J. Polym. Sci.: Part B: Polym. Phys.* **1990**, *28*, 743–761.
25 S.-T. Lee, J. A. Biesenberger, *Polym. Eng. Sci.* **1989**, *29*, 782–790.
26 R. W. Foster, J. T. Lindt, *Polym. Eng. Sci.* **1990**, *30*, 621–634.
27 N. S. Ramesh, D. H. Rasmussen, G. D. Campbell, *Polym. Eng. Sci.* **1994**, *34*, 1685–1697.
28 J. S. Colton, N. P. Suh, *Polym. Eng. Sci.* **1987**, *27*, 485–492.
29 A. L. Yarin, D. Lastochkin, Y. Talmon, Z. Tadmor, *AIChE J.* **1999**, *45*, 2590–2605.
30 R. E. Newman, R. H. M. Simon, *73rd AIChE Meeting, Chicago*, **1980**.
31 H. J. Yoo, C. D. Han, *Polym. Process Eng.* **1984**, *2*, 129–151.
32 M. Amon, C. D. Denson, *Polym. Eng. Sci.* **1984**, *24*, 1026–1034.
33 R. W. Foster, J. T. Lindt, *Polym. Eng. Sci.* **1989**, *29*, 178–185.
34 C.-T. Yang, T. G. Smith, D. I. Bigio, C. Anolick, *AIChE J.* **1997**, *43*, 1861–1873.
35 P. Nising, T. Zeilmann, T. Meyer, *Chem. Eng. Technol.* **2003**, *26*, 599–604.
36 J. P. Skilbeck, US 5 380 822, **1995**.
37 W. R. King, F. R. Feder, J. Naveau, J. M. Mlinaric, US 5 478 509, **1995**.
38 J. Barwich, O. Aydin, U. Erhardt, W. Holtrup, US 5 772 851, **1998**.
39 S. Carrá, M. Morbidelli, E. Santacesaria, G. Niederjaufner, *J. Appl. Polym. Sci.* **1981**, *26*, 1497–1510.
40 J. S. Vrentas, J. L. Duda, H. C. Ling, *J. Appl. Polym. Sci.* **1985**, *30*, 4499–4516.
41 K. Ravindranath, R. A. Mashelkar, *Chem. Eng. Sci.* **1988**, *43*, 429–442.
42 S. Alsoy, J. L. Duda, *AIChE J.* **1998**, *44*, 582–590.
43 H. B. Copelin, US 4 306 058, **1981**.
44 J. T. McQueen, S. Krupinski, EP 0 798 314, **1997**.
45 R. E. Neuman, US 4 361 462, **1982**.
46 S. Kuorinski, D. Desroches, EP 855 410 A1, **1997**.
47 Y. Fujii, C. Oda, US 4 361 462, **1982**.
48 D. Todd, R. H. M. Simon, J. Lane, J. A. Biesenberger, in *Devolatilization of Polymers*, Ed. J. A. Biesenberger, Hanser Publishers, Munich, **1983**.
49 W. R. King, F. R. Feder, J. Naveau, J. M. Mlinaric, US 5 478 509, **1995**.
50 F. Balestri, I. Borghi, A. D. Dall, C. Buonerba, S. Matarrese, A. Mattiussi, EP 352 727, **1990**.
51 P. L. Maffetone, G. Astarita, L. Cori, L. Carnelli, F. Balestri, *AIChE J.* **1991**, *37*, 724–734.
52 G. Ianniruberto, P. L. Maffetone, G. Astarita, in *Polymer Devolatilization*, Ed. R. J. Albalak, Marcel Dekker, New York, **1996**.
53 R. J. Nichols, J. T. Lindt, in *Polymer Devolatilization*, Ed. R. J. Albalak, Marcel Dekker, New York, **1996**.
54 J. Curry, F. Brauer, in *Polymer Devolatilization*, Ed. R. J. Albalak, Marcel Dekker, New York, **1996**.
55 P. Galli, G. Vecellio, *J. Polym. Sci.: Polym. Chem.* **2004**, *42*, 396–415.
56 D. Kukulj, R. G. Gilbert, in *Polymeric Dispersions: Principles and Applications*, Ed. J. M. Asua, NATO

ASI Series, Vol. 35, Kluwer Academic Publishers, **1997**.

57 R. Salazar, D. Alvarez, P. Ilundain, L. Da Cunha, M. J. Barandiaran, J. M. Asua, *Progress in Polymer Colloids* **2003**, *124*, 116–120.

58 E. J. Ellis, L. Mager, GB 2 113 694, **1983**.

59 P. J. Reilly, US 458 508, **1984**.

60 L. Da Cunha, J. M. Asua, Spanish Patent Application, 200 101 018, **2001**.

61 G. S. Misra, U. D. N. Bajpai, *Prog. Polym. Sci.* **1982**, *8*, 61–131.

62 A. S. Sarac, *Prog. Polym. Sci.* **1999**, *24*, 1149–1204.

63 P. Ilundain, D. Alvarez, L. Da Cunha, R. Salazar, M. J. Barandiaran, J. M. Asua, *J. Polym. Sci. Part A: Polym. Chem. Ed.* **2002**, *40*, 4245–4249.

64 I. A. Maxwell, E. M. F. J. Verdurmen, A. I. German, *Makromol. Chem.* **1992**, *193*, 2677–2695.

65 P. Ilundain, L. Da Cunha, R. Salazar, D. Alvarez, M. J. Barandiaran, J. M. Asua, *J. Appl. Polym. Sci.* **2002**, *83*, 923–928.

66 D. Becchi, E. Montesoro, L. M. Saija, C. Sempio, EP 0767180, **1997**.

67 L. Da Cunha, P. Ilundain, R. Salazar, D. Alvarez, M. J. Barandiaran, J. M. Asua, *Polymer* **2001**, *42*, 391–395.

68 J. Tangpakdee, M. Mizokoshi, A. Endo, Y. Tanaka, *Rubber Chem. Technol.* **1998**, *71*, 795–802.

69 P. Ilundain, D. Alvarez, L. Da Cunha, R. Salazar, M. J. Barandiaran, J. M. Asua, *J. Polym. Sci. Part A: Polym. Chem. Ed.* **2003**, *41*, 3744–3749.

70 I. A. Maxwell, B. R. Morrison, D. H. Napper, R. G. Gilbert, *Macromolecules* **1991**, *244*, 1629–1640.

71 P. Ilundain, D. Alvarez, L. Da Cunha, R. Salazar, M. J. Barandiaran, J. M. Asua, *Ind. Eng. Chem. Res.* **2004**, *43*, 1244–1250.

72 G. P. Quadri, *Ind. Eng. Chem. Res.* **1998**, *37*, 2850–2863.

73 N. Frössling, *Beitr. Geophys.* **1938**, *52*, 170–216.

74 P. Lal, S. Kumar, S. N. Upadilla, Y. D. Upadhya, *Ind. Eng. Chem. Res.* **1988**, *27*, 1246–1259.

75 Y. Kawase, M. Moo-Young, *Chem. Eng. J.* **1987**, *36*, 31–40.

76 P. Sanger, W.-D. Deckwer, *Chem. Eng. J.* **1981**, *22*, 179–186.

77 F. Grisafi, A. Brucato, L. Rizzuti, *Can. J. Chem. Eng.* **1998**, *76*, 446–455.

78 J. I. Kim, J. W. Vanderhoff, A. Klein, *J. Appl. Polym. Sci.* **1995**, *58*, 279–290.

79 J. I. Kim, J. W. Vanderhoff, A. Klein, *Polymer Testing* **1995**, *14*, 289–306.

80 R. K. S. Chan, C. B. Patel, R. Gupta, C. H. Worman, R. E. Grandin, *J. Macromol. Sci. – Chem.* **1982**, *A17*, 1045–1064.

81 D. Meier, H. J. Warnecke, J. Prüss, *Chem. Eng. J.* **1997**, *67*, 45–53.

82 C. B. Patel, R. Houck, R. Gupta, R. K. S. Chan, *J. Polym Sci., Polym. Chem. Ed.* **1979**, *17*, 3775–3779.

83 S. M. Englund, *Chem. Eng. Prog.* **1981**, *77*, 55–59.

84 G. Mehos, D. Quick, *Separation Science and Technology* **1994**, *29*, 1841–1856.

85 S. Omi, H. Endo, S. Shibata, M. Iso, H. Kubata, *Contemp. Top. Polym. Sci.* **1984**, *4*, 909–921.

86 P. J. Patel, W. Miringoff, US 4 031 056, **1977**.

87 Hoechst Aktiengessellschaft, French Patent 2 318 176, **1975**.

88 T. Yoichi, Y. Masahiro, EP 0 890 582 A1, **1999**.

89 S. Takao, JP 4 239 506, **1992**.

90 B. P. Chandra, J. C. Greaves, V. G. Lovelock, US 4 283 526, **1981**.

91 F. Bötsch, C. Heinze, H. Wolff, US 4 158 092, **1979**.

92 E. D. Hornbaker, A. Nugent, C. P. Loechelt, US 4 184 924, **1980**.

93 R. G. Nickerson, R. T. Bouchard, P. J. Hurtubise, US 4 168 373, **1979**.

94 T. Kano, H. Okada, S. Masuko, H. Itagaki, US 4 526 656, **1985**.

95 R. S. Miller, F. J. Doyle, US 4 130 527, **1977**.

96 R. Scheiler, J. Walther, K. H. Shoenberg, P. Rauth, US 4 007 022, **1977**.

97 K. Nakano, M. Fukuda, M. Shinkai, S. Aruga, US 4 369 092, **1983**.

98 H. Grosse-Wormann, B. Hentschel, P. Rauth, K. H. Schoenberg, B. Terwiesh, H. Winter, US 4 017 445, **1977**.
99 L. C. Baughman, US 3 930 931, **1976**.
100 R. Salazar, PhD Thesis, University of the Basque Country, Spain, **2003**.
101 K. Rostock, J. Birke, R. Wiedholz, J. Bauer, T. Balwe, US 4 224 437, **1980**.
102 O. Zajicek, J. Kirschner, H. Klapp, R. Rompeltien, US 4 256 470, **1981**.
103 A. M. von Arx, R. Studer, J. M. S. Lorteije, P. L. J. Swinkels, EP 1 024 152, **2000**.
104 J. Dobbelaar, W. Hübinger, P. Keller, B. Stanger, US 6 096 858, **2000**.
105 P. Kelly, EP 0 650 977 A1, **1995**.
106 M. A. Taylor, US 4 529 753, **1985**.
107 R. A. Marshall, D. K. Parker, J. W. Hershberger, US 5 498 693, **1996**.
108 G. L. Burroway, D. K. Parker, J. R. Purdon Jr., US 5 268 431, **1995**.
109 G. C. Overbeek, Y. W. Smak, US 5 292 660, **1994**.
110 P. F. Nicks, M. A. Carver, J. M. Relton, WP 91/15 520, **1991**.
111 A. Nuber, A. Sanrer, D. Urban, US 4 795 802, **1995**.
112 R. Racz, *Macromol. Symp.* **2000**, *155*, 171–180.
113 S.-L. Chen, T. H. Wong, R. N. Gujarathi, US 5 516 818, **1996**.
114 K. Nanri, T. Hirabayashi, Y. Yamamoto, R. Amilla, H. Fujima, JP 200 364 125, **2003**.
115 K. Jackson, J. L. Fulton, in *Supercritical Fluid Cleaning*, Ed. J. McHardy, S. P. Sawan, W. Noyes Publisher, New York, **1998**.
116 M. Kemmere, M. Cleven, M. van Schilt, J. Keurentjes, *Chem. Eng. Sci.* **2002**, *57*, 3929–3937.
117 M. Kemmere, M. van Schilt, M. Cleven, A. van Herk, J. Keurentjes, *Ind. Eng. Chem. Res.* **2002**, *41*, 2617–2622.
118 M. Bouaifi, G. Hebrad, D. Bastoul, M. Roustan, *Chem. Eng. Process.* **2001**, *40*, 97–111.
119 J. F. Perez, O. C. Sandall, *AIChE J.* **1974**, *20*, 770–775.
120 R. S. Albal, Y. T. Shah, A. Schumpe, N. L. Carr, *Chem. Eng. J.* **1983**, *27*, 61–80.
121 H. Yagi, F. Yoshida, *Ind. Eng. Chem., Process Des. Dev.* **1975**, *14*, 488–493.
122 T. Poizat, C. Jallut, A. Accary, J. Lieto, *Chem. Eng. J.* **1992**, *48*, 41–48.

19
Nano- and Microstructuring of Polymers[1]

Christiane de Witz, Carlos Sánchez, Cees Bastiaansen, and Dirk J. Broer

19.1
Introduction

Nano- and microstructures are essential for the growth of well-established technologies, such as microelectronics, but also for the development of emerging new ones in the fields of biology, nanotechnology, and information and communication technology. Since the early 1970s the production of microelectronic silicon chips has been a driving force of paramount importance for the development of micropatterning techniques. Traditionally, this field has focused on optical lithography in which a polymer film is structured using light. The patterned polymer is then used to implement and integrate the functional components of the microchip in the silicon wafer. The semiconductor industry focuses on reducing the feature size of transistors and on reaching higher integration densities for faster computation, higher performance, and cheaper production [1]. Besides microelectronics, miniaturization and high-throughput production of complex structures are also required for electro-optical and electronic devices (LCD, LED, FET), nano- and micro-electromechanical systems (MEMS), denser memories, and biosensors or biological arrays [2]. New patterning techniques and materials need to be explored in order to sustain the historical trend in size reduction in the semiconductor industry and in order to make real advances. For this, polymers are an extremely attractive platform due to their ease of processing and their tailorability, which allows fine-tuning of the properties for each application. As an example, the possibility of patterning polymers with special functionalities such as electrical, (semi-)conductivity, electroluminescence, piezoelectric, and/or dielectric properties is important for the development of plastic electronics. Well-defined structured polymer surfaces also find their application in optical and electro-optical devices [3–8]. For instance, special reflectors, made of a polymer film with a well-designed relief structure and coated with a thin-film metallic mirror, redistribute incoming

1) The symbols used in this chapter are listed at the end of the text, under "Notation".

Handbook of Polymer Reaction Engineering. Edited by T. Meyer, J. Keurentjes

ISBN: 3-527-31014-2

light in a very controlled way and are widely used in reflective and transflective liquid crystal displays. Periodic relief structures are also widely employed in transmissive gratings to control light in transmission or reflection [7, 8]. First, we will briefly review different strategies for the patterning of polymer structures and, subsequently, we will focus on new patterning techniques such as photoembossing.

19.1.1 Patterning Techniques

Photolithographic techniques make use of electromagnetic (EM) radiation (UV, DUV, EUV or X-ray photons) to generate a latent image in a photopolymer [9]. The image of a mask can be transferred, either by putting it in direct contact with the photoresist and subsequently irradiating with collimated light or by projecting the image with the help of appropriate optics. Methods involving multibeam holography have also been explored [10]. Independently of the irradiation method, the EM radiation exposure induces a change in solubility in the exposed areas. A subsequent development step in a solvent removes material locally and a relief structure is obtained. This technique suffers from certain intrinsic limitations such as diffraction or, in the case of projection lithography, the depth of focus. The resolution of optical lithography increases if the wavelength of the EM radiation is reduced. The use of KrF lasers (248 nm) allows the routine fabrication of features in the order of 200 nm. In order to reduce the size of these features, shorter wavelengths need to be used. ArF excimer lasers working at 193 nm can lead to resolutions of 150 nm. Shorter wavelengths such as 157 nm (F2 excimer lasers) can be employed. New grades of modified fused silica have been developed showing weak absorption in this wavelength region, making possible the production of lithographic masks. Calcium fluoride meets also the requirements for implementing the optics for the illumination and projection systems [11]. More problematic is the use of EUV and X-ray lithography, due to the lack of transparent materials to generate the masks and the necessary optics for the lithographic process. One possibility is to use masks with openings in a stencil instead of the conventional lithographic masks consisting of metallic (for example, chromium) patterns supported on a transparent substrate. Reflection-instead of transmission-based optics also needs to be implemented. Besides the typical problems of UV lithography, damage to the mask is also induced during exposure. Although the use of these high-resolution lithographies (EUV and X-ray) has been demonstrated in a laboratory scale, their use in mass production is still restricted because of their high cost. In all the cases described the development of appropriate polymeric photoresists is of crucial importance. For example, the resolution can be enhanced by using resists with a nonlinear response that make possible the fabrication of features with a size below the diffraction limitation of the photolithographic technique [12, 13].

Electron beams (EBs) can also be employed for the fine-scale (tens of nanometers) patterning of polymer structures, because of their short (de Broglie) wavelength [14]. Here the charged character of the writing particles limits the resolution because of interparticle electrostatic interactions that produce scattering and therefore lack of resolution. Another important drawback of this technique is that the patterning is performed with a writing procedure and, consequently, the production of structures is laborious and expensive; this restricts their use to very specific applications. Some projection systems are being investigated in attempts to obtain a high-throughput production of patterned structures by EB lithography [15]. Other related techniques involve ion or uncharged particle beams [16].

Other non-photolithographic techniques, such as imprinting (or embossing) or cast molding, have had a large technological impact in different industrial sectors. In the case of imprinting, a rigid master is pressed against an unstructured polymer film [17]. Afterward the master is released and its inverse replica is obtained. Some popular applications of this imprinting technique are the holograms such as those used in safety features in the security industry (credit cards, passports, and money) and compact disks (CDs) [18]. It is also possible to obtain these inverse replicas by casting a polymer precursor onto the master; after curing this is released to obtain the desired inverse structure. These techniques can reach resolutions of tens of nanometers. The aspect-ratio of the replicated structures with high throughput and reproducibility is limited to values of around 1, although higher values have been reached. Among the physical origins of this limitation are the adhesion forces between the mold and the inverse replica, and their mechanical and thermal properties.

Nanomachining and nanomanipulation techniques using scanning probe microscopes (SPMs) have been shown to be feasible for the patterning of polymers on a nanometer scale [19]. Polymer films have been modified by using high-force AFM. In this case the tip indents into the material, generating a hole in it. Surface profiles consisting of not only lines and dots but also more complex three-dimensional structures in a gray-scale fashion have been demonstrated. Attempts to overcome the serial nature of this AFM-based lithography have also been made ("Millipede" project from IBM) using arrays of independent cantilevers (up to 1024 tips). In this case the hole formation rate (bit rate) is thermally assisted by heating the tip. Densities of storage up to 200 Gbytes inch^{-2} [31 Gbytes cm^{-2}] have been achieved using this system. The movement of individual polymeric chains has also been controlled, showing the power of this new nanolithography. Other nanomanipulation techniques under development involve highly focused beams (optical tweezers) able to control the position of polymeric microbeads [20].

The techniques described above are known as top-down approaches that make use of available technologies to reduce the size of the features of the patterned structures. On the other hand, bottom-up techniques make use of the self-assembly concept to build up structures over several length scales. For example, periodic templates are generated using polymer microbeads of monodispersed size [21]. The steric repulsion between beads promotes the formation of controlled three-dimensional packed crystalline structures. These templates, or equivalent ones made of inorganic materials, can be used subsequently to generate the inverse polymeric opal also [22]. Another self-organization approach involves phase separation of block copolymers [23]. In this case the different natures of the blocks can induce their segregation, giving way to well-defined, long-range structures (lamellas, spheres, cylinders) with sizes that range from several to hundreds of nanometers, well controlled by the length of the different blocks. Although they are intrinsically attractive, these techniques have important limitations for the advanced patterning of complex structures because of the difficulty of controlling the size, position, shape, and directionality of the growth structure.

Soft lithography is being investigated extensively for the large-scale generation of relief structures in metallic substrates. In microcontact printing, which is a subtechnique of soft lithography, an elastomeric stamp with a relief structure is used. The stamp is impregnated with a monolayer of a reactive ink containing, for example, thiols or silanes. The relief structure is then brought into conformal contact with a noble metal substrate (silicon or silica may also have been used). In the contact areas, a reaction occurs between the ink and the substrate, and an ink monolayer is obtained which acts as an etch-resist [24, 25]. Although this procedure is intended to be used to pattern metal layers, it has also been employed in the structuring of polymers. For example, the selective growth of polymer in the reacted areas has been demonstrated [26]. The reacted monolayer on the substrate can also act as a surface tension pattern to deposit polymers selectively.

We have briefly reviewed some of the most significant available technologies for generating patterns in polymers. Each of them has specific advantages and deficiencies with respect the feature size, processability, high-throughput capability, cost effectiveness, and functionality of the final patterned structure. These and other new techniques need to be developed further in order to overcome existing problems.

19.1.2 Photoembossing

As an example of a patterning technique in polymers, we will discuss in more detail photoembossing, a photolithographic technique for the generation of polymeric relief structures. In this particular case, the relief structure develops as a result of a material flux induced by a local polymerization. The material flux deforms the surface of the film, which is permanently fixed by the photoinitiated crosslinking process without the interference of a solvent development step [9, 27, 28]. Therefore this principle is attractive in an economic sense for applications that require the structuring of large surfaces.

The maskwise or holographic exposure of a thin film consisting of a blend of monomers in the presence of a photoinitiator induces a flux of material to the exposed (high UV intensity) areas; this may be accompanied by a counterflux of other material to the dark (low UV intensity) regions [29–32]. When the flux and counterflux are equally balanced in volume, a modulation in composition is obtained, whereas the surface of the film remains flat. This process is used successfully, for instance, to make volume holograms based on refractive index modulation [33, 34]. If there is an unequal volume flux, the surface of the film will become modulated because of a volume expansion of the exposed area and a corresponding volume reduction of the dark areas, or vice versa. This principle is applied to make surface holographic gratings. The process in itself is understood and can be described quantitatively in terms of the change in the spectrum of chemical potentials of the various components during the UV exposure and polymerization [35].

Enhancement of this effect can be achieved with a blend that consists of a polymer and a polyfunctional monomer. In that case the polymer forms a stationary phase whereas the monomer diffuses to the exposed areas, resulting in pronounced surface profiles. A general drawback is that the structures form during exposure. This causes changes of the light path during exposure because the deforming surfaces refract the light in a continuously changing way during the formation of the surface relief. It is possible to overcome this problem by creating a latent image in the film during which the formation of a surface relief is minimized. The structure is then formed in a subsequent processing step, for example heating. This is established by using a blend of a polymer and a monomer which has low monomer mobility during the formation of the latent image because of its high viscosity or because of its glassy state. Diffusion and polymerization are then largely inhibited. The exposed areas contain a high concentration of reactive particles, such as free radicals, that are confined to a very localized space. Only when being heated the mobility is greatly increased and diffusion and polymerization are enabled. This then very rapidly deforms the surface to its desired shape, controlled by diffusion parameters and free surface energy. The process, which we will refer to further as photoembossing, is shown schematically in Figure 19.1. The blends may also contain a thermal initiator that initiates polymerization in the dark area simultaneously or later when heated at higher temperatures, thus fixing the surface structure. An additional, but very important, advantage is that contact masking can be applied because the film is solid and does not stick and thus cause deterioration in the mask and the film.

An interesting feature of this process is that by the use of multiple lithographic exposures before the thermal development of the structures, complex surface profiles can be generated. For this multiple exposure technique the fact that the surface profile is not formed before the thermal treatment is optimally beneficial, since the second masking step can also be performed on a solid, flat surface, enabling easy mask alignment. In the rest of the chapter we will discuss structure formation during single and multiple lithographic or holographic steps, and show some examples of photoembossed structures. A newly developed method to enhance the aspect ratio of the relief structures will also be shown.

19.2 Materials and their Photoresponsive Behavior

A typical formulation for the production of photoembossed structures contains a polymer base material, a polyfunctional monomer, a photoinitiator, and, optionally, a thermal initiator. For the coating process, solvents are added. The blend is composed so that after evaporation of the solvent a solid, tack-free film is formed, al-

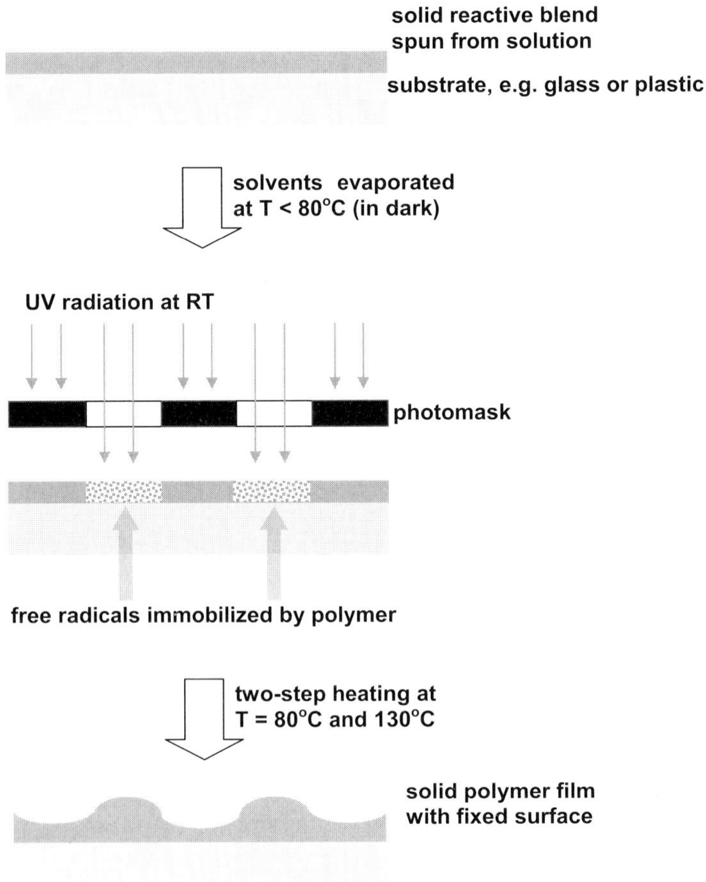

Fig. 19.1. Schematic representation of the photoembossing process.

lowing contact mask exposure. An example of a blend consists of poly(benzyl methacrylate) as the base polymer and di-pentaerythritol tetraacrylate as the monomer. The polymer/monomer ratio is typically 1:1. Poly(benzyl methacrylate) is a solid polymer and has a glass transition of around 54°C. It blends easily and homogeneously with the monomer. Polyfunctional acrylates like di-pentaerythritol tetraacrylate are viscous liquids and basically act as plasticizers for the polymer. If the monomer concentration is too high, the blend becomes tacky. Typically, thin films are formed by spin coating or by a doctor blade and the solvents are removed by heating for 10 min at a temperature well below the decomposition temperature of the initiators: in the case described, this is 80°C.

Figure 19.2 shows the differential scanning calorimetry (DSC) traces of a sample after application as a thin film in the sample holder and drying at 80°C to remove the solvent. From DSC the sample appears stable up to 120°C, above which the thermal initiator activates the thermal polymerization of the acrylate monomer,

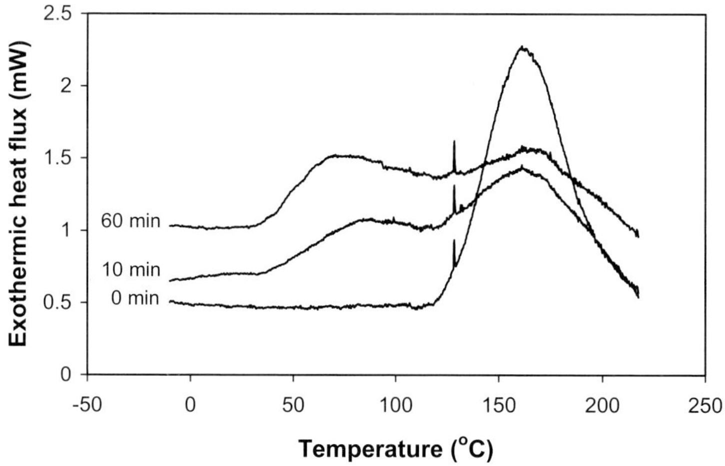

Fig. 19.2. DSC traces of the reactive photoembossing mixture (pre-heated in the dark at 80°C to remove solvents) measured after 0, 10 and 60 min of pre-exposure with a UV lamp (365 nm, 5 mW cm^{-2}).

which reveals itself as the large exothermic peak in the DSC thermogram. When the sample is exposed to UV light before heating, polymerization already starts above 30°C. An isothermal DSC scan for 60 min during the UV exposure at 20°C showed no significant conversion of the acrylate monomer. This indeed shows that monomer mobility at this temperature is very limited, inhibiting further polymerization. Only when heated above 30°C, monomers are gaining diffusional freedom in order to allow propagation of the polymerization. Upon further heating, monomer conversion proceeds up to a temperature of around 80–90°C above which a small decrease is seen, most probably due to a diminishing concentration of free radicals due to termination. At temperatures above 120°C the polymerization propagation rate increases again because of free radical generation from the thermal initiator.

19.3 Single-exposure Photoembossing

The process of latent activation as demonstrated by photo-DSC is also exactly what happens when a thin film of the sample is exposed to UV through a mask. In that case a latent image is formed at the exposed sites consisting of free radicals of the photoinitiator with only minor conversion of the acrylate groups. Only after heating does further polymerization take place. The typical processing sequence has been shown in Figure 19.1. First a thin film is spun on a glass substrate. The film thickness is typically 3 μm after evaporation of the solvent by moderate heating at 80°C without premature polymerization. Cooled down to room temperature, the film is exposed to 365 nm light from a mercury lamp (365 nm bandpass filter, 2.5 mW cm^{-2}) through a mask. After this exposure the surface deformation is present, but is very small in relation to the desired structure. The exposed area contains a high concentration of free radicals, immobilized by the viscous matrix. By heating the film to 80°C, well above the glass transition temperature, monomer mobility is enhanced and the polymerization reaction sets in. This polymerization induces diffusion of the monomer to the exposed areas without noticeable counterdiffusion of polymeric material. This process swells the exposed areas, whereas the unexposed

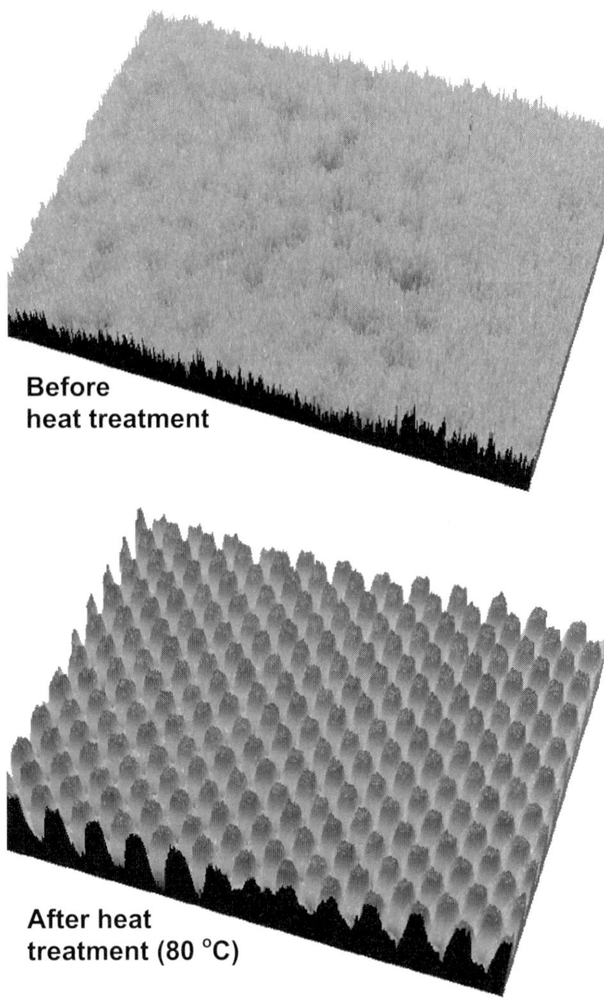

Fig. 19.3. Interferometric microscope picture of a sample surface exposed with a UV dose of 0.2 J cm^{-2} through a checkerboard mask (feature size 10 μm) before and after heat treatment (5 min, 80°C).

areas are depleted. Figure 19.3 shows interferometric microscope picture of a surface before and after the heating step. Before heating, but after UV exposure, the surface remains relatively unstructured and flat. Directly upon heating, the surface develops itself into the relief that was imposed by the checkerboard mask. Note that in these figures the aspect ratio is heavily exaggerated: in a film of 3 μm high, differences of 1 μm are easily feasible. The structure obtained is fixed permanently either by a flood UV exposure or by heating to 130°C, at which temperature the thermal initiator finishes off the polymerization reaction.

The surface profile that is created is determined by a set of parameters. First of all, of course, the mask dimensions are important, but exposure conditions (dose, temperature) and heat treatment conditions have also been proven to matter. Some examples of surface structures obtained by various mask configurations are shown in Figures 19.4 and 19.5. The structure size can be varied between approximately 1 and 20 μm. Below these dimensions the structures become small and the process is hindered by the normal limitations of lithographic processes, such diffraction. At larger dimensions the diffusion lengths of the monomer become too great, resulting in high edges and deeper central parts of the relief (compare, for instance, the two top AFM pictures in Figure 19.4), a phenomenon that is perfectly understood from the models available for this process [35]. The shape can also be affected to some extent by the exposure dose, as can be seen by comparing the two lowest AFM structures in Figure 19.4, in which an overdose of UV light has reshaped the structures that are obtained.

The influence of the exposure energy on the height of the structures is demonstrated in Figure 19.6. This figure again demonstrates that surface deformation of the samples takes place only to a minor extent when they are kept at room temperature, even for a prolonged period of 24 h. The minor relief formation under these conditions is practically independent of the UV dose, but during heating, in this case in one step directly to 130°C, the structures develop and the UV dose appears to be of great importance. Too low energy gives only a small effect, for the obvious reason that photoinitiator conversion is low. A maximum in structure height is obtained at UV doses between 100 and 200 mJ cm^{-2}. At higher doses the structure height decreases again. This is explained by the fact that at doses that are too high (overexposure) some cross-exposure of the intended dark area cannot be avoided. At the same time the exposed area becomes too crosslinked, thus hindering further diffusion in a later stage of the process.

Next to the UV dose, the periodicity also seems to have an influence on the feature's height, as is demonstrated in Figure 19.7. In general, smaller dimensions lead to a lower relief height. This can be understood in terms of monomer deficit at smaller pitches. At greater periodicities, for instance above 20 μm, the structures have some problems filling the whole exposed area, and swelling occurs predominantly near the edges. The diffusion in the reacting medium apparently occurs typically on a length scale of some micrometers. Further optimization outside this preferred region is still possible, and an optimum balance between diffusion and polymerization reaction can be found at other diffusion and curing temperatures. However, comparing the results of the single-step (130°C) baking procedure with

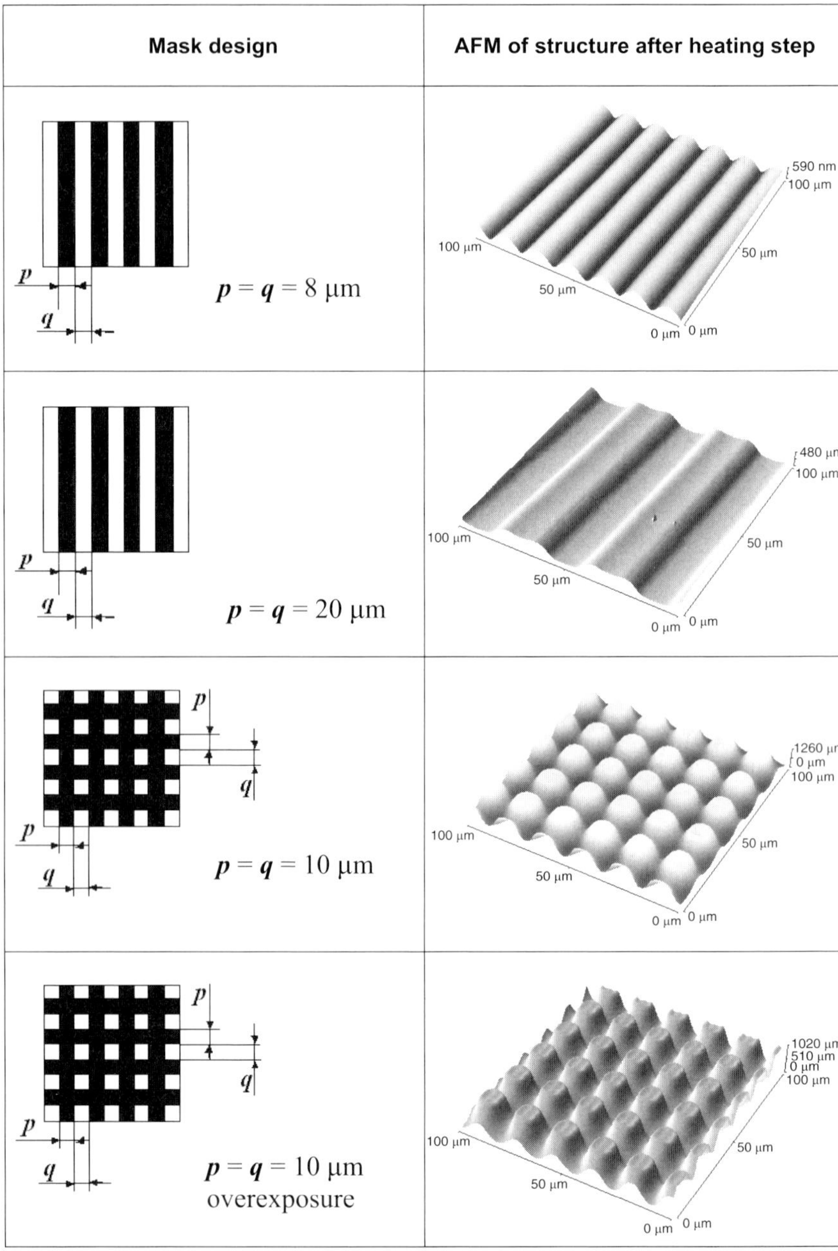

Fig. 19.4. Examples of mask designs and the surface relief that they generate after UV exposure and thermal treatment (10 min, 80°C; 10 min, 130°C).

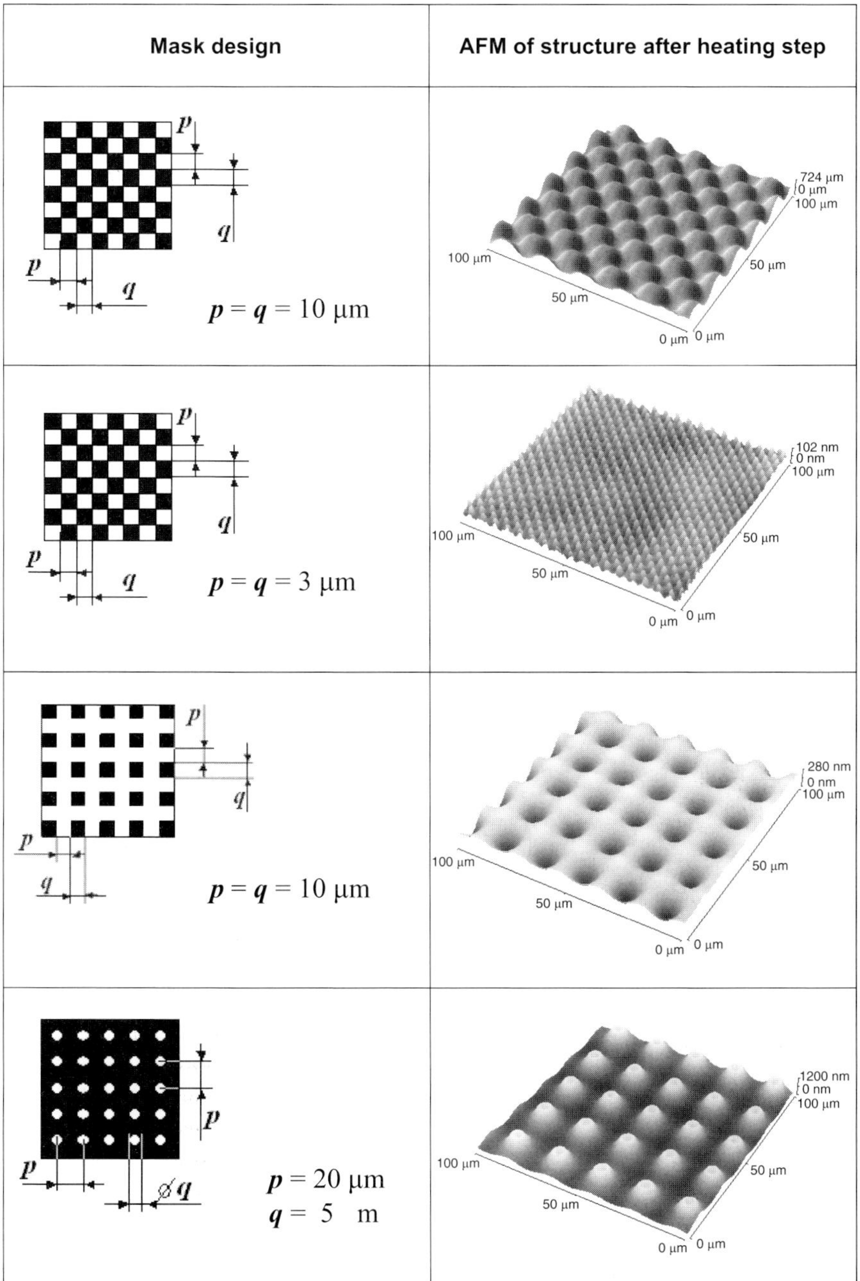

Fig. 19.5. Examples of mask designs and the surface relief that they generate after UV exposure and thermal treatment (10 min, 80°C; 10 min, 130°C).

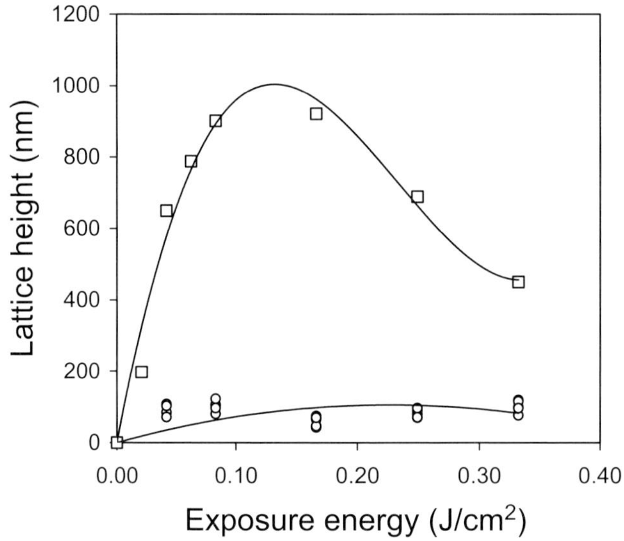

Fig. 19.6. Surface relief heights after exposure through a line mask with a periodicity of 20 μm and an aperture of 50% after storage in air at RT for 25 h (○) and after a subsequent heating step at 130°C in air for 5 min (□).

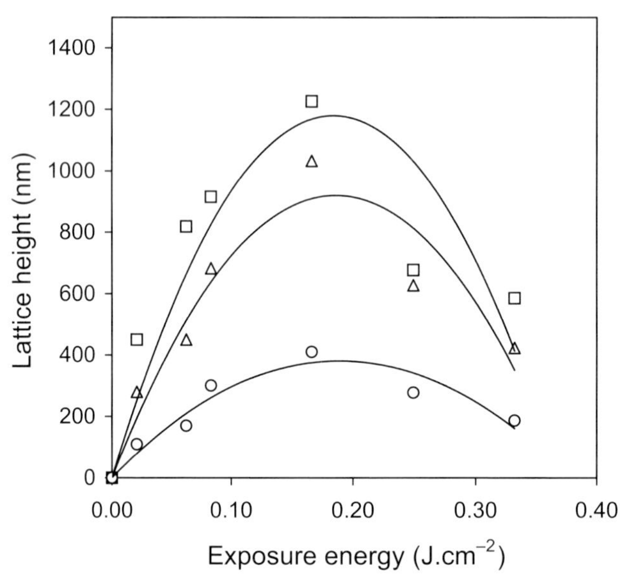

Fig. 19.7. Surface relief heights after exposure through a line mask with periodicities of 5 μm (○), 10 μm (Δ), and 20 μm (□). The films were exposed at RT and heated to 80°C for 5 min, and subsequently to 130°C for another 5 min.

the dual-step (80 and 130°C, respectively) reveals that the one-step process is about as effective as the dual-step procedure. This leads to the conclusion that the balance in diffusion and polymerization kinetics is in favor of diffusion.

19.4 Dual-exposure Photoembossing

Figure 19.3 has already shown that the surface of the reactive polymer film remains almost flat before the heating step. This is confirmed in Figure 19.6, which shows that the height of the structures before the heating step, even after prolonged storage at room temperature, remains below 100 nm. The fact that the pristine structures are small means that the optical behavior of the initially flat film is hardly affected by the exposure. This is very relevant for controlling structure shape during the exposure step. A surface that deforms redirects the rays of the masked image, or in other words the surface profiles start acting as a lens. But the fact the surface profile remains almost unaffected is of even greater importance if one wants to employ a second exposure superimposed on the first one. The procedure to perform this double exposure can be as follows. A first exposure though mask 1 is given at a dose somewhat smaller than the dose corresponding to the maximum profile as shown in Figures 19.6 and 19.7. Then mask 1 is removed and mask 2 is brought into place while the film is maintained at room temperature. A second exposure is performed with a dose equal to the dose of the first exposure step. The film now contains the latent images of two masks which will develop during the heat-diffusion step at 80°C or higher. Figure 19.8 shows some examples of the masks that are used and the resulting surface profiles obtained. The structures that were presented individually in Figures 19.3 and 19.4 are now superimposed, one on top of the other. The AFM pictures show that the formation of complex structures is very readily possible and that the second structure is not hindered in its resolution by the presence of the first structure. In the experiments shown, equal doses for the two mask exposure steps are used. Of course, one has the freedom to choose them differently and play with the dominance of each mask step. In the case where UV doses are selected at the maximum sensitivity for a single exposure or higher dose, it must be realized that local overexposure of doubly exposed areas then lead to smaller features as well.

19.5 Complex Surface Structures from Interfering UV Laser Beams

Patterned irradiation can also be obtained by means of holographic exposure. In this case two beams (or possibly three or more) coming from a laser are made to interfere in the area of the sample. In order to do this, a conventional holographic setup is employed. The *s*-polarized UV beam (351 nm) from an Ar laser is split in two equal-intensity beams. These are made to overlap again, obtaining as a result a

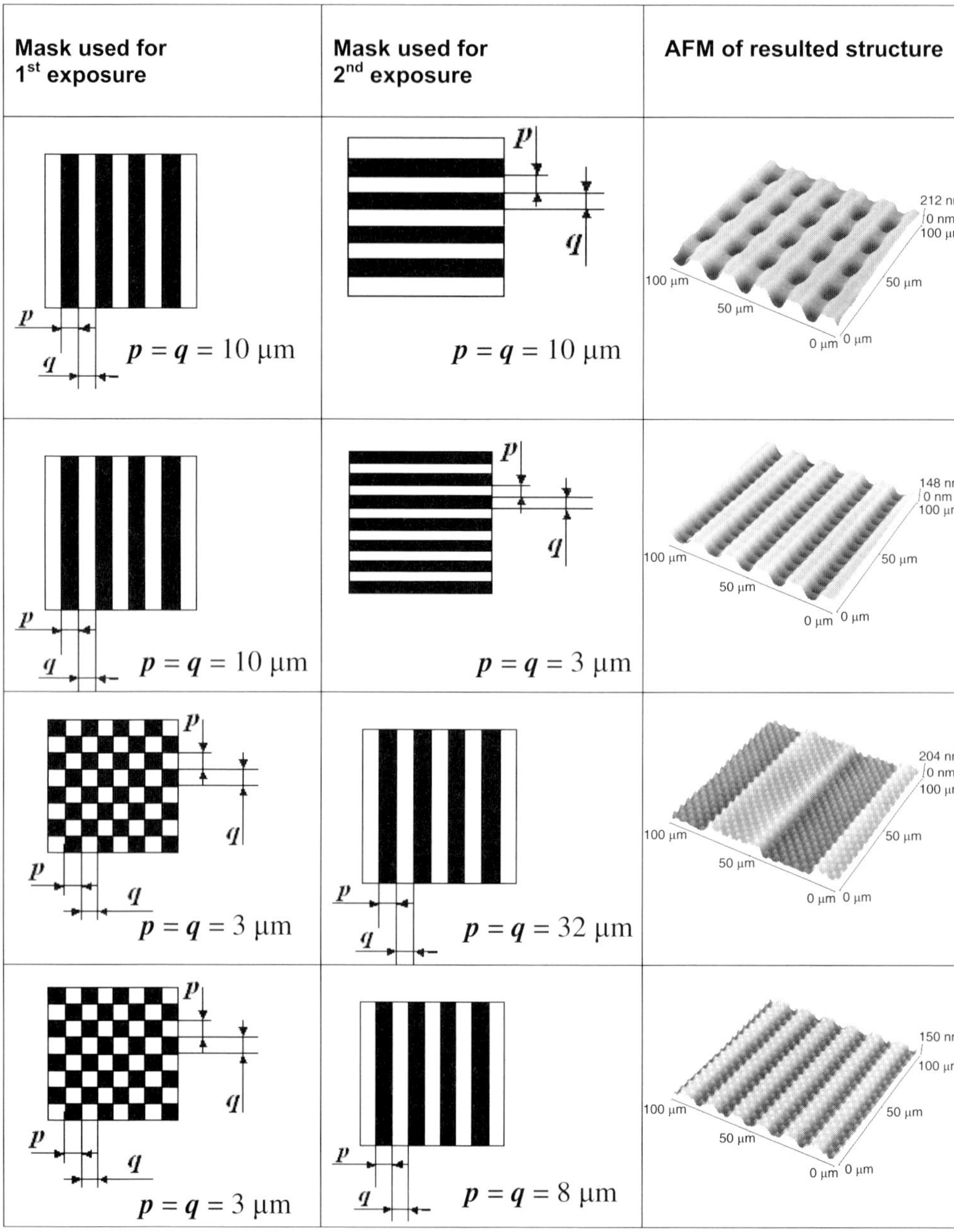

Fig. 19.8. Surface relief structures as measured by AFM after two mask exposures of equal UV doses.

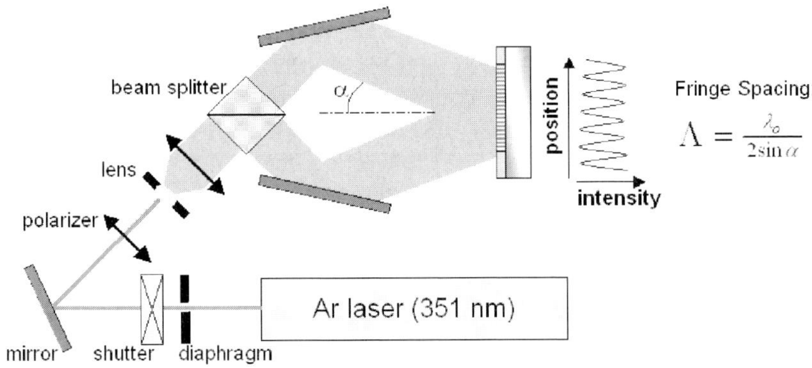

Fig. 19.9. Schematic representation of the holographic setup.

sinusoidal spatial modulation of the light intensity in the interference region (Figure 19.9). The period of the modulation can be adjusted by changing the angle between the two beams. In Figure 19.10 a sinusoidal surface relief grating is shown with a periodicity of around 1 μm, such as is obtained by exposing the sample to a holographic interference and subsequently heating to 80°C as described above.

Multiple holographic exposures can also be performed. As an example, Figure 19.11(a) shows a beat structure obtained by sequential exposures with two holographic gratings with close periods. First a holographic grating with a period of 1.5 μm is recorded. Afterward a second one with a period of 1.62 μm is registered. The subsequent heating step reveals a beat structure with a period of about 20 μm, the same as the beat that is obtained by addition of the two recorded sinusoidal

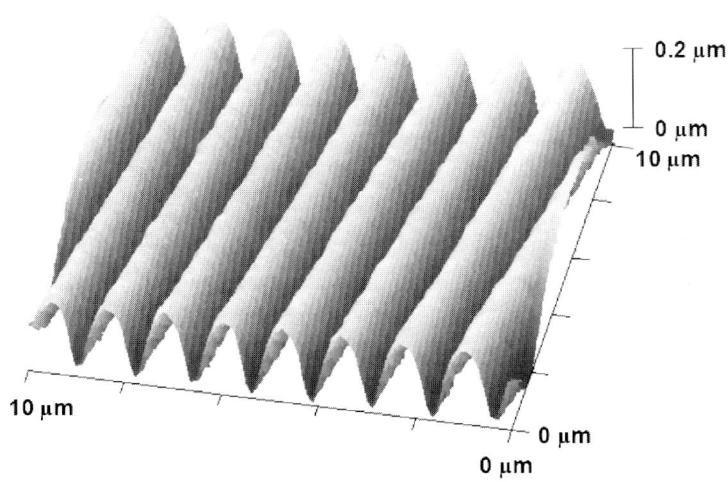

Fig. 19.10. Surface relief grating after holographic exposure and thermal treatment (10 min, 80°C; 10 min, 130°C).

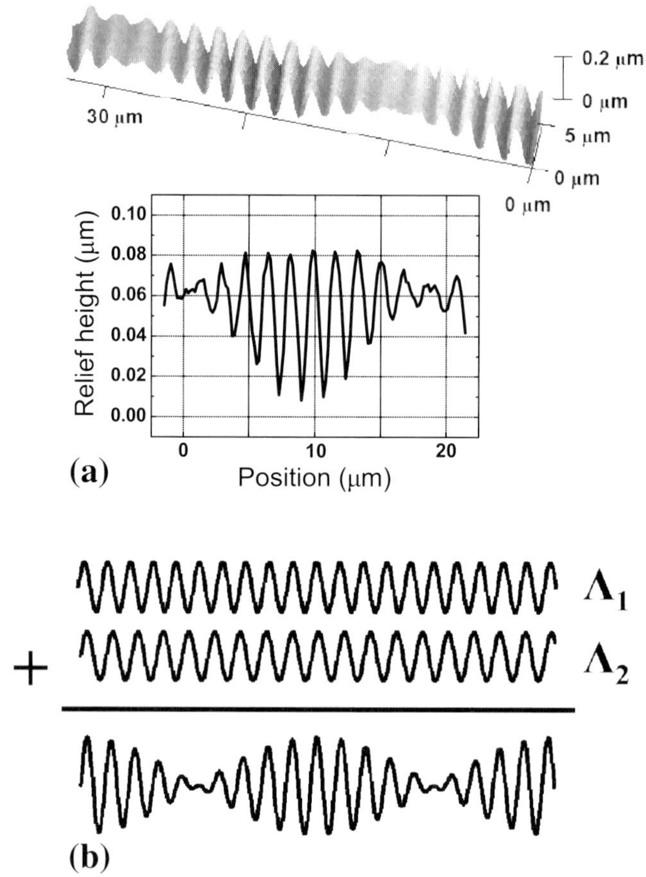

Fig. 19.11. (a) Surface relief beat structure due to sequential double holographic exposure with two grating interference patterns of different periods and equal doses and thermal treatment (10 min, 80°C; 10 min, 130°C). (b) Schematic representation of the result of the addition of two waves of equal amplitude but different periods, Λ_1 and Λ_2.

intensity profiles, as shown in Figure 19.11(b). In Figure 19.12 we see the result of exposing a film with two gratings perpendicular to each other, giving a square lattice.

19.6
Surface Structure Development under Fluids

The formation of surface relief structures starting with a flat film involves an increase in surface, which is counteracted by the free surface energy. Consequently

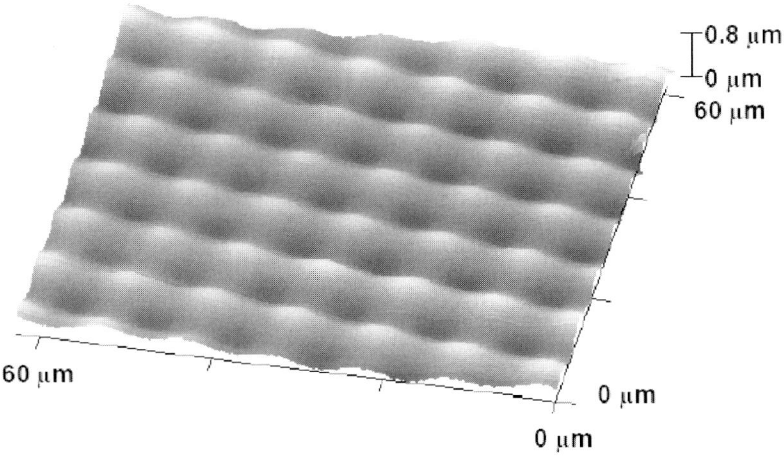

Fig. 19.12. Surface relief structure due to sequential double holographic exposure with two orthogonal grating interference patterns of equal UV doses and thermal treatment (10 min, 80°C; 10 min, 130°C).

the surface energy is a limiting factor for the generation of these relief structures [28]. A new process has been developed to reduce interfacial interactions, favoring the formation of new surface and achieving in this way enhanced aspect ratios in the photoembossed structures. The use of a contact fluid during the development of the relief structure enhances the aspect ratio of the photoembossed structures by a factor almost 2. The effect of this process is shown in Figure 19.13. Irradiation of the film is performed through a lithographic mask (transparent and dark lines

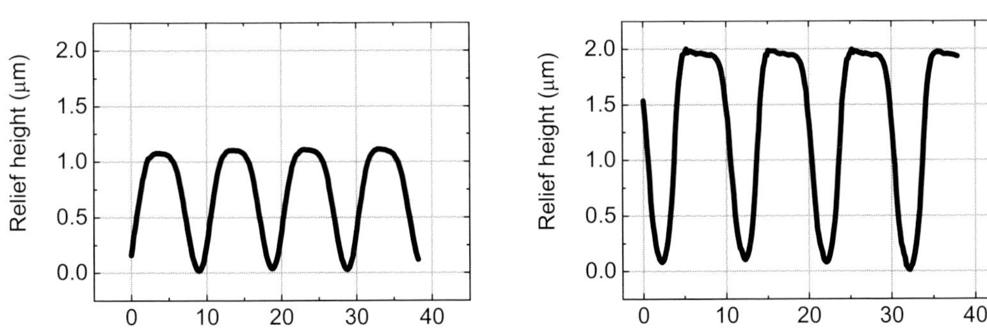

Fig. 19.13. Surface relief heights after exposure through a line mask with a periodicity of 10 μm and an aperture of 50% after development: (a) without silicon oil; (b) with silicon oil on top.

with a period of 10 μm). After the exposure and before the baking step, half of the sample is covered with a drop of silicon oil, which wets the photosensitive film but does not diffuse into it; the other half of the sample remains uncovered. The baking is then performed at 80°C for 10 min, after which the sample is cooled to room temperature. The silicon oil is removed by rinsing the whole film with heptane. It has been checked by infrared spectroscopy of the film and of the liquids employed that neither the silicon oil nor the heptane swells or dissolves the photosensitive film. The photoembossed structure is fixed by a second heating step at 130°C for another 10 min. Figure 19.13 shows the surface profile of (left) the part cured in air and (right) the part cured with silicon oil on top and subsequently rinsed with heptane. The relief height is noticeably higher (almost double) in the part cured in contact with the fluid. The differences in shape of the developed surfaces are also remarkable. Whereas the surface baked in air shows a rounded top part, as driven by the minimization of the surface area under the action of the surface tension, the one baked in contact with the fluid shows a flatter top part and sharper edges. In addition, the top flat part corresponds quite well to the irradiated region of 5 μm. By using this special process, relief structures that reproduce the features of the mask with improved fidelity have been obtained. The research presented in this paragraph is patent pending (inventors: Dick Broer, Carlos Sánchez, and Cees Bastiaansen; applicant and owner: Dutch Polymer Institute (DPI)).

19.7 Conclusion

The generation of latent images by adjusting the monomer mobility during the UV exposure step and the subsequent development of surface structures by polymerization at higher temperatures provide accurate control over polymer surfaces. The fact that the surface deforms only after the exposure step prevents structure blur by refraction and diffraction defects, and enables multiple exposures, creating complex surfaces by multiple mask steps before heat development, or by combined holographic exposures. Reduction of interfacial interactions makes possible to obtain higher aspect ratios and more sharply defined features in the photoembossed structures.

Acknowledgments

We acknowledge Mrs. G. N. Mol, Mr. H. Nulens, Sasha Alexeev (NT-MDT) and Dr. B. van de Zande for their help, with the DSC measurements, the AFM measurements, and the interferometric microscope images. The research of Carlos Sánchez, Cees Bastiaansen, and Dirk Broer forms part of the research program of the Dutch Polymer Institute (DPI), project DPI#298.

Notation

Acronyms

AFM	atomic force microscopy
CD	compact disk
DPI	Dutch Polymer Institute
DSC	differential scanning calorimetry
DUV	deep UV
EB	electron beam
EUV	extreme UV
FET	field-effect transistor
LCD	liquid crystal display
LED	light-emitting diode
MEMS	micro-electromechanical system
SPM	scanning probe microscope

References

1 T. Ito, S. Okazaki, *Nature*, **2000**, *406*, 1027.
2 Y. Xia, J. A. Rogers, K. E. Paul, G. M. Whitesides, *Chem. Rev.*, **1999**, *99*, 1823.
3 P. Ehbets, H. P. Herzig, P. Nussbaum, P. Blattner, R. Dändliker, *Appl. Opt.*, **1995**, *34*, 2540.
4 D. C. Flanders, *Appl. Phys. Lett.*, **1983**, *42*, 492.
5 X. T. Li, A. Natansohn, P. Rochon, *Appl. Phys. Lett.*, **1999**, *74*, 3791.
6 Y. B. Boiko, V. S. Solovjev, S. Calixto, D. J. Lougnot, *Appl. Opt.*, **1994**, *33*, 787.
7 S. T. Han, Y. L. Tsao, R. M. Walser, M. F. Becker, *Appl. Opt.*, **1992**, *31*, 2343.
8 D. Schertler, N. George, *Appl. Opt.*, **1999**, *38*, 291.
9 H. J. Levinson, *Principles of Lithography*, SPIE, Washington, **2001**.
10 M. Campbell, D. N. Sharp, M. T. Harrison, R. G. Denning, A. J. Turberfield, *Nature*, **2000**, *404*, 53.
11 A. K. Bates, M. Rothschild, T. M. Bloomstein, T. H. Fedynyshyn, R. R. Kunz, V. Liberman, M. Switkes, *IBM J. Res. Dev.*, **2001**, *45*, 605.
12 B. H. Cumpston, S. P. Ananthavel, S. Barlow, D. L. Dyer, J. E. Ehrlich, L. L. Erskine, A. A. Heikal, S. M. Kuebler, I.-Y. Sandy Lee, D. McCord-Maughon, J. Qin, H. Röckel, M. Rumi, X. L. Wu, S. R. Marder, J. W. Perry, *Nature*, **1999**, *398*, 51.
13 W. H. Teh, U. Durig, G. Salis, R. Harbers, U. Drechsler, R. F. Mahrt, C. G. Smith, H. J. Güntherodt, *Appl. Phys. Lett.*, **2004**, *84*, 4095.
14 M. Aktary, M. O. Jensen, K. L. Westra, M. J. Brett, M. R. Freeman, *J. Vac. Sci. Technol. B*, **2003**, *21*, L5.
15 S. D. Berger, J. M. Gibson, R. M. Camarda, R. C. Farrow, H. A. Huggins, J. S. Kraus, J. A. Liddle, *J. Vac. Sci. Technol. B*, **1991**, *9*, 2996.
16 W. He, D. B. Poker, K. E. Gonsalves, N. Batina, *Microelectron. Eng.*, **2003**, *65*, 153.
17 M. Heckele, W. K. Schomburg, *J. Micromech. Microeng.*, **2004**, *14*, R1.
18 M. T. Gale, *J. Imag. Sci.*, **1997**, *41*, 211.
19 D. Wouters, U. S. Schubert, *Angew. Chem. Int. Ed.*, **2004**, *43*, 2480.

20 D. G. Grier, *Nature*, **2003**, *424*, 810.
21 S. H. Park, D. Qin, Y. Xia, *Adv. Mater.*, **1998**, *10*, 1028.
22 D. Wang, F. Caruso, *Adv. Mater.*, **2001**, *13*, 350.
23 G. Widawski, M. Rawiso, B. François, *Nature*, **1994**, *369*, 387.
24 B. Michel, A. Bernard, A. Bietsch, E. Delamarche, M. Geissler, D. Juncker, H. Kind, J. P. Renault, H. Rothuizen, H. Schmid, P. Schmidt-Winkel, R. Stutz, H. Wolf, *IBM J. Res. Dev.*, **2001**, *45*, 697.
25 Y. Xia, G. M. Whitesides, *Angew. Chem. Int. Ed.*, **1998**, *37*, 550.
26 T. Kratzmüller, D. Appelhans, H.-G. Braun, *Adv. Mater.*, **1999**, *11*, 555.
27 C. Croutxé-Barghorn, D. J. Lougnot, *Pure Appl. Opt.*, **1996**, *5*, 811.
28 O. Sakhno, T. Smirnova, *Optik*, **2002**, *113*, 130.
29 T. J. Bunning, L. V. Natarajan, V. Tondiglia, R. L. Sutherland, D. L. Vezie, W. W. Adams, *Polymer*, **1995**, *36*, 2699.
30 C. F. van Nostrum, R. J. M. Nolte, D. J. Broer, T. Fuhrman, J. H. Wendorff, *Chem. Mater.*, **1998**, *10*, 135.
31 D. J. Broer, J. Lub, C. F. van Nostrum, M. M. Wienk, *Recent Res. Dev. Polym. Sci.*, **1998**, *2*, 313.
32 C. M. Leewis, D. P. L. Simons, A. M. De Jong, D. J. Broer, M. J. A. De Voigt, *Nucl. Instr. Meth. B*, **2000**, *651*, 161.
33 R. T. Ingwall, D. Waldman, *Holographic Data Storage*, Springer, New York, **2000**.
34 B. M. Monroe, W. K. Smothers, D. E. Keys, R. R. Krebs, D. J. Mickish, A. F. Harrington, S. R. Schicker, M. K. Armstrong, D. M. T. Chan, C. Weathers, *J. Imag. Sci. Technol.*, **1991**, *35*, 19.
35 C. M. Leewis, PhD thesis, Eindhoven University of Technology, Eindhoven, **2003**.

20
Chemical Analysis for Polymer Engineers[1]

Peter Schoenmakers and Petra Aarnoutse

20.1
Introduction

Polymer reaction engineers have basically two different needs for polymer analysis [1]: they want to follow the polymerization process, and to characterize the resulting polymer. These two types of analysis can be classified as process analysis and polymer analysis, respectively; different objectives are associated with each of them (Table 20.1). Process analysis enables the chemical engineer to monitor the polymerization process. For example, (s)he may want to follow the monomer conversion or the polymer molecular weight during the reaction. Process control can be realized if the results of the process monitoring are translated into specific actions. Polymer analysis concerns the product that results from the polymerization process. For example, the physical and chemical (molecular) properties may be compared with given specifications. The results of such analyses may also be used to tune the polymerization process, either for future (batch) reactions or during a continuous polymerization process. Arguably, the ultimate objective of the polymer reaction engineer is to optimize the polymerization process, using either type of measurement.

The different sets of objectives and requirements for the two types of analysis imply that they are performed in quite different manners. Process analyses are preferably carried out on-line (in the reactor) or at-line (in the immediate vicinity of the reactor). Polymer analyses are typically performed in the laboratory. It is also typical for different analytical techniques to be used (Figure 20.1). These will be discussed in somewhat greater detail in subsequent sections of this chapter. We will first discuss the process analysis techniques (bottom left in the figure) in Section 20.2. The polymer analysis techniques will be discussed in Section 20.3.

It should be noted that Figure 20.1 deals with *chemical* analysis – as does the remainder of this chapter. Physical measurements of, for example, the melt viscosity of a polymeric product are very common (at-line) process monitoring tools, but they are beyond the scope of the present chapter.

1) The abbreviations used in this chapter are listed at the end of the text, under "Notation".

Handbook of Polymer Reaction Engineering. Edited by T. Meyer, J. Keurentjes

ISBN: 3-527-31014-2

Tab. 20.1. Objectives of the two types of chemical analysis for polymer reaction engineering.

	Process analysis	*Polymer analysis*
Engineering objectives	process monitoring process control process optimization	product characterization product control process optimization
R&D objectives	polymerization mechanism reaction kinetics	reaction mechanisms structure–property relationships
Analytical objectives	fast response high precision (repeatability) representative data	quality and amount (detail) of information (often) high accuracy (correct information)

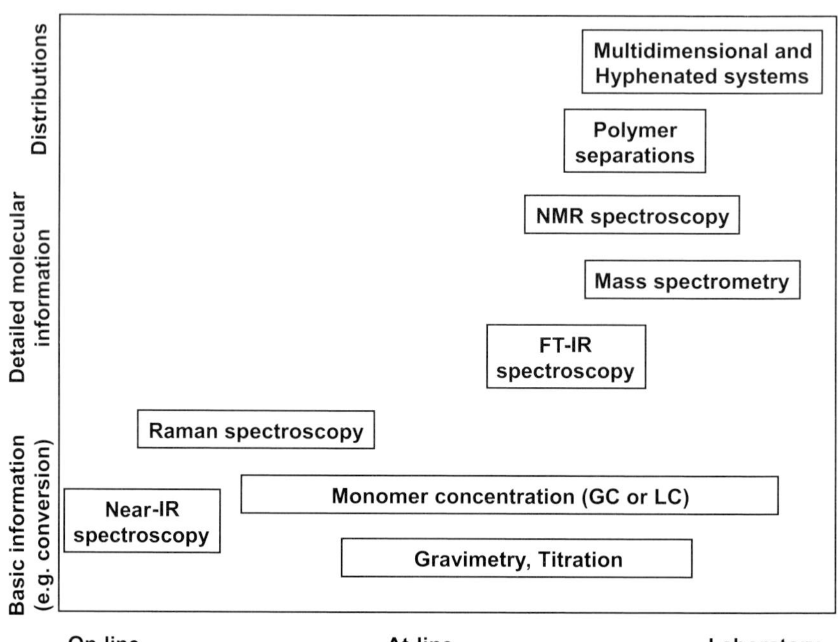

Fig. 20.1. The main analytical techniques used in polymer reaction engineering. A distinction can be made between on-line (process) and off-line (product) analysis (left to right in the figure). On-line techniques typically provide limited information, whereas more detailed characterizations can be performed in the laboratory (bottom to top in the figure).

Moreover, this is not meant to be an exhaustive treatment of all the possible methods of chemical analysis. For example, a very exciting possibility for the direct monitoring of free-radical polymerization reactions in a research environment is electron spin resonance (ESR) spectroscopy [2]. This allows a direct measurement of the concentration of radicals in the reaction mixture. In many cases catalysts are used in polymerization reactions, and determination of the fate of the catalyst during the reaction and in the product polymer may require specific analytical methods. For example, a variety of techniques exist for determining residual traces of metal catalysts in polymers. These include X-ray fluorescence characterization for solid polymers, and atomic absorption (AAS) and emission (AES) spectrometry for polymer solutions. The treatment in this chapter stops at the primary product of the polymerization reaction. Detailed studies on the structure and morphology of the ultimate polymeric product are assumed to be the domain of polymer technologists, rather than polymer reaction engineers. Methods for analyzing additives are also beyond the scope of the present chapter.

20.2 Process Analysis

Analyses during a polymerization reaction can be carried out with the aim of understanding the process (for example, the polymerization mechanism and the kinetics) or with the aim of process control. The former concerns research measurements, for which the amount of information obtained and the accuracy of the data are the prevailing criteria. The latter concerns true process analysis and the response time is a prevailing criterion.

In principle, all kinds of spectroscopic techniques lend themselves to on-line measurements. Only a very few are practical. Although low-field NMR has been used to measure various material properties by applying empirical relationships, NMR is still not a realistic proposition for on-line measurements. Ironically, FTIR spectroscopy suffers from too much sensitivity. Typically, good spectra can be obtained only from very thin polymeric films (or solutions). Attenuated total reflection (ATR) probes, in which only a fraction of the IR light penetrates a very short distance into the sample, reduce the problem of excessive sensitivity. However, they aggravate the problems of variations in the baseline and nonlinear response. The latter problem also obstructs the use of UV spectrometry for monitoring polymerization reactions. Of the remaining options, near-infrared (NIR) and Raman spectroscopy are the most attractive.

20.2.1 Near-infrared Spectroscopy

Near-infrared (NIR) spectra are not very informative. In comparison with conventional (mid-)IR spectra, they are much less useful for identifying unknown compounds and for diagnosing minor variations in the sample. NIR spectra contain fewer bands and much less resolution. Also, the absorption coefficients (sensitivity) are much lower in the NIR wavelength range (typically 800–1200 nm). In contemporary applications of NIR spectroscopy, the lower sensitivity is turned into an advantage: it makes it much easier to deal with the presence of strong IR absorbers, such as water, and it helps to keep the response strictly linear. Also, the spectra of individual compounds of a mixture turn out to be perfectly additive in NIR spectroscopy. Finally, unlike mid-IR spectra, NIR spectra typically show plenty of baseline, which is essential for quantitative analysis. Thus NIR spectroscopy is perfect for on-line analysis in almost all respects. The lack of resolution and of clearly defined bands for specific components or functionalities remains a distinct disadvantage.

The combination of (near-)perfect linearity and spectral additivity and limited specificity and resolution makes NIR a perfect match for multivariate calibration methods or "chemometrics". In this case calibration is not based on a single wavelength (band) per component or functional group (one variable, thus univariate), but on many wavelengths or entire spectra (many variables, thus multivariate). The resulting spectrum is thought to be a linear combination of the individual spectra. In most cases, the spectrum corresponding to the desired property is not known. For example, the number of hydroxyl groups in a polymer can be characterized using NIR and multivariate calibration [3]. This implies that the NIR analysis must be calibrated using a training set of samples with known concentrations. The latter could be a set of (commercially) available reference materials or, more commonly, a set of samples that are analyzed in parallel using an absolute ("primary") method. The latter must be reliable and validated. Building good calibration models is a critical step in the process. If the sample exceeds the boundaries of the training set, the NIR results are highly unreliable and may be grossly incorrect. In case of systematic deviations from the training set (for example, when a different catalyst is used in the polymerization process), recalibration is necessary. The advantages of NIR in combination with multivariate calibration (rapid on-line measurements) must be balanced against the disadvantages of indirect calibration.

NIR has been used to monitor the polymerization of acrylic acid [4], for the solution polymerization of methyl methacrylate (MMA) in toluene [5], for following the emulsion polymerization of MMA and butyl acrylate [6], and for monitoring the copolymerization of MMA and *N*,*N*-dimethylacrylamide (Figure 20.2) [33]. The density of linear low-density polyethylene was monitored using NIR and a partial least-squares (PLS) calibration model [7].

20.2.2 In-situ Raman Spectroscopy

Raman spectroscopy is based not on absorption of light, but on inelastic scattering. The wavelength of the scattered light is slightly changed because of its interaction

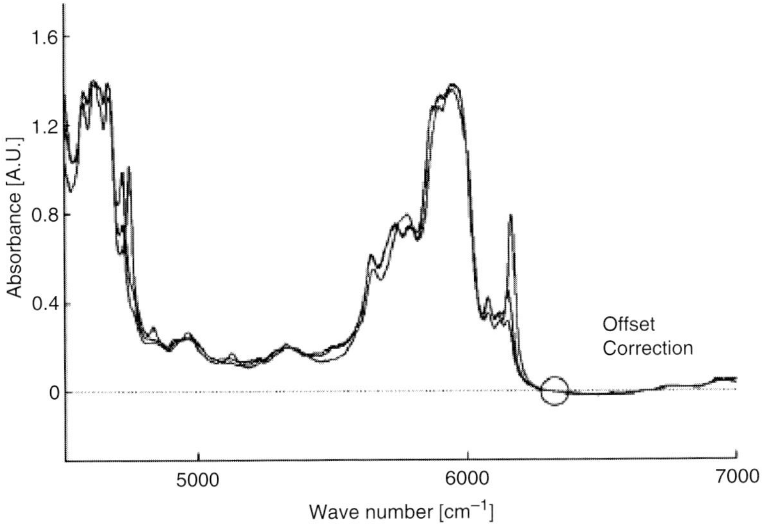

Fig. 20.2. Use of NIR spectroscopy in-line reaction monitoring of copolymerization of methyl methacrylate and *N,N*-dimethylacrylamide [33]. Spectrometer: BOMEM MB 155; the probe was connected to the spectrometer using optical fibers and a BOMEM optical interface. The figure shows calibration samples.

with molecules. The information obtained from Raman spectroscopy is in some ways similar to that from IR spectroscopy (wavelength shifts in the micrometer range) and in some ways very different (different selection rules). For example, water is an infamous interference in mid-IR spectroscopy, but it is not detectable by Raman spectroscopy. Raman spectroscopy shares some of the advantages of NIR spectroscopy. It combines a relatively low sensitivity with a good linearity, rendering it compatible with multivariate calibration. An advantage of Raman in comparison with NIR is the much greater spectral resolution (narrower spectral bands).

The greatest disadvantage of Raman spectroscopy in the analysis of polymers and the monitoring of polymer reactions is the occurrence of fluorescence, which manifests itself as disturbances in the spectral background. Raman spectroscopy is a relatively insensitive technique and because in light-scattering measurements (other than absorption measurements) the sensitivity increases with the intensity of the light source, lasers are commonly used in Raman spectroscopy. This also implies that the incident wavelength is very well defined. In general, fluorescence effects are strongest at short wavelengths. Traditionally, Raman spectroscopy is performed with lasers in the visible range as primary light sources. In FT-Raman near-IR lasers are employed, which give rise to much less fluorescence (Figure 20.3). An attractive compromise is the use of a 785 nm laser, which reduces fluorescence while offering greater sensitivity than NIR lasers.

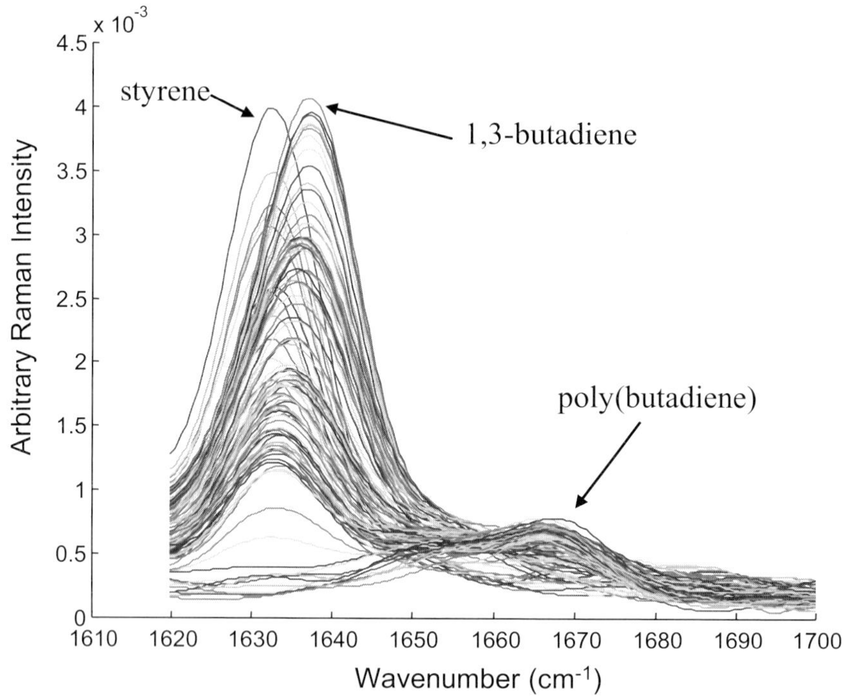

Fig. 20.3. FT-Raman spectra recorded during copolymerization of 1,3-butadiene and styrene [34]. Spectrometer: Bruker IFS66 FTIR/FT-Raman spectrometer with Nd:YAG laser (1.5 W, 1064 nm), N_2-cooled Ge detector. Early stages of the reaction (only polybutadiene homopolymer formed).

20.2.3
At-line Conversion Measurements

Although chromatography is an incredibly versatile (set of) technique(s) for performing analytical separations and quantitative analyses, the typical nature of a polymerization mixture, containing both high molecular weight polymers and low molecular weight monomers, causes serious complications. Polymer-containing samples cannot be injected easily into simple gas chromatographs, because the injector or, worse, the column will be seriously contaminated. Although several elegant solutions exist (see Section 20.3.1.1), these cannot easily be implemented and applied in at-line situations.

Liquid chromatography systems can deal more easily with the combination of large and small molecules. Size exclusion chromatography (SEC; also known as gel permeation chromatography, GPC) is especially suited for the analysis of polymers, which are separated on the basis of their size in solution. Therefore, SEC is

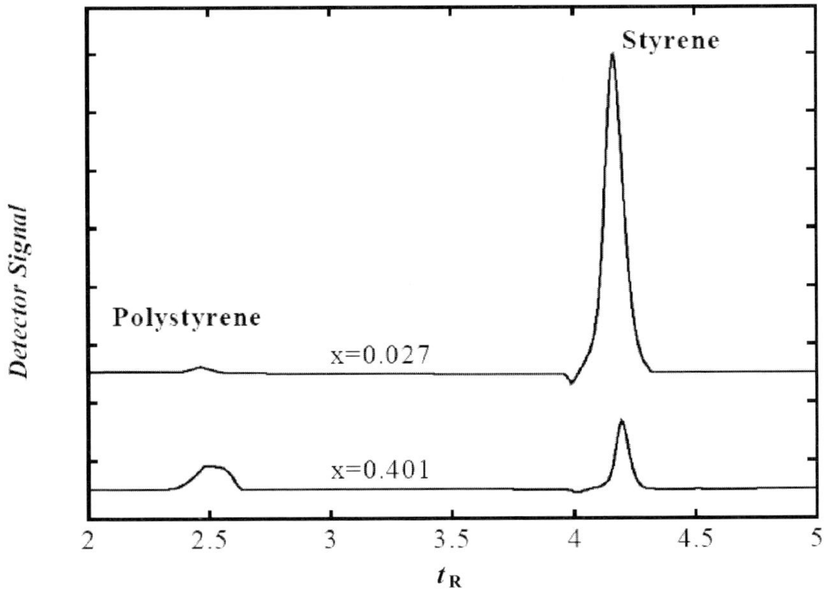

Fig. 20.4. Use of SEC for conversion measurements. Chromatograms of two polystyrene (Acros Organics, $M_w = 1.4 \times 10^5$ g mol^{-1}, PDI = 2.6) calibration samples (conversion $x = 0.027$ and $x = 0.401$). Reprinted from Ref. [8].

often used for measuring molecular weight distributions (see Section 20.3.3.1). The mechanism underlying SEC separations concerns the partial exclusion of larger molecules from pores of the packing material. By inserting a column with very small pores (an "oligomer column") in the separation system, it is possible to create sufficient selectivity between the monomers and the smallest oligomers present in the reaction mixture. If suitable detectors (sufficiently sensitive and, ideally, linear) can be found for both the polymer and the monomer, then the conversion can be elegantly measured by on-line SEC. The polymerization of styrene is a good example (Figure 20.4) [8]. In this case, UV detection can be used conveniently.

There has been a recent trend toward performing fast SEC separations [9], inspired by the developments in combinatorial (polymer) chemistry and the associated high-throughput experimentation. Fast SEC allows molecular weight distributions to be measured in one or two minutes. From both a practical and a theoretical point of view, the possibilities for fast SEC are much greater than was commonly believed until very recently. Fast SEC offers less resolution than conventional SEC. It only offers a rough picture of the MWD and it can only be applied to broadly distributed polymers. Yet the developments in fast SEC make it a feasible technique for determining molecular weight distributions at-line.

20.3
Polymer Analysis

20.3.1
Basic Laboratory Measurements

20.3.1.1 Conversion

In the laboratory, residual monomer concentrations are typically measured by some kind of chromatographic analysis; whether this is gas chromatography (GC) or liquid chromatography (LC) depends largely on the type(s) of monomers used in the polymerization process. Monomers that are gases or highly volatile liquids (for example, alkenes, vinyl chloride, acrylonitrile) or UV-inactive (for example, acrylates) are not readily compatible with LC. On the other hand, nonvolatile monomers (for example, diacids, diamines, bisphenol A) are not amenable to GC.

When GC is being used, possible contamination of the system (injector and column) with nonvolatile polymeric material is the main issue. One option is to allow direct introduction of a polymer-containing sample (solution), but to confine the polymer to a removable part ("liner") of the injection system. On some injection systems (specifically programmable-temperature vaporizers, or PTVs), a series of injections may be performed before the quantitative analysis of monomer is jeopardized and the liner needs to be replaced. Such an approach may be acceptable if

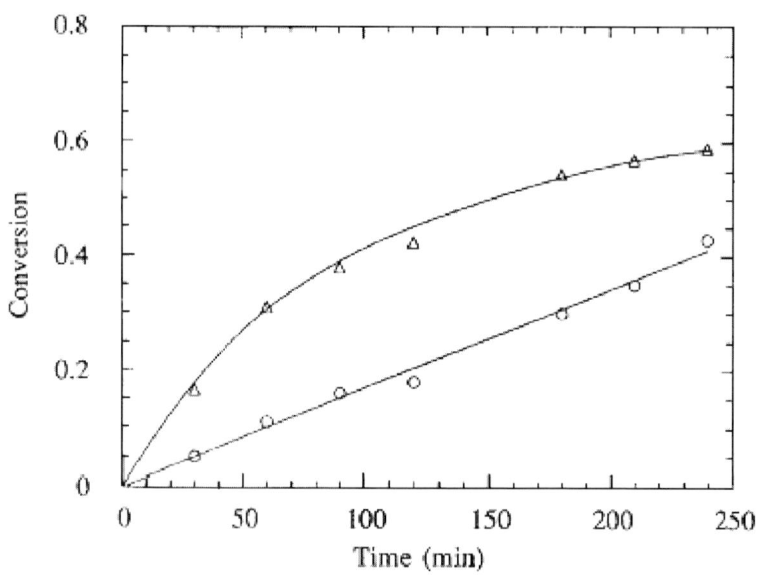

Fig. 20.5. Individual conversion profiles of methyl methacrylate (△) and vinylidene chloride (○) in the batch solution copolymerization of methyl methacrylate and vinylidene chloride in dimethylformamide, estimated by GC measurements of the vapor composition of the reactor head space [35].

the total number of analyses that need to be performed is fairly small. Another, much more sophisticated, option is the on-line coupling of an LC separation of polymer and monomer (for example, using size exclusion chromatography; see Section 20.2.3) with a GC analysis [10]. A fully automated system may be built that allows large numbers of samples to be analyzed. However, LC-GC systems are complicated to develop and to use in practice. High-level expertise is required if one is to opt for such a solution.

An easier solution, that also allows complete automation, is head-space GC. An example is shown in Figure 20.5. This technique is based on measuring volatile components from the vapor phase above a nonvolatile matrix. In this case a polymer-containing sample in a vial is placed in the system. The sample may be a solid or a (viscous) liquid. In this case, the presence of a large amount of volatile solvent may be a disadvantage. Water, especially, may cause insurmountable problems. In head-space GC the vapor phase may be analyzed after equilibrium with the sample has been established (static head-space), or the sample may be purged continuously and all volatile materials may be trapped at the column inlet, to be released from there at the beginning of the analysis ("purge and trap" or dynamic head-space). Head-space analysis (especially in the static mode) requires careful calibration and validation, but once it is in operation fully automatic analyses can be performed with relative ease.

20.3.2 Detailed Molecular Analysis

20.3.2.1 FTIR Spectroscopy

Infrared spectroscopy has long been one of the stalwarts of polymer analysis. One reason is that polymers and IR spectroscopy are highly compatible. As a case in point, thin films of polystyrene have been used for many years as a reference sample for calibrating the wavelength scale of conventional IR spectrometers. In most laboratories, conventional ("dispersive") IR spectrometers have been replaced by contemporary Fourier transform instruments. The latter typically yield a higher resolution, a higher sensitivity, and a greater wavelength accuracy. The advent of FTIR has spurred on the development of highly sensitive measurements on small samples (for example, FTIR microscopy) and of very fast measurements.

The range of information that can be obtained from FTIR measurements is summarized in Table 20.2. FTIR is still often used in polymer analysis, because the measurements are easy and fast and because many different types of polymeric samples can be measured with a minimum of sample preparation. However, FTIR spectroscopy also has its weak points. The resolution in the spectra is usually not limited by the quality of the spectrometer, but by the properties of the sample. In samples of polymers many bands usually overlap and, while band assignments are usually possible, the quantification of overlapping bands is difficult and prone to errors. One significant source of error is variation in the baseline of FTIR spectra. Such variations can be attributed to the fact that in FTIR spectrometers the baseline and the sample are not measured simultaneously. Also, spectral artefacts, such

Tab. 20.2. Information that may be obtained from FTIR spectroscopy.

Type of information	*Comments*
Chemical composition	clear evidence on presence or absence of specific functional groups rapid confirmation of structure chemical composition of copolymers, but quantification can be difficult
End groups	useful only for low MW polymers
Characterization of polymers in many different forms	direct measurements on polymer films thicker objects though ATR or specular reflection (surface characterization) easy characterization of powders (DRIFT) IR microscopy for small spots imaging techniques are emerging

as scattering by the sample, may result in changes in the baseline. Other problems are possible nonlinearity in absorbance (or reflectance) versus concentration curves and the large differences in sensitivity for different absorption bands. In fact, near-IR spectroscopy is often preferred for routine quantitative measurements, because fewer problems are encountered with the linearity and with establishing the baseline.

20.3.2.2 NMR Spectroscopy

Nuclear magnetic resonance spectroscopy probes the spins of specific atomic nuclei in a strong magnetic field. The spin can be changed at very specific (resonance) frequencies, which depend on the chemical environment of the nucleus. Thus, NMR spectra provide very detailed chemical information on the structure of molecules (Table 20.3). The differences between the spin energy levels (with and against the magnetic field) are very small. This implies that the spins are almost evenly distributed across the different levels (Boltzmann's law). Therefore, NMR is an insensitive technique. The difference between the energy levels, and thus the difference between the populations of the different levels, and hence the sensitivity of the NMR measurements, can be increased by increasing the field strength. Therefore there is a continuous trend toward ever-higher field strengths (and ever-costlier instruments) in NMR spectroscopy. This is especially the case since (in solution NMR) the resolution also increases with increasing field strength. The two most common nuclei that are studied in NMR spectroscopy are protons (^{1}H) and carbon (^{13}C). The former has a natural abundance of almost 100%, but ^{13}C represents only about 1% of the carbon population. If only because ^{12}C cannot be observed by NMR spectroscopy, the sensitivity of "carbon-NMR" is significantly lower than that of "proton-NMR".

NMR spectroscopy yields quantitative data on the number of protons or (with some precautions) carbon atoms that have a certain chemical environment. This can be directly translated into chemical composition data (see Figure 20.6). For ex-

Tab. 20.3. Information that may be obtained from NMR spectroscopy.

Type of information	*Comments*
Chemical composition	highly detailed information on chemical structure and composition
	accurate quantitative data (% of specific protons; % of specific carbon atoms) at levels above about 1%
Molecular architecture	good, quantitative information on average chain regularity (tacticity)
	information on branching (at high branching levels)
	easy distinction between random and block copolymers; some information on average block length
Physical structure	information on network mobility
	some information on polymer morphology
Polymers in various forms	solution NMR gives the best resolution
	solid-state NMR (powders, particles, or small objects) routinely possible
	polymer melts studied occasionally

ample, the numbers of aromatic and aliphatic protons in a copolymer of styrene and isobutene unambiguously define the chemical composition. Much more detailed information is revealed by NMR spectroscopy about the specific structure of the polymer. Many different measurement techniques have been developed that reveal intricate details about the molecular structure [11], but these are beyond the scope of the current chapter.

NMR spectroscopy can be used to study polymers in solution (solution NMR) or as such (solid-state NMR). In the latter case, the resolution is typically much lower, mainly because of sample anisotropy. To improve the resolution in solid-state NMR the sample can be spun at a high frequency (in the kilohertz range) and at a specific ("magic") angle (54°) relative to the magnetic field. The sensitivity can be increased by transferring magnetization from ^1H to ^{13}C nuclei. The resulting technique is referred to as cross-polarization magic-angle spinning NMR (CP-MAS-NMR).

20.3.2.3 Mass Spectrometry

Whereas (FT)IR and NMR spectroscopy have been essential tools for polymer analysis for decades, mass spectrometry (MS) only became truly relevant in the later 1990s, and the number of mass spectrometers in polymer analysis laboratories is now increasing rapidly. Our current ability to characterize (very) large molecules successfully can be seen as a small miracle. In many cases, we are now able to ionize polymer molecules and to transport large molecular ions through magnetic or electric fields with great integrity. The possibilities of contemporary MS are fantastic. However, one must be aware that there are still severe limitations. The possibilities and limitations of polymer MS will be briefly reviewed in this section.

MS has been used in polymer analysis for many years, but these techniques

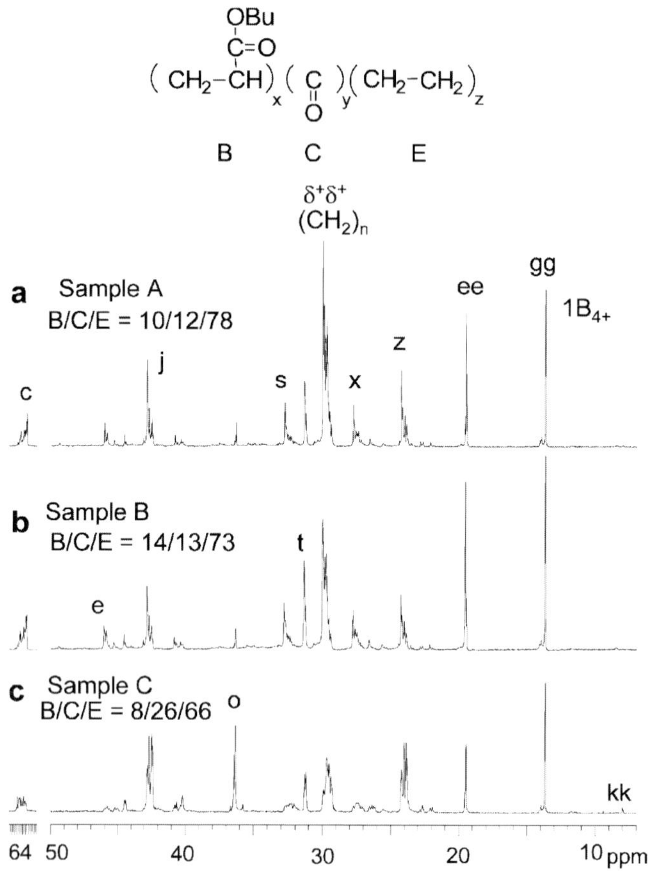

Fig. 20.6. 188 MHz ^{13}C NMR spectra of poly(*n*-butyl acrylate-*co*-carbon monoxide-*co*-ethylene) samples with different monomer compositions (indicated in the figure) [36].

have relied on destruction of the polymer into smaller pieces or fragments. Pyrolysis-GC-MS is a conventional method for polymer analysis and characterization. The pyrolysis stage yields small, uncharged fragments of the polymer, such as residual monomers, which can be separated by gas chromatography (GC) and identified by MS. Pyrolysis-GC-MS yields useful information on the composition of (co)polymers, the presence of additives, and suchlike. However, the results are difficult to quantify. Similarly, in secondary-ion mass spectrometry (SIMS), the polymer is first fragmented using a beam of primary ions. The resulting fragments are charged and these fragments are subsequently analyzed by MS. SIMS can yield very useful (organic chemical) information on polymer surfaces (static SIMS), or it can yield depth profiles of materials (dynamic SIMS). In these conventional applications of MS in polymer engineering, only information on molecular fragments is obtained, but none on the intact polymer molecules.

The main breakthrough in polymer MS has been the development of soft ionization methods. This implies that molecular ions are obtained and characterized. During the formation of molecular ions, small ions (such as H^+) may be abstracted from the molecule, or adducts may be formed with small ions (for example, Na^+), but *no fragmentation* occurs and the resulting ion is highly representative of the original polymeric molecule. Two important soft ionization techniques have emerged in recent years, namely electrospray ionization (ESI) and matrix-assisted laser desorption ionization (MALDI).

Electrospray ionization (ESI) ESI requires a polymer solution to be pumped into the mass spectrometer, either as such or, conveniently, after a liquid-chromatographic separation. In the latter case, a narrow, pre-separated fraction of the polymer is introduced into the mass specrometer, which greatly enhances the chances of obtaining useful mass spectra. The liquid is sprayed into the ionization chamber under the simultaneous action of an electric field of several kilovolts. The solvent should have a significant polarity and some ionic additives (salt or buffer) are typically present. It is not strictly necessary for the polymer itself to be dissolved in this polar solvent. Good results have been obtained by adding a separate (immiscible) stream of solvent to the polymer solution through a T-piece just before the ESI interface [12].

The exact mechanism by which charged molecular ions are ultimately obtained from the polymer solution has not yet been elucidated unequivocally. The spray is formed at atmospheric pressure, but then passed into the vacuum part of the ionization chamber, where the volatile components evaporate. This may result in a concentration of (nonvolatile) ions in very small droplets, which then become unstable and explode to yield a series of individual ions, including the charged polymers. An alternative mechanism, which seems to be supported by the fact that the polar, salt-containing solvent does not need to be miscible with the polymer solution, entails the complete evaporation of all solvents, followed by ionization of the polymer molecules in the gas phase.

In any case, molecular ions are formed that contain one ("singly charged ions") or more charges. The fact that multiply charged ions are being formed makes it easier to analyze molecular ions of high molecular weight, because the behavior of ions in mass spectrometers is determined by the mass-to-charge ratio (m/z) rather than by the absolute mass. However, the presence of multiply charged ions may significantly complicate the resulting mass spectra and their interpretation.

Figure 20.7 shows an example of the application of ESI-MS for polymer analysis. Table 20.4 summarizes the information that can be obtained using this technique.

Matrix-assisted laser desorption ionization (MALDI) In MALDI, the polymer is mixed intensively (usually from a solution) with a matrix, and possibly with other additives such as a salt. It is then deposited as a small spot on a target (MALDI plate). The deposited mixture is irradiated by a laser pulse. The matrix is selected such that it strongly adsorbs the laser light, heats up very rapidly, and induces the transfer of a single charge (usually an adduct ion) to the polymer molecule. The

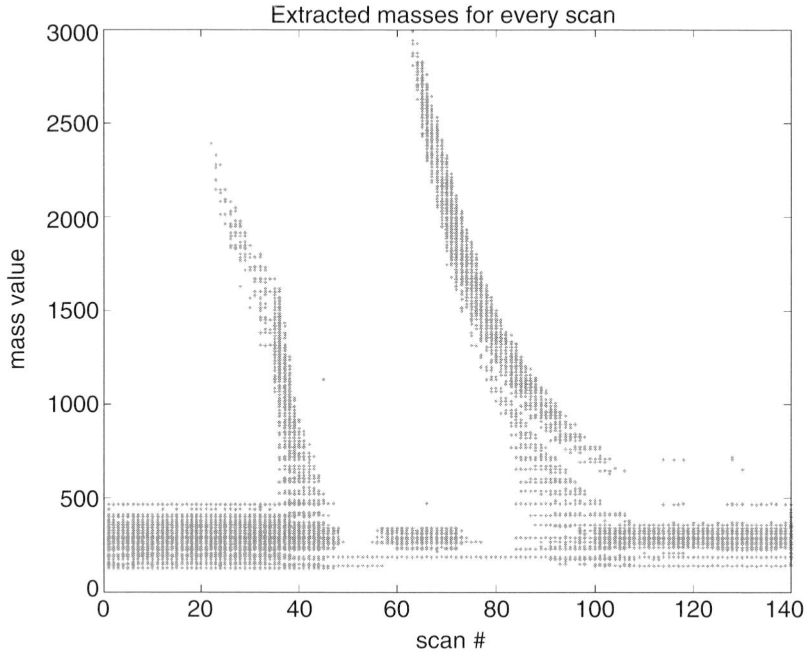

Fig. 20.7. Characterization of perfluoropolyethers by on-line LC-ESI-MS [12]. LC conditions: Column: Nucleosil silica (5 μm), 150 mm × 2.1 mm i.d.; mobile phase: 0.1–20% methyl-*t*-butyl ether in 1,1,2-trichlorotrifluoroethane in 15 min; mass spectrometry: Micromass time-of-flight MS operated in negative-ion mode; capillary voltage: 3 kV; desolvation temperature: 350°C; source temperature: 120°C; cone-gas flow rate: 30 L h^{-1}; desolvation gas flow rate: 350 L h^{-1}. ESI coniditions: 20 mL min^{-1} 50:50 isopropanol/water added post-column; cone voltage: 50 V.

Tab. 20.4. Information that may be obtained from ESI-MS.

Type of information	***Comments***
Absolute mass of molecular ions	limited to polar polymers with molecular weight below ca. 25 kDa structural formulas may be obtained in the case of high-resolution MS
Molecular-weight distribution	only for (homo)polymers with a very narrow distribution in terms of molecular weight, chemical composition, and type of functionality prior separations (e.g., SEC, LC) allow analysis of more broadly distributed polymers by ESI-MS
Combined mass of end groups (homopolymers)	can be obtained from series of MS peaks (one of more different end groups) may be complicated by adduct ions may be facilitated by high-resolution MS

resulting spectra are quite variable in terms of the observed ions and their intensities, but representative spectra can be obtained by summing or averaging the spectra resulting from a large number of pulses. Only singly charged ions are observed and there is usually little or no fragmentation, although MALDI is often found to be slightly "less soft" (that is, more likely to induce fragmentation) than ESI [13].

Polymers that are amenable to MALDI must have some degree of polarity – polyethene and polypropene are still essentially incompatible with the technique. Another essential requirement is that the polymers be narrowly distributed in terms of molecular weight, chemical composition, and functionality. If the molecular weight distribution is not very narrow, then the smallest molecules will be over-represented in the resulting mass spectrum ("discrimination"). If one type of molecule is abundant, it may dominate the entire mass spectrum, completely suppressing the ionization of polymeric components present in lower concentrations ("ion suppression"). Selective ionization may also be observed in the case of variations in functional groups or end groups.

MALDI cannot easily be coupled on-line to separation techniques such as liquid chromatography (LC), but it can be elegantly coupled off-line. Because of the high sensitivity of MALDI, one drop of LC effluent may be sufficient to prepare excellent MALDI spots.

Table 20.5 summarizes the information that can be obtained using this technique.

Types of mass analyzers There are quite a few different types of mass analyzers. Conventional high-resolution MS exploits magnetic-sector instruments. Quadrupole instruments have become the affordable standard for the analysis of fairly small components (up to about 1000 Da), although ion-trap systems have some specific advantages, such as the possibility of performing high-sensitivity tan-

Tab. 20.5. Information that may be obtained from MALDI-MS.

Type of information	*Comments*
Absolute mass of molecular ions	limited to polymers with a certain minimum polarity, up to quite high molecular weights, provided that the sample is very narrowly distributed
Molecular weight distribution	only for (homo)polymers with a very narrow distribution in terms of molecular weight, chemical composition, and type of functionality
	prior separations (e.g., SEC, LC; usually carried out off-line) allow analysis of more broadly distributed polymers by MALDI-MS
Combined mass of end groups (homopolymers)	can be obtained from series of MS peaks (one of more different end groups)
	may be complicated by adduct ions or fragmentation
	may be facilitated by high-resolution MS

dem MS (MS-MS) or multiple MS (MS^n) experiments with relative ease. The big brother of the ion-trap instrument, the Fourier transform ion–cyclotron resonance (FT-ICR-MS, or simply FT-MS) spectrometer, offers extremely high resolution. It has been gaining popularity, mainly thanks to the great advances made in protein MS since the mid-1990s. However, use of FT-MS instruments is still the prerogative of the elite few.

In contrast, another type of mass analyzer that was quite uncommon until recently is now proliferating rapidly in the field of polymer MS. Time-of-flight (ToF) analyzers are based on the principle of measuring the time a specific ion requires to travel a given length (in vacuum). The flight time is directly related to the m/z value. The mass range of ToF analyzers is not fundamentally limited. Thus, polymers of very high molecular weight can be analyzed after soft ionization. ToF analyzers are now the preferred type of instrument for the characterization of (natural and synthetic) macromolecules, in combination both with ESI and with MALDI. ToF combines a high sensitivity (due to a favorable duty cycle) with a broad mass range and a high spectral resolution and accuracy. High-resolution versions of ToF analyzers exploit the so-called reflectron mode, in which the length of the flight path is doubled. In addition, ToF-MS instruments have become much more accessible and much more affordable in recent years. An example of the MALDI-ToF-MS technique is shown in Figure 20.8.

20.3.3
Polymer Distributions

There are many ways to measure the average properties of a polymeric sample. These will not be discussed extensively in this chapter. Among the techniques used to measure average molecular weights are those that yield number averages (M_n), such as end-group titrations and end-group analysis by NMR. Such techniques work best at relatively low molecular weights. Static light scattering yields a direct estimate of the weight-average molecular weight (M_w). It works best for relatively high molecular weights (above ca. 20 kDa). Measurement of the intrinsic viscosity may be used to obtain the viscosity-average molecular weight (M_v). If the entire molecular weight distribution is measured, then all the desired averages and the sample polydispersity ($PDI = M_w/M_n$) can be readily calculated.

20.3.3.1 Molecular Weight Distributions

The most important molecular distribution is the molecular weight distribution (MWD) or, equivalently, the molar mass distribution (MMD). Size exclusion chromatography (SEC; also known as gel permeation chromatography, GPC) is the outstanding technique for measuring the MWD. In SEC, the retention time is related to the molecular weight by constructing a suitable calibration curve (Figure 20.9) based on the retention times (or volumes) of a set of narrowly distributed standards of known molecular weight. SEC separates on the basis of the size of molecules in solution (the hydrodynamic volume), rather than on the molecular weight. Thus, to measure accurate (in SEC terminology, "absolute") molecular weights, it

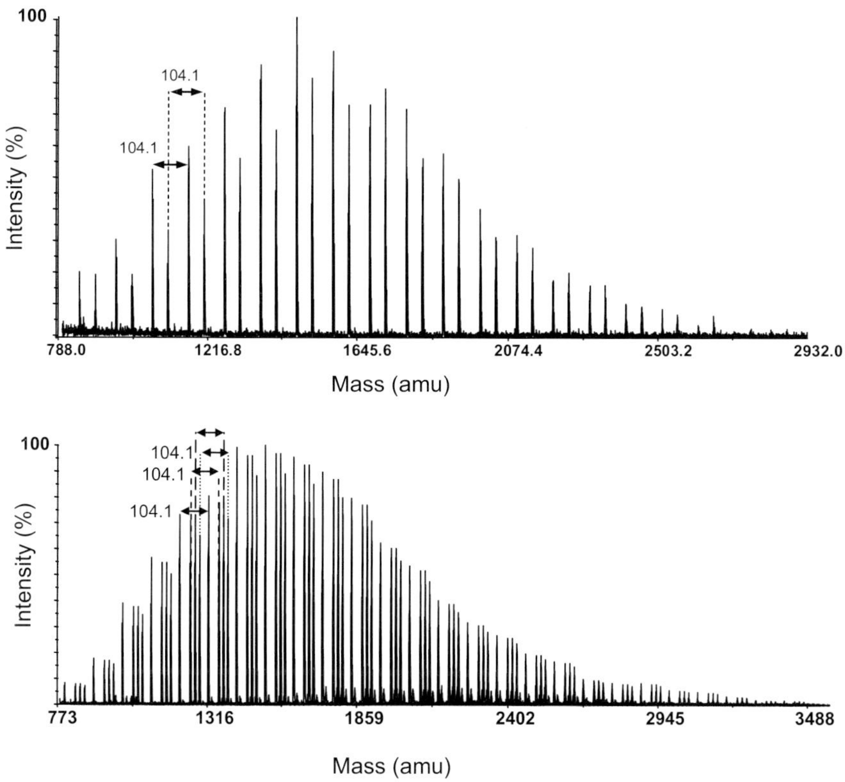

Fig. 20.8. MALDI-TOF-MS spectra of polystyrenes, reflective positive-ion mode, DCTB as matrix and Ag^{+} as a cationization agent. Top: PS with two different end groups or adduct ions. The two different series of peaks differ by 104.1 amu, the mass of one repeat unit styrene. Bottom: PS with four different end groups [37].

is necessary that the calibration standards be chemically identical to the sample polymer. Polystyrene standards should be used to construct a calibration curve for polystyrene samples, PMMA standards for PMMA samples, and so on.

Unfortunately, suitable standards are not available for all polymers. For a number of homopolymers suitable standards are commercially available. However, for copolymers it is notoriously difficult to determine absolute molecular weights by SEC. Even if copolymer standards were available, there are many different molecules (different combinations of molecular weight and chemical composition) that exhibit the same hydrodynamic volume. Also, the size of polymer molecules in solution is affected significantly by other properties, for example the degree of branching. For these reasons, SEC is often used to measure molecular weight distributions *relative* to different standards, such as polystyrene. Such relative measurements may be perfectly adequate for many purposes. For example, for monitoring polymerization reactions and product specifications the precision (repeatability) of the data tends to be much more important than their accuracy.

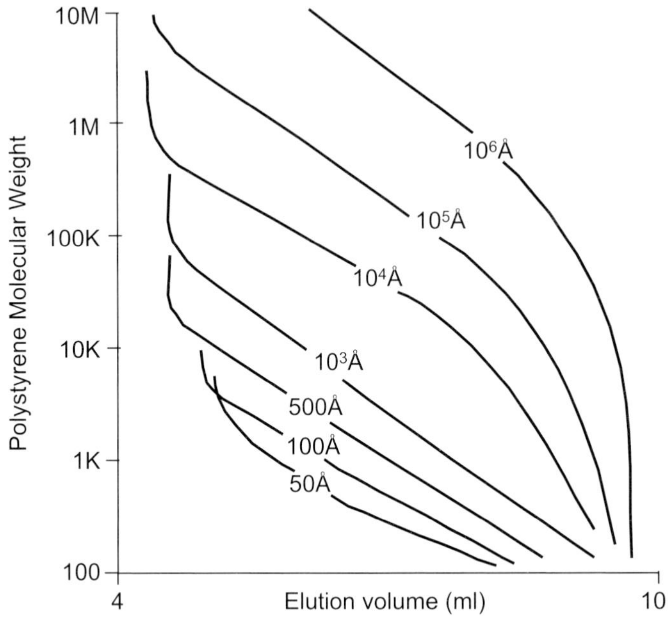

Fig. 20.9. SEC calibration curves for a series of columns with different pore sizes (Polymer Laboratories, Church, Stretton, Shropshire, UK; www.polymerlabs.com/gpc). Different columns show good selectivity (shallow curves) across different molecular weight ranges.

Detectors for size exclusion chromatography The most common detectors used in SEC are the UV absorbance detector (for polymer molecules that possess chromophores) and the (differential) refractive index detector (DRI or RI) for polymers that do not (see Table 20.6). The latter is less convenient to use in practice, and less sensitive. Therefore, the evaporative light-scattering detector (ELSD) is used increasingly for characterizing non-UV-active polymers at low concentrations (for example, polymeric contaminants or polymers separated by comprehensive two-dimensional liquid chromatography). A serious disadvantage of the latter detector is that its response increases exponentially with increasing polymer concentration and is, therefore, highly nonlinear. The response of the ELSD also tends to vary more strongly with the molecular weight of the sample polymer than those of the UV and DRI detectors.

A very important option is the combination of SEC with techniques that allow direct measurement of molecular weight, such as viscometry or light scattering (Table 20.7). This – in principle – eliminates the need for narrow standards of exactly the same polymer, and it alleviates the requirements on the SEC system, because slight variations in retention times due to variations in the flow rate or interactions with the column can be negotiated. This is true for viscometric detection, provided that accurate Mark–Houwink constants (K and α) are available to trans-

Tab. 20.6. General (concentration-sensitive) detectors for size exclusion chromatography.

Detector	*Abbrev.*	*Strong points*	*Weak points*
UV absorbance	UV	sensitive highly linear	only applicable for UV-absorbing polymers
(Differential) refractive index	(D)RI	universal (all polymers)	not very sensitive strong solvent signal sensitive to temperature variations needs reference cell
Evaporative light scattering	ELSD	universal (all polymers) highly sensitive	nonlinear response response depends on polymer structure and molecular weight

Tab. 20.7. Specific (molar mass sensitive) detectors for size exclusion chromatography.

Detector	*Abbrev.*	*Strong points*	*Weak points*
Viscometry	Vis	direct measurement of intrinsic viscosity absolute MW using Mark–Houwink constants or the universal calibration principle	applicable only for $M >$ (ca.) 5000 accuracy relies on accuracy of Mark–Houwink constants or on quality of SEC measurements
Light scattering low angle right angle triple angle multi-angle	LALS[a] RALS TrALS MALS	direct measurement of M_w (for $M >$ (ca.) 10 000) multi-angle LS provides direct measurement of the root-mean-square radius (for $M >$ (ca.) 50 000)	sensitive to calibration parameters (especially the refractive index increment dn/dc) tedious calibration procedure
Mass spectrometry electrospray ionization[b]	ESI	absolute molecular-weight measurement information on end groups	only applicable to polar polymers multiple ionization complicates spectra best for $M <$ (ca.) 20 000
Matrix-assisted laser desorption/ ionization[c]	MALDI	absolute molecular weight measurement information on end groups applicable to narrowly distributed polymers up to very high M	only applicable to moderately or highly polar polymers (very) narrowly distributed samples are required (in terms of MWD, FTD, etc.)

[a] Sometimes called low-angle laser light scattering (LALLS). Analogously, RALLS, TrALLS and MALLS may be found.
[b] Electrospray ionization mass spectrometry can be coupled on-line with SEC. However, this does require a miniaturized SEC system (flow rate $< 100\ \mu L\ min^{-1}$) or a flow splitting device after the SEC column.
[c] MALDI is usually combined off-line with SEC. Samples from the effluent can be collected manually or by using a fraction collector. Devices to facilitate deposition directly from the SEC system onto a MALDI target plate are commercially available.

late the measured intrinsic viscosity ($[\eta]$) for the specific polymer (in the specific solvent used and at the specific temperature of operation) into the molecular weight through $M = K[\eta]^{\alpha}$. When measuring the light scattered by the polymer solution to obtain the (weight-average) molar mass, the refractive index increment ($\mathrm{d}n/\mathrm{d}c$) of the specific polymer in the specific solvent (at the temperature of operation) must be accurately known. Both an on-line viscometer and a light-scattering instrument must be calibrated using known standards.

Both viscometry and light scattering require an independent measurement of the concentration of the polymer. For this purpose, one of the concentration-sensitive detectors (most often refractive index detection) is used in combination with the molecular weight selective detector. Using a combination of several detectors is common practice in SEC. A particularly powerful combination is to use refractive index detection (to measure concentration), viscometry (to measure the intrinsic viscosity), and light-scattering detection (to measure molecular weight), all together. A RALS detector is typically used instead of a MALS detector. Therefore, the root-mean-square radius of the polymer in solution cannot be measured directly and the molecular weight of large polymer molecules ($M > 100\,000$) requires an iterative correction procedure.

The on-line combination of SEC with electrospray ionization MS (SEC-ESI-MS) and the off-line combination of SEC with matrix-assisted laser desorption/ionization (SEC//MALDI-MS) are relatively recent possibilities. For the time being, SEC-ESI-MS is limited to polar polymers of relatively low molecular weight. In contrast, SEC//MALDI-MS is applicable to almost all polymers (polyethylene and polypropylene being the most notable exceptions), up to very high molecular weights. The main requirement is that the polymer samples subjected to MALDI should be narrowly distributed in terms of molecular weight, chemical composition, and functionality. The major problems of MALDI are discrimination and ion suppression. Discrimination implies that the sensitivity (response) of the MALDI-ToF-MS system varies with properties (for example, molecular weight, functionality) of the polymer molecules. Ion suppression occurs when a particular type of ion is present in a very high concentration and the sensitivity of less prevalent ions is (dramatically) reduced. Because MALDI requires narrowly distributed samples and because it is a very sensitive analytical technique, its (off-line) combination with chromatographic separations is highly attractive.

20.3.3.2 **Functionality-type Distributions**

Size exclusion chromatography can be used to separate polymer molecules according to their size in solution, and size can be converted to molecular weight by calibration. SEC cannot be used to separate polymers according to chemical composition or functionality. For this purpose, interactive liquid chromatography (*i*-LC) may be used. In *i*-LC the molecules of the analyte polymers interact with the mobile phase and with the stationary phase in the column. Usually, thermodynamic equilibrium is reached at any point in the column. The distribution of the polymer molecules across the two phases can then be characterized by a distribution coefficient ($K = c_s/c_m$, where c_s is the concentration of the polymer in the stationary phase and c_m is its concentration in the mobile phase). The retention factor (k) is proportional to the distribution coefficient and to the phase ratio [Eq. (1), where V_R is the retention volume, t_R the retention time, V_0 is the column hold-up volume, t_0 the column hold-up time, V_s the volume of stationary phase in the column, and V_m the volume of mobile phase.]

$$k = \frac{V_R - V_0}{V_0} = \frac{t_R - t_0}{t_0} = K\frac{V_s}{V_m} \qquad (1)$$

Retention in *i*-LC is thus directly related to the thermodynamic distribution coefficient. K can be expressed in terms of fundamental thermodynamic properties, that is, the partial molar free energy associated with the transfer of one mole of analyte from the mobile phase to the stationary phase (Δg), the corresponding partial molar enthalpy, and the corresponding entropy effect (Δs), according to Eq. (2), where R is the gas constant and T the absolute temperature.

$$RT \ln K = -\Delta g = -\Delta h + T\Delta s \qquad (2)$$

Enthalpic (heat) effects arising from molecular interactions are reflected in Δh. SEC is a strictly entropic process; that is, $\Delta h = 0$ and temperature has no significant effect on the elution volume.

Separation between chemically different polymers can be achieved if different parts of the molecules (different co-monomers, end groups, functional groups) exhibit different interactions with the mobile phase and the stationary phase in the column. We can rewrite Eq. (2) for a homopolymer as Eq. (3), where we assume that the partial molar free energy is built up from p contributions of monomeric units (p being the degree of polymerization) and the sum of all contributions from end groups (or functional groups).

$$RT \ln K = -\Delta g = -p\Delta g_{\text{monomer}} - \sum \Delta g_{\text{end groups}} \qquad (3)$$

Because p is a large number for high M_r polymers, reasonable distribution coefficients (and, therefore, reasonable retention factors) can usually be obtained only if $\Delta g_{\text{monomer}} \approx 0$.

A special case is the situation in which $\Delta g_{\text{monomer}} = 0$. In this case the distribution coefficients and chromatographic retention factors are determined only by the functional groups and they are independent of the chain length (p). This situation is known as liquid chromatography at the critical conditions or, simply, as critical chromatography. It is eminently suitable for separating polymers based on functionality. Figure 20.10 shows examples of the separation of functional poly(methyl methacrylate)s [14, 18]. All PMMA molecules without OH groups are eluted with a retention time of around 4 min in Fig. 20.10(a), irrespective of the molecular weight and (possible) other end groups present (provided that the Δg values for all the end groups other than OH are either negligibly small or very similar). The polymers with one OH end group are all eluted at around 5 min and the bifunc-

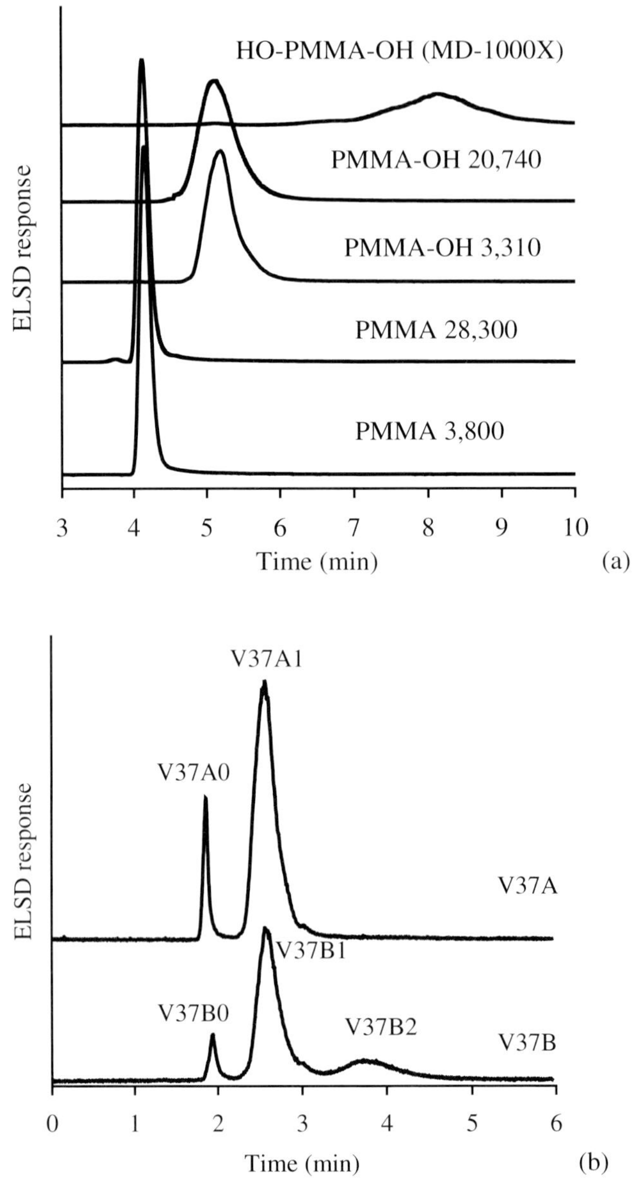

Fig. 20.10. (a) Separation of end-functional poly(methyl methacrylate)s based on the number of end groups. Column: 150 mm long × 4.6 mm i.d., home-packed with Hypersil silica (3 μm particles, 100 Å pore size); mobile phase: 43% acetonitrile in dichloromethane; temperature: 25°C; flow rate: 0.5 mL min^{-1}; injection volume: 10 μL; sample concentration: 1 mg mL^{-1} in dichloromethane; detector: evaporative light scattering [14]. (b) Separation of end-functional PMMAs prepared by RAFT polymerization [18]. Mobile phase: 40% acetonitrile in dichloromethane; other conditions as in (a).

tional ("telechelic") PMMAs are eluted at around 8 min in Fig. 20.10(a). The elution profiles of real samples can be translated into a functionality-type distribution (FTD), provided the detector is suitably calibrated [15–17].

The separation shown in Fig.10(a) has proven to be quite robust. However, this is not usually the case for critical separations. Indeed, for carboxyl-functionalized poly(**n**-butyl acrylates) it proved much more difficult to achieve genuine critical chromatography [19]. Critical liquid chromatographic separations are not always easy, but they can be highly rewarding, especially for determining FTDs.

Mass spectrometry (ESI-MS and MALDI-ToF-MS) is very useful for obtaining information on the total masses of various end group combinations present. In this case Eq. (4) holds, where M_{ion} is the observed mass for the polymeric ion, p is the degree of polymerization, M_{mon} is the mass of the monomeric unit, M_{adduct} is the mass of the adduct ion (typically added as a salt to the matrix/polymer mixture) and $\sum M_{\text{endgroup}}$ is the sum of the masses of the end groups.

$$M_{\text{ion}} = pM_{\text{mon}} + M_{\text{adduct}} + \sum M_{\text{endgroup}} \tag{4}$$

Very accurate and precise values of M_{mon} and M_{endgroup} can be obtained by MALDI. Complicating factors are:

- uncertainty in the true end group mass, which may equal $\sum M_{\text{endgroup}}$, but also $\sum M_{\text{endgroup}} + nM_{\text{mon}}$ (where n is an integer);
- the (broad) isotope pattern of large polymers, which, if not properly accounted for, may cause errors in the values obtained for $\sum M_{\text{endgroup}}$;
- the possibility that several different end group combinations yield exactly or nearly the same value for $\sum M_{\text{endgroup}}$.

It is very difficult to obtain quantitative FTDs by MS. Because the response factors are usually different for polymers with different end groups, highly specific reference materials (of accurately known composition) would be required.

In some cases capillary (zone) electrophoresis can be elegantly used to separate polymers according to functionality, and to obtain accurate FTDs. This is especially the case if the end groups are charged (in a suitable buffer solution) or if they can be derivatized to yield ionizable groups. Under suitable conditions, the effect of the molecular charge on the electrophoretic mobility can be much greater than the effect of the molecular weight [20].

20.3.3.3 Chemical Composition Distributions (CCDs)

For a copolymer, Eq. (3) becomes Eq. (5), where the subscript i depicts the different monomeric units.

$$RT \ln K = -\Delta g = -\sum_i p_i \Delta g_{\text{monomer},\,i} - \sum \Delta g_{\text{end groups}} \tag{5}$$

It is difficult to achieve $\Delta g_{\text{monomer}, i} = 0$ for one particular monomer (critical chromatography) by carefully adjusting the ratio of two components of the mobile phase (a strong eluent and a weak eluent or "nonsolvent"). It is impossible to find conditions that are critical for several different monomers simultaneously. Therefore, critical chromatography is much more useful for determining the FTDs of functional (homo)polymers than for determining the CCDs of copolymers. In the latter case, two options are open. One is to find conditions at which the separation is critical toward one type of monomer, while the second monomeric unit does not show any interaction, so that it is eluted under SEC conditions. Such conditions have been applied to block copolymers. The block for which critical conditions are maintained is made "invisible" and the separation reflects the block length distribution of the second block [22]. There is some discussion in the literature on whether any effect of the "invisible" block on the retention of the polymer molecule can be avoided [21–23].

The alternative (and more common) way to analyze copolymers is to resort to gradient elution. In this case the composition of the mobile phase is changed over time. At the initial composition, both monomers are highly retained (negative Δg). When the eluent becomes stronger, the critical conditions for one of the monomeric units will be approached. At a later point in time, this will be the case for the second monomer. In this way, a blend of polymers can be separated into its constituents. Copolymers will be eluted according to their composition (Figure 20.11). In principle, gradient elution liquid chromatography can be used to obtain CCDs of copolymers. Again, proper calibration of the detector is a significant issue if quantitative data are to be obtained.

i-LC separations can also be coupled to spectrometers and other highly informative detection devices. The situation is similar, but not identical, to that described for SEC. In many cases, gradient elution is used: that is, the solvent changes as a function of time. In that case hyphenation between *i*-LC and viscometry or light scattering is horribly difficult. Gradient elution is not used with such devices. Thus, it is difficult to determine the (average) molar mass as a function of the chemical composition $\langle M_r \rangle(\varphi_{\text{pol}})$. *i*-LC//MALDI-ToF-MS is a feasible technique; *i*-LC-Vis and *i*-LC-LS are not realistic options. SEC-FTIR and SEC-NMR can – in principle – be used to obtain the average composition as a function of molar mass $\langle \varphi_{\text{pol}} \rangle(M_r)$. In principle it is possible to combine viscometry or light scattering with critical chromatography. However, since the latter technique is more practical for relatively low molar masses and the detection devices are most suitable for high masses, this is not a perfect match.

Both LC-FTIR and LC-NMR can be applied in combination with solvent gradients. In both cases there are some complicating factors. In the case of FTIR, only solvent elimination interfaces can realistically be used. This implies that the effluent from the LC is sprayed into a (heated) evaporation chamber and that the nonvolatile analyte polymers are deposited on a suitable substrate (for example, a germanium disc). By moving the spray or the substrate, the entire chromatogram can be recorded. Some authors have programmed the deposition conditions to obtain optimal results for gradient elution LC-FTIR. However, as was mentioned ear-

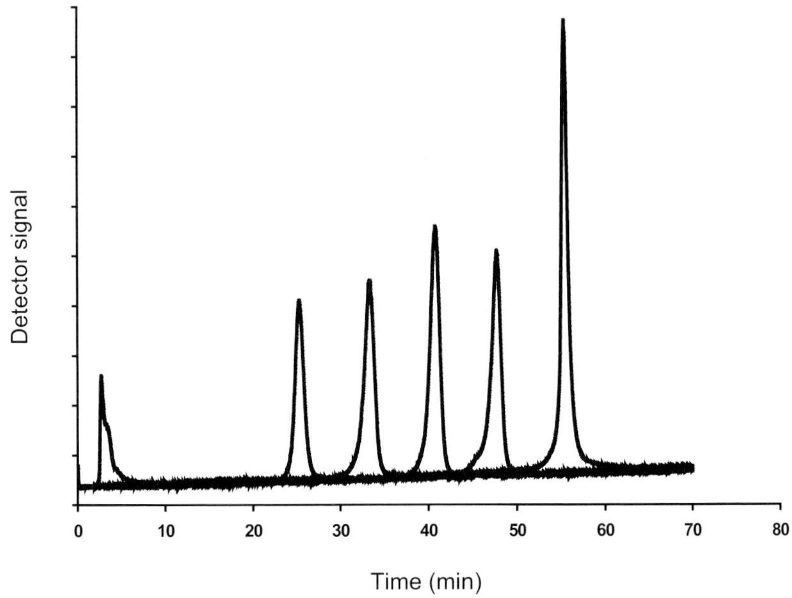

Fig. 20.11. Gradient elution liquid chromatography of a mixture ("blend") of a number of copolymers. Column: Supelcosil Discovery C_{18}, 150 mm long × 2.1 mm i.d.; particle size: 5 μm; pore diameter: 180 Å; temperature: 25°C; flow rate: 0.2 mL min^{-1}; injection volume: 5 μL; sample concentration: 1.5 mg mL^{-1}; gradient from 5 to 95% THF in acetonitrile. Peaks from left to right: poly(methyl methacrylate) homopolymer, 20% polystyrene-*co*-MMA, 40% PS-*co*-MMA, 60% PS-*co*-MMA, 80% PS-*co*-MMA, PS homopolymer.

lier (Section 20.3.2.1), it is not easy to obtain accurate quantitative results for the copolymer composition using FTIR. In LC-NMR a solvent gradient causes severe complications associated with the suppression of the solvent signal. While suppression techniques for gradient elution LC have been developed and applied successfully, the interferences in the spectrum are more serious than they are in isocratic separations. In either case, LC-FTIR or LC-NMR, the amount of additional information obtained is limited. The LC retention axis contains information on the polymer composition. The information present in the spectra is related to this. Although additional information on structural aspects may be obtained from both FTIR and NMR spectra, the two information dimensions are far from orthogonal.

This is fundamentally different for the combination of *i*-LC with MS, either on-line (most commonly using LC-ESI-MS) or off-line (most commonly using LC//MALDI-ToF-MS). In that case, the LC axis contains mainly structural information, whereas the MS axis provides information on the molar mass. A disadvantage of this combination is that fractions resulting from the *i*-LC separation are expected to be narrow in terms of their chemical composition distribution, but may be quite broad in terms of their molar mass distribution. In critical or pseudo-critical *i*-LC the very purpose of the separation is to elute all the different molar masses as one peak. Such broad fractions are not really compatible with MS. As mentioned in section 20.3.2.3, biased results are anticipated from the MS analysis of broadly distributed samples.

Better results are anticipated if fractions that are narrow both in chemical composition and in molar mass are subjected to mass spectrometry. Such fractions can be obtained from two-dimensional separations (see Section 20.3.3.5).

20.3.3.4 **Degree of Branching Distributions**

When polymer molecules are branched, their size in solution decreases. This is progressively the case with an increasing degree of branching. This also implies that molecules of equal size in solution (and thus equal SEC retention times) can have different molecular weights, depending on the degree of branching. SEC alone is not capable of revealing both the MWD and the degree of branching distribution (DBD), because the two distributions are confounded. Other information is required to yield information on the two parameters simultaneously. This extra information may be obtained from light scattering or from viscosity data, and both of these techniques can be readily combined on-line with SEC, yielding so-called hyphenated systems – that is, SEC-Vis (size-exclusion chromatography with viscometric detection) and SEC-LS (light-scattering detection). Static light scattering is most commonly coupled with SEC, but in recent years the on-line coupling of SEC and dynamic light scattering has become more readily available. The instrumentation for static light scattering is typically distinguished by the type and/or number of measurement angles. Thus, we have low-angle light scattering (LALS) and right-angle light scattering (RALS), as well as multi-angle light scattering (MALS) and triple-angle light scattering (TrALS). Often an additional "L" is added to the abbreviations, signifying the use of a laser source (for example, multi-angle laser light scattering, MALLS). In any case (viscometry or light scattering) a concentration-sensitive detector, typically a refractive index (RI) detector, is also installed on-line. Also quite common is the use of both viscometry and light scattering in combination with refractive index (RI) detection and SEC. Such a system with three detectors is known as a TripleSEC system.

Branching parameters g'_{vis} are obtained from the ratio of the intrinsic viscosity of the branched polymer $[\eta]_{\text{br}}$ (measured by viscometry) to that of the corresponding linear reference polymer $[\eta]_{\text{lin}}$ (eluting at the same time) [Eq. (6)], or from the molecular weight of a linear reference polymer (as measured by light scattering) divided by the molecular weight of the branched polymer [Eq. (7), where α is the Mark–Houwink coefficient].

$$g'_{vis} = \frac{[\eta]_{br}}{[\eta]_{lin}} \tag{6}$$

$$g'_{LS} = \left[\frac{M_{lin}}{M_{br}}\right]^{\alpha} \tag{7}$$

The average number of branches in polymer molecules as a function of the molecular weight can be estimated from Eqs. (6) or (7) and the Zimm–Stockmayer equation [24].

20.3.3.5 Complex Polymers (Multiple Distributions)

We have already encountered several examples of confounded distributions. SEC may not suffice to obtain the MWD of a copolymer, because the molecules eluted at any one time may have different molecular weights and different compositions resulting in equal hydrodynamic volumes. Likewise, in the SEC separation of a branched polymer the molecules eluted at any one time may have different molecular weights and different degrees of branching, again resulting in equal hydrodynamic volumes. Just as the characterization of polymer distributions necessitates polymer separations, the characterization of two-dimensional polymer distributions necessitates two-dimensional polymer separations. Only if two distributions are fully independent do two separate one-dimensional separations suffice. This is the case if every chemical composition fraction exhibits the same MMD and every molar mass fraction exhibits the same CCD. Because this is not usually the case, one two-dimensional separation usually reveals (much) more information than two one-dimensional separations.

Comprehensive two-dimensional liquid chromatography Two-dimensional liquid chromatographic separations can be performed in the linear ("heart-cut") format or in the comprehensive mode. In the former case, one fraction (or a few fractions) is (are) isolated from the sample and these are subsequently subjected to a second separation. An advantage of this approach is that the specific fraction(s) can be subjected to two (lengthy) high-resolution separations. A great disadvantage is that only one or a few small fractions of the sample are extensively characterized. In comprehensive two-dimensional liquid chromatography the entire sample is subjected to two different separations. The word "comprehensive" is justified if the final (two-dimensional) chromatogram is representative of the entire sample [25]. The recommended notation for linear ("heart-cut") two-dimensional liquid chromatography is LC-LC, whereas comprehensive two-dimensional liquid chromatography is commonly denoted by LC $\times$ LC [25].

In the case of polymer separations, the MMD is usually one of the distributions of interest. The second most important distribution is usually either the CCD or the functionality-type distribution (FTD). This implies that SEC and *i*-LC are attractive candidates for the two dimensions in comprehensive two-dimensional liquid chromatography. These two techniques can in principle be coupled in two different orders (either LC $\times$ SEC or SEC $\times$ LC, with the first dimension listed first). LC $\times$ SEC has a number of prevailing advantages [26]. These include (i) the possibility of performing high-resolution (gradient) LC in the first dimension, (ii) the finite time of analysis in the second dimension, (iii) the greater choice of detectors (because the separation in the second dimension is isocratic), (iv) the possibility of changing the first-dimension LC conditions without the need to re-optimize the second-dimension conditions, (v) the relatively high concentration in the first-dimension *i*-LC system (not easily overloaded) and the relatively low concentration in the second-dimension SEC system (more easily overloaded), and (vi) the greatly reduced chance of "breakthrough". If the first dimension were SEC, the sample (fraction) transferred to the second-dimension LC would be dissolved in a very strong solvent, creating a great danger of detrimental "breakthrough" peaks [27]. A disadvantage is that the resolution in the second (fast-SEC) dimension is limited, but the series of advantages prevails. Therefore LC × SEC is now the commonly employed technique.

If we are to maintain the separation (resolution) that has been achieved in the first dimension in the eventual LC × SEC chromatogram, we need to collect a large number of fractions. To maintain a reasonable overall analysis time, this implies that the second-dimension separation should be fast and that the resolution that can be obtained in this second dimension is limited. There have been significant developments toward fast SEC in recent years [9, 28] Moderate-resolution SEC can be performed within one or two minutes. If we want to collect 100 fractions from the first dimension, this implies that typical LC × SEC analysis times are of the order of two to three hours. Indeed, these are the analysis times commonly encountered in practice.

If we wish to transfer the entire first-dimension fraction to the second dimension, then the first-dimension column should have a much smaller internal diameter than the second-dimension column. Either a "miniaturized" first-dimension column can be used in combination with a conventional second-dimension column, or a conventional first-dimension column can be used in combination with a wide-bore ("maxiaturized") second-dimension column. Both approaches have been demonstrated successfully. The first approach requires (very) much less solvent, produces correspondingly less waste, and is compatible with most existing LC detectors, including the molar-mass selective detectors (viscometry, light scattering) that are of great interest in polymer separations. The second approach puts fewer demands on the chromatographic (first-dimension) system in terms of extra-column band broadening and it yields larger separated fractions for subsequent off-line analysis by other methods (for example, NMR).

Figure 20.12(a) shows an outline of a typical LC × SEC system and Figure 20.12(b) shows an enlarged representation of the switching valve. A comprehensive two-dimensional liquid chromatography system typically consists of two liquid chromatographs that are interfaced by means of a switching valve. In the case of LC × SEC the first dimension often features a gradient elution system; that is, the composition of the eluent can be programmed during the run. The valve is configured so that while one fraction is being analyzed, the next fraction is being collected (Figure 20.12(b)).

Figure 20.13 features a contemporary example of an LC × SEC separation [26]: in this two-dimensional separation of a series of copolymer "standards" of known molar mass and chemical composition, the first (gradient elution LC) dimension shows a high resolution, whereas the resolution in the second (SEC) dimension is adequate. The two separations are seen to be nearly orthogonal (that is, separation in the first dimension is (nearly) completely based on the chemical composition, whereas that in the second dimension is based on molecular size).

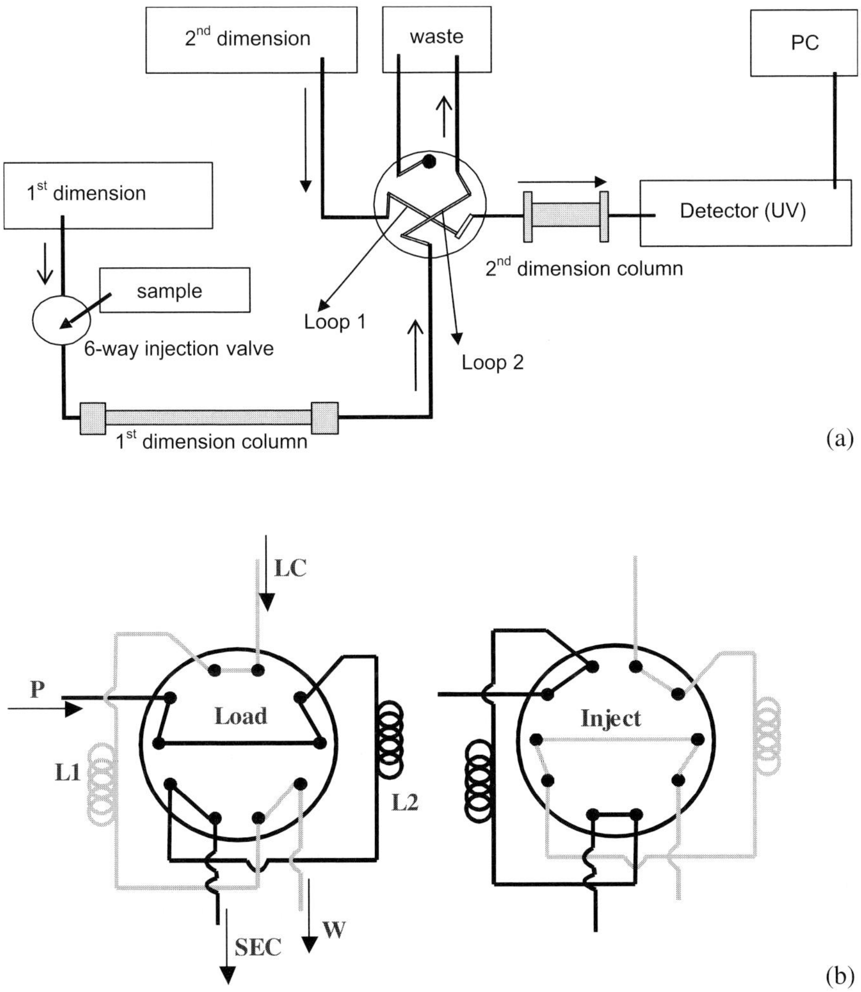

Fig. 20.12. Instrumentation for comprehensive two-dimensional liquid chromatography: (a) schematic of the complete instrument; (b) preferred configuration of a 10-port switching valve. Reprinted from Ref. [26] with permission.

Comprehensive two-dimensional liquid chromatography has seen a strong increase in popularity and in the number of applications in recent years. LC × SEC has been applied to a large number of problems in polymer science. For example, the technique has been used to provide a detailed analysis of polystyrene–poly(methyl methacrylate) diblock copolymers [29], to analyze well-defined star polylactides [30], and to study to the grafting reaction of methyl methacrylate onto EPDM [31] or onto polybutadiene [32].

The main bottlenecks for the proliferation of LC × SEC (and other two-

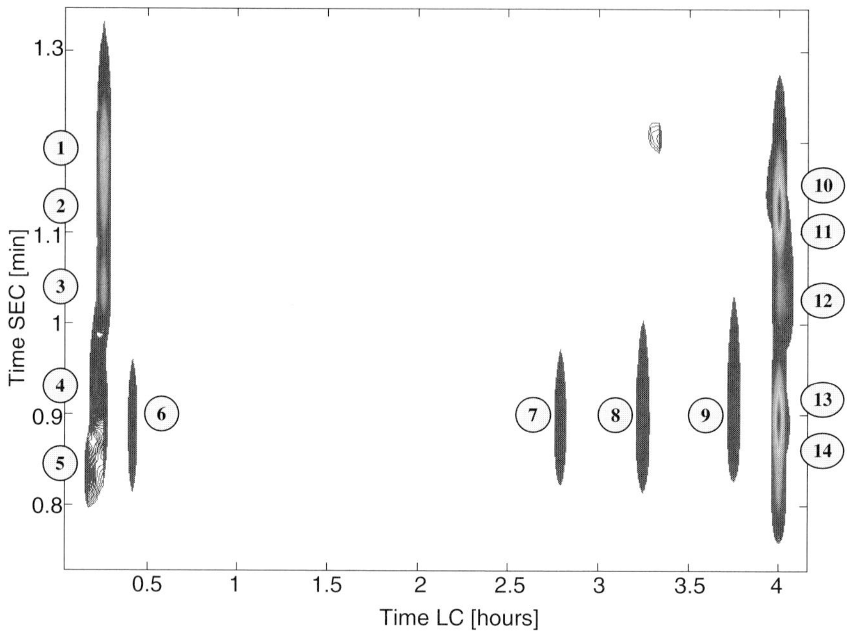

Fig. 20.13. LC × SEC-ELSD contour plot of a mixture of homo- and copolymeric reference materials: PMMA 2900 (1), 6950 (2), 28 300 (3), 127 000 (4), 840 000 (5); S-*co*-MMA 20% S (6), 40% S (7), 60% S (8), 80% S (9); PS 2450 (10), 7000 (11), 30 000 (12), 200 000 (13), 900 000 (14). LC (first dimension): C_{18} column; flow: 4 μL min^{-1}; gradient: 5–70% THF in acetonitrile 0–300 min (40°C). SEC (second dimension): mixed C column; flow: 0.6 μL min^{-1} THF. Figure reprinted from Ref. [26] with permission.

dimensional polymer separations) are now in the development of suitable calibration procedures and the associated software. Because suitable hardware for LC × SEC is already commercially available, significant progress is expected in this direction.

Notation

Acronyms

ATR	attenuated total reflection
CC	critical (liquid) chromatography (or liquid chromatography at the critical conditions of adsorption)
CCD	chemical composition distribution
CP-MAS-NMR	cross-polarization magic-angle spinning NMR
DRI	differential refractive index
ELSD	evaporative light-scattering detector
ESI	electrospray ionization

FTD	functionality-type distribution
FT-ICR-MS	Fourier transform ion–cyclotron resonance mass spectrometry
FTIR	Fourier transform infrared
GC	gas chromatography
GPC	gel permeation chromatography (same as SEC)
i-LC	interactive liquid chromatography
IR	infrared
LA(L)LS	low-angle (laser) light scattering
LC	liquid chromatography
LC × LC	comprehensive two-dimensional liquid chromatography
LC × SEC	comprehensive two-dimensional (liquid chromatography) × (size exclusion chromatography)
LS	light scattering
MA(L)LS	multi-angle (laser) light scattering
MALDI	matrix-assisted laser desorption ionization
MS	mass spectrometry
MMD	molar mass distribution (equivalent to MWD)
MWD	molecular weight distribution (equivalent to MMD)
NIR	near infrared
NMR	nuclear magnetic resonance
RA(L)LS	right-angle (laser) light scattering
RI	refractive index
SEC	size exclusion chromatography
ToF	time-of-flight
TrA(L)LS	triple-angle (laser) light scattering
Triple SEC	SEC with light-scattering, viscometry, and refractive index detection
Vis	viscometry

References

1 J. A. Biesenberger, D. H. Sebastian, *Principles of Polymerization Engineering*, Kruger, Malabar, **1993**.

2 M. Buback, M. Egorov, T. Junkers, E. Panchenko, *Macromol. Rapid Commun. 25* (**2004**) 1004.

3 S. C. Rutan, O. E. de Noord, R. R. Andrea, *Anal. Chem. 70* (**1998**) 3198.

4 N. D. Othman, G. Fevotte, D. Peycelon, J.-B. Egraz, J.-M. Suau, *AIChE J. 50* (**2004**) 654.

5 A. Cherfi, G. Fevotte, C. Novat, *J. Appl. Polym. Sci. 85* (**2002**) 2510.

6 R. A. M. Vieira, C. Sayer, E. L. Lima, J. C. Pinto, *J. Appl. Polym. Sci. 84* (**2002**) 2670.

7 M. Watari, H. Higashiyama, N. Mitsui, M. Tomo, Y. Ozaki, *Appl. Spectrosc. 58* (**2004**) 248.

8 H. A. Lousberg, H. C. J. Hoefsloot, H. F. M. Boelens, P. J. Schoenmakers, A. K. Smilde, *Int. J. Polym. Anal. Char. 7* (**2002**) 76–92.

9 H. Pasch, P. Kiltz, *Macromol. Rapid Commun. 24* (**2003**) 104–108.

10 J. Blomberg, P. J. Schoenmakers, N. van den Hoed, *J. High Resolut. Chromatogr. 17* (**1994**) 411.

11 P. A. Mirau, Nuclear magnetic resonance spectroscopy, in: R. F. Brady (ed.), *Comprehensive Desk Reference of Polymer Characterization*

and Analysis, Oxford University Press, New York, NY, USA, **2003**, pp. 181–219.

12 F. P. Fitzpatrick, H.-J. Ramaker, P. J. Schoenmakers, R. Beerends, M. Verheggen, H. J. A. Phillipsen, *J. Chromatogr. A 1043* (**2004**) 239.

13 X.-L. Jiang, P. J. Schoenmakers, X.-W. Lou, V. Lima, J. L. M. van Dongen, J. Brokken-Zijp, *Anal. Chem. 75* (**2003**) 5517.

14 X.-L. Jiang, V. Lima, P. J. Schoenmakers, *J. Chromatogr. A 1018* (**2003**) 19.

15 R. Peters, Y. Mengerink, S. Langereis, M. Frederix, H. Linssen, J. van Hest, Sj. van der Wal, *J. Chromatogr. A 994* (**2002**) 327.

16 Y. Mengerink, R. Peters, C. G. de Koster, Sj. van der Wal, H. A. Claessens, C. A. Cramers, *J. Chromatogr. A 914* (**2001**) 131.

17 X.-L. Jiang, A. van der Horst, V. Lima, P. J. Schoenmakers, *J. Chromatogr. A*, submitted for publication.

18 X.-L. Jiang, P. J. Schoenmakers, X.-W. Lou, V. Lima, J. L. M. van Dongen and J. Brokken-Zijp, *Anal. Chem. 75* (**2003**) 5517.

19 X.-L. Jiang, P. J. Schoenmakers, J. L. M. van Dongen, X.-W. Lou, V. Lima, J. Brokken-Zijp, *J. Chromatogr. A*, **1055** (**2004**) 123.

20 K. Oudhoff, *Capillary Electrophoresis for the Characterization of Synthetic Polymers*, Ph.D. thesis, University of Amsterdam, **2004**.

21 J. Falkenhagen, H. Much, W. Stauf, A. H. E. Muller, *Macromolecules 33* (**2000**) 3687.

22 W. Lee, D. Cho, T. Chang, K. J. Hanley, T. P. Lodge, *Macromolecules 34* (**2001**) 2353.

23 H. J. A. Philipsen, *J. Chromatogr. A 1037* (**2004**) 329.

24 S. Grcev, P. J. Schoenmakers, P. D. Iedema, Polymer *45* (**2004**) 39.

25 P. J. Schoenmakers, P. Marriott, J. Beens, *LC-GC Europe 16* (**2003**) 335.

26 A. van der Horst, P. J. Schoenmakers, *J. Chromatogr. A 1000* (**2003**) 693.

27 X. Jiang, A. van der Horst, P. J. Schoenmakers, *J. Chromatogr. A 982* (**2002**) 55–68.

28 S. T. Popovici, W. Th. Kok, P. J. Schoenmakers, *J. Chromatogr. A* (**2004**) DOI information: 10.1016/j.chroma.2004.05.099.

29 H. Pasch, K. Mequanint, J. Adrian, *e-Polymers* (**2002**) Paper No. 5, CODEN: EPOLCI CAN 137:125589 AN 2002:181118 CAPLUS.

30 T. Biela, A. Duda, S. Penczek, K. Rode, H. Pasch, *J. Polym. Sci. Part A: Polym. Chem. 40* (**2002**) 2884.

31 A. Siewing, J. Schierholz, D. Braun, G. Hellmann, H. Pasch, *Macromol. Chem. Phys. 202* (**2001**) 2890.

32 A. Siewing, B. Lahn, D. Braun, H. Pasch, *J. Polym. Sci. Part A: Polym. Chem. 41* (**2003**) 3143.

33 C. P. Beyers, H. F. M. Boelens, L. Klumperman, J. A. Westerhuis, *Appl. Spectrosc. 58* (**2004**) 863.

34 F. Bandermann, I. Tausendfreund, S. Sasic, Y. Ozaki, M. Kleimann, J. A. Westerhuis, H. W. Siesler, *Macromol. Rapid Commun. 22* (**2001**) 690.

35 O. Kammona, E. G. Chatzi, C. Kiparissides, *J. Macromol. Sci. Rev. Macromol. Chem. Phys. C39*(1), 57–134 (**1999**).

36 F. J. Wyzgoski, P. L. Rinaldi, E. F. McCord, M. A. Stewart, D. R. Marshall, *Macromolecules 37* (**2004**) 846.

37 B. B. P. Staal, Ph.D. thesis, TU Eindhoven, **2005**.

21
Recent Developments in Polymer Processes[1]

Maartje Kemmere

21.1
Introduction

Currently, many industrial polymers are being produced in organic solvents. A typical production scheme for conventional polymer processes is shown in Figure 21.1, for which three sub-processes can be distinguished: the polymerization of the monomers, the shaping of the raw polymer, and the post-processing of the polymer. A major drawback of polymer processes in organic solvents is the inefficient removal and recovery of the solvents and monomers in each of the three sub-processes. Often the solvent recovery requires more process steps and energy than the actual production process, as indicated in Figure 21.1. For example, in the production of butadiene rubber a typical production process applies a recycle of approximately four tons of solvent per ton of polybutadiene produced [1]. Another example is the production of elastomers such as EPDM (ethylene–propylene–diene copolymer), for which typically 20 wt.% of polymer is dissolved in an excess of hexane. Annually, these conventional polymer processes add substantially to the total amount of discharged volatile organic compounds (VOCs) such as hexane and tol-

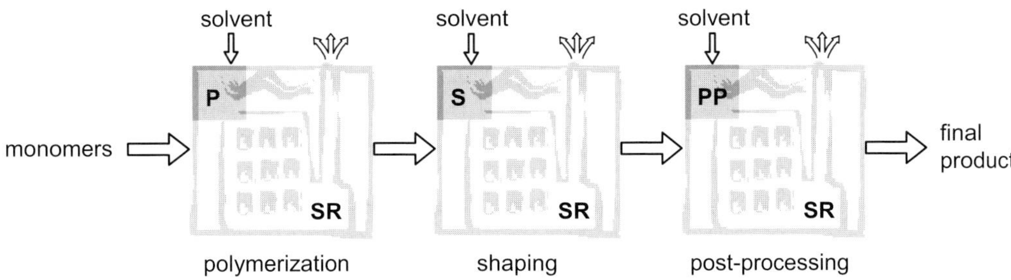

Fig. 21.1. Conventional polymerization (P), shaping (S) and post-processing (PP) steps based on organic solvents, including significant solvent recovery steps (SR).

1) The symbols used in this chapter are listed at the end of the text, under "Notation".

Handbook of Polymer Reaction Engineering. Edited by T. Meyer, J. Keurentjes

ISBN: 3-527-31014-2

uene. Approximately 20 million tons of VOCs are emitted into the atmosphere each year as a result of industrial activities [2].

For these reasons, increasing environmental awareness demands more sustainable polymer processes. Currently, several technologies are being investigated to develop cleaner processes. These include the reduction and/or recycling of solvents by closed-loop operation as well as switching to solvent-free processes, for instance melt-phase polymerization [3]. However, these methods have their limitations, because many industries require process liquids for operations such as reactions, separations, and processing steps. Therefore, the possibilities of alternative green solvents to replace organic solvents are being explored, for which important examples are aqueous solvents, ionic liquids, fluorous phases, and supercritical or dense-phase fluids [3, 4]. Obviously, each of these approaches exhibits specific advantages and potential drawbacks. Ionic liquids (room-temperature molten salts), for example, have a vapor pressure that is negligible. Because ionic liquids are non-volatile, their commercial application would significantly reduce the present VOC emission. In general, ionic liquids can be used in existing equipment at reasonable capital costs [5]. Nevertheless, the cost price of the ionic liquid itself is substantial. In addition, the separation of ionic liquids from a process stream is another important point of concern. With respect to dense-phase fluids, supercritical water has been shown to be a very effective reaction medium for oxidation reactions [6, 7]. Despite extensive research efforts, however, corrosion and investment costs still form major challenges in these processes due to the extreme operation conditions required (above 647 K and 221 bar). Particularly, supercritical carbon dioxide has made its appearance as an alternative to organic solvents, because of its tunable nature, its excellent wetting characteristics, and its low viscosity [3].

Besides alternative reaction media, process and product research has resulted in a number of new classes of polymers, such as dendritic and conductive polymers. At the same time the application field of polymers has been extended, ranging nowadays from bulk materials such as polyethylene and polypropylene all the way to specialty products, for example for biomedical applications. These developments ask for well controlled polymerization processes. In this respect ultrasound has proven to be a clean and safe technology for generating radicals in situ during the polymerization reaction.

In this chapter supercritical carbon dioxide will be discussed in more detail as one of the most promising green solvents for polymerizations. Subsequently, the possibilities and challenges of applying ultrasound in polymer processes are described. Finally, a short overview of the application potential of polymer processes based on supercritical carbon dioxide or ultrasound technology will be given.

21.2
Polymer Processes in Supercritical Carbon Dioxide

A supercritical fluid is defined as a substance for which the temperature and pressure are above their critical values (see Figure 21.2A) [8, 9]. Above the critical

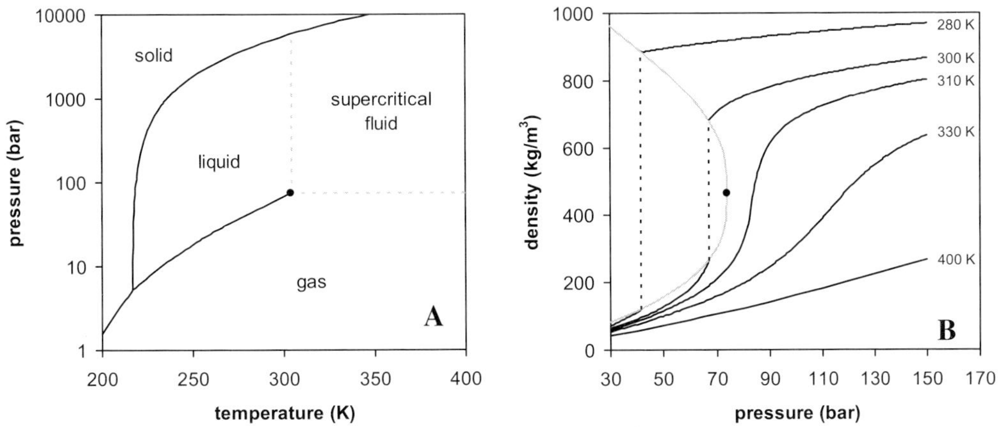

Fig. 21.2. Projections of the phase diagram of carbon dioxide A) in the pressure–temperature plane and B) in the density–pressure plane [8, 9].

temperature, the line representing vapor and liquid coexistence has disappeared. Therefore, supercritical fluids can be regarded as "hybrid solvents" because the properties can be tuned from liquid-like to gas-like without crossing a phase boundary by simply changing the pressure or the temperature. The unique properties of supercritical fluids provide opportunities for a variety of industrial processes. In Table 21.1 [10–13], the critical properties are shown for some components, of which carbon dioxide and water are the most frequently used as supercritical fluids. In polymerization processes, supercritical ethylene and propylene are also applied, where they act both as a solvent and as the reacting monomer [14].

With respect to the development of sustainable polymer processes, supercritical carbon dioxide ($scCO_2$) has significant potential as a substitute for organic solvents. In addition to the environmental benefits, $scCO_2$ has interesting physical and chemical properties from an engineering point of view. These include its relative chemical inertness, readily accessible critical point, and highly tunable solvent behavior. Moreover, carbon dioxide is nontoxic and nonflammable. In Figure 21.2B, the density of CO_2 at different temperatures is plotted as a function of pressure. The graph clearly shows that the modification of the density requires only small changes in temperature or pressure just above the critical point. In general terms, $scCO_2$ is a solvent with a low viscosity, high diffusion rates, and no surface tension [15]. The viscosity is in the range of 0.02–0.1 mPa s, where liquids have viscosities of approximately 0.5–1.0 mPa s and gases approximately 0.01 mPa s, respectively. Diffusivities of solutes in supercritical carbon dioxide are higher than in liquid solvents by a factor of up to 10. Binary diffusion coefficients of various substances in $scCO_2$ have been determined experimentally as a function of CO_2 density [16]. Because of the high diffusivity compared with ordinary solvents, $scCO_2$ is often associated with high mass and heat transfer. Additionally, the

Tab. 21.1. Physical properties of various solvents [10–13].

Solvent	T_c *[K]*	P_c *[bar]*	*Polarizability* α *[10^{-19} m^3]*	*Dipole moment* μ *[10^{-30} C m]*	*Quadrupole moment Q* *[3.16 × 10^{-17} $J^{1/2}$ $m^{5/2}$]*
Methane	190.4	46.0	26	0.0	
Ethane	305.3	48.7	45.0	0.0	−0.7
Propane	369.8	42.5	62.9	0.1	
Ethylene	282.4	50.4	42.3	0.0	
Propene	364.9	46.0	62.6	0.4	
Ethyne	308.3	61.4	33.3	0.0	+3.0
Dimethyl ether	400.0	52.4	51.6	1.3	
Sulfur hexafluoride	318.7	37.6	54.6	0.0	
Difluoromethane	351.6	58.3	24.8	2.0	
Trifluoromethane	299.3	48.6	26.5	1.7	
Tetrafluoromethane	227.6	37.4	28.6	0.0	
Difluororethane	386.7	45.0	41.5	2.3	
Hexafluoroethane	293.0	30.6	47.6	0.0	−0.7
Carbon dioxide	304.1	73.8	27.6	0.0	−4.3
n-Hexane	507.5	30.1	118.3	0.0	
Cyclohexane	553.5	40.7	109	0.3	
Diethyl ether	466.7	36.4	87.3	1.3	
Methanol	512.6	80.9	32.3	1.7	
Acetone	508.1	47.0	63.3	2.9	

above-mentioned properties are strongly pressure-dependent in the vicinity of the critical point, making $scCO_2$ a highly tunable solvent.

For application of $scCO_2$ as a medium in polymer processes, it is important to have insight into the specific system involved. In Section 21.2.1 a brief overview of the interactions of carbon dioxide with polymers and monomers will be given, including solubility in CO_2 (Section 21.2.1.1), sorption and swelling of polymers (Section 21.2.1.2) as well as phase behavior (Section 21.2.1.3) of polymer–monomer–CO_2 systems. Subsequently, an overview of processes involving polymerization (Section 21.2.2) as well polymer processing (Section 21.2.3) in $scCO_2$ will be discussed.

21.2.1
Interactions of Carbon Dioxide with Polymers and Monomers

The thermodynamic properties of pure substances and mixtures of molecules are determined by intermolecular forces acting between the molecules or polymer segments. By examining these potentials between molecules in a mixture, insight into the solution behavior of the mixture can be obtained. The most commonly occurring interactions are dispersion, dipole–dipole, dipole–quadrupole, quadrupole–quadrupole, and specific interactions. For small molecules, the contribution of each interaction to the intermolecular potential energy $\Gamma_{ij}(r, T)$, is given by Eq. (1) [12], where α is the polarizability, μ is the dipole moment, Q is the quadruple moment and SI represents specific interactions such as complex formation or hydrogen bonding.

$$\Gamma_{ij}(r,T) \approx -\left[C_1\frac{\alpha_i\alpha_j}{r^6} + C_2\frac{\mu_i^2\mu_j^2}{r^6kT} + C_3\frac{\mu_i^2Q_j^2}{r^8kT} + C_4\frac{Q_i^2Q_j^2}{r^{10}kT} + \text{SI}\right] \quad (1)$$

The interactions work over different distances, with the longest range for dispersion and dipole interactions. Note that the dispersion interaction depends on the polarizability alone, and not on the temperature. Consequently, an increased polarizability of the supercritical solvent is expected to decrease the pressures needed to dissolve a nonpolar solute or polymer. Furthermore, at elevated temperatures, the configurational alignment of directional interactions such as dipoles or quadrupoles is disrupted by the thermal energy, leading to a nonpolar behavior. Hence, it may be possible to dissolve a nonpolar solute or a polymer in a polar supercritical fluid. However, to obtain sufficient density for dissolving the solutes at those elevated temperatures, substantially higher pressures need to be applied. Additionally, specific interactions such as complex formation and hydrogen bonding can increase the solvent strength of the supercritical fluid. These interactions are also highly temperature-sensitive.

21.2.1.1 Solubility in Carbon Dioxide

The solvent strength of carbon dioxide for solutes is dominated by a low polarizability and a strong quadrupole moment. Consequently, carbon dioxide is difficult to compare with conventional solvents, due to this ambivalent character. With its low polarizability and nonpolarity, carbon dioxide is similar to perfluoromethane, perfluoroethane, and also methane. These fluids are very weak solvents for many components. Additionally, the acidity of carbon dioxide increases the solvent strength for weakly basic solutes. In general, carbon dioxide is a reasonable solvent for small molecules, both polar and nonpolar. For many components, with the exception of water, complete miscibility can be obtained at elevated pressures. However, the critical point of the mixture, which is the lowest pressure at a given temperature where CO_2 is still completely miscible, rises sharply with increasing molecule size. Consequently, most of the larger components and polymers exhibit a very limited solubility in carbon dioxide. Polymers that do exhibit a high solubility in carbon dioxide are typically characterized by a flexible backbone and high free volume (hence, a low T_g), weak interactions between the polymer segments, and a weakly basic interaction site such as a carbonyl group [17–20]. Examples of polymers that are soluble in carbon dioxide and incorporate these characteristics include poly(alkene oxides), perfluorinated poly(propylene oxide), poly(methyl acrylate), poly(vinyl acetate), poly(alkylsiloxanes), and poly(ether carbonate).

To illustrate the influence of the carbonyl group on the solubility of the polymer in carbon dioxide, Figure 21.3 shows the difference in cloud-point pressures for polymethyl acrylate (PMA) and polyvinyl acetate (PVA) [17]. Structurally, these polymers are very similar, but the carbonyl group in PVA is more easily accessible to CO_2, making it more susceptible to complex formation. As a result, the cloud-

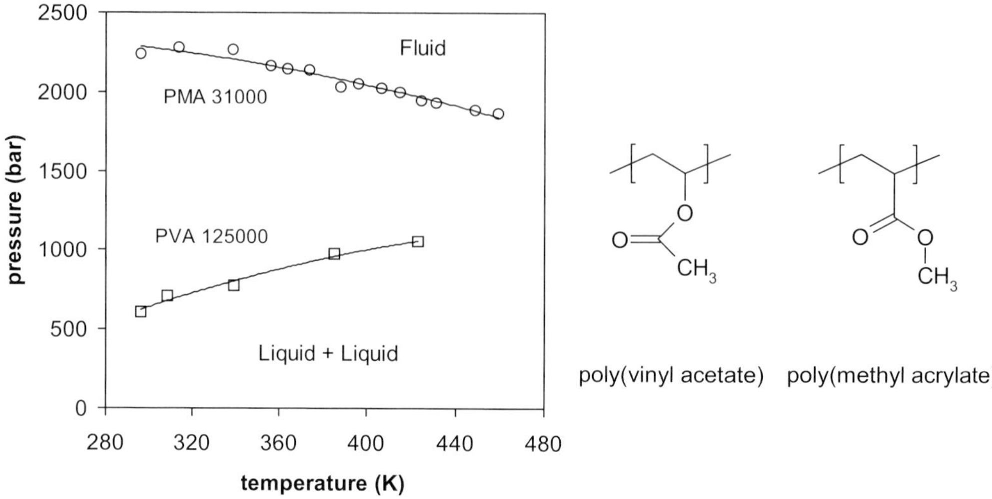

Fig. 21.3. Cloud-point curves of carbon dioxide–poly(methyl acrylate) and carbon dioxide–poly(vinyl acetate) with polymer concentrations of ~5 wt.% [17].

point pressure of PVA is up to 1500 bar lower than that of PMA, even though the molecular weight of PVA is four times higher than that of PMA, and its glass transition temperature is 21 K higher.

21.2.1.2 **Sorption and Swelling of Polymers**

Although the solubility of polymers in CO_2 is typically very low, the solubility of carbon dioxide in many polymers is substantial. When the polymer comes into contact with $scCO_2$, the sorption of carbon dioxide by the polymers and the resultant swelling of the polymer influence the mechanical and physical properties of the polymer. The most important effect is plasticization, that is, the reduction of the glass transition temperature (T_g) of glassy polymers. The plasticization effect, characterized by an increased segmental and chain mobility as well as by an increase in interchain distance, is largely determined by polymer–solvent interactions and solvent size [21]. The molecular weight of the polymer is of little influence on the swelling once the entanglement molecular weight has been exceeded. The influence of specific interactions of carbon dioxide with various groups in the polymer on the swelling isotherms is shown in Figure 21.4 [22]. These polymers all have a T_g of approximately 378 K. Carbon dioxide shows little interaction with the phenyl groups in polystyrene (PS), resulting in limited swelling. Substituting one of the CH groups in styrene with a nitrogen atom renders the polymer somewhat basic, thereby increasing the swelling level as well as the swelling rate. Building carbonyl groups into the polymer has an even stronger effect. In the case of poly(methyl methacrylate) (PMMA), the swelling level is doubled as compared to PS. Besides the enhancing interactions with carbon dioxide, the sorption and swell-

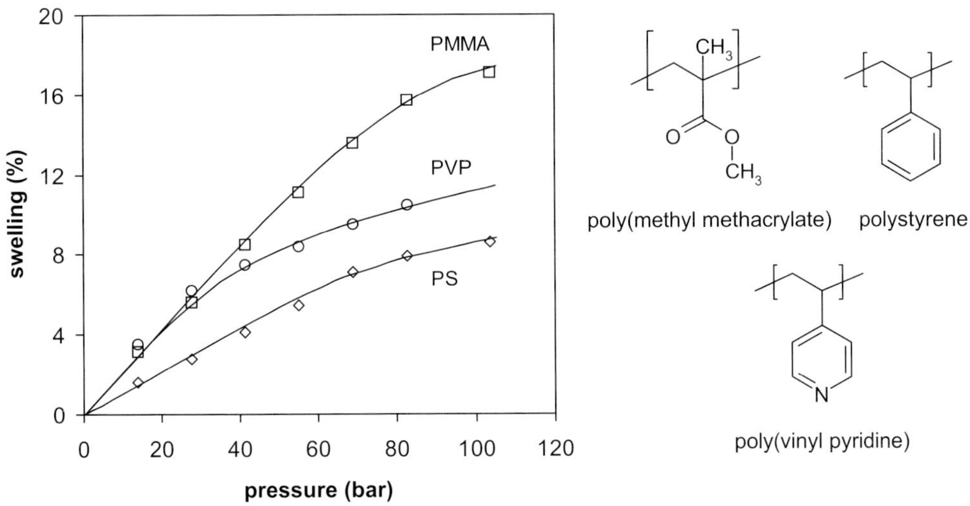

Fig. 21.4. Swelling isotherms of poly(methyl methacrylate), polystyrene and poly(vinyl pyridine) in carbon dioxide at 308 K [22].

ing can also be increased by lowering the interchain interactions of the polymer. This is illustrated by comparison of commercial semi-crystalline low-density polyethylene (ldPE) [23] (40% crystallinity) with highly branched amorphous polyethylene [24] ($<2\%$ crystallinity). Although both polymers lack specific interactions with carbon dioxide, the highly branched PE shows a strongly enhanced swelling (up to 25 vol.%) in carbon dioxide, whereas the ldPE exhibits a low maximum swelling (4 vol.%) at 323 K [25]. This is caused by the absence of crystallinity and the increased free volume introduced by branching.

To study the CO_2-induced plasticization, different techniques can be applied, for example, gas sorption and polymer swelling [26, 27]; in-situ FTIR spectroscopy [28, 29], NMR spectroscopy [30], and dynamic light scattering [31]. Two new techniques for simultaneous sorption and swelling measurement have been proposed, based on detection of the volume change of the polymer sample measured by mercury displacement [32], and on gravimetric measurement [33], respectively.

The sorption and swelling of a wide variety of polymers in $scCO_2$ have been reported in the literature. For example, Liau and McHugh [34] have investigated the swelling of PMMA and sorption of CO_2 in PMMA at temperatures ranging from 313 to 343 K and at pressures up to 272 bar by using a camera to record the change in length of the polymer sample. Wissinger and Paulaitis [26] have reported swelling and sorption in glassy polymer–gas systems for CO_2 with polycarbonate (PC), PMMA, and PS at temperatures between 306 to 338 K and pressures up to 100 bar. Sada et al. [35] have studied the sorption and diffusion of CO_2 in glassy polymer films of PS and PC over a temperature range of 293 to 313 K and pressures up to 30 bar by a pressure decay method. Berens et al. [36] have investigated the sorption behavior of the glassy polymers poly(vinyl chloride) (PVC), PC, PMMA, and PVA at 298 K up to 700 bar in the presence of liquid CO_2 through a simple gravimetric procedure. Additionally, sorption and swelling data at 308 K have been reported by Zhang et al. [37] and up to 100 bar for PMMA, PS, polyvinylpyridine and polyisoprene as well as for block copolymers of styrene–methyl methacrylate, styrene–vinylpyridine and styrene–isoprene, respectively. Otake et al. [31] have measured the swelling of a 50 nm monodisperse PS latex in water by CO_2 by dynamic light scattering at 298 K at pressures up to 350 bar.

In general, the sorption and swelling of polymers by CO_2 are crucial effects in designing polymer processes based a high-pressure technology, because important properties such as diffusivity, viscosity, glass transition, melting point, compressibility, and expansion will change. The plasticization effect of CO_2 facilitates mass transfer properties of solutes in and out of the polymer phase, which leads to many applications: increased monomer diffusion for polymer synthesis, enhanced diffusion of small components in polymers for impregnation and extraction purposes, polymer fractionation, and polymer extrusion.

21.2.1.3 **Phase Behavior of Monomer–Polymer–Carbon Dioxide Systems**

One of the requirements for the development of new polymer processes based on $scCO_2$ is knowledge about the phase behavior of the mixture involved, which enables one to tune the process variables properly to achieve maximum process efficiency. Important parameters in the phase behavior of the system are the solvent quality, the molecular weight, chain branching, and chemical architecture of the polymer, as well as the effect of end groups, and addition of a co-solvent or an anti-solvent. The literature available on the phase behavior of polymers in supercritical fluids has been reviewed extensively by Kirby and McHugh [38].

The usual phase equilibrium issue in polymer systems consists of determining whether phase separation occurs, and if it does, then what the phase compositions are. Although measuring is still the most accurate way at present to obtain information about the phase behavior [39–42], it is rather time-consuming. Most of the experimental work described in the literature has focused on polymer–solvent systems rather than on non-solvents. For example, binary systems of linear and branched polyethylenes in ethylene have been measured [40, 43]. In addition, the effect of carbon dioxide as a non-solvent on polyolefin–solvent systems has been studied [44–47].

When thermodynamic models are used to correlate and predict phase equilibria, the experimental effort can be reduced substantially. Modeling of polymer–supercritical fluid mixtures is particularly challenging, as the model has to be able to describe the behavior of a mixture containing molecules covering a wide range of molecular weights, which is often referred to as the "asymmetry" of the system. Additionally, the proximity to the critical point is difficult to model. There exist several thermodynamic models suitable for the description of phase behavior of polymer systems in supercritical media; see Chapter 3 for a detailed description of the existing models. However, the Sanchez–Lacombe equation of state (eos) and statistical associating fluid theory (SAFT) eos are the most commonly used.

The Sanchez–Lacombe model [48–50] is a lattice fluid model in which each component is divided into parts that are placed in a lattice. The different parts are allowed to interact with a mean-field intermolecular potential. By introducing an appropriate number of vacant sites (holes) in the lattice, the correct solution density can be obtained. SAFT [51–53] is based on the perturbation theory. The principle of perturbation theory is that first a model is derived for some idealized fluid with accurately known properties, called the "reference fluid". Subsequently, the properties of this model are related to those of a real dense fluid. By expanding this reference fluid into power series over a specified parameter, the power terms can be regarded as "corrections" or "perturbations" for the reference fluid as compared to reality. Obviously, the more the reference model approaches reality, the smaller the corrections are. Therefore, the key issue for applying perturbation theory is deriving the most suitable reference fluid.

In general, the choice of a certain model to describe the phase behavior of a polymer–monomer–CO_2 system is far from arbitrary, and depends on, among other factors, the particular components involved and the required detail of the description. The Sanchez–Lacombe eos has the advantage that it is very tractable and it is suitable for interpolation of data. On the other hand, the Sanchez–Lacombe model gives a poor description of systems containing specific interactions, because it is a mean-field theory with mixing rules assuming a random mixture. For that reason specific interactions can only be introduced through a temperature-dependent interaction parameter. Modeling with SAFT provides a more rigorous approach, although it is more computationally intensive. Many non-idealities can be incorporated in calculations with this eos. However, similarly to Sanchez–Lacombe, description of the binary polymer–CO_2 system can be troublesome using SAFT. As an example the SAFT eos and the Sanchez–Lacombe eos have been compared for the poly(ethylene-*co*-propoylene)–ethylene–CO_2 system [54]. Both models give a qualitative description of the increase in ethylene–PEP cloud-points upon addition of carbon dioxide as anti-solvent. For a quantitative correlation with the Sanchez–Lacombe model, a temperature-dependent interaction parameter is required. Additionally, the parameter is determined for each individual cloud-point composition. For the SAFT model, one temperature-dependent interaction parameter suffices for all compositions. For this particular system, the description of the phase behavior with the SAFT model is in better agreement with the experimental data than the description with the Sanchez–Lacombe model (see Figure 21.5).

21.2.2
Polymerization Processes in Supercritical Carbon Dioxide

With respect to the possibilities of carbon dioxide as a reaction medium for polymerization reactions, extensive reviews have been published by Scholsky [55], Shaffer et al. [56], Kendall et al. [57], and Cooper [58]. Typically, a division is made between homogeneous and heterogeneous polymerization in $scCO_2$.

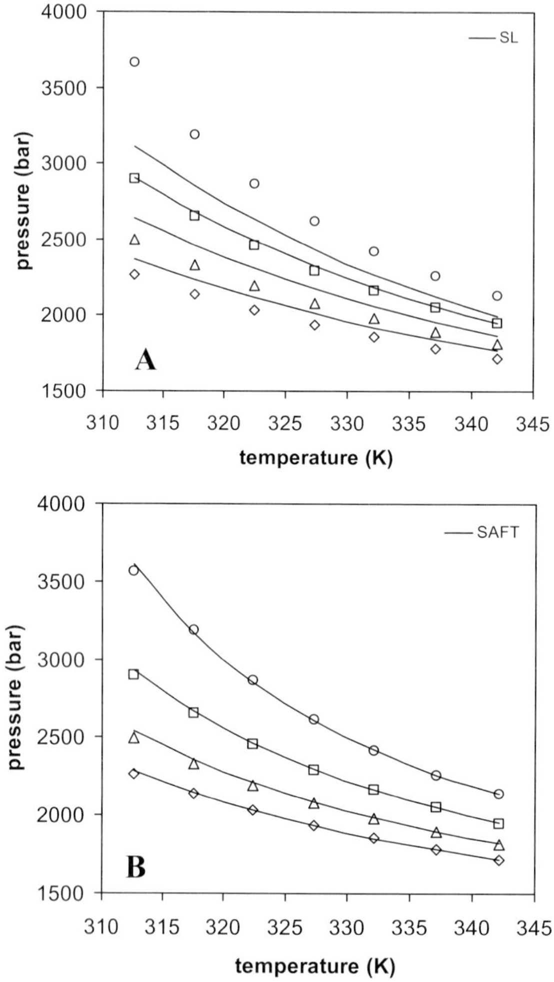

Fig. 21.5. Sanchez–Lacombe (A) and SAFT (B) correlation of the cloud-point curves for the ternary system ethylene–carbon dioxide–PEP containing 10 wt.% PEP, as a function of temperature and mass fraction of carbon dioxide: ◇, 0.05; Δ, 0.10; □, 0.15; ○, 0.20 [54].

Both chlorofluorocarbons (CFCs) and carbon dioxide appear to be very good solvents for amorphous, low-melting fluoropolymers. Since environmental restrictions have limited the use of CFCs drastically, carbon dioxide has become a highly viable alternative solvent for the production of amorphous fluoropolymers [59]. Examples of polymerization of fluorinated monomers in a homogeneous reaction medium of $scCO_2$ are the polymerization of fluorinated acrylates (see, for example Refs. 59–61, fluoroalkyl-derivatized styrene [62], fluorinated vinyl and cyclic ethers [63], and the telomerization of 1,1-difluoroethylene [64]. Other options to run a polymerization under homogeneous conditions are to increase the temperature and pressure until a one-phase system is obtained [65] or to add a co-solvent.

Except for certain fluoropolymers and a few other examples of polymers that are soluble in it, as given in Section 21.2.1.1, carbon dioxide exhibits a strong anti-solvent effect for most high molecular weight polymers. This implies that the majority of the polymerization reactions, such as dispersion, emulsion, suspension, and precipitation polymerization, involve heterogeneous polymerization. Examples of dispersion polymerization in $scCO_2$ are the polymerization of MMA [66] and styrene [67]. Beckman and co-workers have reported the inverse emulsion polymerization of acrylamide in $scCO_2$ [68]. Moreover, it is possible to synthesize well-defined porous polymers by templating in $scCO_2$–water emulsions [69]. Macroporous beads of trimethylolpropane trimethacrylate (TRIM) can be synthesized by means of suspension polymerization [70]. Much work has been performed on precipitation polymerization, such as the polymerization of vinyl ethers and oxetanes [63], the continuous polymerization of vinylidene fluoride as well as of acrylic acid [71, 72], MMA polymerization initiated by ultrasound [73], the catalytic polymerization of olefins [74, 75] and the enzymatic polymerization to polycaprolactone [76]. An interesting extension in the application of $scCO_2$ for polymerization purposes is the use of CO_2 both as a reactant and as a solvent in a copolymerization with cyclohexene oxide [77].

From an engineering point of view several issues have to be addressed in order to design polymerization processes based on $scCO_2$. The previous sections have emphasized the importance of the phase behavior of the system as well as the sorption and swelling of the polymer. Additionally, in catalytic polymerization processes the solubility and recycle of the catalyst are important. Moreover, when the polymer-shaping step is to be included in the same equipment or the purification of the polymer product, this will set additional requirements on the design. Obviously, mass and heat transfer have to be investigated. In general, supercritical fluids have gas-like viscosities and diffusion rates, which allow for fast mass transfer. In addition, the thermal diffusivity of supercritical CO_2 is significantly higher than for gaseous CO_2. The need for forced convection in order to facilitate sufficient mass and heat transfer during the polymerization process has to be determined for each particular type of polymerization. If the process involves a precipitation polymerization, for instance, the effect of polymer deposition on heat transfer surfaces forms a potential point of concern. Given the scale of most commercial polymerization processes, a translation from a batchwise operation as it is usually performed on a laboratory scale toward a continuous process forms another important consideration in the development of polymerization processes based on $scCO_2$. As an example, it has been suggested that the process concept of the catalytic polymerization of olefins in $scCO_2$ should include a loop type of apparatus [78], as described by Ahvenainen et al. [79] (see Figure 21.6). Next to the requirements mention above, this reactor design allows adjustment of the residence time to relatively long values in order to obtain a reasonable conversion.

Although many new polymerization reactions have been discovered in $scCO_2$ since the early 1990s, the development of the engineering aspects of these pro-

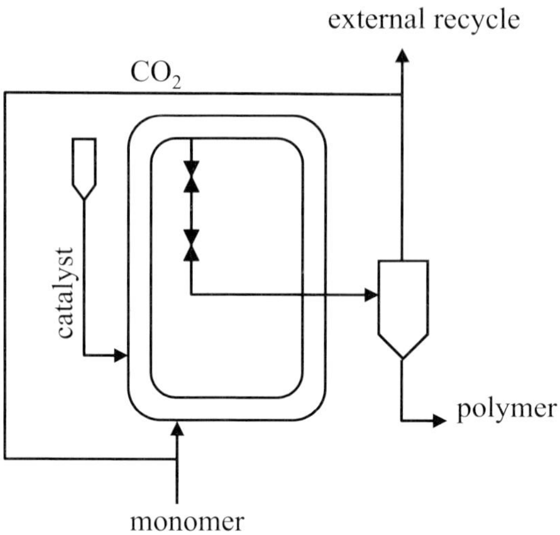

Fig. 21.6. Proposed process concept of a loop reactor used for the catalytic polymerization of olefins in $scCO_2$ [77].

cesses and the step toward pilot scale and industrial application had only just been initiated in the last few years. The commercialization by DuPont of the production of Teflon® (polytetrafluoroethylene, PTFE) in $scCO_2$ is a promising example [80].

21.2.3 Polymer Processing in Supercritical Carbon Dioxide

Applications of supercritical fluids in polymer processing can be divided into three areas: applications where carbon dioxide does not interact with the polymer; and processing of swollen, or dissolved, polymers. Schematically, this is shown in Figure 21.7 [81, 82]. In the first of these areas, when no interaction with the polymer occurs, which is typically the case for crystalline and rigid polymers, supercritical carbon dioxide can be used as a cleaning solvent. Applications in this category include precision cleaning of surfaces, degreasing, particle removal, and dry cleaning. In particular, dry cleaning is being investigated for widespread industrial application, because traditional processes based on perchloroethylene are under pressure due to health and environmental issues. Current research in this area is focused on the development of low-cost nonfluorous CO_2-philic surfactants [18, 19] and on the development of membrane systems for the economical reuse of carbon dioxide [83].

In the case of moderate interactions of carbon dioxide with the polymer, substantial levels of sorption and swelling can occur. The swelling of the polymer significantly enhances diffusivities of solutes in the polymer matrix, allowing for rapid extraction, impregnation, shaping, and blending processes. Extraction applications include removal of residual monomers, solvents, and catalysts [84, 85]. An impor-

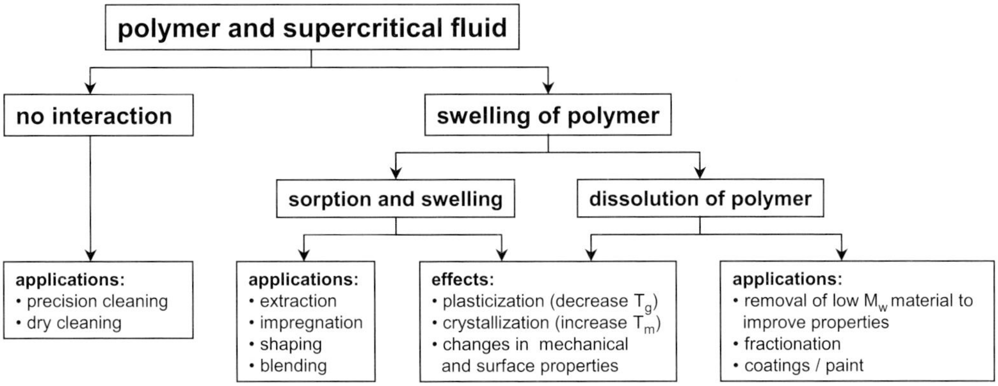

Fig. 21.7. Interactions of supercritical fluids with polymers, and their applications in polymer processes [81].

tant application of impregnation of polymers is the dyeing of textiles and fibers [86, 87]. Other impregnation applications include impregnation of polymer matrices with drugs for the production of controlled drug delivery devices [88]. Advantages in this process are the low processing temperatures and the lack of organic solvents, which need to be removed in post-processing steps. Applying $scCO_2$ for shaping purposes is advantageous because the material properties can be tuned by varying the supercritical solvent density. This enables the removal of the solvent without passing a liquid–gas interface, which would cause the structures formed to collapse [89]. With respect to blending, an interesting application is in-situ polymerization in an existing polymer matrix after impregnation with a monomer and an initiator to form polymer blends [90]. This enables the formation of polymer blends that cannot easily be obtained by conventional methods, for example due to large differences in the melting temperatures of the pure polymers.

Applications in the third category, where the polymer is dissolved in the supercritical fluid, include coating processes as well as fractionation. For example, various groups have worked on the RESS (rapid expansion from supercritical solution)-based process for spray coating using supercritical carbon dioxide as a solvent [91–94]. The main advantage of these processes is the reduction of VOC emissions. Methods of applying coatings from CO_2 involve dip or spray coatings, solutions of coatings in a CO_2 atmosphere and dispersion coatings. An example that has already been commercially developed is the Unicarb™ spray coating process [95], which can use highly viscous coatings and results in a narrow droplet size distribution. A potential drawback of spray coating processes using supercritical carbon dioxide is the fact that most coating materials have a low solubility in supercritical carbon dioxide. Consequently, co-solvents such as methanol are needed to enhance solubilities. An alternative to the addition of co-solvents is the use of stabilizers [96]. Typical stabilizer structures involve graft copolymers, perfluoropolyether acid and polyether carbonates.

An extensive review on polymer processing using supercritical fluids by Kazarian [21] includes the applications mentioned above. To illustrate the possibilities of polymer processing with $scCO_2$ from an engineering point of view, two important applications will be discussed in more detail, namely extraction (see Section 21.2.3.1) and impregnation (Section 21.2.3.2).

21.2.3.1 **Extraction**

Commercial-grade polymers contain a range of additives and contaminants such as residual monomer, solvent, and catalyst residue. Before application of the polymer in the end-product, manufactures need to remove these impurities from it, for instance to meet environmental regulatory requirements or to improve the physical polymer properties [58]. Supercritical fluid extraction (SFE) has many advantages over conventional cleaning techniques, such as a shorter extraction time, adjustable solvent strength by tuning the pressure, a wide range over which the extraction temperature can be varied, and less energy consumption [15].

A number of studies have been reported in supercritical extraction of polymer additives, extraction of monomers and oligomers, and removal of residual solvent from polymer foams [97], of which some examples will be given here. SFE can be achieved in two ways: statically or dynamically. In a static extraction, the extraction vessel is pressurized to the desired pressure with the extracting fluid and subsequently left for a certain length of time, whereas during dynamic extraction the supercritical fluid passes continuously over the sample, extracting soluble compounds and depositing them in a suitable solvent or on a solid trap. Cotton et al. [98] used dynamic extraction for the extraction of Irgafos 168 and Irganox 1010 from a commercial-grade polypropylene matrix. The rate of extraction can be hampered by two factors: the diffusion of the additive through the matrix, or the solubility of the extracted material in the supercritical fluid, can be limited. Similar results have been obtained by Lou et al. [99] for the supercritical extraction of polyethylene by applying the two-film theory. In general, if diffusion through the polymer matrix is the rate-limiting step in the extraction process, an enhanced extraction temperature will increase the extraction rate, because the diffusion coefficient increases with temperature. Moreover, the plasticizing effect will significantly enhance the diffusion rates. If the rate-limiting step is the solubility of the solute (extracted material) in the supercritical fluid, the extraction rates can be enhanced by either increasing the pressure (solvent strength) or by increasing the fluid flow rate [58].

Another example of supercritical fluid extraction is the extraction of liquid crystalline 4,4′-dibutylazobenzene from a polystyrene matrices [100]. Furthermore, Sekinger et al. [101] have investigated the feasibility of using SFE for extracting additives from styrene–butadiene rubber (SBR) by determining the effects of density, temperature, flow rate, and contact time on the extraction efficiency. Kemmere et al. [85] have developed a post-polymerization process which reduces the amount of residual monomer in latexes using $scCO_2$. Typically, the method comprises a counterflow process, in which part of the residual monomer is converted by the increased diffusion inside the polymer particles due to the swelling by $scCO_2$. In

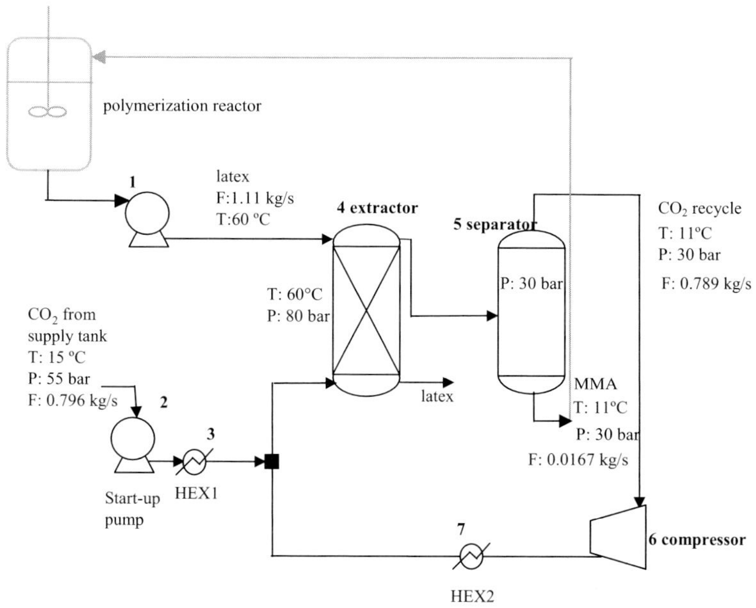

1. Latex pump	**2. Centrifugal pump**	**3. Heat exchanger 1**	**4. Extractor**	**5. Separator**	**6. Compressor (screw)**	**7. Heat exchanger 2**
flow $4.0\ m^3/hr$	flow $3.45\ m^3/hr$	heating duty 80.5kW	height 4.25 m	height 3.38 m	total power 53.66 kW	heating duty (-)51.57 kW
head of feet 2592.0	head of feet 995.7	U $850.0\ W/m^2K$	diameter 0.60 m	diameter 1.12 m	U	$850.0\ W/m^2K$
power 12.5 kW	power 3.44 kW	Tm 69.12 K	wall thickness 0.024 m	wall thickness 0.045 m		T_m 57.00 K
		required area 1.4				required area 4.52

Fig. 21.8. Process flow sheet for the removal of residual monomer from latex-products using $scCO_2$ [84].

addition, the amount of residual monomer is further reduced by the extraction capacity of $scCO_2$ [102]. Figure 21.8 shows the flow diagram of the extraction process. A viability study, including equipment sizing and economic evaluation, has shown that the removal of residual monomer from latex-products using $scCO_2$ in principle yields a process which is both technically and economically feasible, and capable of meeting future requirements [85]. Other examples of extrapolation of SFE results obtained on a laboratory scale toward production equipment are described by Brunner [103] and Perrut and co-workers [104].

21.2.3.2 **Impregnation and Dyeing**

The same properties that make $scCO_2$ attractive for extraction purposes can be exploited in the impregnation and dyeing of polymeric materials [105]. By applying supercritical conditions, it is possible to impregnate polymers with pharmaceuticals, flavors, and fragrances, or with additives such as pigments, stabilizers, and plasticizers. A wide variety of polymer systems have been impregnated using $scCO_2$, for example PMMA [106], PS, PE, PC, PVC, and polypropylene (PP) [107], and polydimethylsiloxane (PDMS) [108, 109], as well as poly(ethylene terephthalate) (PET) [110, 111].

The key parameter that determines the feasibility of the impregnation process is the equilibrium distribution or partition coefficient of the solute that is being impregnated into the polymer. There are two different methods of supercritical fluid impregnation of additives into polymer products. The first involves diffusion of a compound that is soluble in the supercritical fluid. The impregnation takes place when a polymer matrix is contacted with the supercritical fluid containing the solute. On depressurization, the CO_2 flows out of the polymer matrix, leaving the solute trapped in it [112]. For example, Ma and Tomasko [113] have investigated the impregnation of high-density polyethylene (hdPE) with the nonionic surfactant *N,N*-dimethyldodecylamine-*N*-oxide using $scCO_2$. They have observed that the penetration and absorption into the polymer produces no structural changes or losses of mechanical strength. However, as a result of the swelling by $scCO_2$, the wetting properties of hdPE are changed permanently, in contrast to conventional techniques using an aqueous solution. In the second route to impregnation, the compound typically has a low solubility in the supercritical phase; dyeing is a typical example (see, for example, Refs. 106, 107, 111, 114).

As described earlier in Section 21.2.3, impregnation applications based on $scCO_2$ technology include polymer blending and the production of polymeric drug delivery devices. However, the largest application of impregnation with $scCO_2$ at present is the dyeing of textiles and fibers. The main motivation is the replacement of large amounts of water in the dyeing process with benign $scCO_2$, which can reduce the wastewater in the textile industry significantly [85, 86]. The ease of recovering the carbon dioxide by reducing the pressure and recycling it without the necessity to clean the carbon dioxide provides a strong environmental advantage. For these reasons, there has been a significant effort to commercialize supercritical fluid dyeing of PET [112], for which the first pilot plant has been built [115].

21.3
Ultrasound-induced Radical Polymerization

For the development of sustainable polymer processes, ultrasound is an interesting technology, as it allows for polymerizations without the use of initiator. The radicals are generated in situ by cavitation events [116, 117], which make possible a clean and intrinsically safe polymerization reaction. As a result of the high strain rates outside the bubble, cavitation can also induce chain scission [118, 119], which provides an additional means to control the molecular weight of the polymer produced. In Sections 21.3.1 and 21.3.2 the physical background of ultrasound-induced cavitation and radical formation will be described. Subsequently (see Section 21.3.3), an overview of the several types of ultrasound-induced polymerizations will be given, namely bulk, precipitation, and emulsion polymerization.

Finally, in Section 21.3.4, the breakage of polymer chains by cavitation will be discussed, including the possibility of synthesizing block copolymers.

21.3.1 Ultrasound and Cavitation in Liquids

Ultrasound passes through an elastic medium as a longitudinal wave, which is a series of alternating compressions and rarefactions. This means that liquid is displaced parallel to the direction of motion of the wave. Ultrasound comprises sound waves typically in the range of 20 kHz to approximately 500 MHz. The frequency (f) and the acoustic amplitude ($P_{A,\max}$) are the most important properties to characterize the pressure wave. The variation of the acoustic pressure (P_A) of an ultrasound wave as a function of time (t) at a fixed frequency is described by Eq. (2) [120].

$$P_A = P_{A,\max} \cdot \sin(2 \cdot \pi \cdot f \cdot t) \tag{2}$$

The use of ultrasound can be divided into two areas: low-intensity, high-frequency ultrasound (2–500 MHz, 0.1–0.5 W cm^{-2}) and power ultrasound with a high intensity and a low frequency (20–900 kHz, >10 W cm^{-2}); see Table 21.2. The first does not alter the state of the medium through which it travels and is commonly used for nondestructive evaluation and medical diagnosis [121]. This type of ultrasound cannot be used for reactions. Contrarily, power ultrasound uses the energy to create cavitations, which involve the formation, growth, and implosive collapse of microscopic bubbles in a liquid. These bubbles are generated when the "negative" pressure during the rarefaction phase of the sound wave is sufficiently great to disrupt the liquid. The implosive collapse of the bubbles can produce extreme temperatures and pressures locally for very short times, due to compression of the gas phase inside the cavity [122, 123]. These hot-spots can lead to irreversible changes such as the formation of excited states, bond breakage, and the generation of radicals. Power ultrasound is applied for cleaning purposes, treatment of kidney stones, plastic welding, and chemical reactions [124].

Tab. 21.2. Overview of different types of ultrasound, including the various applications.

Power ultrasound, 20–100 kHz	*Sonophoresis, 20 kHz, low power*	*Therapeutic ultrasound, 1 MHz, high power*	*Nondestructive ultrasound, >2 MHz, low power*
Sonochemistry Welding Cleaning Cell disruption Sterilization Kidney stones	transdermal drug delivery	therapeutic massage controlled release	flaw detection medical diagnoses

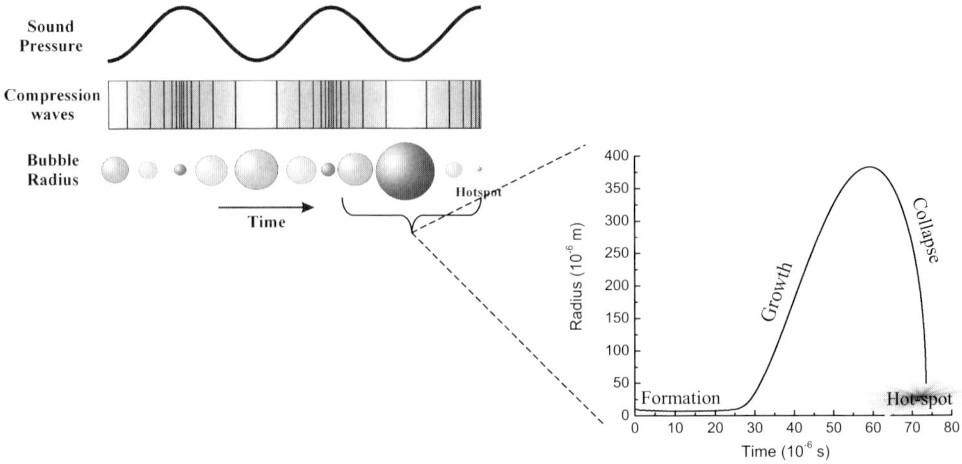

Fig. 21.9. Schematic representation of bubble growth and collapse in water at ambient pressure irradiated with ultrasound and the resulting hot-spot due to adiabatic compression. The radius of the cavitation bubble as a function of time has been calculated using the dynamic bubble model based on the Rayleigh–Plesset equation [136].

Sonochemistry comprises the chemical effects that are induced by power ultrasound. The ultrasound waves cause the molecules to oscillate around their mean position in a liquid. During the positive-pressure cycle the distance between the molecules decreases, while during the negative-pressure period the distance increases. At a sufficiently high intensity a critical distance between the molecules is exceeded during the negative-pressure period, and a cavity is formed [125]. Due to the presence of nuclei such as dissolved gases and solid impurities, cavities are formed at far lower sound pressures than theoretically predicted [126]. After the formation of a cavity, a critical acoustic pressure has to be overcome to initiate the explosive growth of this bubble. This explosive growth is followed by an implosive collapse (see Figure 21.9). During this collapse, the contents of the bubble are heated almost adiabatically, which leads to local short-lived hot-spots in the liquid. Depending on the specific conditions, bubble wall velocities, and pressures and temperatures in the bubble, can increase up to 1500 m s^{-1}, 200 bar and 5000 K, respectively [127].

The acoustic pressure amplitude determines the growth of a cavitation bubble and consequently the chemical effects upon collapse. The amplitude of the pressure wave can be measured directly with a hydrophone or calculated using a calorimetric method [128, 129], by which it is possible to determine the ultrasound power (Q_{US}) that is transferred to the liquid. With the ultrasound power, the density of the liquid (ρ), the speed of sound in the medium (v), and the surface area of the ultrasound source (A_{US}), the acoustic amplitude can be calculated according to Eq. (3). The ultrasound intensity is the power input divided by the surface area of the source [130].

$$P_{A,\max} = \sqrt{2 \cdot \rho \cdot v \cdot \frac{Q_{US}}{A_{US}}} = \sqrt{2 \cdot \rho \cdot v \cdot I_{US}} \tag{3}$$

The Blake threshold pressure (P_B) describes the critical pressure to initiate the explosive growth of a cavitation bubble [Eq. (4)] [131].

$$P_B = P_0 - P_v + \frac{4}{3} \cdot \sigma \cdot \sqrt{\frac{2}{3} \cdot \frac{\sigma}{\left(P_0 + 2 \cdot \frac{\sigma}{R_0} - P_v\right) \cdot R_0^3}} \tag{4}$$

Equation (4) assumes that the external pressure (P_0), the vapor pressure (P_v), the surface tension (σ), and the equilibrium radius of the bubble (R_0) determine the negative pressure in the liquid required to start an explosive growth of a cavity [132]. The Blake threshold pressure is based on a static approach and is only valid when the surface tension dominates all dynamic effects, such as mass transfer and viscosity. The influence of the medium characteristics and the physical conditions on the implosion can be calculated with a dynamic bubble model [120, 133]. The dynamic movement of a bubble in a sound pressure field can be described with the Rayleigh–Plesset equation. By combining the Rayleigh–Plesset equation with a mass and energy balance over the bubble, the temperature and pressure in the bubble can be calculated [134–136]. The model also describes the dynamic movement of the bubble wall, which results in a calculated radius of the cavitation bubble as a function of time (see Figure 21.9). The explosive growth phase and the collapse phase of the bubble can be distinguished clearly. Moreover, when dynamic effects are more important than the surface tension, the cavitation threshold can be calculated with the dynamic model, while the Blake threshold pressure cannot be used in these conditions.

21.3.2
Radical Formation by Cavitation

Because of the extreme conditions during a cavitation event, radicals can be formed. Several parameters affect cavitation and thereby the polymerization reaction, since the radical formation rate is directly influenced by the cavitational collapse. The number of radicals formed due to sonification is a function of the number of cavities created and the number of radicals that are formed per cavitation bubble. The bubble wall velocity during collapse and the hot-spot temperature determine the rate at which radicals are formed, both inside and outside a single bubble. These two parameters depend on the physical properties of the liquid as well as on the physical and chemical processes occurring around the cavity. The most important properties and processes occurring in a cavitation bubble are depicted schematically in Figure 21.10. The number of cavities is determined, for instance, by the impurities in the liquid, the static pressure, the ultrasound intensity, and the vapor pressure. This emphasizes the complexity of the influences on the overall

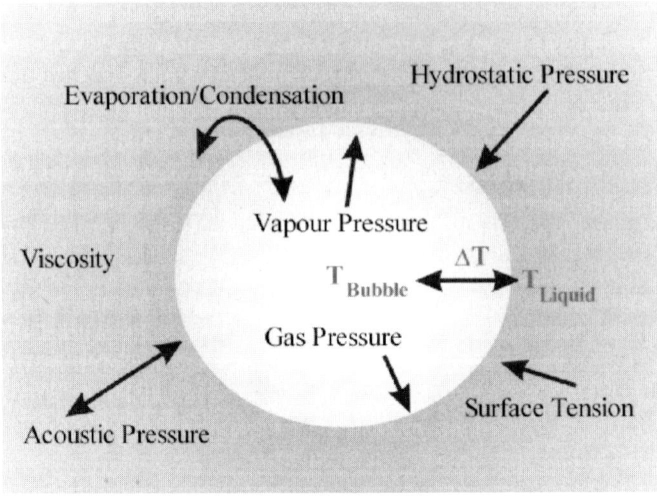

Fig. 21.10. Schematic representation of a cavitation bubble, including the processes and physical properties that determine the cavitation process.

radical formation rate [137, 138]. In the following sections, the effect of the most important parameters will be discussed.

Reaction temperature When the reaction temperature is altered, the liquid properties will change. Although these properties (viscosity, surface tension, sound velocity, vapor pressure, and so on) all have an influence on the chemical effect of cavitation, the change in vapor pressure dominates the other liquid properties. As the temperature is raised, the vapor pressure in the bubble is increased, which cushions the implosion of the cavity. This results in a lower local temperature inside the cavity at higher overall temperatures. Consequently, fewer radicals are formed per cavitation bubble. On the other hand, a higher vapor pressure can lead to easier bubble formation due to the decrease in the cavitation threshold. In most cases, however, an increase in reaction temperature will result in an overall decrease in the radical formation rate. Therefore, ultrasound-induced reactions exhibit the opposite behavior to common radical reactions [139].

Static pressure A high static pressure can prevent the formation of cavitation bubbles, as the Blake threshold pressure increases. This implies that fewer or no cavitations are formed at higher static pressures. To counteract this effect a higher acoustic pressure is required, which will result in a more violent collapse of a cavitation bubble.

Viscosity At increasing viscosities the growth and collapse of a cavitation bubble is retarded, due to the higher drag of the liquid. At a certain viscosity the drag force becomes too high and the bubble has insufficient time to grow to a critical radius.

If this radius is not reached, no collapse will occur. Additionally, the higher viscosity slows down the collapse, which makes it possible for heat to be transferred to the liquid. Due to this heat transfer, lower hot-spot temperatures are reached and subsequently fewer radicals are being generated.

Ultrasound intensity At first the radical formation rate will increase to a maximum with increasing ultrasound intensity [137]. This is caused by the higher cavitation intensity per bubble and the larger number of cavitation bubbles. At ultrasound intensities that are too high, however, a cloud of cavitation bubbles is formed near the ultrasound source, due to which the pressure wave is no longer transmitted efficiently to the liquid. As a result, the cavitation intensity and consequently the radical formation rate decrease with a continued increase in intensity [137]. An optimum radical formation rate with ultrasound intensity can thus be found.

Ultrasound frequency The frequency of ultrasound has a significant effect on the cavitation process. At very high frequencies (>1 MHz), the cavitation effect is reduced as the inertia of a cavitation bubble becomes too high to react to fast-changing pressures. Most ultrasound-induced reactions are therefore carried out at frequencies between 20 and 300 kHz. Typically a frequency of 20 kHz is used for ultrasound-induced polymerizations and polymer scission reactions [140].

21.3.3 Cavitation-induced Polymerization

Generally, free-radical polymerization consists of four elementary steps: initiation, propagation, chain transfer, and termination. When ultrasound is used to initiate polymerization, radicals can be formed both from monomer and from polymer molecules. This implies that due to radical formation by polymer scission an additional elementary step is involved in ultrasound-induced polymerization, as indicated in Figure 21.11.

The radicals from polymer and monomer are generated by two different mechanisms. The monomer molecules are dissociated by the high temperatures inside the hot-spot, whereas the polymer chains are fractured by the high strain rates outside the bubble [141]. The majority of the radicals in an ultrasound-induced polymerization reaction originate from the polymer chains [142]; see Figure 21.12. It has to be noted that the radicals are only formed in the immediate vicinity of the ultrasound source where cavitation occurs. Subsequently, these radicals are dispersed throughout the reactor. In the literature several types of ultrasound-induced polymerizations have been reported, namely bulk, precipitation, and emulsion polymerization.

21.3.3.1 Bulk Polymerization

Most ultrasound-induced bulk polymerizations are performed at room temperature [143]. This low temperature is chosen because radical formation induced by

Initiation

$$\text{M} + \text{Cavitation} \longrightarrow 2\,\text{R}\bullet \qquad R_i = 2\,k_{dmon}\,[M]$$

Propagation

$$\text{M}_n\bullet + \text{M} \longrightarrow \text{M}_{n+1} \qquad R_p = k_p\,[M][M_n\cdot]$$

Chain Transfer

to monomer

$$\text{M}_n\bullet + \text{M} \longrightarrow \text{M}_n + \text{M}\bullet \qquad R_{ctm} = k_{ctm}\,[M][M_n\cdot]$$

to polymer

$$\text{M}_n\bullet + \text{M}_m \longrightarrow \text{M}_n + \text{M}_m\bullet \qquad R_{ctp} = k_{ctp}\,[M_n\cdot][M_m]$$

Termination

by combination

$$\text{M}_n\bullet + \text{M}_m\bullet \longrightarrow \text{M}_{n+m} \qquad R_{tc} = k_{tc}\,[M_n\cdot][M_m\cdot]$$

by disproportionation

$$\text{M}_n\bullet + \text{M}_m\bullet \longrightarrow \text{M}_n + \text{M}_m \qquad R_{td} = k_{td}\,[M_n\cdot][M_m\cdot]$$

Polymer Scission

$$\text{M}_n + \text{Cavitation} \longrightarrow \text{M}_m\bullet + \text{M}_{n\text{-}m}\bullet \quad R_d = 2\,k_{dpol}\,[M_n]$$

Fig. 21.11. Reaction mechanism of ultrasound-induced radical polymerization, assuming intrinsic polymerization and avoiding thermal initiation.

ultrasound appears to be more efficient at lower temperatures [128]; see also Section 21.3.2. In contrast to conventional thermal initiators such as potassium persulfate, ultrasound can initiate a polymerization reaction at ambient temperatures. It should be noted, however, that the propagation rate increases with an increasing temperature. Therefore, a suitable well-tuned system has to be chosen to optimize ultrasound-induced polymerization. In the literature, several bulk polymerizations initiated by ultrasound have been reported; examples are the polymerization of methyl methacrylate [116, 144, 145], ethyl methacrylate [146], *n*-butyl methacrylate [117, 141], and styrene [147]. Since the basic reaction kinetics are well known [144, 148] the ultrasound-induced polymerization of MMA is by far the most studied system. Typically, the resulting molecular weight of the PMMA obtained is in the range 1×10^5–6×10^5 g mol^{-1}.

An important parameter during ultrasound-induced bulk polymerizations is the viscosity. As the reaction proceeds, the polymer chains formed cause a dramatic increase in the viscosity, resulting in slower growth and collapse of the cavity. As

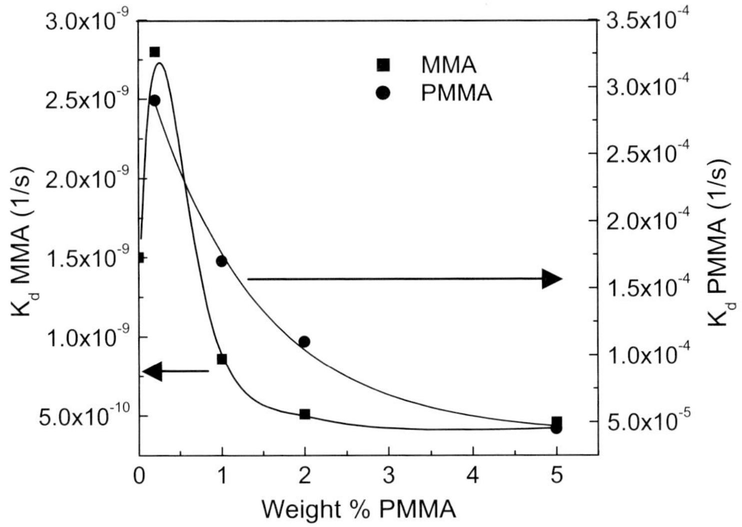

Fig. 21.12. Radical formation rate constants from MMA and PMMA as a function of the weight percentage of polymer dissolved at a temperature of 293 K and an ultrasound intensity of 62 W cm^{-2} [142].

the cavitations become less effective, the radical formation rate both from monomer and from polymer will decrease. Subsequently, the reaction rate decreases as shown in Figure 21.13. At a conversion of approximately 20% the collapse is no longer sufficiently strong to induce hot-spot temperatures that are able to generate monomeric radicals. Moreover, the strain rates outside the collapsing bubble are not large enough to produce polymeric radicals by polymer scission [116]. As a consequence, the polymerization ultimately stops at this conversion, which represents a serious drawback for the development of ultrasound-induced bulk polymerization toward industrial application.

21.3.3.2 **Precipitation Polymerization**

As described in Section 21.3.3.1, ultrasound-induced bulk polymerizations are limited to relatively low conversions, because a strong increase in viscosity upon reaction hinders cavitation. In order to obtain higher conversions, precipitation polymerization forms a potential solution. Because the polymer produced is precipitated from the reaction medium, the viscosity and consequently the radical formation rate are expected to remain virtually constant. From this perspective, liquid carbon dioxide is a suitable reaction medium, because most monomers have a high solubility in CO_2, whereas it exhibits an anti-solvent effect for most polymers. Moreover, CO_2 is regarded as an environmentally friendly compound, which is nontoxic, nonflammable, and naturally abundant. Since higher pressures are required for CO_2 to act as an anti-solvent [56, 149], the possibility of ultrasound-induced cavitation in pressurized carbon dioxide systems has been studied [73, 150].

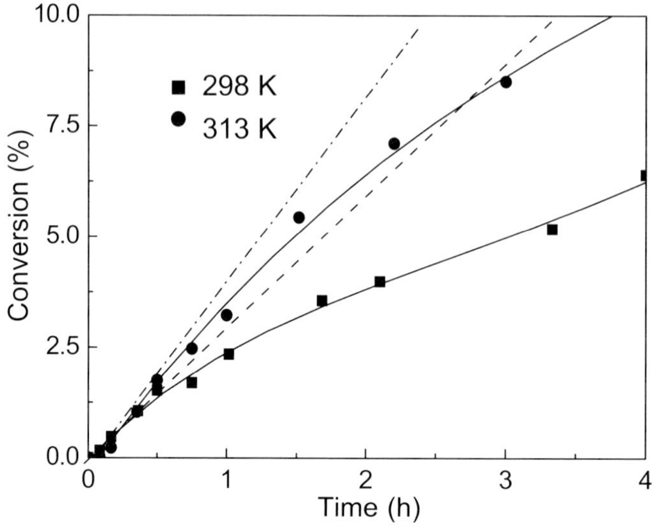

Fig. 21.13. Conversion development of an ultrasound-induced bulk polymerization of MMA at 298 and 313 K [150]. The broken lines indicate the initial reaction rate.

During pressurization of a liquid, the Blake threshold pressure [Eq. (4)] increases, which implies that higher acoustic pressures are needed to produce cavitations. Obviously, no cavitation occurs when the Blake threshold pressure exceeds the maximum acoustic pressure that can be applied with the currently available equipment. The vapor pressure of the liquid, however, can counteract the static pressure (see Figure 21.14). As a result of this, the Blake threshold is reduced in liquids with a high vapor pressure, such as CO_2, ethylene, and ammonia. This enables cavitation at increased static pressures [73, 150]. Additionally, the dynamic movement of the bubble in pressurized CO_2 has been calculated using the dynamic bubble model based on the Rayleigh–Plesset equation. According to the calculations, the bubble exhibits a similar movement to that of water at ambient pressure. To validate these simulations, cavitation experiments have been performed in pressurized CO_2–MMA systems using the radical scavenger 1,1-diphenyl-2-picrylhydrazyl [151]. The radical formation rate appears to be in the order of 1.5×10^{-4} s^{-1}, from which no noticable difference occurs upon varying the MMA–CO_2 ratios. Moreover, ultrasound-induced polymerizations of MMA in CO_2-expanded MMA have resulted in high molecular weight polymers [151].

21.3.3.3 **Emulsion Polymerization**

The generation of radicals by ultrasound can also be applied in emulsion polymerization, which comprises a free-radical polymerization in a heterogeneous reaction system, yielding submicron polymer particles in a continuous aqueous phase. Ultrasound can be applied for emulsification purposes as well as at higher conversions in the emulsion polymerization process.

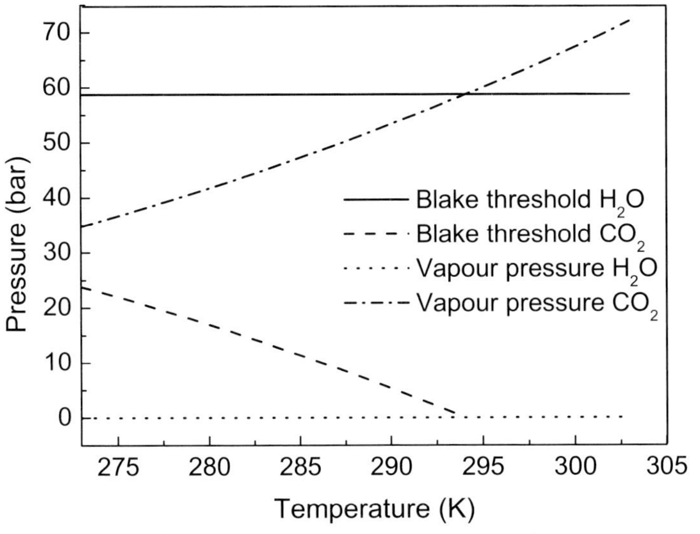

Fig. 21.14. Calculated Blake threshold and vapor pressure for water and carbon dioxide at 58.2 bar [73].

At the beginning of the polymerization, emulsification and nucleation govern the course of the process. The monomer droplets have to be small enough to provide a negligible resistance to monomer transport from the droplets through the aqueous phase to the growing polymer particles. Only in the case of sufficient emulsification, can intrinsic polymerization kinetics be assumed. This implies that in conventional emulsion polymerization vigorous stirring is required to disperse the monomer droplets. When ultrasound is applied as an energy source for emulsification, a very uniform emulsion of small monomer droplets is obtained without additional stirring. The high implosion velocity of the cavitation bubbles ensures that the monomer droplets are well dispersed. A special case involves mini- or microemulsion polymerization. Depending on the surfactant system used and the emulsification induced by ultrasound, the monomer droplets become so small that they can serve as nucleation loci. In ordinary emulsion polymerization, the main locus of nucleation is within the monomer-swollen micelles. Typically, polymer particles produced by ultrasound-induced emulsion polymerization have a diameter of approximately 50 nm [152, 153].

During emulsion polymerization induced by ultrasound, the redundancy of initiator is advantageous in terms of process control and safety. Moreover, initiator residues do not contaminate the product. In contrast to ultrasound-induced bulk polymerizations, high conversions can be obtained in ultrasound-induced emulsion polymerization [154, 155]. Since a heterogeneous reaction system is involved, in which the polymer is insoluble in the continuous aqueous phase, the viscosity of the water phase does not increase upon reaction. The cavitation events occur in the continuous phase, producing radicals mainly from water and surfactant molecules

[156]. Subsequently, the oligomeric radicals enter the monomer-swollen polymer particles and continue to polymerize. Examples of ultrasound-induced emulsion polymerization are described for styrene [153, 155, 157], methyl methacrylate [152, 156, 158–161], and *n*-butyl acrylate [162] and the copolymerization of vinyl acetate and butyl acrylate is also reported [154].

21.3.4 Cavitation-induced Polymer Scission

In terms of product properties, the molecular weight distribution is an important characteristic of polymers. In the polymer industry a post-processing step is often applied to alter the molecular weight of the polymers, for example for the peroxide-induced degradation of polypropylene [163]. In this process, fracture of the polymer chain occurs at a random site. An alternative method is ultrasound-induced polymer scission, which involves a much better-controlled, nonrandom process [119]. This enables relatively straightforward production of the desired molecular weight as well as the formation of block copolymers, as described in Section 21.3.5.

It has been shown that ultrasound-induced polymer breakage is a direct consequence of cavitation because, under conditions that suppress cavitation, no degradation is observed. In this nonrandom scission process the polymer is fractured at the center of the chain [164, 165]. This is clearly shown in Figure 21.15, in which a polymer with a molecular weight of 90 kg mol^{-1} is produced from a polymer with an initial molecular weight of 180 kg mol^{-1}. It is often thought that the extreme temperatures inside the bubble upon implosion contribute to the degradation.

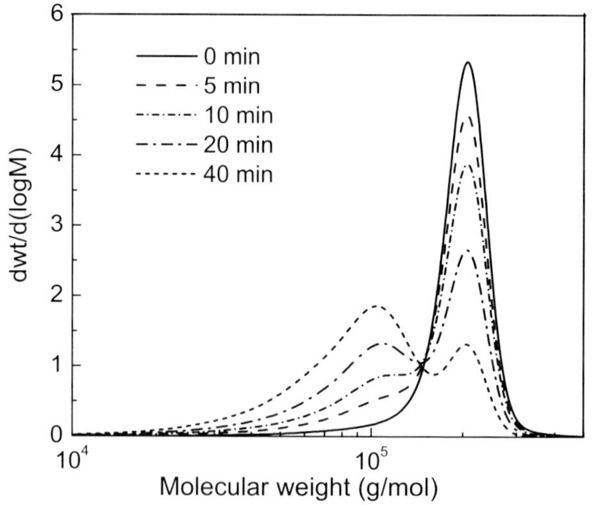

Fig. 21.15. Molecular weight distributions of an ultrasound-induced polymer scission of PMMA into MMA with an initial M_n of 18.0×10^4 g mol^{-1} [151].

However, temperature degradation implies a random process and thus does not explain the scission in the middle of the chain.

Ultrasound-induced chain fracture arises from the high strain rates on the polymer chain upon implosion. The nonrandom fracture in the middle of the chain by cavitation can only occur when the polymer chain is in a nonrandom conformation and thus completely stretched [166, 167]. In the absence of an external force a polymer chain in solution is randomly coiled. Upon bubble collapse, the entire molecule will move along with the fluid. However, due to the velocity profile near the cavities, friction between the polymer chain and the liquid will occur. Under sufficiently strong flow conditions, the solvent drag force causes extension of the polymer molecule and finally full stretching. If the polymer chain is stretched, the maximum stress due to the drag force will be in the center of the polymer chain, which is in analogy with flow-induced polymer scission [168]. If the drag force on the stretched polymer molecule exceeds the bond strength, scission will occur in the middle of the chain; otherwise the chain is not fractured and a limiting molecular weight (M_{lim}) is reached [169]. This limiting molecular weight is no longer fractured as the forces acting on the chain are smaller than the force required to break a bond.

The kinetics of ultrasonic scission can be directly ascribed to the implosion velocity of the cavitation bubbles and the polymer that is fractured. The scission rate and the limiting molecular weight are thus influenced by the ultrasonic wave, the solvent properties, and the polymer structure. The influence of ultrasound intensity, liquid viscosity, and liquid temperature on the degradation rate and the limiting molecular weight have been studied by Price et al. [170]. A higher intensity results in a faster scission and a lower limiting molecular weight. Lower temperatures and lower viscosities have similar influences on the scission rate and M_{lim}. These three effects can be ascribed to the higher implosion velocity, which induces higher strain rates around the bubble and therefore lead to a lower limiting molecular weight and a higher scission rate.

The initial molecular weight of the polymer influences neither the implosion velocity of a cavity nor the limiting molecular weight that is reached. However, the molecular weight has a large influence on the fracture rate of the polymer chains. A higher scission rate is observed for polymers with a higher initial molecular weight [170]. This is a consequence of the higher solvent drag force on the longer chains. As a result, narrow molecular weight distributions can be produced by ultrasound-induced polymer scission, because a high molecular weight polymer is fractured faster. Independently of the initial molecular weight, a similar limiting molecular weight is obtained for different experiments, as the final conditions are equal after several breakage events of the higher molecular weight polymers [167].

21.3.5 Synthesis of Block Copolymers

An interesting application of ultrasound-induced scission and polymerization is the synthesis of block copolymers, which are used in many applications where dif-

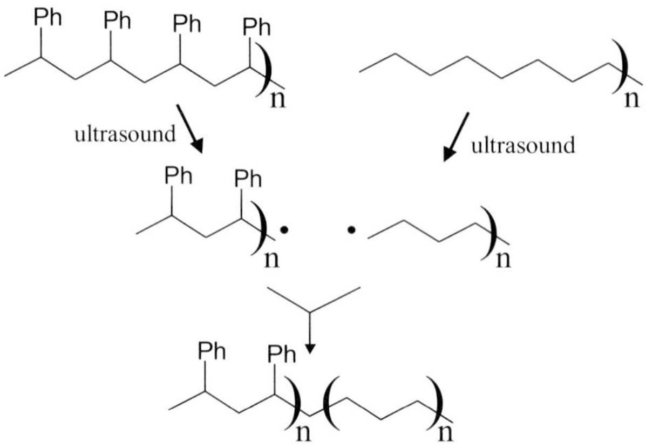

Fig. 21.16. The production of a block copolymer from two homopolymers with ultrasound [117].

ferent polymers are connected to yield a material with hybrid properties [171], for instance as a compatibilizing agent between immiscible polymers [172]. Anionic and living radical polymerization reactions are the most important techniques for synthesis of block copolymers. Alternatively, ultrasound can be used to produce these block copolymers, for which two different methods are applicable.

The first route to synthesis of block copolymers by ultrasound starts with the dissolution of a homopolymer in a different monomer [173]. Subsequently, ultrasonic scission of the polymer chains generates polymeric radicals, which initiate the polymerization reaction with the monomer present. In this way ultrasound provides the controlled formation of block copolymers, examples of which are polyethylene with acrylamide [173], and PMMA with styrene [174]. Secondly, dissolving two different homopolymers in a nonreactive solvent can lead to block copolymers as well (see Figure 21.16). In this case, the polymeric radicals generated have to undergo termination by cross-combination [170, 175]. If no cross-combination occurs, the original homopolymers are reproduced again. The advantage of the second method is that block copolymers can be produced from homopolymers of which the polymer–monomer systems are immiscible. Using this approach the synthesis of the block copolymer of polystyrene and poly(methyl phenyl silane) [170] as well as the block copolymer of poly(vinyl chloride) and poly(acrylonitrile-*co*-butadiene) [171] has been described.

21.4 Concluding Remarks and Outlook for the Future

In this chapter, supercritical carbon dioxide and ultrasound have been discussed as two examples of emerging technologies in polymer processes. Replacing the traditional organic solvents with supercritical carbon dioxide provides possibilities for the development of sustainable polymer processes. Additionally, the tunability of the solvent and the relatively easy separation of the polymer from the reaction medium are important advantages. However, applying $scCO_2$ as a clean solvent in polymer processes is not the simplest route, because it implies, among other things, high-pressure equipment, complex phase behavior, new measurement techniques, and the development of novel process concepts rather than extending the conventional technologies. Currently, there is a lack of integrated tools between the research on a laboratory scale and the industrial-scale application, mainly caused by the absence of pilot-scale facilities. Recent innovative tools, such as supercritical reaction calorimetry [176] on the liter scale, could fill this gap by allowing the determination of engineering, process, scaleup, and safety data [177]. Additionally, several process design calculations have shown that polymer processes based on $scCO_2$ technology can be economically feasible, depending on the value of the product and the process conditions. Moreover, further developments will reduce costs of supercritical application substantially. It is expected that the major application of supercritical carbon dioxide will first be in the food and pharmaceuticals industry because of additional marketing advantages, such as the GRAS (generally regarded as safe) status. Nevertheless, the fact that DuPont is commercializing the production of fluoropolymers in $scCO_2$ illustrates the possibility of applying supercritical fluid technology in polymer processes as well. In addition, the long-existing ldPE tubular process (approximately 2500 bar, 600 K) proves that a high-pressure polymerization process performed on a large scale can survive in a highly competitive field.

With respect to controllability of polymerization processes, ultrasound has significant potential as a clean and safe technology. After the production of most types of polymers, catalyst and initiator residues contaminate the product. Since ultrasound generates the radicals in situ, no initiator or catalyst is required to start a polymerization reaction. In addition, high-temperature free-radical polymerization is emerging in order to avoid the use of initiators that will remain present in the polymer chain at the end of the process [178]. The mechanism of thermal initiation at high temperatures is still not well described. However, in analogy with initiation by ultrasound at low temperatures, high-temperature processes open the route for the sustainable production of polymers composed only of monomer units. An additional advantage of ultrasound is the intrinsically safe operation, because turning off the electrical power supply will immediately stop the radical formation and consequently the polymerization reaction. Besides polymerization, polymer scission can also occur through irradiation with ultrasound, due to the rapid flow around a cavitation bubble. At a sufficiently small chain length, further fracture of the polymer is prevented. In this way ultrasound-induced polymer scission provides an additional means to control the molecular weight of the product. At the moment the major challenge of ultrasound-induced polymer processes is the scaleup, mainly in terms of energy conversion of the ultrasound generator to the probe. Although no large-scale industrial polymerization process based on ultrasound exists yet, commercial applications in other fields such as ultrasound cleaning and sterilization prove that ultrasound is a readily available technique and relatively simple to implement in existing industrial equipment. Still, the application of ultrasound for polymerization purposes requires a thorough, multidisciplinary understanding of both ultrasound parameters and liquid properties, including physics, chemistry, and engineering.

In general, the various subjects in this chapter have been addressed from an engineering point of view, for which industrial application is one of the important issues. Although significant effort is being put into new developments by a large number of research groups, the development trajectory from the concept idea, via the laboratory, bench, and pilot scale, toward industrial implementation is often long and not easy. To reduce the boundaries between the academic approach and industrial practice, collaboration between industrial R&D, research institutes, and universities is essential to reduce costs and to exploit existing know-how and experimental facilities, as well as to reduce the development time. Given the economics of an emerging technology as compared to long-existing processes, it is a challenge to implement new process concepts at reasonable costs. For these reasons, in the short term the number of large-scale industrial polymer processes based on supercritical fluid or ultrasound technology will be limited, for which stimulation from government and research consortia can contribute substantially to facilitate the development trajectory. Nevertheless, the progress made in research today will enable the development of sustainable and well-controlled polymer processes for the future.

Acknowledgments

The author thanks Marc Jacobs and Martijn Kuijpers for their contribution to this chapter.

Notation

A_{US}	surface area of ultrasound source [m^2]
f	frequency [Hz]
I_{US}	ultrasound intensity [W/cm^{-2}]
P_0	external pressure [bar]
P_A	acoustic pressure [bar]
$P_{A,\max}$	maximum acoustic amplitude [bar]
P_B	Blake threshold pressure [bar]
P_c	critical pressure [bar]
P_v	vapor pressure [bar]
Q	quadrupole moment [$J^{1/2}\ m^{5/2}$]
Q_{US}	ultrasound power transferred to liquid [W]
R_0	equilibrium radius of cavitation bubble [m]
T_c	critical temperature [K]
T_g	glass transition temperature [K]

v speed of sound in medium [m s^{-1}]
α polarizability [m^3]
Γ_{ij} potential energy [J]
μ dipole moment [C m]
v speed of sound in medium [m s^{-1}]
ρ density [kg m^{-3}]
σ surface tension [N m^{-1}]

References

1 International Institute of Synthetic Rubber Producers, Inc., Internet publication http://www.iisrp.com/WebPolymers/o1FinalPolybutadieneVer2.pdf.
2 D. T. ALLEN, D. R. SHONNARD, *Green Engineering, Environmentally Conscious Design of Chemical Processes*, Prentice Hall PTR, Upper Saddle River, **2002**.
3 J. M. DESIMONE, *Science*, **2002**, *297*, 799.
4 P. T. ANASTAS, Green chemistry as applied to solvents, in *Clean Solvents, Alternative Media for Chemical Reactions and Processing*, M. A. ABRAHAM, L. MOENS (eds.), ACS Symposium Series no. 819, Washington DC, **2002**.
5 J. F. BRENNECKE, E. J. MAGINN, *AIChE J.*, **2001**, *47*(11), 2384.
6 M. MODELL, *US Patent 4 338 199*, **1982**.
7 T. B. THOMASON, M. MODELL, *Hazard Waste*, **1984**, *1*(4), 453.
8 S. ANGUS, B. ARMSTRONG, K. M. DE REUCK, *International Thermodynamic Tables of the Fluid State. Carbon Dioxide*, Pergamon Press, Oxford, **1976**.
9 R. SPAN, W. WAGNER, *J. Phys. Chem. Ref. Data*, **1996**, *25*, 1509.
10 R. C. REID, J. M. PRAUSNITZ, B. E. POLING, *The Properties of Gases and Liquids*, 4th ed., McGraw-Hill, New York, **1987**.
11 C. F. KIRBY, M. A. MCHUGH, *Chem. Rev.*, **1999**, *99*, 565.
12 J. M. PRAUSNITZ, R. N. LICHTENTHALER, E. G. DE AZEVEDO, *Molecular Thermodynamics of Fluid Phase Equilibria*, 2nd ed., Prentice-Hall, Englewood Cliffs, **1986**.
13 N. AJZENBERG, F. TRABELSI, F. RECASENS, *Chem. Eng. Technol.*, **2000**, *23*(10), 829.
14 H. ORBEY, C. P. BOKIS, C. C. CHEN, *Ind. Eng. Chem. Res.*, **1998**, *37*, 4481.
15 M. A. MCHUGH, V. J. KRUKONIS, *Supercritical Fluid Extractions: Principles and Practice*, 2nd ed., Butterworth-Heinemann, Stoneham, MA, **1994**.
16 J. J. SUAREZ, I. MEDINA, J. L. BUENO, *Fluid Phase Equilibria*, **1998**, *153*, 167.
17 F. RINDFLEISCH, T. P. DINOIA, M. A. MCHUGH, *J. Phys. Chem.*, **1996**, *100*, 15 581.
18 T. SARBU, T. STYRANEC, E. J. BECKMAN, *Ind. Eng. Chem. Res.*, **2000**, *39*, 1678.
19 T. SARBU, T. STYRANEC, E. J. BECKMAN, *Nature*, **2000**, *405*, 165.
20 S. G. KAZARIAN, M. F. VINCENT, F. V. BRIGHT, C. L. LIOTTA, C. A. ECKERT, *J. Am. Chem. Soc.*, **1996**, *118*, 1729.
21 S. G. KAZARIAN, *Polym. Sci., Ser. C*, **2000**, *42*(1), 78.
22 Y. ZHANG, K. K. GANGWANI, R. M. LEMERT, *J. Supercrit. Fluids*, **1997**, *11*, 115.
23 T. HIROSE, K. MIZOGUCHI, Y. KAMIYA, *J. Polym. Sci. B*, **1986**, *24*, 2107.
24 T. J. DE VRIES, *Late Transition State Metal Catalyzed Polymerizations of Olefins in Supercritical Carbon Dioxide*, PhD thesis, Technische Universiteit Eindhoven, Eindhoven, **2003**.
25 M. A. JACOBS, *Measurement and Modeling of Thermodynamic Properties for Processing of Polymers in Supercritical Fluids*, PhD thesis, Technische Universiteit Eindhoven, Eindhoven, **2004**.

26 R. G. Wissinger, M. E. Pauliatis, *J. Polym. Sci. B*, **1987**, *25*, 2497.
27 R. G. Wissinger, M. E. Pauliatis, *Ind. Eng. Chem. Res.*, **1991**, *30*, 842.
28 S. G. Kazarian, M. F. Vincent, F. V. Bright, C. L. Liotta, C. A. Eckert, *J. Am. Chem. Soc.*, **1996**, *118*, 1729.
29 S. G. Kazarian, *Appl. Spectroscopy Rev.*, **1997**, *32*, 301.
30 T. Miyoshi, K. Takegoshi, T. Tereo, *Macromolecules*, **1997**, *30*, 6582.
31 K. Otake, S. E. Webber, P. Munk, K. P. Johnston, *Langmuir*, **1997**, *13*, 3047.
32 M. A. Jacobs, M. F. Kemmere, Th. W. de Loos, J. T. F. Keurentjes, 4th International Symposium on High Pressure Technology and Chemical Engineering, Venice, *Chem. Eng. Trans.*, **2002**, *2*, 779.
33 A. Rajendran, B. Bonavoglia, G. Storti, M. Mazzotti, *Annual AIChE Meeting*, **2003**, San Francisco, paper 95F.
34 I. S. Liau, M. A. McHugh, in *Supercritical Fluid Technology*, J. M. L. Penninger, M. Radosz, M. A. McHugh (eds.), Elsevier Science, Amsterdam, **1985**.
35 E. Sada, H. Kumazawa, H. Yakushiji, Y. Bamba, K. Sakata, S. Wang, *Ind. Eng. Chem. Res.*, **1987**, *26*, 433.
36 A. R. Berens, G. S. Huvard, in *Supercritical Fluid Science and Technology*, American Chemical Society, Washington DC, **1989**, p. 207.
37 Y. Zhang, K. Gangwani, R. M. Lemert, *J. Supercrit. Fluids*, **1997**, *11*, 115.
38 C. F. Kirby, M. A. McHugh, *Chem. Rev.*, **1999**, *99*, 565.
39 J. A. M. van Hest, G. A. M. Diepen, Solubility of naphthalene in supercritical methane, in *The Physics and Chemistry of High Pressures, Symposium Papers*, Society of Chemical Industry, London, **1963**, p. 10.
40 Th. W. de Loos, W. Poot, G. A. M. Diepen, *Macromolecules*, **1983**, *16*, 111.
41 M. A. McHugh, T. L. Guckes, *Macromolecules*, **1985**, *18*, 674.
42 M. Lora, M. A. McHugh, *Fluid Phase Equilib.*, **1999**, *157*, 285.
43 Th. W. de Loos, W. Poot, R. N. Lichtenthaler, *J. Supercrit. Fluids*, **1995**, *8*, 282.
44 A. A. Kiamos, M. D. Donohue, *Macromolecules*, **1994**, *27*, 357.
45 Y. Xiong, E. Kiran, *J. Appl. Polym. Sci.*, **1994**, *53*, 1179.
46 T. M. Martin, A. A. Lateef, C. B. Roberts, *Fluid Phase Equilib.*, **1999**, *154*, 241.
47 T. J. de Vries, P. J. A. Somers, Th. W. de Loos, M. A. G. Vorstman, J. T. F. Keurentjes, *Ind. Eng. Chem. Res.*, **2000**, *39*, 4510.
48 I. C. Sanchez, R. H. Lacombe, *J. Phys. Chem.*, **1976**, *80*, 2352.
49 R. H. Lacombe, I. C. Sanchez, *J. Phys. Chem.*, **1976**, *80*, 2568.
50 I. C. Sanchez, R. H. Lacombe, *Macromolecules*, **1978**, *11*, 1145.
51 S. H. Huang, M. Radosz, *Ind. Eng. Chem. Res.*, **1990**, *29*, 2284.
52 S. H. Huang, M. Radosz, *Ind. Eng. Chem. Res.*, **1991**, *30*, 1994.
53 S. H. Huang, M. Radosz, *Ind. Eng. Chem. Res.*, **1993**, *32*, 762.
54 M. A. Jacobs, T. J. de Vries, M. F. Kemmere, Th. W. de Loos, J. T. F. Keurentjes, *Recent Res. Devel. Appl. Polym. Sci.*, **2002**, *1*, 171.
55 K. M. Scholsky, *J. Supercrit. Fluids*, **1993**, *6*, 103.
56 K. A. Shaffer, J. M. DeSimone, *Trends in Polym. Sci.*, **1995**, *3*(5), 146.
57 J. L. Kendall, D. A. Canelas, J. L. Young, J. M. DeSimone, *Chem. Rev.*, **1999**, *99*, 543.
58 A. I. Cooper, *J. Mater. Chem.*, **2000**, *10*, 207.
59 J. M. DeSimone, Z. Guan, C. S. Elsbernd, *Science*, **1992**, *257*, 945.
60 M. Shaw, J. Ryan, C. Erkey, *Abstr. Am. Chem. Soc.*, **1997**, *214*, 243.
61 M. Poliakoff, S. M. Howdle, M. W. George, High pressure chemical engineering, in *Process Techn. Proc.*, Ph. R. von Rohr, Ch. Trepp (eds.), Elsevier, Amsterdam, **1996**, p. 67.
62 Z. Guan, J. R. Combes, C. S. Elsbernd, J. M. DeSimone, *Polym. Prepr. Am. Chem. Soc.*, **1993**, *34*, 329.
63 M. R. Clark, J. M. DeSimone, *Macromolecules*, **1995**, *28*, 3002.
64 J. R. Combes, Z. Guan, J. M.

DeSimone, *Macromolecules*, **1994**, *27*, 865.

65 S. Beuermann, M. Buback, C. Isemer, A. Wahl, *Macromol. Rapid Commun.*, **1999**, *20*, 26.

66 J. M. DeSimone, E. E. Maury, Y. Z. Menceloglu, J. B. McClain, T. J. Romack, J. R. Combes, *Science*, **1994**, *265*, 356.

67 D. A. Canelas, D. E. Betts, J. M. DeSimone, *Macromolecules*, **1996**, *29*, 2818.

68 F. A. Adamsky, E. J. Beckman, *Macromolecules*, **1994**, *27*, 312.

69 R. Butler, C. M. Davies, A. I. Cooper, *Adv. Mater.*, **2001**, *13*(19), 1459.

70 C. D. Wood, A. I. Cooper, *Macromolecules*, **2001**, *34*, 5.

71 P. A. Carpentier, K. A. Kennedy, J. M. DeSimone, G. W. Roberts, *Macromolecules*, **1999**, *32*, 5973.

72 P. A. Carpentier, J. M. DeSimone, G. W. Roberts, Continuous polymerization in supercritical carbon dioxide, in *Clean Solvents, Alternative Media for Chemical Reactions and Processing*, M. A. Abraham, L. Moens (eds.), ACS Symposium Series No. 819, Washington DC, **2002**.

73 M. W. A. Kuijpers, D. van Eck, M. F. Kemmere, J. T. F. Keurentjes, *Science*, **2002**, *298*, 1969.

74 T. J. de Vries, R. DuChateau, M. A. G. Vorstman, J. T. F. Keurentjes, *Chem. Comm.*, **2000**, 263.

75 T. J. de Vries, M. F. Kemmere, J. T. F. Keurentjes, *Macromolecules*, **2004**, *37*(11), 4241.

76 R. Kondo, K. Toshima, S. Matsumura, *Macromol. Biosci.*, **2002**, *2*(6), 267.

77 M. S. Super, E. Beluche, C. A. Costello, E. Beckman, *Macromolecules*, **1997**, *30*, 368.

78 M. F. Kemmere, T. J. de Vries, M. A. G. Vorstman, J. T. F. Keurentjes, *Chem. Eng. Sci.*, **2001**, *56*, 4197.

79 A. Ahvenainen, K. Sarantila, H. Andtsjo, *WO Patent 92/12 181*, **1992**.

80 G. Parkinson, *Chem. Eng.*, **1999**, *106*(May), 17.

81 B. Bungert, G. Sadowski, W. Arlt, *Ind. Eng. Chem. Res.*, **1998**, *37*, 3208.

82 Y. T. Shieh, J. H. Su, G. Manivannan, P. F. Lee, S. P. Sawan, W. D. Spall, *J. Appl. Polym. Sci.*, **1996**, *59*, 695.

83 Y. W. Chiu, C. S. Tan, *J. Supercrit. Fluids*, **2001**, *21*, 81.

84 S. Alsoy, J. L. Duda, *AIChE J.*, **1998**, *44*, 582.

85 M. F. Kemmere, M. H. W. Cleven, M. A. van Schilt, J. T. F. Keurentjes, *Chem. Eng. Sci.*, **2002**, *57*, 3929.

86 M. R. De Giorgi, E. Cadoni, D. Maricca, A. Piras, *Dyes and Pigments*, **2002**, *45*, 75.

87 M. W. Park, H. K. Bae, *J. Supercrit. Fluids*, **2002**, *22*, 65.

88 O. Guney, A. Akgerman, *AIChE J.*, **2002**, *48*, 856.

89 A. I. Cooper, *Adv. Mater.*, **2003**, *15*(13), 1.

90 J. J. Watkins, T. J. McCarthy, *Macromolecules*, **1994**, *27*, 4845.

91 R. C. Petersen, D. W. Matson, R. D. Smith, *Polym. Eng. Sci.*, **1987**, *27*, 1693.

92 Y. Chernyak, F. Henon, R. B. Harris, R. D. Gould, R. K. Franklin, J. R. Edwards, J. M. DeSimone, R. G. Carbonell, *Ind. Eng. Chem. Res.*, **2001**, *40*, 6118.

93 R. K. Franklin, J. R. Edwards, Y. Chernyak, R. D. Gould, F. Henon, R. G. Carbonell, *Ind. Eng. Chem. Res.*, **2001**, *40*, 6127.

94 J. Jung, M. Perrut, *J. Supercrit. Fluids*, **2001**, *20*, 179.

95 C. Lee, K. L. Hoy, M. D. Donohue, *U.S. Patent 4 923 720*, **1990**; *U.S. Patent 5 027 742*, **1991**.

96 J. J. Shim, M. Z. Yates, K. P. Johnston, *Ind. Eng. Chem. Res.*, **2001**, *38*, 3655.

97 H. J. Vandenburg, A. A. Clifford, K. D. Bartle, J. Carroll, I. Newton, L. M. Garden, J. R. Dean, C. T. Costley, *Analyst*, **1997**, *122*, 101R.

98 N. J. Cotton, K. D. Bartle, A. A. Clifford, C. J. Dowle, *J. Appl. Polym. Sci.*, **1993**, *48*, 1607.

99 X. W. Lou, H. G. Janssen, C. A. Cramers, *J. Microcolumn Separations*, **1995**, *7*(4), 303.

100 M. D. Bermudez, F. J. Carrion-Vilches, G. Martinez-Nicolas, M.

PAGAN, *J. Supercrit. Fluids*, **2002**, *23*, 59.

101 J. K. SEKINGER, G. N. GHEBREMESKEL, L. H. CONCIENNE, *Rubber Chemistry and Technology*, **1996**, *69*, 851.

102 M. F. KEMMERE, M. A. VAN SCHILT, M. H. W. CLEVEN, A. M. VAN HERK, J. T. F. KEURENTJES, *Ind. Eng. Chem. Res.*, **2002**, *41*, 2617.

103 G. BRUNNER, *Nato Science Ser. E, Appl. Sci.*, **2000**, *366*, 517.

104 P. MENGAL, J.-Y. CLAVIER, M. PERRUT, GVC-department, *High Pressure Chemical Engineering*, March 3–5, Karlsruhe, Germany, **1999**, p. 279.

105 A. R. BERENS, G. S. HUVARD, R. W. KORSMEYER, F. W. KUNIG, *J. Appl. Polym. Sci.*, **1992**, *46*, 231.

106 B. L. WEST, S. G. KAZARIAN, M. F. VINCENT, N. H. BRANTLEY, C. A. ECKERT, *J. Appl. Polym. Sci.*, **1998**, *69*, 911.

107 Y. WANG, C. Y. YANG, D. TOMASKO, *Ind. Eng. Chem. Res.*, **2002**, *41*, 1780.

108 P. D. CONDO, S. R. SUMPTER, M. L. LEE, K. P. JOHNSTON, *Ind. Eng. Chem. Res.*, **1996**, *35*, 1115.

109 S. G. KAZARIAN, M. F. VINCENT, B. L. WEST, C. A. ECKERT, *J. Supercrit. Fluids*, **1998**, *13*, 107.

110 W. SAUS, D. KNITTEL, E. SCHOLLMEYER, *Textile Res. J.*, **1993**, *63*(3), 135.

111 J. VON SCHNITZLER, R. EGGERS, *J. Supercrit. Fluids*, **1999**, *16*, 81.

112 S. M. HOWDLE, J. M. RAMSAY, A. I. COOPER, *J. Polym. Sci. B Polym. Phys.*, **1994**, *32*(3), 541.

113 X. MA, D. L. TOMASKO, *Ind. Eng. Chem. Res.*, **1997**, *36*, 1586.

114 S. K. LIAO, P. S. CHANG, Y. C. LIN, *J. Polym. Res. Taiwan*, **2000**, *7*(3), 155.

115 D. KNITTEL, W. SAUS, E. SCHOLLMEYER, *Indian J. Fibre Textile Res.*, **1997**, *22*, 184.

116 P. KRUUS, T. J. PATRABOY, *J. Phys. Chem.*, **1985**, *89*, 3379.

117 G. J. PRICE, *Ultrason. Sonochem.*, **1996**, *3*, S229.

118 G. J. PRICE, *Adv. Sonochem.*, **1990**, *1*, 231.

119 J. NIEZETTE, A. LINKENS, *Polymer*, **1978**, *19*, 939.

120 J. P. LORIMER, T. J. MASON, *Chem. Soc. Rev.*, **1987**, *16*, 239.

121 L. KORNET, A. P. HOEKS, B. J. JANSSEN, J. M. WILLIGERS, R. S. RENEMAN, *J. Hypertension*, **2002**, *20*, 1165.

122 Y. T. DIDENKO, K. S. SUSLICK, *Nature*, **2002**, *407*, 877.

123 K. S. SUSLICK, *Sci. Am.*, **1989**, *260*, 80.

124 A. SHOH, *Ultrasonics*, in *Kirk-Othmer Encyclopedia of Chemical Technology*, Wiley, New York, **1983**.

125 M. P. BRENNER, S. HILGENFELDT, D. LOHSE, *Rev. Modern Physics*, **2002**, *74*, 425.

126 W. P. MASON, *Physical Acoustics; Principles and Methods*, Vol. IB, Academic Press, New York, **1964**.

127 K. S. SUSLICK, Y. DIDENKO, M. M. FANG, T. HYEON, K. J. KOLBECK, W. B. MCNAMARA III, M. M. MDELELENI, M. WONG, *Phil. Trans. Roy. Soc. A.*, **1999**, 1.

128 M. W. A. KUIJPERS, M. F. KEMMERE, J. T. F. KEURENTJES, *Ultrasonics*, **2002**, *40*, 675.

129 J. M. LÖNING, C. HORST, U. HOFFMANN, *7th Meeting of the European Society of Sonochemistry*, **2000**, p. 197.

130 U. HOFFMANN, C. HORST, U. WIETELMANN, S. BANDELIN, R. JUNG, Sonochemistry, in *Ullmann's Encyclopedia of Industrial Chemistry*, Wiley-VCH, Weinheim, **2003**.

131 T. J. LEIGHTON, *The Acoustic Bubble*, Academic Press, London, **1994**.

132 Y. T. SHAH, A. B. PANDIT, V. S. MOHOLKAR, *Cavitation Reaction Engineering*, Plenum, New York, **1999**.

133 S. HILGENFELDT, M. P. BRENNER, S. GROSSMANN, D. LOHSE, *J. Fluid Mech.*, **1998**, *365*, 171.

134 K. YASUI, *J. Acoust. Soc. Am.*, **1995**, *98*, 2772.

135 K. YASUI, *Ultrasonics*, **1998**, *36*, 575.

136 M. W. A. KUIJPERS, M. F. KEMMERE, J. T. F. KEURENTJES, Ultrasound-induced radical polymerization, in *Encyclopedia of Polymer Science and Technology*, John Wiley, New York, **2004**.

137 L. H. THOMPSON, L. K. DORAISWAMY, *Ind. Eng. Chem. Res.*, **1999**, *38*, 1215.

138 J. M. PESTMAN, J. B. F. N. ENGBERTS,

F. de Jong, *Recl. Trav. Chim. Pays-Bas*, **1994**, *113*, 533.

139 M. Ibisi, B. Brown, *J. Acoust. Soc. Am.*, **1967**, *41*, 568.

140 J.-L. Luche, *Synthetic Organic Chemistry*, Plenum, New York, **1998**.

141 G. J. Price, P. J. West, P. F. Smith, *Ultrason. Sonochem.*, **1994**, *1*, S51.

142 M. W. A. Kuijpers, M. F. Kemmere, J. T. F. Keurentjes, *Annual AIChE Meeting*, **2001**, paper 341B.

143 G. J. Price, P. F. Smith, *Polymer*, **1993**, *34*, 4111.

144 P. Kruus, J. A. G. Lawrie, M. L. O'Neill, *Ultrasonics*, **1988**, *26*, 352.

145 G. J. Price, *Chemistry & Industry*, **1993**, 75.

146 N. Erolan, F. Arisan, H. Çatalgil-Giz, *Angew. Makromol. Chem.*, **1995**, *226*, 53.

147 P. Kruus, *Ultrasonics*, **1987**, *25*, 20.

148 D. Feldman, *Polymer News*, **1995**, *20*, 138.

149 P. G. Jessop, W. Leitner, *Chemical Synthesis using Supercritical Fluids*, Wiley-VCH, Weinheim, **1999**.

150 M. W. A. Kuijpers, *Physics and Chemistry of Ultrasound Polymer Reaction Engineering*, PhD thesis, Technische Universiteit Eindhoven, Eindhoven, **2004**.

151 M. F. Kemmere, M. W. A. Kuijpers, L. J. M. Jacobs, J. T. F. Keurentjes, *Macromol. Symp.*, **2004**, *206*, 321.

152 H. C. Chou, J. O. Stoffer, *J. Appl. Polym. Sci.*, **1999**, *72*, 797.

153 S. K. Ooi, S. Biggs, *Ultrason. Sonochem.*, **2000**, *7*, 125.

154 G. Cooper, F. Grieser, S. Biggs, *J. Colloid Interface Sci.*, **1996**, *184*, 52.

155 C. Zhang, Q. Wang, H. Xia, G. Qiu, *Eur. Polym. J.*, **2002**, *38*, 1769.

156 H. C. Chou, J. O. Stoffer, *J. Appl. Polym. Sci.*, **1999**, *72*, 827.

157 S. Biggs, F. Grieser, *Macromolecules*, **1995**, *28*, 4877.

158 M. Bradley, M. Ashokkumar, F. Grieser, *J. Am. Chem. Soc.*, **2003**, *125*, 525.

159 H. C. Chou, W. Y. Lin, J. O. Stoffer, *Polym. Mater. Sci. Eng*, **1995**, *72*, 363.

160 H. C. Chou, Y. B. Liu, O. C. Sitton, J. O. Stoffer, *Polym. Mater. Sci. Eng.*, **1993**, *69*, 376.

161 Y. Liao, Q. Wang, H. Xia, X. Xu, S. M. Baxter, R. V. Slone, S. Wu, G. Swift, D. G. Westmoreland, *Ultrason. Sonochem.*, **2002**, *9*, 151.

162 Q. Wang, H. Xia, Y. Liao, X. Xu, S. M. Baxter, R. V. Slone, S. Wu, G. Swift, D. G. Westmoreland, *Polym. Int.*, **2001**, *50*, 1252.

163 A. V. Machado, J. A. Covas, M. van Duin, *J. Appl. Polym. Sci.*, **2001**, *81*, 58.

164 P. A. R. Glyn, B. M. E. van der Hoff, P. M. Reilly, *J. Macromol. Sci.-Chem. A6*, **1972**, 1653.

165 B. M. E. van der Hoff, C. E. Gall, *J. Macromol. Sci. A*, **1977**, *11*, 1739.

166 G. Madras, S. Kumar, S. Chattopadhyay, *Polym. Degr. Stab.*, **2000**, *59*, 73.

167 M. W. A. Kuijpers, P. D. Iedema, M. F. Kemmere, J. T. F. Keurentjes, *Polymer*, **2004**, *45*, 6461.

168 A. M. Basedow, K. H. Ebert, *Adv. Polym. Sci.*, **1977**, *22*, 83.

169 J. A. Odell, A. Keller, *J. Polym. Sci. B*, **1986**, *24*, 1889.

170 G. J. Price, P. J. West, *Polymer*, **1996**, *37*, 3975.

171 I. W. Hamley, Block copolymers, in *Encyclopedia of Polymer Science and Technology*, John Wiley, New York, **2002**.

172 T. J. Mason, *Sonochemistry, The Uses of Ultrasound in Chemistry*, CRC Press, Cleveland, **1990**.

173 H. Fujiwara, J. Tanaka, A. Horiuchi, *Polymer Bulletin*, **1996**, *36*, 723.

174 G. Schmidt-Naake, M. Drache, M. Weber, *Macromol. Chem. Phys.*, **2002**, *15*, 203.

175 H. Fujiwara, *Polymer Bulletin*, **2001**, *47*, 247.

176 S. Fortini, F. Lavanchy, Th. Meyer, *Macromol. Symp.*, **2004**, *206*, 79.

177 F. Lavanchy, S. Fortini, Th. Meyer, *Organic Process Research Development*, **2004**, *8*(3), 504.

178 P. Nising, Th. Meyer, *Ind. Eng. Chem. Res.*, **2004**, *43*, 7220.

Index

Handbook of Polymer Reaction Engineering. Edited by T. Meyer, J. Keurentjes

ISBN: 3-527-31014-2

d

g

t

u

v